Advanced Sciences and Technologies for Security Applications

Advanced Sciences and Technologies
for Security Applications

The series Advanced Sciences and Technologies for Security Applications focuses on research monographs in the areas of
—Recognition and identification (including optical imaging, biometrics, authentication, verification, and smart surveillance systems)
—Biological and chemical threat detection (including biosensors, aerosols, materials detection and forensics),
and
—Secure information systems (including encryption, and optical and photonic systems).

The series is intended to give an overview at the highest research level at the frontier of research in the physical sciences.

The editors encourage prospective authors to correspond with them in advance of submitting a manuscript. Submission of manuscripts should be made to the Editor-in-Chief or one of the Editors.

Bahram Javidi
Editor

Optical Imaging Sensors and Systems for Homeland Security Applications

With 246 Figures

 Springer

Bahram Javidi
Distinguished Professor of Electrical and Computer Engineering
University of Connecticut
Electrical & Computer Engineering Dept.
371 Fairfield Road, Unit 1157
Storrs, CT 06269-1157, USA
E-mail: bahram@engr.uconn.edu

Library of Congress Control Number: 2005927080

ISBN-10: 0-387-26170-2   e-ISBN: 0-387-28001-4
ISBN-13: 978-0387-26170-6

Printed on acid-free paper.

Printed in the United States of America.   (SPI/MVY)

9 8 7 6 5 4 3 2 1

springeronline.com

Table of Contents

Contributor List ix

Section I.   Active Optical 3D Sensing and 3D Imaging Systems for Homeland Security Applications

1 3D Object Reconstruction and Recognition Techniques Based on Digital Holography
Yann Frauel, Enrique Tajahuerce, Osamu Matoba, Albertina Castro, and Bahram Javidi............................... 1

2 Compression of Encrypted Digital Holograms Using Artificial Neural Networks
Alison E. Shortt, Thomas J. Naughton, Bahram Javidi................ 25

3 Digital Hoplography: Recent Advancements and Prospective Improvements for Applications in Microscopy
Pietro Ferraro, Sergio De Nicola, Giuseppe Coppola 47

4 Hybrid Optical Encryption of a 3D Object by Use of a Digital Holographic Technique
Takanori Nomura 85

5 3D Object Recognition using Gabor Feature Extraction and PCA-FLD Projections of Holographically Sensed Data
Sekwon Yeom, Bahram Javidi.................................. 97

6 Distortion-tolerant 3D Volume Recognition Using X-ray Imaging
Sekwon Yeom, Bahram Javidi, Young Jun Roh, and Hyung Suck Cho 115

7 3D Imaging and Recognition of Microorganism using Single-exposure On-line (SEOL) Digital Holography
Bahram Javidi, Inkyu Moon, Seokwon Yeom and Edward Carapezza 139

Section II. Passive 3D Sensing and 3D Imaging Systems for Homeland Security

8 Integral Imaging Applied to the Digital Reconstruction and Recognition of 3D Scenes
Yann Frauel, Osamu Matoba, Enrique Tajahuerce, and Bahram Javidi 157

9 Real-time Remote Identification and Verification of Objects Using Optical ID Tags
Bahram Javidi 177

Section III. Surveillance and Image Recognition Algorithms for Homeland Security Applications

10 An Adaptive Technique for Minimizing Rate of Sensory Data Transmission in Unmanned Aerial Vehicles
Firooz Sadjadi 185

11 Information Processing Across Distributed and Netted Systems for Security and Surveillance
Abhijit Mahalanobis, Mubarak Shah, Alan van Nevel 205

12 Composite Correlation Filters and Neural Networks for Identification and Pose Estimation
Albertina Castro, Yann Frauel, and Bahram Javidi 225

13 Evolutionary Sensor Fusion for Security
Bir Bhanu and Sohail Nadimi 245

14 The Use of Synthetic Data in Eye/Face Recognition
Behrooz Kamgar-Parsi, Behzad Kabmar-Parsi, Benjamin N. Waber 271

15 Hyperspectral Target Detection based on Kernels
Heesung Kwon and Nasser M. Nasrabadi 287

16 Detecting 3D Location and Shape of Distorted 3D Objects using LADAR Trained Optimum Nonlinear Filters
Seung-Hyun Hong and Bahram Javidi 323

Section IV. Optical Devices and Hardware for Homeland Security Applications

17 Planar Microoptical Systems for Correlation and Security Applications
Stefan Sinzinger, J. Jahns, J. Glueckstad, V. Daria 339

18 Optical Waveguide-mode Resonant Biosensors
D. Wawro, S. Tibuleac, and R. Magnusson.......................... 367

19 Improved Optical Document Security Techniques based on Volume Holography and Lippmann Photography
Hans I. Bjelkhagen 385

Index 401

Contributor List

B. Bhanu
Electrical Engineering & Computer
    Science
University of California at Riverside
Center for Research in Intelligent Systems
Bourns B232, College of Engineering,
    UCR
Riverside, CA 92521

H.I. Bjelkhagen
OpTIC Technium
Centre for Modern Optics
Ffordd William Morgan
St Asaph Business Park
St Asaph LL17 0JD
North Wales, UK

A. Castro
Instituto Nacional de Astrofsica
Optica y Electronica,
Luis Enrique Erro No. 1
Tonantzintla, Puebla
72840, Mexico

E. Carapezza
DARPA Advanced Technology Office
3701 N.Fairfax Drive
Arlington, VA 22203-1714

H.S. Cho
Department of Mechanical Engineering
Korea Advanced Institute of Science and
    Technology (KAIST)
373-1 Kusung-dong,Yusung-ku
Taejeon 305-701, South Korea

G. Coppola
Istituto per la Microelettronica e
    Microsistemi del CNR
Via P.Castellino
111 80134
Naples, Italy

V. Daria
Risoe National Laboratory
Optics and Plasma Research
    Department
P.O. Box 49
DK-4000 Roskilde
Denmark

S. De Nicola
Istituto di Cibernetica CNR
Comprensorio "Olivetti", building 70
Via Campi Flegrei, 34
I 80072 Pozzuoli
Italy

P. Ferraro
Istituto Nazionale di Ottica Applicata
    (INOA)
c/o Istituto di Cibernetica "E.Caianiello"
    del CNR
Via Campi Flegrei, 34
c/o Compr. "Olivetti"
80072 Pozzuoli (Na), Italy

Y. Frauel
IIMAS-UNAM
Cto. Escolar, Ciudad Universitaria
Del. Coyoacan
04510 Mexico, D.F., Mexico

J. Glückstad
Risoe National Laboratory
Optics and Plasma Research Department
P.O. Box 49
DK-4000 Roskilde
Denmark

S.H. Hong
University of Connecticut
Electrical & Computer Engineering Dept.
371 Fairfield Road, Unit 2157
Storrs, CT 06269-2157
USA

T. Nomura
Department of Opto-Mechatronics
Faculty of Systems Engineering
Wakayama University
930 Sakaedani
Wakayama 640-8510 JAPAN

J. Jahns
Jürgen Jahns
Fernuniversitat Hagen
Optische Nachrichtentechnik
Universitatsstr. 27/PRG
58084 Hagen, Germany

B. Javidi
Electrical and Computer Engineering
    Dept.
University of Connecticut
U-1157, Storrs, CT 06269-1157

B. Kamgar-Parsi
Naval Research Laboratory
Washington, DC 20375

H. Kwon
Department of the Army
US Army Research Laboratory
2800 Powder Mill Road
Adelphi, Maryland 20783-1145

R Magnusson
University of Connecticut
Electrical & Computer Engineering Dept.
371 Fairfield Road, Unit 2157
Storrs, CT 06269-2157
USA

A. Mahalanobis
Lockheed Martin MFC
MP 450
5600 Sandlake Road
Orlando , FL 32819
USA

O. Matoba
Dept. of Computer and System
    Engineering
Faculty of Engineering
Kobe University
1-1 Rokkodai-cho, Nada-ku
Kobe 657-8501, Japan

I. Moon
University of Connecticut
Electrical & Computer Engineering
    Dept.
371 Fairfield Road, Unit 2157
Storrs, CT 06269-2157
USA

S. Nadimi
Electrical Engineering & Computer
    Science
University of California at Riverside
Center for Research in Intelligent
    Systems
Bourns B232, College of Engineering,
    UCR
Riverside, CA 92521

N. M. Nasrabadi
Department of the Army
US Army Research Laboratory
2800 Powder Mill Road
Adelphi, Maryland 20783-1145

T.J. Naughton
Department of Computer Science
National University of Ireland, Maynooth
County Kildare, Ireland

Y.J. Roh
Korea Advanced Institute of Science and
   Technology (KAIST)
373-1 Kusung-dong,Yusung-ku
Taejeon 305-701, South Korea

F. Sadjadi
Lockheed Martin Corporation
3400 Highcrest Rd.
Saint Anthony MN 55418

M. Shah
School of Computer Science
University of Central Florida
Orlando, FL 32816

A.E. Shortt
Department of Computer Science
National University of Ireland, Maynooth
County Kildare, Ireland

S. Sinzinger
Technische Universitat Ilmenau
Fakultat für Maschinenbau
Technische Optik
PF 100565
98684 Ilmenau

E. Tajahuerce
Dept. de Ciencies Experimentals
Universitat Jaume I
Campus Riu Sec, s/n, P.O. Box 224
12080-Castellon, Spain

S. Tibuleac
Director of Product Management
Movaz Networks
One Technology Parkway South
Norcross, GA 30092

A. van Nevel
Image and Signal Processing Branch
Sensor and Signals Sciences Division
   Research Dept.
Naval Air Warfare Center, Weapons
   Division
China Lake, CA 93555

B. N. Waber
Information Technology Division
Naval Research Laboratory
Washington, DC 20375
and
Department of Computer Science
Boston University
Boston, MA 02215

D. Wawro
CEO/President
Resonant Sensors Incorporated
202 E. Border St., #201
Arlington, Texas 76010

S. Yeom
University of Connecticut
Electrical & Computer Engineering Dept.
371 Fairfield Road, Unit 2157
Storrs, CT 06269-2157

3D Object Reconstruction and Recognition Techniques Based on Digital Holography

Yann Frauel,[1] Enrique Tajahuerce,[2] Osamu Matoba,[3] Albertina Castro,[4] and Bahram Javidi[5]

[1]IIMAS, Universidad Nacional Autónoma de México, Mexico
[2]Departament de Ciencies Experimentals, Universitat Jaume I, Spain
[3]Department of Computer and System Engineering, Kobe University, Japan
[4]Instituto Nacional de Astrofísica, Óptica y Electrónica, Mexico
[5]Department of Electrical and Computer Engineering, University of Connecticut, USA

1.0 Introduction

Optical techniques and algorithms have proven to be particularly suitable to the realization of fast and efficient security and information processing systems.[15,16,21,32] Recent advances in electrooptical devices and components have allowed researchers to take advantage of the properties of free-space optics such as massive parallelism and high space-bandwidth product.[1,7] One of the most promising applications of optical processing is the recognition of objects or images. In particular, the properties of coherent light allow an instant computation of Fourier transforms with a single lens.[10] This explains why optical correlation techniques have been widely studied for pattern recognition.[13,14,26,31,33] However, these techniques have been developed in order to deal with two-dimensional (2D) images and they cannot be applied easily to three-dimensional (3D) objects.

The first problem of 3D object recognition is that one has to acquire the depth information of an object or a scene. This can be done using various methods.[9,17] The easiest one consists in taking several conventional pictures of an object from different viewpoints or with different viewing angles. These pictures can be recorded by several cameras or by a single moving camera.[25,28] They can also be taken with a microlens array as in the case of integral photography.[20,23,24] These techniques only provide indirect information about the depth through changes of parallax. Another method to find the depth information involves the projection of fringes onto the object and the analysis of their deformation.[5] This analysis introduces some additional complexity.

A more direct technique is to use a rangefinder system to measure the 3D shape of the object.[12] However, this measurement has to be done point by point.

On the contrary, holography is a more natural technique since it is an extension of the photographic technique to 3D objects. A hologram records – in one single acquisition – the complete 3D information of a scene because it records the phase information of the optical beam along with its magnitude.[4] Moreover, an interesting approach consists in using digital holography to acquire these data.[11,29] In this case, the holograms are recorded by a CCD camera and the 3D object can be reconstructed numerically in a computer. This technique avoids the analog recording of the hologram and the corresponding chemical or physical development. Since the information is directly available in digital form, it can be used to reconstruct the object in a computer and to compute digital correlations. In this contribution, we will first describe the technique of phase-shift digital holography that we use for recording the information about the 3D objects. We will describe the methods used to reconstruct the diffraction volume of the object from digital holograms. Then we will show how it is possible – with some limitations – to achieve 3D object recognition by direct correlation of digital holograms. Next, we will describe a modified technique that allows us to obtain a shift-invariant recognition. Finally, we will explain how to extend the technique to perform a 3D recognition in the presence of distortions of the object such as out-of-plane rotation or longitudinal shift.

1.1 Phase-shift Digital Holography

1.1.1 Recording of the Hologram

1.1.1.1 Experimental Setup

Our optical setup (Fig. 1.1) uses a linearly polarized argon laser tuned to 515 nm. Its beam is split into a reference path and an object path. On both paths, the beams are expanded and spatially filtered in order to obtain two uniform plane waves. The reference beam passes through a quarter- and a half-wave plates whose axes are either parallel or orthogonal to the polarization of the beam. Depending on the relative orientations of the fast and slow axes of each plate, it is possible to achieve a phase retardation of 0, $\pi/2$, π, or $3\pi/2$. The object beam illuminates the 3D object, which then scatters light in the direction of the camera. This scattered light is added coherently to the phase-modulated reference beam and the interference pattern is detected by a CCD camera. This geometry is similar to a Mach-Zehnder interferometer.

1.1.1.2 Recording of a Digital Hologram

Let us call $O(x, y, z)$ the complex amplitude of the beam scattered by the object at location (x, y, z). This complex field contains both the magnitude and

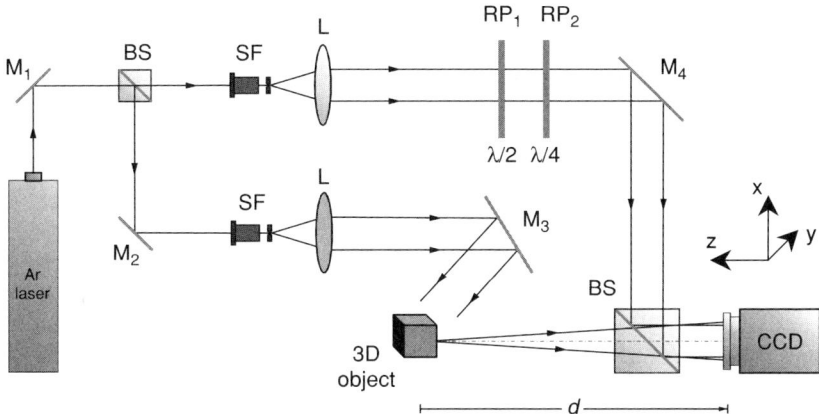

Figure 1.1. Experimental setup–M mirror, BS beamsplitter, SF spatial filter, L lens, RP retardation plate.

the phase of the optical wave. The complex amplitude of the object beam in the plane of the CCD sensor is a sum of spherical waves emitted by all the points of the 3D object. Assuming that this plane is located at $z = 0$, and in the Fresnel approximation, this sum can be written as follows:

$$
O(x,\, y,\, 0) = \frac{i}{\lambda} \iiint\limits_{Object} O(x',\, y',\, z') \frac{1}{z'} \exp\left(-i\frac{2\pi}{\lambda} z'\right) \times
$$

$$
\times \exp\left(-i\pi \frac{(x-x')^2 + (y-y')^2}{\lambda z'}\right) dx'\, dy'\, dz'. \tag{1}
$$

We can also express this complex amplitude in the form

$$
O(x,\, y,\, 0) = A_O(x,\, y) \exp\left[i\varphi_O(x,\, y)\right] \tag{2}
$$

where $A_O(x,\, y)$ and $\varphi_O(x,\, y)$ are the magnitude and the phase of the object wave in the plane of the camera. Similarly, the complex amplitude of the reference wave in the same plane can be written as

$$
R_{\Delta\varphi}(x,\, y,\, 0) = A_R(x,y) \exp[i(\varphi_R(x,y) + \Delta\varphi)] \tag{3}
$$

where $\Delta\varphi$ is the phase retardation introduced by the two waveplates.

The intensity interferogram recorded by the camera is given by

$$
\begin{aligned}
I_{\Delta\varphi}(x,\, y) &= |O(x,\, y,\, 0) + R_{\Delta\varphi}(x,\, y,\, 0)|^2 \\
&= |O(x,\, y,0)|^2 + |R_{\Delta\varphi}(x,\, y,\, 0)|^2 + \\
&\quad + O(x,\, y,\, 0) R_{\Delta\varphi}{}^*(x,\, y,\, 0) + O^*(x,\, y,\, 0) R_{\Delta\varphi}(x,\, y,\, 0).
\end{aligned} \tag{4}
$$

This interferogram is also called a digital hologram. Indeed, if it is illuminated by the reference wave $R_{\Delta\varphi}(x, y, 0)$, the third term provides a reconstruction of the original object beam. However, the three other terms give undesirable information. In conventional holographic techniques, the hologram is usually recorded with a sufficient angle between the reference and the object beams. In this way, the readout of the hologram angularly separates the different terms. In the case of digital holography, the relatively large size of the pixels of available CCD cameras (typically $10\,\mu$m) limits the spatial resolution of the interferogram and therefore imposes that the two beams be almost parallel to each other. During readout, all the beams overlap which prevents the separation of the reconstructed object beam from the unwanted beams.

1.1.1.3 Phase-shift Technique

In order to avoid the problem of overlapping of the terms, we need to extract the value of the object amplitude $O(x, y, 0)$ from the interferogram. This is not possible with only one interferogram but it is possible if we record several interferograms with different values of the phase retardation of the reference beam $\Delta\varphi$.[2,34] As mentioned before, we use the values 0, $\pi/2$, π, or $3\pi/2$. Using Eqs. (2) and (3) in Eq. (4) we obtain

$$\begin{aligned}
I_{\Delta\varphi}(x, y) = {}& A_O^2(x, y) + A_R^2(x, y) + \\
& + 2A_O(x, y)A_R(x, y)\cos[\varphi_O(x, y) - \varphi_R(x, y) - \Delta\varphi].
\end{aligned} \tag{5}$$

In our case, the reference beam is actually a plane wave that is orthogonally incident onto the camera sensor. Consequently, its magnitude $A_R(x, y)$ and phase $\varphi_R(x, y)$ are constant over the sensor and can be replaced with 1 and 0 respectively without loss of generality. We can then rewrite Eq. (5) for the four values of phase retardation:

$$\begin{cases}
I_0 = A_O^2(x, y) + 1 + 2A_O(x, y)\cos\left(\varphi_O(x, y)\right) \\
I_\pi = A_O^2(x, y) + 1 - 2A_O(x, y)\cos\left(\varphi_O(x, y)\right) \\
I_{\pi/2} = A_O^2(x, y) + 1 + 2A_O(x, y)\sin\left(\varphi_O(x, y)\right) \\
I_{3\pi/2} = A_O^2(x, y) + 1 - 2A_O(x, y)\sin\left(\varphi_O(x, y)\right).
\end{cases} \tag{6}$$

It is then easy to show that the magnitude of the object wave can be found by

$$A_O(x, y) = \frac{1}{4}\sqrt{\left(I_0 - I_\pi\right)^2 + \left(I_{\pi/2} - I_{3\pi/2}\right)^2} \tag{7}$$

and its phase by

$$\varphi_O(x, y) = \text{Arctan}\left(\frac{I_0 - I_\pi}{I_{\pi/2} - I_{3\pi/2}}\right) \tag{8}$$

1.1.2 Reconstruction of Views of the Object

The phase-shift technique described in the previous paragraph provides the object wave $O(x, y, 0)$ as defined in Eq. (2). By extension, we give the name of digital hologram to this object wave. Indeed, it allows us to reconstruct the 3D object without parasitic images. This reconstruction is made possible by numerically computing the reversed propagation of the light from the plane of the camera (where the so-called hologram is located) to a plane in the middle of the object. This method actually provides an accurate view of the parts of the object that are in the chosen plane only. The areas that stand in different planes will be blurred. This allows us to get information about the depth of the object by focusing specifically in various planes. However, the depth resolution ("focus depth") is typically much lower than the lateral resolution. If the depth of the object is less than this focus depth, it is possible to see clearly the whole object by reconstructing a view in the median plane of the object. In this case we will call this plane "the plane of the object."

The simulated propagation is computed by using the Fresnel–Kirchhoff integral:

$$O(x, y, d) = O(x, y, 0) * h_d(x, y) \tag{9}$$

where

$$h_d(x, y) = \frac{i}{\lambda d} \exp\left(-i \frac{2\pi}{\lambda} d\right) \exp\left[-i\pi \frac{(x^2 + y^2)}{\lambda d}\right] \tag{10}$$

is the point-spread function of the free space, λ denotes the wavelength of the beam and the symbol * stands for the 2D convolution.

One efficient way of computing Eq. (9) is to use fast Fourier transforms in order to compute the convolution:

$$O(x, y, d) = FT^{-1}\{FT[O(x, y, 0)].FT[h_d(x, y)]\} \tag{11}$$

Of course, we only know a sampled version of $O(x, y, 0)$. In this case, it can be shown[22] that, aside from the computation speed, the formula used in Eq. (11) has another advantage: it keeps the same sampling step for the origin and the destination planes. This is particularly useful for object recognition because it means that no change in scale is introduced when the distance of the object is modified.

A property of holography is that each point of the hologram records light coming from the whole object. It is thus possible to reconstruct a view of the object by using only a partial window extracted from the hologram. Of course, this method yields a loss of information. In the case of Eq. (11), the sampling step remains constant while the number of pixels is reduced. Hence, this window extraction in the hologram results in a window extraction in the reconstruction plane. In other words, only a part of the object field is obtained.

This technique is therefore usable only when the size of the object is less than the size of the window. In this case, it is possible to reconstruct different perspectives of the object by using different windows of the hologram (Fig. 1.2). However, the application of Eq. (9) simulates a propagation orthogonally to the hologram, so that each window provides a reconstruction of a different part of the object plane (Fig. 1.2a). In order to have all the reconstructed perspectives centered at the same point, we need to multiply each window by a linear phase factor. This multiplication is equivalent to reading the holographic window with a tilted plane wave (Fig. 1.2b). In mathematical terms, in Eq. (9) we replace $O(x, y, 0)$ with $W_{\alpha, \beta}(x, y) \exp[-i2\pi (\alpha x + \beta y)/\lambda d]$, where $W_{\alpha, \beta}(x, y)$ is the window centered at the coordinates (α, β) in the hologram. The angle of view with respect to the orthogonal axis is then $(\alpha/d, \beta/d)$.

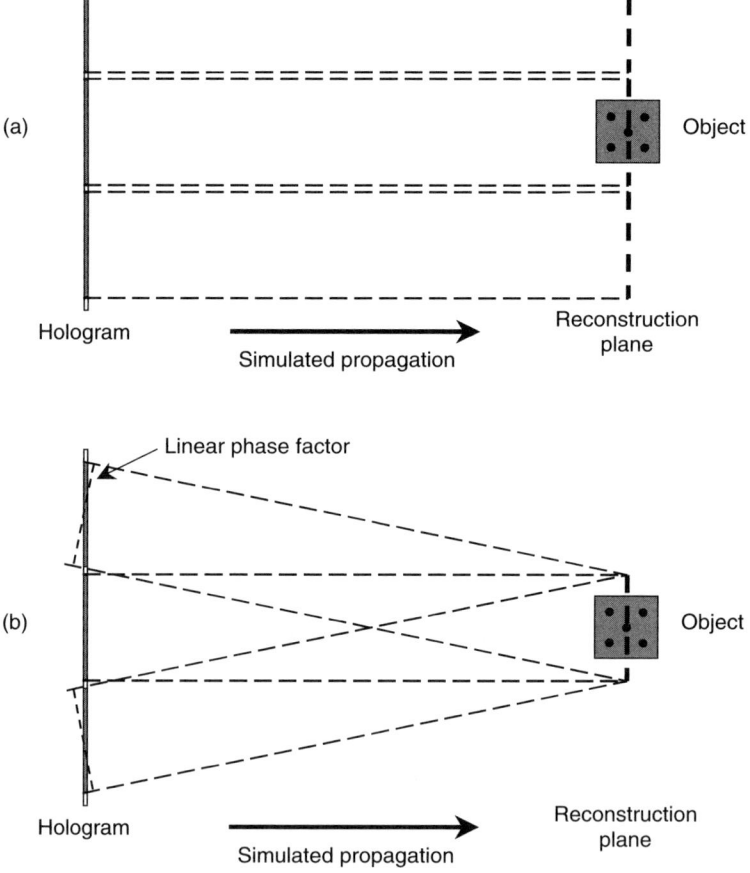

Figure 1.2. Reconstruction using windows in the hologram: (a) Without phase factor, (b) With a linear phase factor to recenter the reconstructed view.

1.1.3 Enhancement of the Reconstructed Images

Since the digital holograms are recorded with coherent light, they are marred by a speckle pattern. This spatially random noise pattern is caused by the interference of the light emitted by every point of the rough surface of the object. The speckle phenomenon occurs in the hologram but also in any reconstruction plane, including the plane(s) of the object. This noise affects both the amplitude and the phase of the beam. It is not only bothering for the visual aspect of the reconstructed images but also for the recognition applications. Indeed, these high-frequency spatial variations – especially in the phase – dramatically reduce the correlation lengths in every direction and therefore strongly reduce the tolerance to distortions and displacements of the objects. This means that an object can only be recognized if it is replaced exactly in its original position. Moreover, another object, even though its global structure is identical to the one of the reference object, is likely to have a different micro-structure and therefore not to be recognized.

In order to improve the correlation results, we have to get rid of the phase of the complex field and to keep the amplitude information only. We then remove the remaining amplitude speckle in two steps. First, we replace each block of 8×8 pixels of the image with its average amplitude. The size of the reconstructed image is therefore reduced by a factor 8 in both directions. This also allows us to reduce the computation time. However, some amplitude variations remain after this process and we remove them by performing a median filtering over 7×7 pixels. Namely, for each pixel we classify the values of the 49 closest neighbors in an ascending order and we keep the 25th value. Figure 1.3 presents reconstructed images of a die both before and after removing the speckle pattern. The levels have been scaled to enhance the contrast.

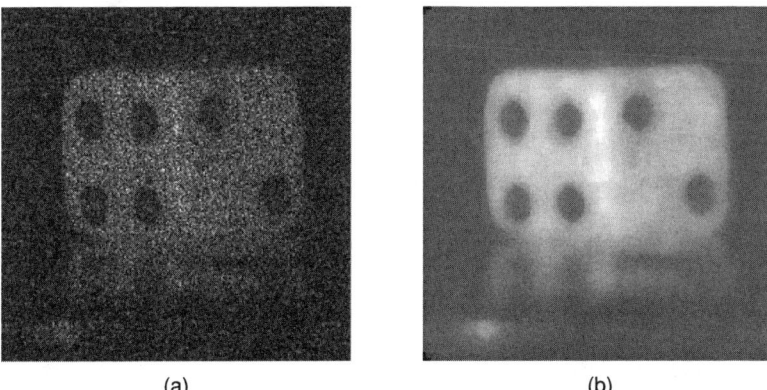

(a)   (b)

Figure 1.3. Image of a die reconstructed from a digital hologram- (a) before and (b) after filtering the speckle pattern.

1.3 Correlation of Holograms

1.3.1 Object Recognition

As can be seen in Eq. (1), the hologram contains the 3D information about the object. Therefore, it is possible to compare two 3D objects by correlating their holograms directly, without the need to reconstruct views of the objects. We use a matched filter approach. The filter is the digital hologram of the reference object $O_R(x, y, 0)$. This filter is compared to the digital hologram of the object to be tested $O_T(x, y, 0)$ through a digital correlation:

$$Cor_{RT} = O_R(x, y, 0) \otimes O_T(x, y, 0) \qquad (12)$$

where the symbol $\otimes$ stands for the correlation operation. This correlation can be efficiently computed by using Fourier transforms:

$$Cor_{RT} = FT^{-1}\{FT[O_R(x, y, 0).FT^*[O_T(x, y, 0)]]\} \qquad (13)$$

As mentioned previously, due to the speckle pattern, this correlation is very sensitive to displacements of the object and can thus be used to detect very small changes in the location or in the shape of an object.

We conducted a 3D object recognition experiment using two models of cars with an approximate size of $25 \times 25 \times 45$ mm. They were located at a distance $d = 865$ mm from the CCD detector. The digital holograms contain 256×256 pixels. In Fig. 1.4 are presented the reconstructions of the objects obtained using Eq. (9). In Fig. 1.5a we show a plot of the autocorrelation of the object in Fig. 1.4a performed by autocorrelating its digital hologram. Figure 1.5b shows the crosscorrelation of the 3D object in Fig. 1.4a with that in Fig. 1.4b obtained by crosscorrelation of the digital holograms. Both plots are normal-

(a)   (b)

Figure 1.4. Reconstructed images of (a) the reference object and (b) the input object.

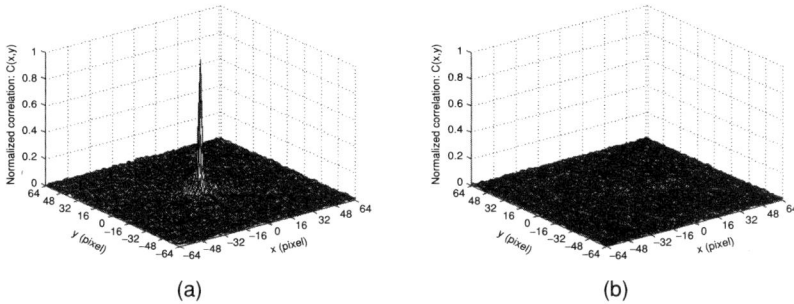

(a)   (b)

Figure 1.5. (a) Autocorrelation and (b) cross-correlation of the digital holograms of the objects represented in Fig. 1.1.4.

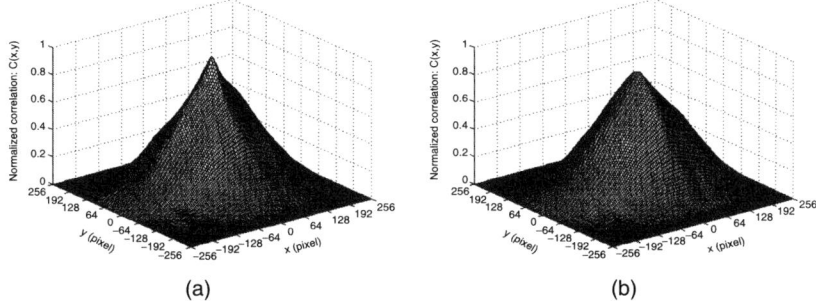

(a)   (b)

Figure 1.6. Conventional 2D correlation of the objects in Fig. 1.4 taken as 2D intensity images: (a) Autocorrelation and (b) cross correlation.

ized to the same value. As a comparison, we present in Fig. 1.6 the conventional 2D autocorrelation and crosscorrelation, using the objects of Fig. 1.4 as 2D intensity images. It can be seen that the 3D correlation with the holograms is much more discriminant and sensitive to displacements of the object.

1.3.2 Measurement of Small Rotations

Instead of using the entire holograms, it is also possible to measure the correlation of sub-windows of the holograms properly modified by a linear phase factor. This is equivalent to comparing different perspectives of the objects and therefore allows us to evaluate small rotations of the reference object.

In order to illustrate this property, we use as a reference the hologram corresponding to the object shown in Fig. 1.4a. We take as an input a hologram of the same object slightly rotated. Both holograms have 2028×2044 pixels. We then compare 256×256 pixel windows in these holograms. In the reference one, the window is centered, whereas on the input one the window is moved across the hologram. Figure 1.7a presents the maximum

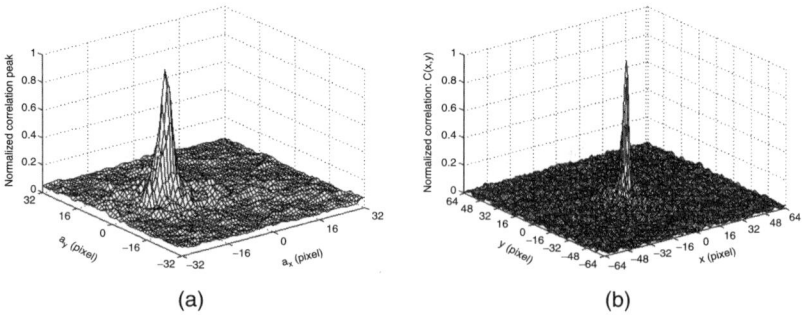

(a)   (b)

Figure 1.7. Correlation of the hologram windows corresponding to the object in Fig. 1.1.4a, and a rotated version of the same object- (a) Correlation peak versus displacement of the window and (b) Correlation plane for the displacement giving the maximum peak.

value of the correlation plane versus the displacement (α, β) of the latter window. A peak is obtained for $\alpha = -12$ and $\beta = -2$, which corresponds to an angle of view $(0.007°, 0.001°)$. Figure 1.7b shows the correlation plane for this particular perspective of the 3D object.

1.4 Shift-invariant 3D Object Recognition

1.4.1 Principle

The problem with the technique described in the previous section is that the recognition can only take place when the input object is almost in the same location and with the same orientation as the reference object. In this section we describe a different technique that is able to deal with translations and limited rotations of the objects.

Once the holograms have been obtained, we use Eq. (9) to calculate the complex wavefronts generated by the 3D objects in a set of planes parallel to the output plane within the Fresnel approximation. As mentioned in Section **1.1.3**, we keep only the amplitude information while dismissing the phase information. In contrast with the technique described in the previous section, instead of using the digital holograms directly, we first evaluate the amplitude distribution of the objects in the 3D object space. This approach allows us to apply 3D correlation techniques to the 3D amplitude distribution generated by the objects. As described in Section **1.1.2**, it is also possible to reproduce the amplitude distribution in planes tilted with respect to the output plane by using partial information from the digital holograms. Therefore, not only the three Cartesian coordinates of the reference in the 3D input space but also the relative out-of-plane rotation of the target with respect to the reference can be obtained. Also, by filtering the phase information, we achieve a system less sensitive to noise and fast fluctuations of the complex distribution produced by the rough surfaces of the objects. With this new method we are able to achieve full 3D shift invariance.

As explained in Section **1.1.2**, the digital holograms allow us to reconstruct the light field at any plane orthogonal to the output plane, including those planes containing the 3D objects, as shown in Fig. 1.8. This permits us to reconstruct the 3D light distribution generated by the scene in a volume with a depth limited only by the Fresnel approximation. In each reconstruction plane at a distance d given by Eq. (9), regions of the 3D object such that $z = d$ result focused, but other areas of the object appear defocused in the same plane. The set of different 2D distributions with variable propagation distances d constitutes a 3D function that contains information about the location of 3D objects in the 3D scene under consideration. Similar 3D objects will generate similar 3D light distributions in volumes located at the same relative distance from the object. Therefore, it is possible to show that this 3D transformation represents a linear shift-invariant operation. Thus, application of correlation techniques to the resulting 3D distribution provides information about the presence and position of the reference 3D object in the 3D input scene.

Actually some preliminary operations are performed before evaluating the correlation in order to improve the efficiency of the technique. First, the filtering operations described in Section **1.1.3** are applied to the reconstructed complex amplitudes. Second, in order to reduce the computation time, instead of computing a 3D correlation between the two 3D functions generated from the digital holograms, we only consider as reference function the 2D distribution corresponding to the reconstruction in the median plane of

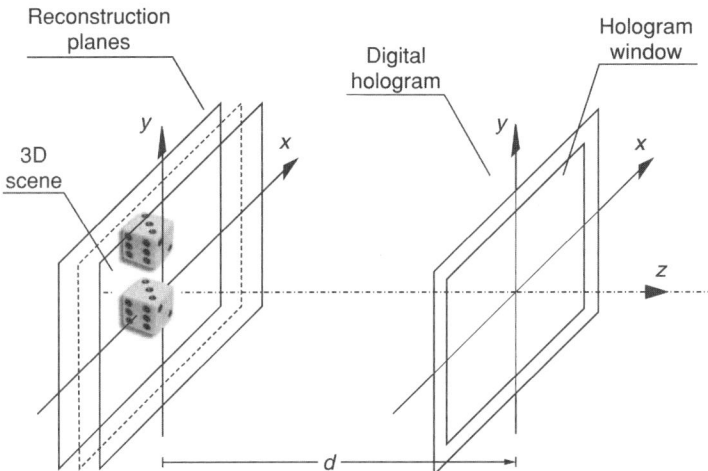

Figure 1.8. Reconstruction of the amplitude distribution in different planes from the Fresnel digital hologram of a 3D object.

the reference object. In this way, this 2D reference function can be sequentially correlated in the computer with the different 2D functions that characterize the 3D input scene by computing simple 2D correlations. Finally, in order to increase the discrimination capability, we use a phase-only filter to perform the correlation.[13] The filter is constructed using only the phase of the complex conjugate of the Fourier transform of the reference function. The result, after computing the correlation for different 2D reconstructions of the irradiance distribution associated to the input, constitutes a 3D correlation volume. By analyzing the correlation peaks in this volume it is possible to determine the 3D position of the reference in the input scene.

1.4.2 Experimental Results

An experiment was performed to detect the presence and 3D position of a 3D reference in an input scene. The reference object was a die with a lateral size equal to 4.6 mm. The center of the die was located at a distance $d_1 = 345$ mm from the camera sensor. In Fig. 1.9(a) a picture of the irradiance distribution at this distance d_1 is shown corresponding to the location of the reference object. It was obtained by applying Eq. (11) to the digital hologram.

As input scene, we use two similar dice located at different distances. One was located at the same position than the reference, the other had its center located at a different axial distance from the output plane and was displaced transversally and rotated a small angle around the vertical axis. The 3D amplitude distribution generated by the input scene was obtained by recording a digital hologram and using Eq. (11) to generate a set of 2D distributions with different propagation distances d. This set of 2D functions constitutes the 3D distribution to be used in the recognition step. We show the irradiance distribution generated by the 3D input scene at only two planes, those corre-

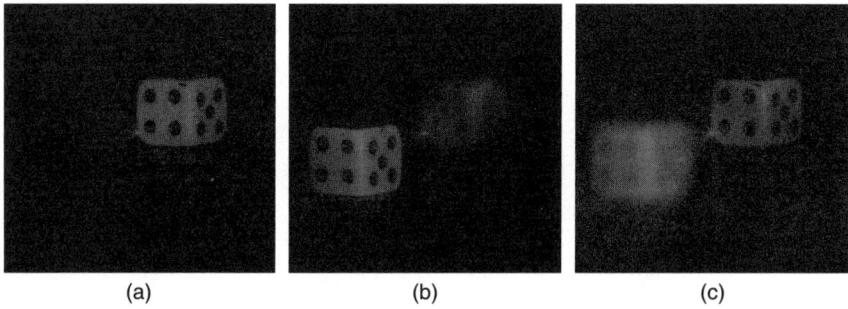

(a)   (b)   (c)

Figure 1.9. Computer reconstructions of the 3D reference object and the 3D input scene for two focusing distances: (a) reference object, (b) input objects for $z = 315$ mm, and (c) for $z = 345$ mm.

sponding to distances $z = 315$ mm and $z = 345$ mm in Fig. 1.9(b) and 1.9(c), respectively. The 2D distributions in Fig. 1.9 are just three elements of the set of 2D functions characterizing the 3D reference and input scene.

In Fig. 1.10, we show a plot of the correlation between the reference and two different 2D sections of the 3D input amplitude distribution. We selected the same reconstructions depicted in Fig. 1.9, which give us two local maxima of the correlation. Figure 1.10(a) corresponds to the correlation between the reference in Fig. 1.9(a) with the 2D amplitude distribution in Fig. 1.9(b), while Fig. 1.10(b) shows the result obtained by correlation of the distributions in Fig. 1.9(a) and 1.9(c). Both plots, normalized to the same value, show a clear maximum at the locations of the 3D reference. The maximum in Fig. 1.10(a) is lower than that in Fig. 1.10(b) due to the small rotation of the reference. It can be shown that the height of this correlation peak can be increased by reconstructing the amplitude distribution generated by the 3D input scene at planes tilted with respect to the output plane. This can be done by using partial windows in the corresponding Fresnel digital hologram, as was explained in Section **1.1.2** (Fig. 1.2). The locations of the maxima determine the transversal locations of the objects in the input scene with respect to the original position of the reference. The distances z for which we obtain the maximum values of the correlation peak, $z = 315$ mm and $z = 345$ mm, determine the axial position of the two objects. With the information obtained, it is possible to localize the 3D reference in the 3D input space. In this way, we see that one of the objects is located at the same position than the original reference object. The second one, similar to the reference but displaced and rotated, is located at 3D coordinates given by $(-7.72, -3.35, -30)$ mm with respect to the original position of the reference.

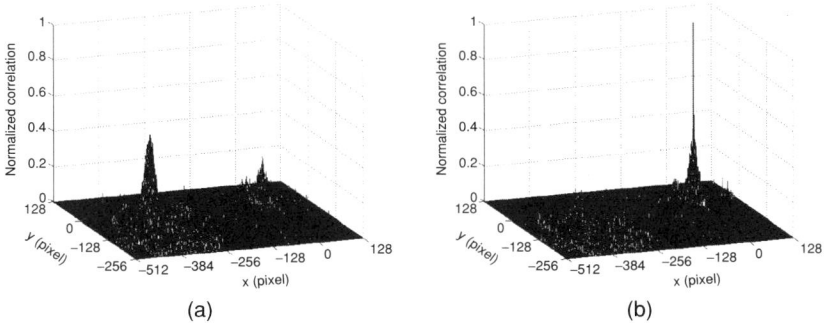

Figure 1.10. Correlation of the irradiance distribution associated to the reference with that of the input scene for different distances: (a) Correlation of the distribution in Fig. 1.9(a) with that in Fig. 1.9(b) and (b) correlation of Fig. 1.9(a) with Fig. 1.9(c).

1.5 Distortion-tolerant 3D Object Recognition

In the previous section, it has been mentioned that the recognition can tolerate slight out-of-plane rotations of the object. However, this tolerance is limited to very small angles, typically less than $1°$. In this section, we show how it is possible to make the 3D recognition really robust to distortions by using nonlinear composite filters.[3,17] Such a filter is a combination of several matched filters corresponding to various distortions of the 3D reference object. As an example, we demonstrate tolerance to out-of-plane rotations and longitudinal shifts. The same approach could be generalized, for instance to achieve tolerance to in-plane rotation or scaling. We also describe how to complement the composite filters with a neural network in order to further enhance the distortion tolerance.

1.5.1 Composite Correlation Filters

First of all, we need to dismiss the phase information and to improve the quality of the reconstructed views of the object by ridding them of the speckle noise as described in section **1.1.3**. In the previous section, we performed a nonlinear correlation by using a phase-only filter. Now we use a more general case of nonlinear correlation called kth law.[14] Specifically, we compute the Fourier transforms of the filtered images and raise their Fourier amplitudes to the kth power while retaining their Fourier phase. The case $k = 1$ corresponds to a linear correlation; the case $k = 0$ corresponds to the phase-only filter we utilized in the previous section. For the rest of this chapter, we will use $k = 0.1$, which is not a phase-only filter but is still a strongly nonlinear correlation and is therefore highly discriminant.

As previously, our reference object is a die. Our holograms have 2028×2044 pixels but we only use 1024×1024 pixels windows in order to reconstruct the views of the objects. Here we want to achieve rotation tolerance. Therefore, we record 19 holograms of the reference die with several out-of-plane rotations. For each new hologram, the die is rotated roughly $0.5°$ around the axis, which is orthogonal to Fig. 1.1. The overall rotation angle is around $9°$. Although we only use out-of-plane rotation around the vertical axis, our approach is easy to generalize to any axis of rotation. For every hologram, we reconstruct the corresponding image in the plane of the object. These 19 images are our nontraining true targets (Images #1–19). In order to study the robustness of the object recognition, we also record holograms of the die with a very different illumination (Image #20) and in a different 3D position (Image #21). For this latter case, we will present the correlation in the best focus plane. Finally, we need several false targets to test the discrimination of our filter. Hence, we use seven various objects (Images #22–28) which are completely different from the die.

First, we synthesize a nonlinear filter with only one view of the reference die (the "training" image). This view is reconstructed from the same hologram we

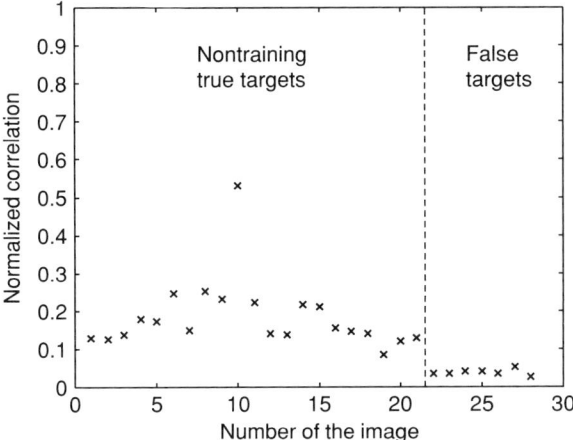

Figure 1.11. Correlation results obtained with various test objects. Filter made from one single view.

used for reconstructing Image #10, but with a different window. Figure 1.11 presents the values of the correlation peaks for all the images. It can be seen that – except for the one that is very close to the training image – the true targets are barely distinguished from the false targets.

In order to improve the recognition range, we construct a synthetic discriminant function filter[3,17] with several reconstructed views of our reference object. Specifically, we normalize the energy of these views and combine them linearly in the Fourier domain in such a way that the resulting composite filter produces the same output peak for each training image.[8] Since we want to achieve rotation tolerance, we have to include in the training images different perspectives of the die. We therefore use three different holograms for reconstructing the views. These holograms are the ones corresponding to Images #3, #10, and #16. They were recorded with the die rotated $-3°$, $0°$, and $+3°$, respectively, compared to the orientation for the previous filter. With each of these holograms, we use three different windows: one centered window and one laterally shifted window in both directions. These windows allow us to reconstruct views of the die with a regular angle change of $0.6°$. We thus have $3 \times 3 = 9$ different views of the reference object with which we make our composite filter. Figure 1.12 shows the results of the correlations with the test images. The filter has been designed in order to obtain an output peak of 1 for the training targets. For the nontraining images, we obtain obviously an output lower than 1. However, it can be seen that it is easy to discriminate true targets from false targets by using a threshold, for instance at 0.2. When comparing Figs. 1.11 and 1.12, it is clear that using a composite filter enhances the recognition of the out-of-plane rotated object.

As we mentioned before, even when testing an object that is similar to the reference, we only get a high output peak when the reconstruction plane is

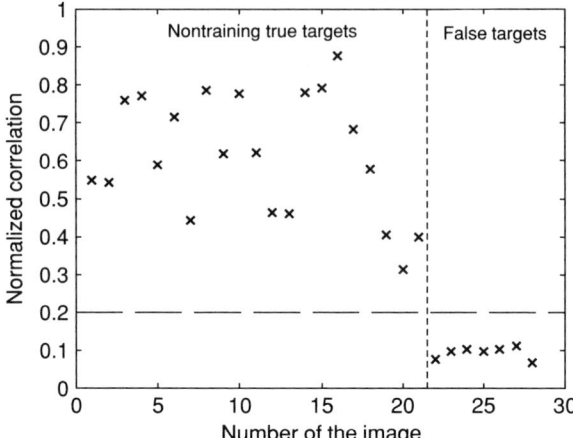

Figure 1.12. Correlation results obtained with various test objects. Filter made from 9 views based on 3 different holograms.

the plane of the object. In some cases it is useful to lower the sensitivity to longitudinal shifts along the z-axis. This allows a reduction of the number of reconstructions that have to be computed in order to recognize the object. Therefore, we design a new composite filter, which includes defocused images of the reference object. Namely, we use again the hologram corresponding to Image #10 and also the same three reconstruction windows as above. However, for each window, in addition to reconstructing the image of the object in the focus plane, we also reconstruct views with a defocus of -20 mm, -10 mm, 10 mm, and 20 mm. We finally obtain $3 \times 5 = 15$ images with which we make the composite filter. We then test the filter with the hologram corresponding to Image #21. The evolution of the output peak value versus longitudinal shift along the z-axis can be seen in Fig. 1.13 for both this new filter and a filter made with the three focused images only. It appears that the new filter is less sensitive to longitudinal shift in the reconstruction of the image. Indeed, the recognition of the die is achieved over a wider range of longitudinal shifts along the z-axis. Moreover, Fig. 1.14 gives the correlation values for all the test images. It appears that the performance of this filter is comparable to the one made with three different holograms (Fig. 1.12), but this time we only need one single hologram to construct the filter.

1.5.2 Neural Network

As we have seen above, the rotation tolerance obtained with a composite filter is significantly greater than with a simple filter. Yet, it remains limited to a few degrees. A natural idea to enlarge the tolerance angle would be to add more

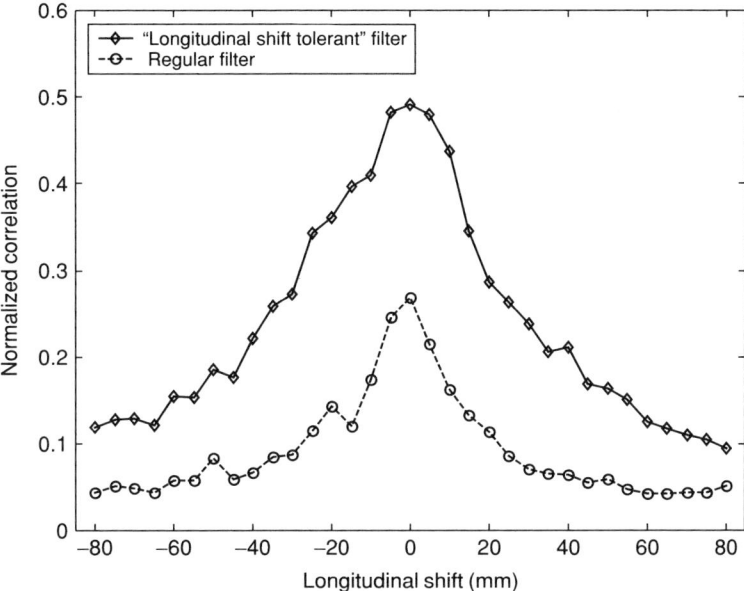

Figure 1.13. Value of the output peak versus longitudinal shift along the z-axis. Comparison between a filter made with focused images ("regular filter") and a filter including defocused images ("longitudinal shift tolerant filter").

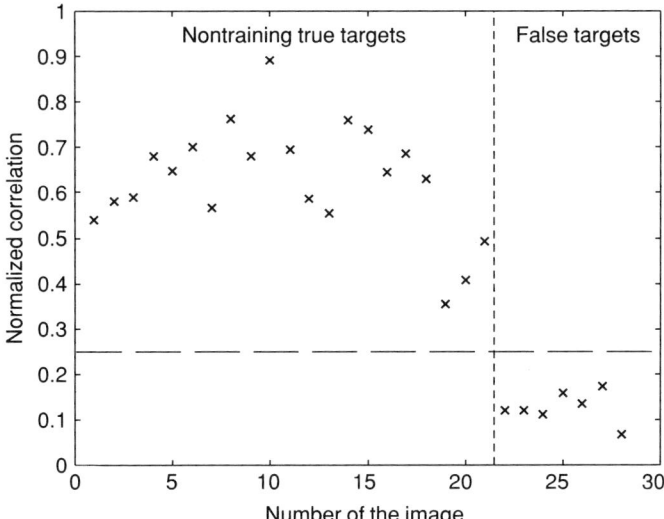

Figure 1.14. Correlation results obtained with various test objects. Filter made from 15 focused and defocused views based on one single hologram.

images during the construction of the composite filter. However this would actually blur the filter and lower its discrimination. Therefore, in order to generalize the recognition of the object to any arbitrary rotation and, in addition, to estimate the out-of-plane orientation of the object, we can think of using a bank of composite filters.

In the previous subsection, we have seen that a single composite filter is able to recognize an object within a 10° rotation angle. In order to be able to deal with a 360° rotation of the object, we record 36 holograms of the die with a 10° rotation step. Each of these holograms is used to construct a composite filter following the technique described above for the longitudinal shift tolerant filter. We thus construct 36 rotation-tolerant composite filters that allow us to recognize the die with any rotation angle.[6] However, it is difficult to estimate the die's orientation because of the similarity of its shape after rotation of 90°. For instance, Figure 1.15 provides the values of the correlation peaks for the 36 filters when presenting a die in a particular position. It can be seen that, besides the correct filter (filter #15), other filters tend to give a significant correlation peak, especially filters #6, #24, and #33, which correspond to rotations of −90°, +90°, and +180°, respectively. Because of this, and as the height of the correlation peak varies with different input holograms, it is difficult to determine a threshold value in order to decide whether one particular filter gives a correct detection or not. As a test, we use twenty holograms of the die with various orientations and also seven false targets. Figure 1.16 (a) shows the recognition error rates (non detections and false alarms) versus the threshold applied to all the filters. The minimum error rate is around 5%.

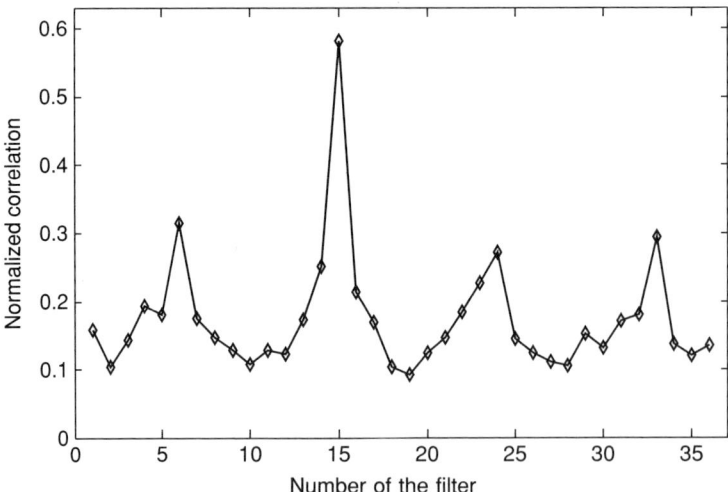

Figure 1.15. Values of the correlation peaks for all of the filters when presented with the die in the orientation corresponding to filter #15.

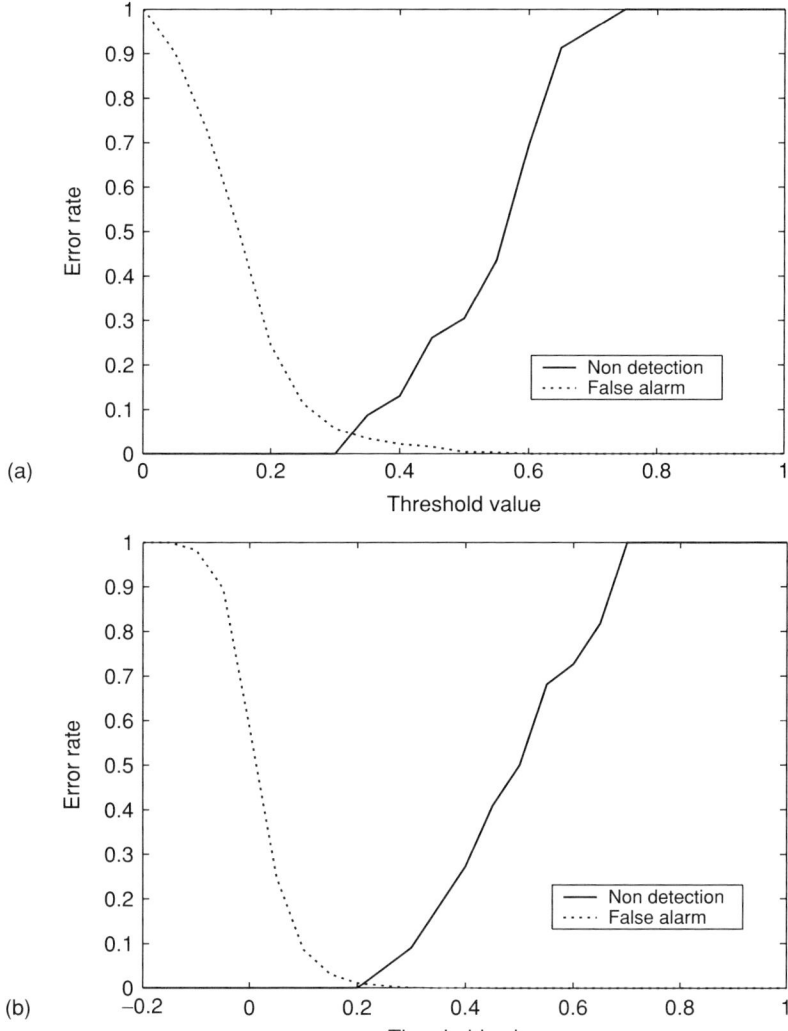

Figure 1.16. Error rates for the recognition and orientation estimation of a die with a 360° rotation: (a) using a bank of composite filters and (b) using a bank of filters and a neural net.

In order to improve the classification, we feed the output of the filter bank to a neural network composed of 36 linear neurons.[6] Our aim is that these neurons correspond to the different orientations of the die (with a 10° step) and that each particular neuron will respond only when the presented image has the correct orientation. In order to determine the weights of this layer, we provide 144 training images with the corresponding desired results. The training images are views of the die reconstructed from the same 36 holograms we used for

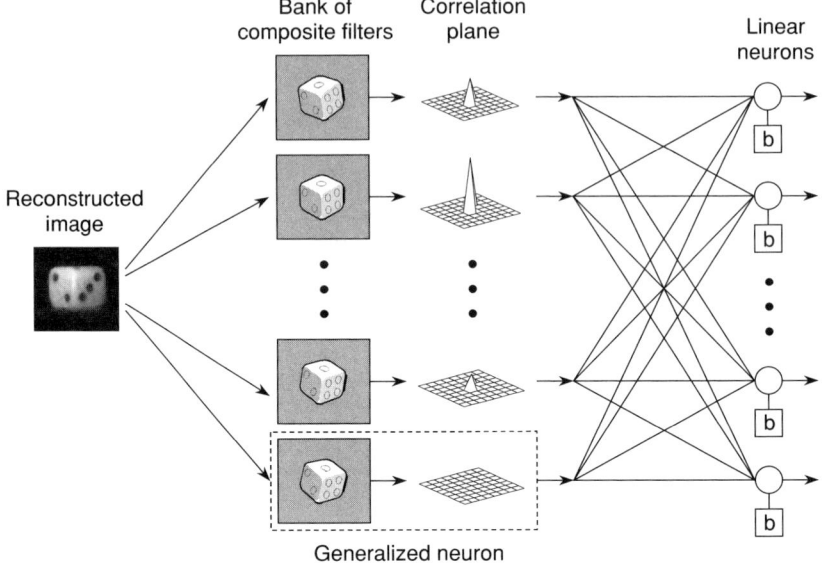

Figure 1.17. Two-layer neural network for recognition and orientation estimation of a die. *b* denotes bias.

making the composite filters. We actually construct four images from each hologram using windows that are different from the ones we used for the composite filters. The responses we desire for these input images are "1" for the output neuron corresponding to the correct orientation and "0" for the other neurons. We can compute the values of the correlation peaks provided by the filters for every image. This gives us the training input vectors for the second layer. Knowing these training vectors and the desired outputs, we can directly compute the weights and the biases that minimize the mean square error.[27] Once the network is designed, we present as inputs the same 27 reconstructed test images we used when testing the first layer. Figure 1.16 (b) shows the error rates of the classification versus the threshold value. The minimum error is now less than 1%. We are thus able to recognize and find the orientation of the object with any rotation.

1.6 Conclusion

In this chapter, we have described how to use digital holograms to reconstruct various views of a 3D object and how to utilize either the holograms themselves or the reconstructed views in order to perform 3D recognition.

The reconstruction technique is based on a numerical computation of the Fresnel diffraction from the digital hologram. The distance of propagation can be chosen arbitrarily and it is thus possible to reconstruct a whole diffraction volume. Moreover, by selecting partial windows of the hologram for the reconstruction, small changes in the angle of view can be obtained. Proper filtering methods allow the removal of the speckle noise on the reconstructed images.

Concerning the 3D recognition, we have first described a technique in which the recognition is carried out by a digital matched filter method applied directly to the holographic information. This technique can measure accurately very small orientation changes of the 3D object. However, because of its high sensitivity, it is unable to deal with larger rotations or translations of the object.

Therefore, we have described a different technique in order to achieve full shift-invariance and to reduce the sensitivity to noise generated by the rough surfaces of the 3D objects. In this second approach, the correlation operation is no longer applied to the holograms but rather to the 3D irradiance distributions generated by the reference and input objects in the object space. These 3D irradiance distributions are generated from single digital holograms. The method allows one to perform 3D correlations in order to recognize the presence and 3D position of the reference in the 3D input scene. In a first experiment, the method has been simplified by evaluating only 2D correlations between 2D sections of the light distributions generated by the 3D reference and the 3D input scene.

Finally, we have extended the previous technique to achieve distortion-tolerance in the recognition process. We have described how to construct non-linear composite filters to take into account distortions of the reference object. These composite filters are made with several views of the object obtained from one or several digital holograms. As an example, we have demonstrated some tolerance to out-of-plane rotation and to longitudinal shift along the z-axis. We have also complemented the composite filters with a neural network in order to improve the rotation tolerance even more. With this combination of techniques, we have been able to achieve a full 360° rotation tolerance. These same techniques can be applied to other kinds of distortions.

The presented experimental results have proven the benefits of applying digital holography to 3D object reconstruction and recognition. Digital holograms can either be used directly to perform 3D recognition, or they can serve for digitally reconstructing the 3D irradiance distributions in the object space. These distributions can then also be used for 3D recognition. In this latter case, it is possible to achieve a shift-invariant and distortion-tolerant recognition.

References

[1]   An X, Psaltis D, and Burr GW. (1999). "Thermal fixing of 10,000 holograms in LiNbO$_3$: Fe." *Appl. Opt.*, 38:386–393.

[2]   Bruning JH, Herriott DR, Gallagher JE, Rosenfeld DP, White AD, and Brangaccio DJ. (1974). "Digital wavefront measuring interferometer for testing optical surfaces and lenses." *Appl. Opt.*, 13:2693–2703.

[3] Casasent D. (1984). "Unified synthetic discriminant function computational formulation." *Appl. Opt.*, 23:1620–1627.

[4] Caulfield HJ. (1979). *Handbook of Optical Holography*. Academic, London.

[5] Esteve-Taboada JJ, Mas D, and García J. (1999). "Three-dimensional object recognition by Fourier transform profilometry." *Appl. Opt.*, 38:4760–4765.

[6] Frauel Y and Javidi B. (2001). "Neural network for three-dimensional object recognition based on digital holography." *Opt. Lett.*, 26:1478–1480.

[7] Frauel Y, Pauliat G, Villing A, and Roosen G. (2001). "High-capacity photorefractive neural network implementing a Kohonen topological map." *Appl. Opt.*, 40:5162–5169.

[8] Frauel Y, Tajahuerce E, Castro M-A, and Javidi B. (2001). "Distortion-tolerant 3D object recognition using digital holography." *Appl. Opt.*, 40:3887–3893.

[9] Frauel Y, Tajahuerce E, Matoba O, Castro A, and Javidi B. (2004). "Comparison of passive ranging integral imaging and active imaging digital holography for three-dimensional object recognition." *Appl. Opt.*, 43:452–462.

[10] Goodman JW. (1968). *Introduction to Fourier Optics*. McGraw-Hill, New York.

[11] Goodman JW and Lawrence RW. (1967). "Digital image formation from electronically detected holograms." *Appl. Phys. Lett.*, 11:77–79.

[12] Guerrero-Bermudez J, Meneses J, and Gualdrón O. (2000). "Object recognition using three-dimensional correlation of range images." *Opt. Eng.*, 39:2828–2831.

[13] Horner JL and Gianino PD. (1984). "Phase-only matched filtering." *Appl. Opt.*, 23:812–816.

[14] Javidi B. (1989). "Nonlinear joint power spectrum based optical correlation." *Appl. Opt.*, 28:2358–2367.

[15] Javidi B, ed. (2002). *Image Recognition and Classification: Algorithms, Systems, and Applications*. Marcel-Dekker, New York.

[16] Javidi B and Horner JL. (1994). *Real-time Optical Information Processing*. Academic, Orlando.

[17] Javidi B and Okano F, eds. (2002). *Three-Dimensional Television, Video, and Display Technologies*. Springer-Verlag, Berlin.

[18] Javidi B and Painchaud D. (1996). "Distortion-invariant pattern recognition with Fourier-plane nonlinear filters." *Appl. Opt.*, 35:318–331.

[19] Javidi B and Tajahuerce E. (2000). "Three-dimensional object recognition using digital holography." *Opt. Lett.*, 25:610–612.

[20] Lippmann G. (1908). "La photographie intégrale." *Comptes-rendus de l'Académie des Sciences*, 146:446–451.

[21] MacAulay AD. (1991). *Optical Computers Architectures*. John Wiley, New York.

[22] Mas D, Garcia J, Ferreira C, Bernardo LM, and Marinho F. (1999). "Fast algorithms for free-space diffraction patterns calculation." *Opt. Commun.*, 164:233–245.

[23] Matoba O, Tajahuerce E, and Javidi B. (2001). "Real-time three-dimensional object recognition with multiple perspectives imaging." *Appl. Opt.*, 20:3318–3325.

[24] Okoshi T. (1971). *Three-dimensional Imaging Techniques*. Academic, New York.

[25] Pu A, Denkewalter R, and Psaltis D. (1997). "Real-time vehicle navigation using a holographic memory." *Opt. Eng.*, 36:2737–2746.

[26] Refrégiér Ph, Laude V, and Javidi B. (1994). "Nonlinear joint transform correlation: An optimum solution for adaptive image discrimination and input noise robustness." *Opt. Lett.*, 19:405–407.

[27] Ritter H, Martinetz T, and Schulten K. (1992). *Neural Computation and Self-organizing Maps*. Addison-Wesley, New York.

[28] Rosen J. (1998). "Three-dimensional joint transform correlator." *Appl. Opt.*, 37:7538–7544.

[29] Schnars U and Jüpter W. (1994). "Direct recording of holograms by a CCD target and numerical reconstruction." *Appl. Opt.*, 33:179–181.

[30] Tajahuerce E, Matoba O, and Javidi B. (2001). "Shift-invariant three-dimensional object recognition by means of digital holography." *Appl. Opt.*, 40:3877–3886.

[31] VanderLugt AB. (1964). "Signal detection by complex spatial filtering." *IEEE Trans. Inf. Theory IT*, 10:139–145.

[32] VanderLugt AB. (1992). *Optical Signal Processing*. John Wiley, New York.

[33] Weaver CS and Goodman JW. (1966). "A technique for optically convolving two functions." *Appl. Opt.*, 5:1248–1249.

[34] Yamaguchi I and Zhang T. (1997). "Phase-shifting digital holography." *Opt. Lett.*, 22:1268–1270.

Compression of Encrypted Digital Holograms Using Artificial Neural Networks

Alison E Shortt[1], Thomas J Naughton[1], and Bahram Javidi[2]

[1]Department of Computer Science, National University of Ireland, Maynooth, County Kildare, Ireland tom.naughton@may.ie
[2]Department of Electrical and Computer Engineering University of Connecticut, 260 Glenbrook Road, U-2157 Storrs, CT 06269-2157, USA bahram@engr.uconn.edu

2.0 Introduction

An important aspect of security and defense is information gathering, dissemination, processing, and analysis. Central to this is the encryption and decryption of messages for storage and transmission. Although public key cryptosystems are the state of the art currently, there is a place for private key systems in cases where hardware implementation permits very high throughputs. Optical implementation is one such candidate that promises huge throughputs.[1-22] Optics has some very promising scalability advantages over purely electronic systems as, in principle, the size of the key can be increased without increasing the encryption or decryption time. Furthermore, optics is perfectly suited to scenarios where message distortion in the encryption/decryption process is permissible in order to increase efficiency. In such scenarios, the secure transmission of image information, for example, compression and encryption/decryption, go hand in hand.

Digital holography,[23-31] and particularly phase-shift interferometry (PSI),[29-31] can record high quality representations of both the amplitude and phase of complex-valued optical wavefronts, and has been proposed for three-dimensional (3D) object recognition and 3D display applications[32-38]. Recently, digital holography has been used in the encryption of two-dimensional images[12-14] and 3D objects.[20-22]

In this chapter, the complex-valued encrypted holographic pixels are quantized nonuniformly using an unsupervised artificial neural network (unsupervised ANN) to achieve lossy data compression. Two important differences between digital hologram compression and conventional image compression[39], are that our holograms store 3D information in complex-valued pixels, and their inherent speckle content which gives the holograms a white-noise appearance. Holographic speckle is difficult to remove since it actually carries 3D information. Its presence causes lossless data compression techniques to perform badly, therefore, lossy compression techniques are necessary for effective compression of 3D digital holograms.[36]

Quantization in holograms,[40,41] and compression of real-valued[42] and complex-valued[21,36,37,43] digital holograms has received some attention to date. Some studies have also been performed on the decrypted-domain effects of perturbations, including quantization, in the encrypted domain.[44,45] This introduces a third reason why compression of digital holograms differs from compression of digital images; a change locally in a digital hologram will, in theory, affect the whole reconstructed object. Furthermore, when gauging the errors introduced by lossy compression, we are not directly interested in the defects in the hologram itself, only how compression noise affects the quality of reconstructions of the compressed 3D object.

We used PSI to create our in-line digital holograms.[32,34] These holograms were encrypted by perturbing the Fresnel diffraction of the 3D objects with a random phase mask. We simulated this encryption step in software.[22] The dimensions of each encrypted hologram are 1024×1024 pixels. Encrypted digital holograms have been successfully quantized previously. We extend these results[21] by choosing nonuniform distributions of quantization values. We describe these nonuniform quantization techniques and present experimental results to justify our final choice of a Kohonen competitive neural network. We consider each complex value as a vector of length two and use the unsupervised ANN to locate the most suitable clusters in the encrypted digital hologram data. We then quantize our encrypted holograms with the centers of these clusters. We use a reconstructed-object-plane RMS metric to quantify the quality of our decompressed and decrypted holograms.

The structure of the chapter is as follows. In Section 2.1, we outline how we perform encryption of the Fresnel propagation of 3D objects using a random phase mask, and how the complex wavefront is subsequently captured using PSI. In Section 2.2, the decryption and reconstruction steps are explained. In Section 2.3, we examine the amenability of encrypted digital holograms to lossless compression using four well-known techniques. In Section 2.4 we discuss two types of Kohonen ANN that we used to quantize our 3D digital hologram data. We assess the performance of the nonuniform quantization techniques in Section 2.5 and find one that best suits our hologram data. We then apply this lossy technique of quantization to the real and imaginary encrypted components of each holographic pixel in Section 2.6. In this section too, we quantify quantization error by measuring deformation in the decrypted and reconstructed 3D object intensities, and finally conclude in Section 2.7.

2.1 Digital Hologram Encryption

The encrypted complex-valued holograms can be captured using an optical setup (shown in Fig. 2.1) based on a Mach-Zehnder interferometer architecture. [32,34] A linearly polarized Argon ion (514.5 nm) laser beam is divided into object and reference beams, both of which are spatially filtered and expanded. The first beam illuminates the 3D object placed at an approximate distance $d_1 + d_2 = 350\,\mathrm{mm}$ from a 10-bit 2028×2044 pixel Kodak Megaplus CCD camera. A random phase mask is placed a distance d_1 from the 3D object. Due to free-space propagation, and under the Fresnel approximation, [46,47] the signal at the detector plane $H_\mathrm{E}(x, y)$ is given by the superposition integral

$$
\begin{aligned}
H_\mathrm{E}(x, y) = \; & \frac{-\mathrm{i}}{\lambda\, d_2} \exp\left(\mathrm{i}\frac{2\pi}{\lambda} d_2\right) \int \int_{-\infty}^{\infty} \exp\left[\mathrm{i}\Phi(x', y')\right] \\
& \times A_\mathrm{M}(x', y') \exp\left[\mathrm{i}\phi_\mathrm{M}(x', y')\right] \\
& \times \exp\left\{\mathrm{i}\frac{\pi}{\lambda\, d_2}[(x - x')^2 + (y - y')^2]\right\} \mathrm{d}x'\mathrm{d}y'
\end{aligned}
\tag{1}
$$

where A_M and ϕ_M are the amplitude and phase, respectively, of the signal in the plane of, but immediately before, the random phase mask Φ. $H_\mathrm{E}(x, y)$ will have both its amplitude and phase modulated by the mask and will have a dynamic range suitable for capture by a CCD camera. The reference beam passes through half-wave plate RP_1 and quarter-wave plate RP_2. This linearly polarized beam can be phase-modulated by rotating the two retardation plates.

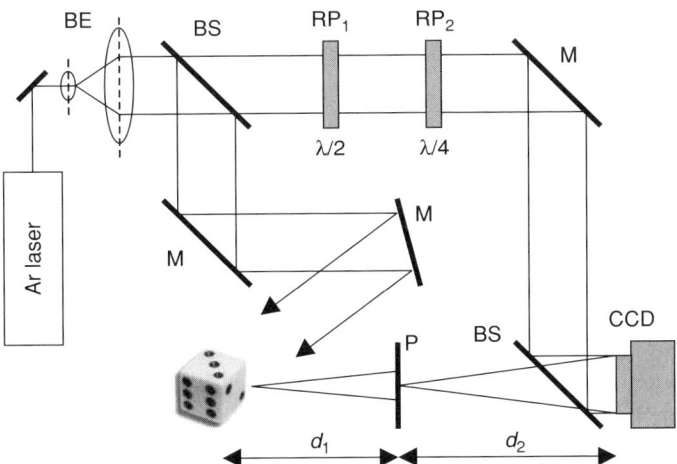

Figure 2.1. Experimental setup for three-dimensional object encryption using phase-shift digital holography: BE, beam expander; BS, beam splitter; M, mirror; RP, retardation plate; P, phase mask.

Through permutation of the fast and slow axes of the plates we can achieve phase shifts of 0, $-\pi/2$, $-\pi$, and $-3\pi/2$. The reference beam combines with the light diffracted from the object and forms an interference pattern in the plane of the camera. At each of the four phase shifts we record an interferogram. Using these four real-valued images, the complex camera-plane wavefront can be approximated to good accuracy using PSI.[32,34]

In this system, the encryption key is $(\boldsymbol{\Phi}, x, y, d_2, \lambda, e_x, e_y, d_1)$, consisting of the random phase mask, its position in 3D space, the wavelength of the illumination, the dimensions of the detector elements (for a pixilated device), and the distance between the mask and the notional center of the object, respectively. This key is also exactly the decryption key: a means of decrypting and reconstructing an arbitrary view of the 3D object encoded in the hologram.

2.2 Decryption and Reconstruction

The decryption and reconstruction of the digital hologram can be carried out optically or digitally. The hologram is propagated a distance d_2 to plane P and decrypted by multiplying it with the phase mask. It is reconstructed through further Fresnel propagation to focus in any chosen plane in the range $d_1 \pm D$.

A decrypted digital hologram contains sufficient amplitude and phase information to reconstruct the complex field $U(x, y, z)$ in a plane in the object beam at any distance z from the camera. Like traditional holography,[47] different angles of view of the object can be reconstructed using different windowed subsets of the hologram. These views are obtained by multiplying the decrypted and reconstructed object by a suitable linear phase factor[32] within the angular range of the hologram. The number of possible viewing angles is dependent on the ratio of the window size to the full CCD sensor dimensions. Our CCD sensor is approximately $18.5 \, \text{mm} \times 18.5 \, \text{mm}$ and so a 1024×1024 pixel window has a maximum lateral shift of 9 mm across the face of the CCD sensor.[34] So the range of viewing angles that are possible with an object placed $d = 350 \, \text{mm}$ from the camera is ± 0.74 deg. Smaller windows will permit a larger range of viewing angles at the expense of image quality at each viewpoint.

The intensity images of two of the objects used for the experiments in this chapter are shown in Fig. 2.2. These images were reconstructed from digital holograms that were created using a similar setup to that shown in Fig. 2.1, without the phase mask positioned in plane P.[32,34] Both objects are approximately $5 \, \text{mm} \times 5 \, \text{mm} \times 5 \, \text{mm}$ in size and were positioned 323 mm (for the die) and 390 mm (for the bolt) from the camera. We use our reconstructions later in the chapter to quantify lossy compression errors.

By digitally encrypting the holograms that were captured without a random phase mask,[22] we achieve added flexibility and security,[22] while still accommodating the possibility for a real-time optical reconstruction.[16,37,48] Figure 2.3 shows the 1024×1024 pixel phase mask used in our experiments.

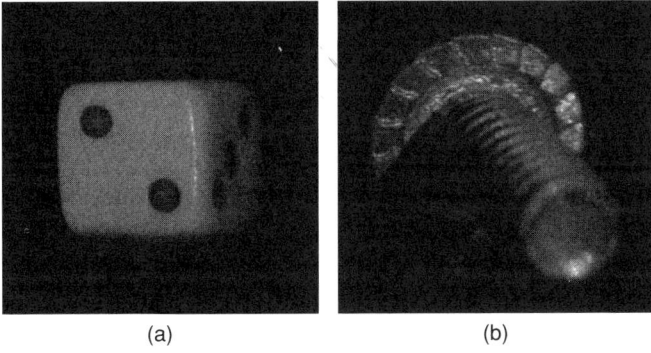

(a)   (b)

Figure 2.2. Objects used in the study: (a) die, (b) bolt.

It contains values chosen with uniform probability from the range $[0, 2\pi)$ using a pseudo-random number generator. The position of the phase mask is illustrated in Fig. 2.1 and the ratio of the distances $d_1 : d_2$ is 35 : 65. In Fig. 2.4 we show the amplitude and phase of the bolt hologram before encryption, and after encryption as described by (1). In Fig. 2.5 we show the results of reconstructing an encrypted digital hologram with and without the phase mask used in the encryption step.

2.3 Lossless Compression of Encrypted Digital Holograms

In order to motivate the need for lossy compression techniques, the digital holograms were treated as binary data streams and compressed using the

Figure 2.3. Example of a random phase mask used in the study.

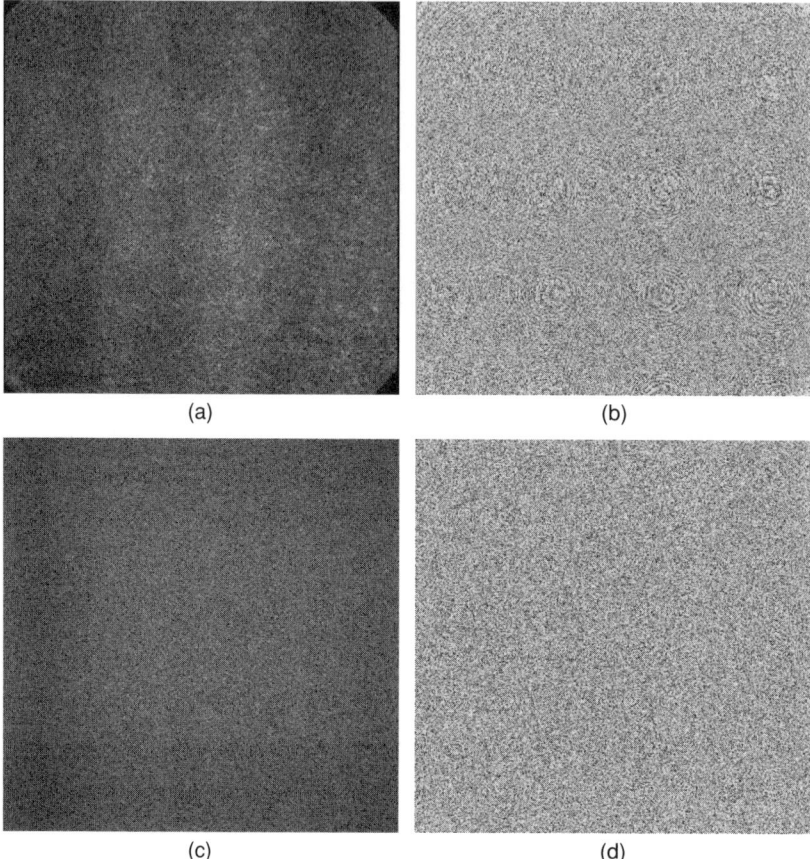

Figure 2.4. The bolt hologram before and after encryption: (a) amplitude, and (b) phase of the original hologram, and (c) amplitude, and (d) phase of the encrypted hologram.

lossless data compression techniques of Huffman,[49] Lempel–Ziv (LZ77),[50] Lempel-Ziv–Welch (LZW),[51] and Burrows–Wheeler (BW).[52] Huffman coding,[49] an entropy-based technique, is one of the oldest and most widely used compression methods. Each symbol in the input is replaced by a codeword, with more frequent symbols assigned shorter codewords. The LZ77 algorithm[50] takes advantage of repeated substrings in the input data and replaces variable length strings with a pointer to the previous occurrence of that string. The LZW[51] improves upon LZ77 by maintaining a lookup table of variable sized codewords and is also less biased towards local redundancy. Finally, the BW technique[52] uses a sorting operation to transform its input into a format that can be compressed very effectively using standard techniques (in our particular implementation, Huffman coding).

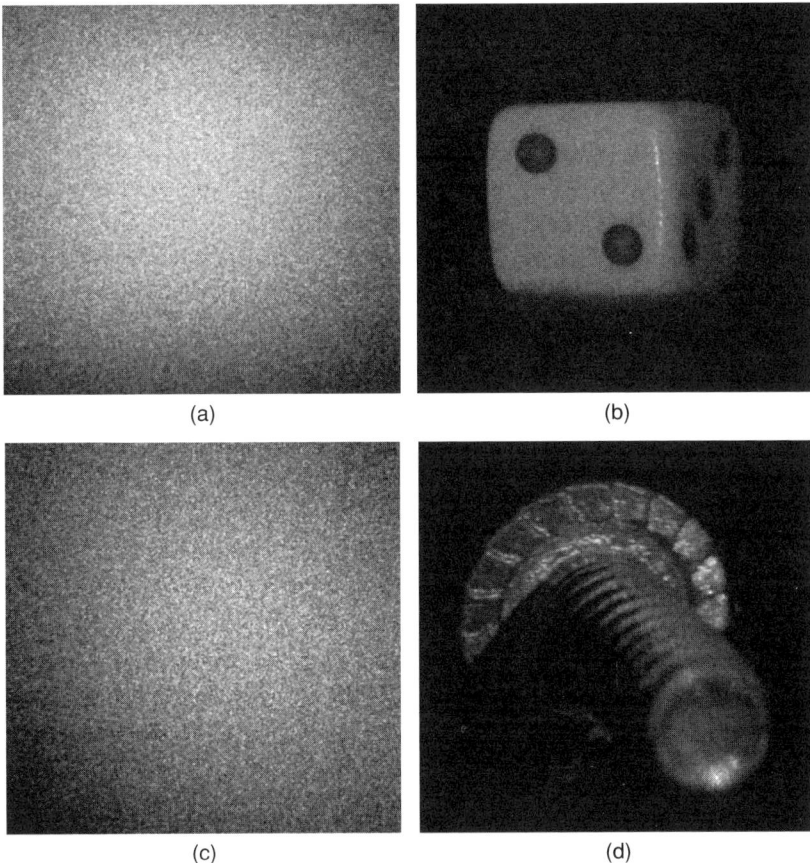

(a)   (b)

(c)   (d)

Figure 2.5. Reconstruction of the die hologram (a) without the phase mask, and (b) with the phase mask, and reconstruction of the bolt hologram (c) without the phase mask, and (d) with the phase mask.

The two digital holograms used in the experiments have dimensions of 1024×1024 pixels, with each pixel storing 8 bytes of amplitude information and 8 bytes of phase information. This amounts to a file size of 16384kB where $1\,\mathrm{kB} = 2^{10}$ bytes. In a previous study of unencrypted digital holograms,[36] lossless techniques have been shown to achieve compression ratios in the range [1.0, 6.66] where compression ratio is calculated by dividing a hologram's uncompressed size by its compressed size.

The two holograms were encrypted with the phase mask shown in Fig. 2.3. For these experiments, unencrypted holograms of the 3D objects were captured optically[32,34] and the encryption steps described in (1) were simulated in software.[22] The four lossless compression techniques were applied to each hologram and the results are shown in Table 2.1. The poor compression ratios

Table 2.1. Lossless compression of encrypted digital holograms; c.r., compression ratio.

Hologram	Size (kB)	LZ77 (kB)	LZW (kB)	Huff. (kB)	BW (kB)	LZ77 c.r	LZW c.r.	Huff. c.r.	BW c.r.
die	4097	3918	5296	3914	4003	1.05	1.00	1.05	1.02
bolt	4097	3918	5297	3915	4003	1.05	1.00	1.05	1.02
Averages:						1.05	1.00	1.05	1.02

testify to the lack of redundancy or structure in the encrypted hologram data, even compared to unencrypted digital holograms. The random phase mask, combined with Fresnel propagation, is very effective at removing apparent structure from the hologram data. With LZW, the compressed sizes were even larger than the uncompressed. In these cases a compression ratio of 1.0 (indicating zero compression) is reported. These results illustrate the urgent need to explore lossy compression techniques suitable for encrypted digital holograms. One such lossy technique that has been successfully applied to 3D digital holograms is quantization.[21,36,37,43]

2.4 ANNs Suitable for Nonuniform Quantization

Artificial Neural Network clustering algorithms have been successfully used for vector quantization and image compression[53,54] in the past. We use the Kohonen competitive network[55] (also known as a vector quantization network) and the self-organizing map (SOM)[55] for quantizing our digital holograms.

The Kohonen competitive neural network[55] consists of two layers, an input layer and a competitive layer. Weight vectors, connecting the input neurons to the output neurons, are initially set to the midpoint of the range of input values. An unsupervised learning strategy allows these weight vectors to learn to cluster the input data naturally without any a priori information. An input vector is randomly chosen and presented to the network. The neuron whose weight vector is closest to the input vector wins the competition. The winning neuron has its weight vectors updated in order to draw it closer to the input vector and the weight vectors of all other neurons are unchanged. This is known as hard competition.

Kohonen desired a characteristic known as equiprobability for his competitive network, whereby an input vector chosen at random from the training set would have equal probability of being close to any of the weight vectors.[56] It was Desieno[57] who proposed a conscience mechanism that not only enforced equiprobability but also fixed the over-clustering problem (the problem of combining a number of diverse clusters into one large cluster) and alleviated the dead neuron problem (where neurons that are positioned far away from the input data may never influence clustering) that were present in the original competitive network. By monitoring the success of all neurons, conscience creates a fatigue effect[55] on neurons that are winning a lot in order to give others a chance. This encourages neurons to spread out into undersampled areas of the input space. This network also has a learning rate associated with it that controls the amount by which the winning weight vectors are updated during learning. Training ceases when the maximum number of epochs is reached, performance has minimized the goal, or the maximum amount of time has been exceeded.

The SOM,[55] based on earlier work by a number of researchers,[58,59] updates the winning neuron's weights and the weights of neurons located in the neighborhood of the winner. This is known as soft competition. The neighboring weight vectors are updated to a lesser degree depending on how far away from the winning neuron they are. Experimentation has shown that the best results are obtained when the neighborhood is large initially and shrinks monotonically over time.[55] This results in a rough global ordering of the input data initially and as the neighborhood shrinks this ordering becomes finely tuned. The SOM network also has a two-layer structure. The competitive layer consists of a grid, usually 2D, of connected neurons. This grid stretches and mutates its shape to arrange its neurons to successfully represent the patterns in the input data. The number of neurons in the grid affects quality of results and training time; more neurons give improved accuracy but increase training time.

During the ordering learning phase of the SOM, the neighborhoods are defined, i.e., neurons arrange themselves so that neurons that are sensitive to similar inputs will be located close together. The learning rate is initially high to allow self-organization. In the tuning phase, the weight vectors are expected to spread out relatively evenly over the input space, while retaining their topological ordering that was found during the ordering phase. This tuning phase generally performs between 10 and 100 times as many steps as the ordering phase.[55] The learning rate should be kept small as the neighborhood will also be small at this stage. The distance function most often used for the SOM is Euclidean distance.

Both the Kohonen competitive and the SOM neural networks are given an initial number of cluster centers and will use as many as required to successfully cluster the input data. A maximum number of epochs is also allocated. Both networks learn the distribution of the input data. In addition to this, SOM learns and preserves the structure of the input space; neighboring neurons represent similar input data and densely populated regions are mapped to larger regions in the output space. In the next section we discuss the results we obtained from evaluating these neural network quantization techniques with our digital hologram data.

2.5 Evaluation of Nonuniform Quantization Techniques

Uniform quantization is the optimal choice when the data values are uniformly distributed. Since our hologram data consists of unevenly distributed complex

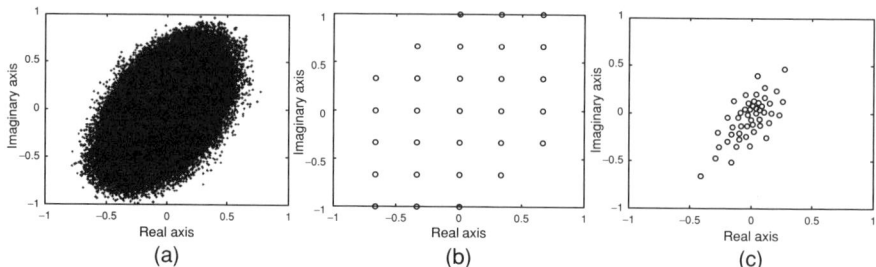

Figure 2.6. Scatter plots of the complex-valued data in the die hologram: (a) before quantization, (b) uniformly quantized with 3 bits per dimension (49 clusters), (c) nonuniformly quantized (k-means) with 49 clusters.

values [see Fig. 2.6(a)], nonuniform quantization techniques are more suitable. Methods for clustering data include discriminant analysis[60] the k-means algorithm,[61-64] competitive neural networks,[55] SOMs,[55] and support vector machines.[65] We applied some of these nonuniform quantization techniques to our original (unencrypted) hologram data and found that the Kohonen competitive neural network performed best.

Initially we looked at the popular k-means clustering algorithm,[61-64] which is suitable for clustering large amounts of data. This algorithm clusters the data by observing similarity. It is an iterative process operating on a fixed number of k clusters (codebook vectors) that attempts to minimize some distance metric between the input vectors (unquantized data) and code-book vectors. In our tests, an input vector of 2048 complex-valued pixels was chosen randomly from the digital hologram. We dealt with empty clusters by repositioning their centers to the data vector that was furthest from it. For the initial codebook we chose k initial cluster centroid positions randomly from the input. The distance measure that we used was the sum of the Euclidean distances, which sets each centroid to the mean of the points in its cluster. Figure 2.6(c) shows the distribution of clusters relative to the hologram data, compared to uniform quantization Fig. 2.6(b). One advantage of k-means nonuniform quantization over uniform is that no codebook vectors are wasted on unpopulated regions. This is quite visible in Fig. 2.6(b) where only 29 of the 49 uniform clusters are actually used.

Our subsequent experiments involved the use of unsupervised ANN techniques (Kohonen competitive and SOM) to quantize hologram data, with k-means used to compare performance. The ANNs were given an initial number of centers before training. Training was then performed for a fixed number of epochs, during which time each network used as many centers as it needed to cluster the input data. Generally, only a subset of the centers would be used, in contrast to k-means where all of the centers are utilized. For the initial codebook all initial cluster centroid positions are set to the midpoint of the input data and these continue to spread out over the input data as training proceeds.

The distance that the codebook vector is moved depends on the learning rate. As explained in Section 2.4, the Kohonen competitive network has both a learning rate and a conscience learning rate, while the SOM has an ordering learning rate and a tuning learning rate.

We performed extensive tests in order to determine the most appropriate ANN parameters for our data. The first set of experiments sought to determine how many training epochs would be required. A 32×32 pixel window of the die hologram was used. For training durations from 2 epochs to 2000 epochs, and for numbers of clusters from 9 to 81, the networks were trained with the hologram window. The trained networks were then used to quantize the full 1024×1024 pixel hologram and the resulting reconstructions U' by numerical propagation were compared with reconstructions U_0 from original (unquantized) versions of the holograms in terms of normalized rms (NRMS) difference between their intensities, defined as

$$
D = \left[\sum_{m=0}^{N_x-1} \sum_{n=0}^{N_y-1} \left\{ |U_0(m,n)|^2 - |U'(m,n)|^2 \right\}^2 \right. \\
\left. \times \left(\sum_{m=0}^{N_x-1} \sum_{n=0}^{N_y-1} \left\{ |U_0(m,n)|^2 \right\}^2 \right)^{-1} \right]^{1/2} ,
\tag{2}
$$

where (m, n) are discrete spatial coordinates in the reconstruction plane, and N_y and N_x are the height and width of the reconstructions, respectively. In order to lessen the effects of speckle noise we examine only intensity in the reconstruction plane and apply a mean filtering operation prior to calculating NRMS. The results for both networks can be seen in Fig. 2.7. Since, for both types of network, in the order of 10^3 epochs produced only marginally better performance than 10^2 epochs, we chose 200 epochs as our default training duration.

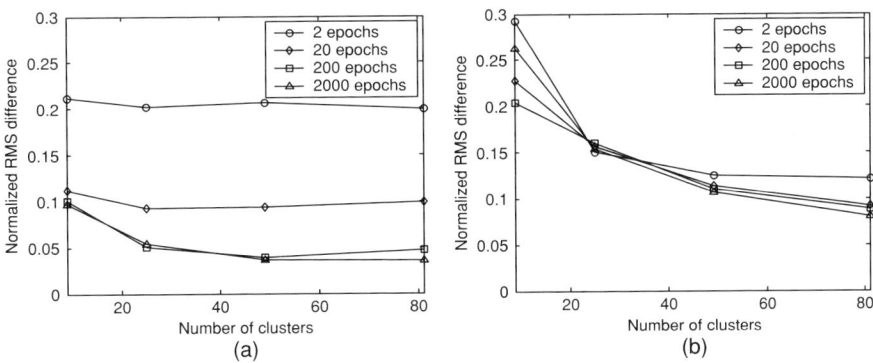

Figure 2.7. Results of experiment using a 32×32 pixel window of the die hologram to determine the required number of epochs for digital hologram compression: (a) Kohonen competitive network, and (b) SOM.

Appropriate learning rates had to be chosen for the ANNs. For these experiments, 128×128 pixel windows from each of the two holograms were used. For several learning rates, and for several numbers of clusters, the networks were trained on the hologram windows. After each training cycle of 200 epochs, the hologram was quantized using the network and the error in the hologram reconstruction measured. The results are shown in Fig. 2.8. For the Kohonen competitive network, the learning rate of 0.1 was deemed the most appropriate. For the SOM, the combination of an ordering phase learning rate of 0.9 and a tuning phase learning rate of 0.1 was favored. For the SOM, the following additional parameter settings were chosen. A topology was chosen that creates a set of neurons that form a hexagonal pattern. We used $(\sqrt{n} \times \sqrt{n})$ as the dimensions of the ith layer, where n was the number of clusters. We employed 1000 ordering phase steps, and set a tuning phase neighborhood distance equal to 1.

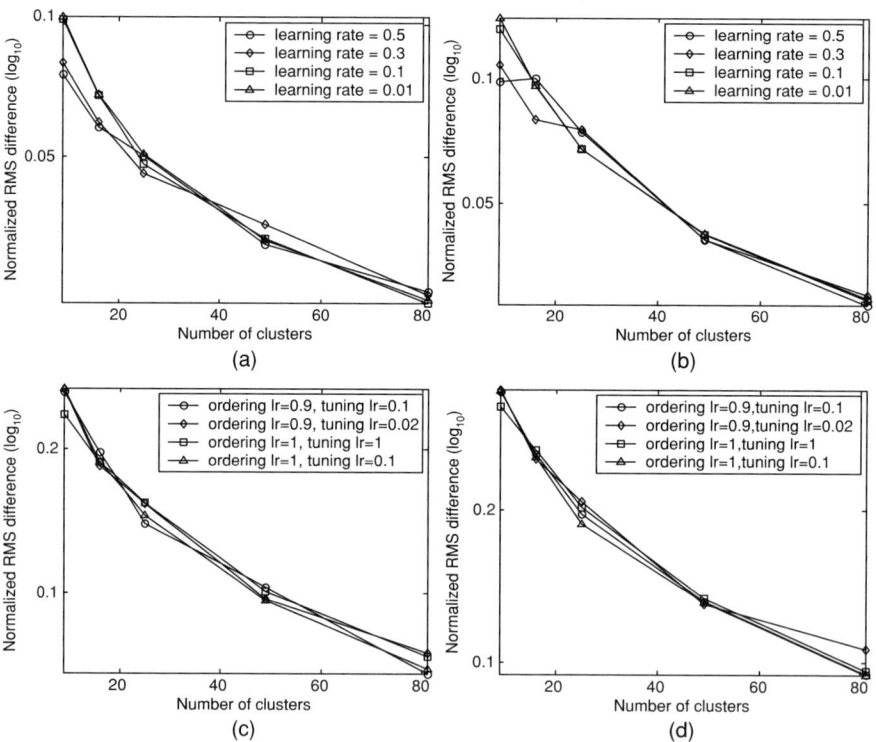

Figure 2.8. Results from experiment to determine required learning rates: Kohonen competitive network with 128×128 pixel window of (a) die hologram, and (b) bolt hologram; and SOM with 128×128 pixel window of (c) die hologram, and (d) bolt hologram. lr: learning rate.

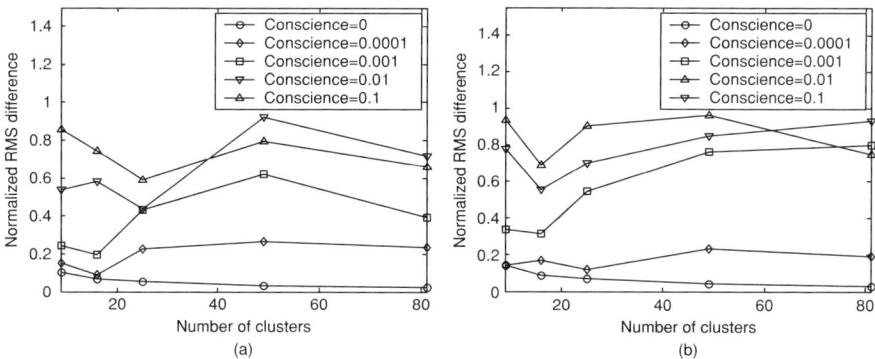

Figure 2.9. Results of experiment to determine the appropriate conscience learning rate for the Kohonen competitive network, with 128×128 pixel windows of (a) die hologram, and (b) bolt hologram.

The Kohonen competitive network also has a conscience learning rate parameter. Experiments were performed to determine the appropriate value for this parameter, the results of which are shown in Fig. 2.9. It was found that all nonzero conscience learning rates were unsuitable for our white-noise-like digital hologram data. In these experiments, the number of neurons was set to be equal to the required number of clusters.

Having determined the appropriate parameters to get the best possible performance out of the two neural networks for our particular holographic data, we applied both networks to the compression of larger, 256×256 pixel, windows of both digital holograms. Figure 2.10 shows the distribution of clusters for both the Kohonen competitive network and the SOM, for the die hologram and with 49 clusters. The Kohonen competitive network seems to

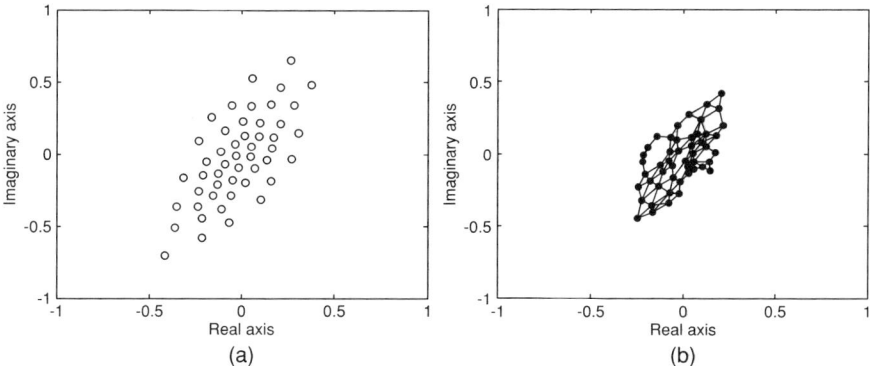

Figure 2.10. Scatter plots of the nonuniformly quantized complex-valued data in the die hologram, quantized with (a) Kohonen competitive, and (b) SOM, both with 49 clusters.

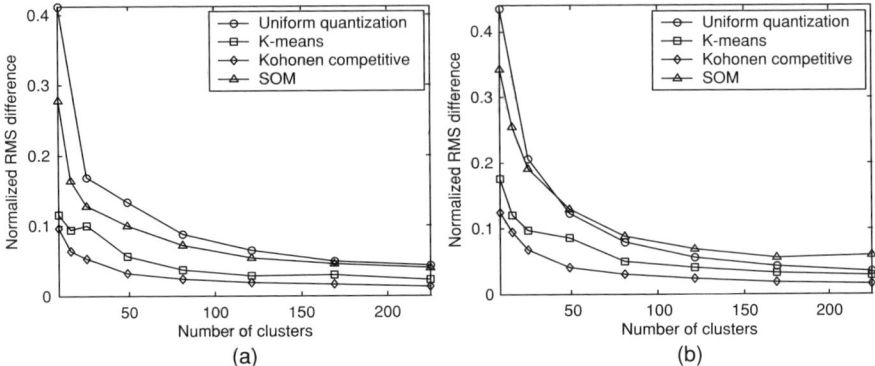

Figure 2.11. NRMS difference in the reconstructed objects plotted against number of clusters: (a) die hologram, and (b) bolt hologram.

allocate its clusters for greater coverage of the hologram data [recall Fig. 2.6(a)] than the SOM.

The resulting NRMS reconstruction errors [calculated using (2)] for various numbers of clusters are compared in Fig. 2.11. Fewer numbers of clusters corresponds to a higher compression ratio. In our experiments, the following learning rates were chosen. For the Kohonen competitive network, a learning rate of 0.1 was used. For the SOM, the combination of an ordering phase learning rate of 0.9 and a tuning phase learning rate of 0.1 was applied. The training for each network was set at 200 epochs. For comparison purposes, Fig. 2.11 also includes the NRMS error for uniform quantization and nonuniform k-means quantization. The k-means algorithm acts as an appropriate benchmark for comparison with the neural network results. For both holograms, k-means clearly performs better than the SOM, which itself is only slightly better than uniform quantization. The Kohonen competitive network consistently beat the other techniques over all trials. Having identified the Kohonen competitive network as being the more appropriate unsupervised ANN for digital hologram compression, we next apply it to larger encrypted digital holograms.

2.6 Quantization of Encrypted Digital Holograms

A uniform quantization technique was used to investigate the loss in reconstruction quality due to quantization in encrypted holograms, and to comparatively evaluate the quality of the results obtained using the Kohonen competitive neural network. The uniform quantization technique linearly rescaled the encrypted holograms to the square in the complex plane $[-1 - i, 1 + i]$ without changing their aspect ratio in the complex plane. The real and imaginary components of each holographic pixel were then quantized. Quantization levels were chosen to be symmetrical about zero; as a result b bits encode $(2^b - 1)$ levels. For example, 2 bits encode levels $\{-1, 0, 1\}$, 3 bits encode levels $\{-1, -2/3, -1/3, 0, 1/3, 2/3, 1\}$, and so on. The combined rescale and quantization operation is defined for individual pixels as

$$H'(x,\ y) = \text{round}[H(x,\ y) \times \sigma^{-1} \times \beta] \times \beta^{-1} \tag{3}$$

and was applied to each pixel $(x,\ y)$ in the encrypted hologram H, where

$$\sigma = \max\{|\min [\text{Im}(H)]|, |\max [\text{Im}(H)]|, \\ |\min [\text{Re}(H)]|, |\max [\text{Re}(H)]|\}, \tag{4}$$

and where $\beta = 2^{(b-1)} - 1$. Here, b represents the number of bits per real and imaginary value, max ($\cdot$) returns the maximum scalar in its argument(s), and round (α) is defined as $\lfloor \alpha + 0.5 \rfloor$. After quantization, each real and imaginary value will be in the range $[-1, 1]$.

Nonuniform quantization was then employed to quantize the encrypted hologram data. The Kohonen competitive neural network was trained on a 128×128 pixel window of encrypted digital hologram data. We used the resulting centers to quantize the full 1024×1024 pixel encrypted digital hologram. Figure 2.12(a) shows a scatter plot of the unquantized 128×128 pixel window of the die hologram that was used to train the ANN. Figure 2.12(b) shows the cluster positions found by the ANN (equivalently, this is a scatter plot of the quantized encrypted data). Figure 2.12(c) shows a scatter plot of the full 1024×1024 pixel hologram that the clusters from Fig. 2.12(b) were applied to. Figures 2.12(d)–(f) show equivalent scatter plots for the bolt hologram.

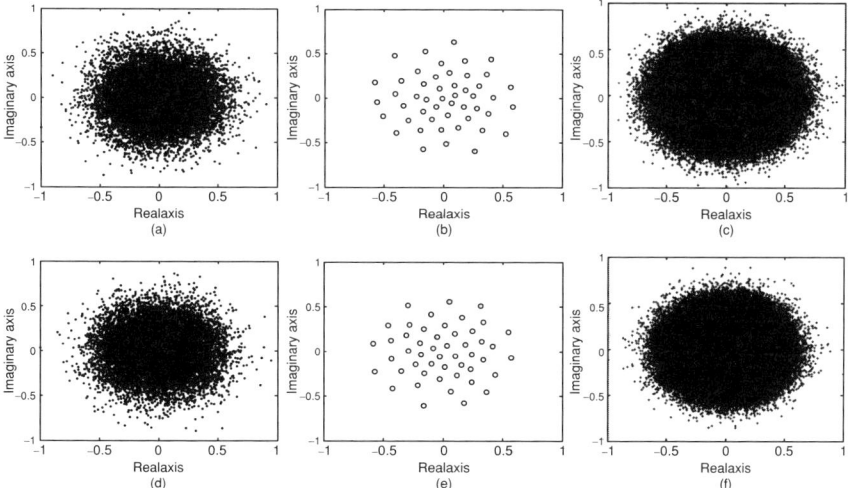

Figure 2.12. Scatter plots of the complex-valued encrypted data in the die hologram: (a) 128×128 window before quantization, (b) 128×128 window nonuniform quantization (Kohonen competitive), (c) 1024×1024 window before quantization; and of the data in the bolt hologram: (d) 128×128 window before quantization, (e) 128×128 window nonuniform quantization (Kohonen competitive), (f) 1024×1024 window before quantization.

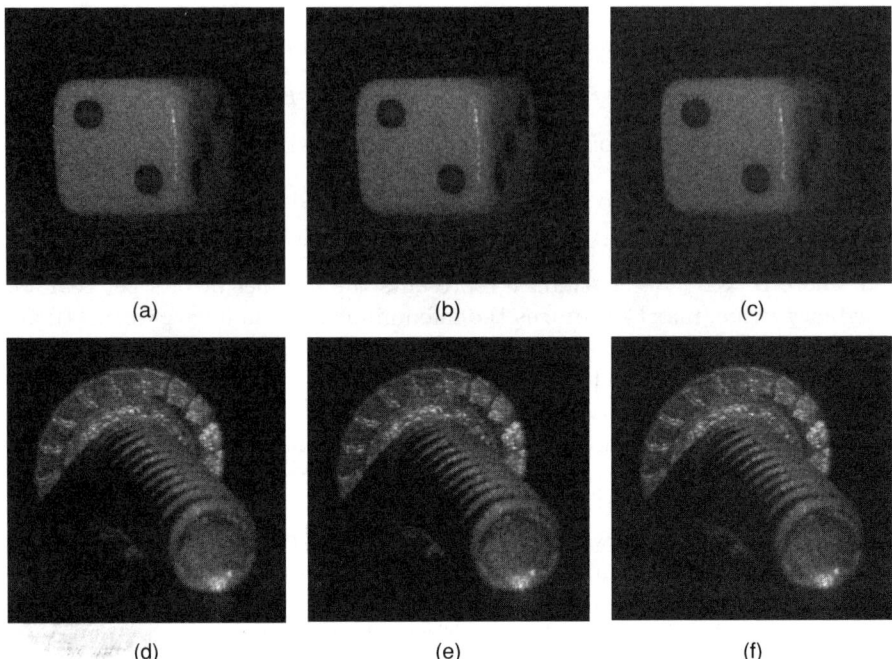

Figure 2.13. Reconstructed objects (with 5 × 5 pixel mean filtering) from encrypted digital holograms nonuniformly quantized (Kohonen competitive) with various numbers of bits of information in each real and imaginary value: die hologram (a) 4 bits, (b) 3 bits, (c) 2 bits; and bolt hologram (d) 4 bits, (e) 3 bits, (f) 2 bits.

Figure 2.13 shows reconstructed object intensities for both objects for selected quantization resolutions. Figure 2.14 shows plots of NRMS difference against number of bits of encrypted holographic data for both uniform quantization and Kohonen competitive quantization. Note from Fig. 2.13 that quantization at 4 bits (with 5 × 5 pixel mean filtering) reveals little visible loss in reconstruction quality, and (from Fig. 2.14) small NRMS errors of 0.02 and 0.01 for the die and bolt, respectively. Figure 2.14 illustrates the consistently lower NRMS error achieved by Kohonen competitive nonuniform quantization over uniform quantization on our encrypted digital holograms. Further evidence of this performance gain achieved with nonuniform quantization is shown in Fig. 2.15 where we see the improved quality in the reconstructed objects using nonuniform quantization compared with uniform quantization. Reductions from 8 bytes to 4 bits, 3 bits, and 2 bits correspond to compression ratios of 16, 21, and 32, respectively.

Ideally, the cluster centers from one hologram could be stored in a lookup table and applied with reasonable results to the quantization of subsequent holograms. (The JPEG algorithm uses a hard-coded lookup table of cosine-domain quantization values arrived at through performance evaluation over a

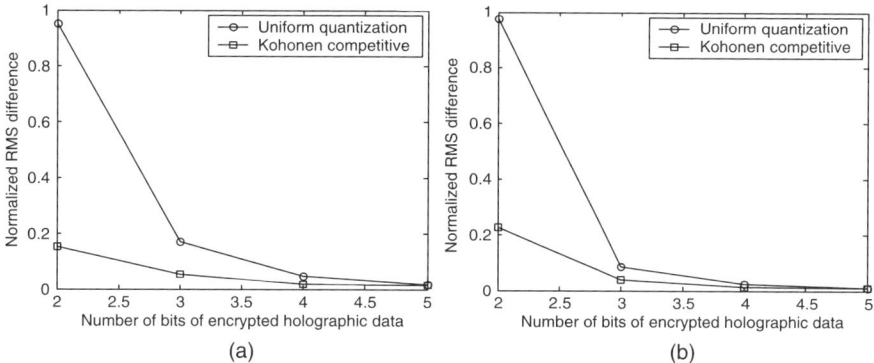

(a)   (b)

Figure 2.14. NRMS intensity difference in decrypted and reconstructed 3D object images plotted against quantization level: (a) die, and (b) bolt.

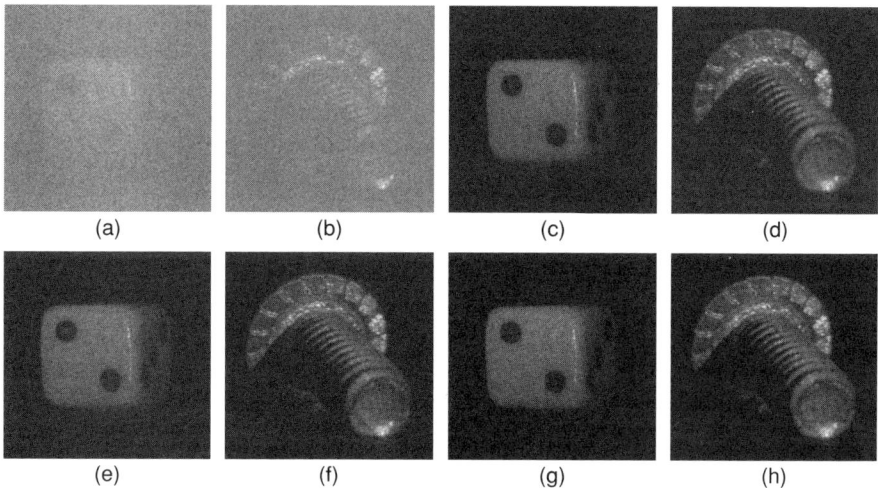

Figure 2.15. Reconstructions from encrypted digital hologram data with uniform quantization (upper row) and Kohonen competitive nonuniform quantization (lower row): (a),(b),(e),(f) 2 bits per real and imaginary value, and (c),(d),(g),(h) 3 bits per real and imaginary value. Mean filtering (5×5 pixel) was applied in each case.

database of sample input images.) We have found that the set of cluster centers we obtained from the Kohonen competitive neural network is very effective when applied in the quantization of a different hologram. This is illustrated in Fig. 2.16, where it can be seen that quantizing the die hologram using the centers obtained by applying Kohonen to the bolt hologram results in comparably low NRMS errors compared to those obtained when applying the centers produced specifically for the die hologram. By using the centers

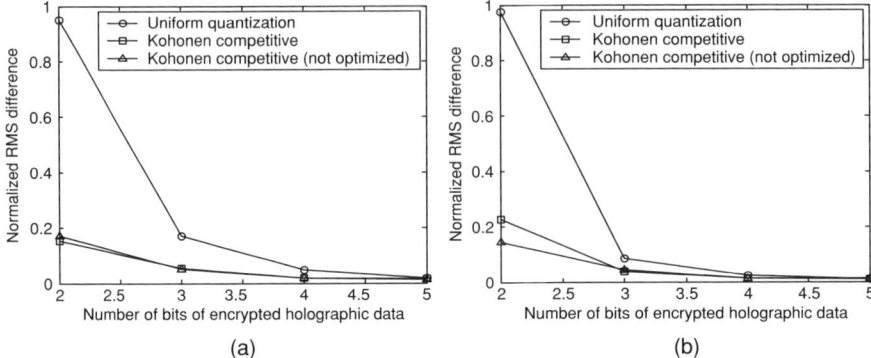

Figure 2.16. NRMS intensity difference in the decrypted and reconstructed 3D objects plotted against quantization level, with uniform quantization and nonuniform quantization (Kohonen competitive): (a) for die hologram, where nonoptimized means using the Kohonen centers from bolt hologram, and (b) for bolt hologram, where nonoptimized means using the Kohonen centers from die hologram.

obtained from the Kohonen competitive network to quantize other encrypted holograms, we have the improved performance of nonuniform quantization combined with the speed advantage of uniform quantization.

2.7 Conclusions

This chapter outlines an optical encryption technique, based on phase-shift digital holography, that is suitable for secure 3D object storage and transmission applications. This technique takes advantage of both the massive parallelism inherent in optical systems and the flexibility offered by digital electronics/software. Both the amplitude and phase of the hologram is encrypted by a phase-only perturbation of the Fresnel diffraction from the 3D object. Therefore, a phase mask is only required for this encryption scheme. Decryption and reconstruction of particular views of the 3D object can be performed optically or electronically. If the incorrect phase mask result is used, the reconstruction will be an unintelligible wavefront. The level of encryption can be increased by the use of multiple keys at different locations. Following encryption the hologram data is in a form suitable for digital electronic storage, transmission, or manipulation.

Lossless and lossy compression techniques were applied to the digital hologram data. Lossless techniques, such as LZ77, LZW, Huffman, and BW, perform very poorly on digital hologram data due to its white noise characteristics. We find that the encrypted digital holograms are compressed even less effectively. We evaluated two ANN-based nonuniform quantization techniques and found that the Kohonen competitive neural network performed best with our digital hologram data. We achieved reduced NRMS error and increased compression ratios using this technique. The Kohonen network was also shown to outperform the popular k-means clustering algorithm. We found that as few as 2 bits in each real and imaginary value (corresponding to a compression ratio of 32) results in good quality decompressed and decrypted 3D object reconstructions. Nonuniform quantization not only performs significant compression itself, it will also reduce the number of symbols (for Huffman) and introduce structure into the bit stream (for LZ77 and LZW) to allow them to perform further compression.

Acknowledgments

The authors wish to thank Enrique Tajahuerce and Yann Frauel for use of their hologram data. The first author wishes to acknowledge support from Enterprise Ireland.

References

[1]   B Javidi and JL Horner. (1994). *Opt. Eng.*, **33**:1752.
[2]   Ph Réfrégier and B Javidi. (1995). *Opt. Lett.*, **20**:767.
[3]   S Fukushima, T Kurokawa and Y Sakai. (1991). *IEEE Photon. Technol. Lett.*, **3**: 1133.
[4]   M Madjarova, M Kakuta, M Yamaguchi and N Ohyama. (1997). *Opt. Lett.*, **22**:1624.
[5]   JF Heanue, MC Bashaw and L Hesselink. (1995). *Appl. Opt.*, **34**:6012.
[6]   RK Wang, IA Watson and CR Chatwin. (1996). *Opt. Eng.*, **35**:2464.
[7]   LG Neto and Y Sheng. (1996). *Opt. Eng.*, **35**:2459.
[8]   B Javidi and E Ahouzi. (1998). *Appl. Opt.*, **37**:6247.
[9]   G Unnikrishnan, J Joseph and K Singh. (1998). *Appl. Opt.*, **37**:8181.
[10] O Matoba and B Javidi. (1999). *Opt. Lett.*, **24**:762.
[11] PC Mogensen and J Glückstad. *Opt. Lett.*, **25**:566.
[12] B Javidi and T Nomura. *Opt. Lett.*, **25**:28.
[13] S Lai and MA Neifeld. *Opt. Commun.*, **178**:283.
[14] E Tajahuerce, O Matoba, SC Verrall and B Javidi. (2000). *Appl. Opt.*, **39**:2313.
[15] E Tajahuerce, J Lancis, B Javidi and P Andrés. (2001). *Opt. Lett.*, **26**:678.
[16] O Matoba and B Javidi. (2002). *Opt. Lett.*, **27**:321.
[17] B Hennelly and JT Sheridan. (2003). *Opt. Lett.*, **28**:269.
[18] NK Nishchal, J Joseph and K Singh. (2003). *Opt. Eng.*, **42**:1583.
[19] NK Nishchal, G Unnikrishnan, J Joseph and K Singh. (2003). *Opt. Eng.*, **42**: 3566.
[20] E Tajahuerce and B Javidi. (2000). *Appl. Opt.*, **39**:6595.
[21] TJ Naughton and B Javidi. (2004). *Opt. Eng.*, **43**.
[22] TJ Naughton and B Javidi. "Encryption and decryption of 3D objects using digital holography". Unpublished
[23] JW Goodman and RW Lawrence. (1967). *Appl. Phys. Lett.*, **11**:77.

[24] T-C Poon and A Korpel. (1979). *Opt. Lett.*, **4**:317.

[25] L Onural and PD Scott. (1987). *Opt. Eng.*, **26**: 1124.

[26] U Schnars and WPO Jüptner. (1994). *Appl. Opt.*, **33**:179.

[27] U Schnars. (1994). *J. Opt. Soc. Am. A.*, **11**:2011.

[28] G Pedrini, P Frning, H Fessler and HJ Tiziani. (1998). *Appl. Opt.*, **37**:6262.

[29] JH Bruning, DR Herriott, JE Gallagher, DP Rosenfeld, AD White and DJ Brangaccio. (1974). *Appl. Opt.*, **13**:2693.

[30] J Schwider, B Burow, KW Elsner, J Grzanna and R Spolaczyk. (1983). *Appl. Opt.*, **22**:3421.

[31] I Yamaguchi and T Zhang. (1997). *Opt. Lett.*, **22**:1268.

[32] B Javidi and E Tajahuerce. (2000). *Opt. Lett.*, **25**:610.

[33] E Tajahuerce, O Matoba and B Javidi. (2001). *Appl. Opt.*, **40**:3877.

[34] Y Frauel, E Tajahuerce, M.-A Castro and B Javidi. (2001). *Appl. Opt.*, **40**:3887.

[35] Y Frauel and B Javidi. (2001). *Opt. Lett.*, **26**:1478.

[36] TJ Naughton, Y Frauel, B Javidi and E Tajahuerce. (2002). *Appl. Opt.*, **41**:4124.

[37] O Matoba, TJ Naughton, Y Frauel, N Bertaux and B Javidi. (2002). *Appl. Opt.*, **41**:6187.

[38] B Javidi and F Okano. (2002). *Three-dimensional Television, Video, and Display Technologies.* Springer, Berlin.

[39] M Rabbani. (1992). "Selected Papers on Image Coding and Compression." *SPIE Milestone Series MS48.* SPIE Press, Bellingham, W.

[40] JW Goodman and AM Silvestri. (1970). IBM J. Res. Develop., **14**:478.

[41] WJ Dallas and AW Lohmann. (1972). *Appl. Opt.* **11**:192.

[42] T Nomura, A Okazaki, M Kameda, Y Morimoto and B Javidi. (2001). *Proc. SPIE,* **4471**:235.

[43] TJ Naughton, JB Mc Donald and B Javidi. (2003). *Appl. Opt.*, **42**:4758.

[44] B Javidi, A Sergent, G Zhang and L Guibert. (1997). *Opt. Eng.*, **36**:992.

[45] F Goudail, F Bollaro, B Javidi and Ph. Réfrégier. (1998). *J. Opt. Soc. Am. A.*, **15**:2629.

[46] JW Goodman. (1996). *Introduction to Fourier Optics.* McGraw-Hill, New York.

[47] HJ Caulfield. (1979). *Handbook of Optical Holography.* Academic, New York.

[48] M Sutkowski and M Kujawinska. (2000). *Opt. Lasers Eng.*, **33**:191.

[49] DA Huffman. (1952). *Proc. IRE*, **40**:1098.

[50] J Ziv and A Lempel. (1977). *IEEE Trans.*, **IT-23**:337.

[51] TA Welch. (1984). *IEEE Comput.*, **17**:8.

[52] M Burrows and DJ Wheeler. (1994). *Digital SRC Report*, **124**:1994.

[53] C Amerijckx, M Verleysen, P Thissen and J-D Legat. (1998). *IEEE Trans. Neural Netw.*, **9**:3.

[54] NM Nasrabadi and Y Feng. (1988). *Int. Conf. on Neural Networks.* San Diego, CA, pp 1–101.

[55] T Kohonen. (1994). *Self-organizing Maps.* Springer-Verlag, Berlin.

[56] R Hecht-Nielsen. (1989). *Neurocomputing.* Addison-Wesley.

[57] D Desieno. (1995). *Proc. Int. Conf. on Neural Networks-I*, pp. 117–124.

[58] S Grossberg. (1976). *Biol. Cybern.*, **23**:121.

[59] C von der Malsburg. (1973). *Kybernetik*, **14**:85.

[60] RA Fisher and J Wishart. (1931). *Proc. Lon. Math. Soc. ser. 2*, **33**:195.

[61] J MacQueen. (1967). *Proc. Fifth Berkeley Symposium on Math., Stat. and Prob.*, Berkeley, Los Angeles, CA p. 281.

[62] EE Hilbert. (1977). *NASA JPL Technical Report*, pp. **77–43**.

[63] Y Linde, A Buzo and RM Gray. (1980). *IEEE Trans. Commun.*, **28**:84.

[64] SP Lloyd. (1982). *IEEE Trans. Inform. Theory*, **28**:129.

[65] VN Vapnik. (1995). *The Nature of Statistical Learning Theory*. Springer-Verlag, New York.

Digital Holography: Recent Advancements and Prospective Improvements for Applications in Microscopy

Pietro Ferraro,[1] Sergio De Nicola,[2] and Giuseppe Coppola[3]

[1]Istituto Nazionale di Ottica Applicata, Sez. di Napoli, Via Campi Flegrei 34, 80078, Pozzuoli (NA), Italy
[2]Istituto di Cibernetica del CNR 'E. Caianiello', Via Campi Flegrei 34, 80078, Pozzuoli (NA), Italy
[3]Istituto per la Microelettronica e Microsistemi del CNR, Sez. di Napoli, Via P. Castellino 111, 80131, Napoli, Italy

3.0 Introduction

Homeland security involves a wide number of technologies involving different scientific disciplines from biology to chemistry, physics, and numerous fields of engineering. New advancements in each field could help existing technologies meet stricter requirements in security, and progress in one or more scientific disciplines may lead to completely new instruments that may address emerging needs that had no previous solutions. Many aspects of security involve imaging, which involves many types of radiation from X-rays to microwaves. Imaging can generally be divided into two main categories based upon the type of radiation used: incoherent or coherent. In the optical spectrum, the special properties of coherent light, which is generated most efficiently by lasers, are used to allow special types of imaging such as holography that are only found in the realm of synthetic aperture radar.

This chapter describes the state-of-the-art of the interferometric imaging method called Digital Holography (DH). Special emphasis will be given to the recent advances and to achievements resulting from research efforts in different applications. Most of the recent results can have important direct or indirect impact on Homeland Security. The aim of the chapter is to furnish people with an up-to-date overview of the most recent advances in the imaging capabilities of digital holography to relate these applications to Homeland Security. New developments in coherent imaging can have significant applications for object recognition, ranging, and 3D scene reconstruction. The fact that DH is a coherent imaging technique is very important for obtaining nanometer scale resolution (i.e., a fraction of the wavelength used for recording the hologram) for quantitative measurement of object 3D profiles and comparing a copy to a master object.

Applications in microscopy are very important because they can have an impact in fingerprint recognition and applications to forensics. Biological sample visualization and study can be performed accurately and in a new way by means of Digital Holography. The chapter will not only describe recent advances but will present examples of applications in digital holography and will show how these results open a new perspective for improvements in the field of optical coherent microscopy and imaging. Although the applications reported do not involve specific security themes, the recent improvements of this coherent technique may be applied in the fields of biology, imaging, etc. where microscopy is required and thus they are certain to have an impact in Homeland Security.

3.1 Digital Holography

Dennis Gabor invented holography in 1948 as a method for recording and reconstructing both the amplitude and the phase of an optical wavefront,[1] with the objective of improving electron microscope images. Holography requires the use of coherent light, which is split into two waves by a beam splitter. One wave illuminates the object and is reflected, scattered from it, or possibly transmitted through it. The second wave is called the reference beam and it interferes with the light from the object in the plane of a recording medium, which traditionally was a photographic plate. The hologram (from the Greek words 'holos' meaning 'whole' or 'entire' and 'graphein' meaning 'to write') contains information about the entire 3D distribution of the optical wave field in the form of an interferometric fringe pattern. The object is reconstructed by illuminating the recorded hologram with a replica of the original recording reference wave.

Holography, through the discovery of holographic interferometry,[2–4] has become a very useful metrological tool in experimental mechanics, biology, fluid dynamics, and nondestructive inspection. One of the main limitations of holography and its related approaches has been the inconvenient chemical procedures connected with photographic recording media. That limitation has been largely overcome for application in metrology by the advent of digital speckle holographic methods.[5] However, most speckle methods, as they have been developed, are not truly holographic since they are not used to reconstruct the object field but rather phase difference between two or more fields from the object under investigation.

The idea of using a computer for reconstructing a hologram was first proposed by Goodman and Laurence and by Kronrod et al.[6,7] The development of computer technology and solid-state image sensors made it possible to record holograms directly on charge coupled device (CCD) cameras.[8,9] This important step enabled full digital recording and reconstruction of holograms without the use of photographic media, and is commonly referred to as digital holography (DH). Replacing the photographic film with a CCD TV camera requires that conventional recording techniques be modified to meet the limitations of these devices, but it does not change the basic purpose of holography which is the reconstruction of the object field from the recorded interference pattern. Indeed, the key step is to calculate quantities related to the object investigation such as the vectorial displacement field due to surface deformation, object shape, refractive index changes in transparent media, particle tracking, and microscopy just to cite a few examples of current research. In these applications, a modification of the state of the object under investigation leads to a modification of the light field scattered, reflected or transmitted by it and to a change of the digitally recorded interference pattern.

In DH, the reconstruction of the object field is performed numerically from the direct recording of the digitized numerical hologram. Since the information of the interfering waves is stored in the form of matrices, the numerical reconstruction process enables full digital processing of the holograms and it offers more possibilities than conventional optical reconstruction. Both amplitude and phase of the reconstructed complex field can be computed. Digital processing allows for subtraction of background noise and for the elimination of the zero-order diffraction term.[10] The parameters governing the reconstruction algorithm can be selected to control and optimize the spatial resolution of the reconstructed object field, thus compensating for lack of spatial resolution of digital cameras in comparison to holographic quality photographic plates. The limitation imposed by the low spatial resolution of CCD camera array compared to that of photographic materials has been widely discussed and various configurations of DH have been proposed and applied in various fields of science and engineering[11] and refs. therein. Furthermore, recent efforts toward developing new optoelectronic devices, such as solid-state pyroelectric sensors for infrared, make it also possible to extend the potential of DH to metrological applications with other than visible light sources.[12]

3.2 Theory and Principle of Operation of Digital Holography

The principle of the optical recording and reconstruction in classical holography can be understood from the setups shown in Fig. 3.1. The reference beam R and object beam O interfere at the plane of the holographic plate with an angle θ between them as shown in Fig. 3.1a. This angle is necessary so that the reconstructed image, shown in Fig. 3.1b, will be angularly separated from the zero-order diffraction and the second image, the so-called twin or conjugate

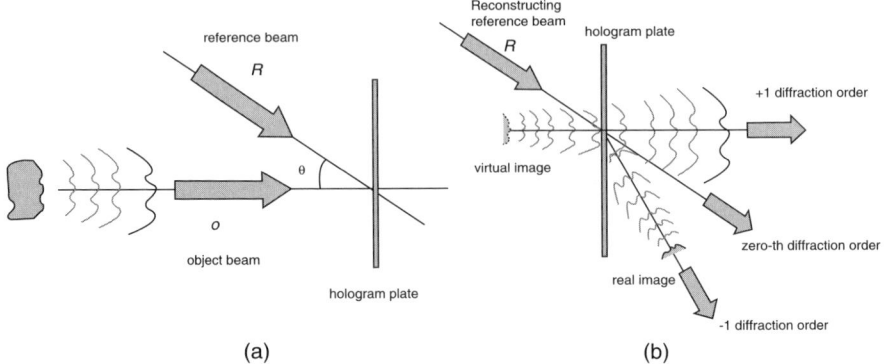

Figure 3.1. Optical configuration for recording (a) for reconstruction and (b) of off-axis holograms.

image. These three diffraction orders propagate at different directions and can be observed separately—a significant improvement over the original in-line configuration developed by Gabor where the zero-order, the real, and the conjugate image overlap.[13]

The intensity distribution $I(x, y)$ across the x-y holographic recording plane can be written as the modulus squared of $O(x, y) + R(x, y)$, namely

$$I(x, y) = |O(x, y) + R(x, y)|^2$$
$$= |R(x, y)|^2 + |O(x, y)|^2 + R^*(x, y)O(x, y) + R(x, y)O^*(x, y) \quad (1)$$

where the symbol $*$ denotes the complex conjugate, $O(x, y) = |O(x, y)|$ $\exp[i\phi_o(x, y)]$ is the complex amplitude of the object wave with real amplitude $|O(x, y)|$ and phase $\phi_o(x, y)$ and $R(x, y) = |R(x, y)| \exp[i\phi_R(x, y)]$ is the complex amplitude of the reference wave with real amplitude $|R(x, y)|$ and phase $\phi_R(x, y)$.

For the reconstruction of the recorded hologram, the interference pattern $I(x, y)$ is illuminated by the reference wave $R(x, y)$, and we have

$$R(x, y)I(x, y) = R(x, y)|R(x, y)|^2 + R(x, y)|O(x, y)|^2$$
$$+ |R(x, y)|^2 O(x, y) + R^2(x, y)O^*(x, y) \quad (2)$$

The first term on the right side of this equation is proportional to reference wave field, and the second one is a spatially varying "halo" surrounding the first term. These two terms constitute the zero-order diffraction (sometimes called the DC term by analogy to electrical current) and the autocorrelation of the object field with itself. The third term represents, apart from a constant factor, an exact replica of the original wavefront $O(x, y) = |O(x, y)|$ $\exp(i\phi_o(x, y))$ and it is usually a virtual image referred to simply as the image of the object. The last term is another copy, the so-called twin, or conjugate, image of the original object wave and it usually appears as a real image.

In photographic holography, the hologram acts as an amplitude transmittance structure that diffracts the reconstructed field from the reference field. In DH, the field at the object plane $\xi - \eta$ is calculated from the optical field $R(x, y)h(x, y)$ at the hologram plane using the scalar diffraction theory of the Fresnel approximation of the Rayleigh–Sommerfield diffraction integral.[14] The reconstructed diffracted field $Q(\xi, \eta)$ in the reconstruction plane $\xi - \eta$ at distance d from the hologram plane can be written in the paraxial approximation (PA) as

$$Q(\xi, \eta) = \frac{1}{i\lambda d} \exp\left(i\frac{2\pi}{\lambda d}\right) \int_{-\infty}^{\infty} \int_{-\infty}^{\infty} R(x, y)I(x, y)$$
$$\exp\left[i\frac{\pi}{\lambda d}\left[(\xi - x)^2 + (\eta - y)^2\right]\right] dxdy \tag{3}$$

Equation (3) provides a numerical reconstruction from the digitized hologram in the paraxial approximation where the x and y values and the corresponding ξ and η values in the reconstructed plane are small compared to the distance d (see Fig. 3.2).

Once the complex field $Q(\xi, \eta)$ has been calculated at distance d, the intensity $I(x, y;d)$ and phase distribution $\phi(x, y;d)$ of the reconstructed image can be determined by the following:

$$I(x, y;d) = |Q(x, y)|^2 \tag{4(a)}$$

$$\phi(x, y;d) = \arctan\frac{\text{Im}[Q(x, y)]}{\text{Re}[Q(x, y)]} \tag{4(b)}$$

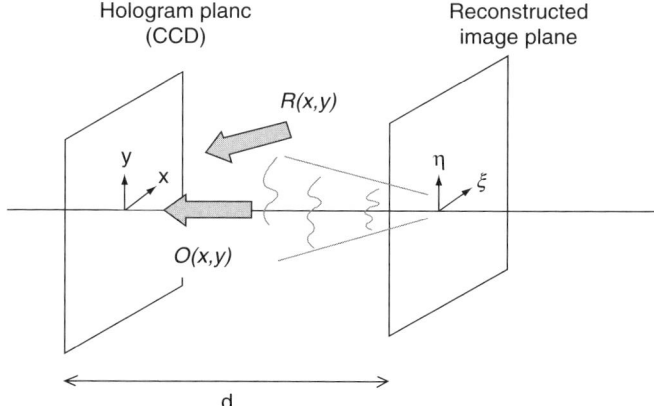

Figure 3.2. Optical set-up in off-axis digital holography.

If the signs of the numerator and denominator are accounted for, Eq. (4b) provides phase values wrapped in the interval $[-\pi, \pi]$. (This is commonly denoted as the arctan2 function.) Well-known unwrapping procedures can be employed to convert the phase modulo -2π into a continuous phase distribution in order to obtain a smooth phase image.[15]

3.2.1 Reconstruction Methods

Different techniques can be adopted for implementing the reconstruction of digital holograms. The following is a description of the main adopted methods.

3.2.1.1 Fresnel Transformation Method (FTM)

The convolution integral given by Eq. (3) can be manipulated to obtain the reconstructed diffracted field $Q(\xi, \eta)$ as a Fresnel transformation of the hologram function. The result is

$$Q(\xi, \eta) = \frac{1}{i\lambda d} \exp\left(i\frac{2\pi}{\lambda}d\right) \exp\left[i\frac{\pi}{\lambda d}(\xi^2 + \eta^2)\right]$$
$$\times \int_{-\infty}^{\infty}\int_{-\infty}^{\infty} R(x, y)I(x, y)\exp\left[i\frac{\pi}{\lambda d}(x^2 + y^2)\right]\exp\left[-i\frac{2\pi}{\lambda d}(\xi x + y\eta)\right]dxdy \quad (5)$$

Equation (5) shows that the reconstruction field is determined essentially by the 2D Fourier transformation of the multiplication of the hologram $I(x, y)$ by the reference wave $R(x, y)$ and the quadratic (or chirp) phase function

$$w(x, y) = \exp\left[i\frac{\pi}{\lambda d}(x^2 + y^2)\right] \quad (5a)$$

Equation (5) can be written in terms of the Fourier integral

$$Q(v_\xi, v_\eta) = \frac{1}{i\lambda d}\exp\left(i\frac{2\pi}{\lambda}d\right)$$
$$\exp\left[i\pi\lambda d(v_\xi^2 + v_\eta^2)\right]\mathcal{J}^{+1}[R(x, y)I(x, y)w(x, y)](v_\xi, v_\eta) \quad (6)$$

where the direct $(+1)$ or inverse (-1) continuous 2D Fourier transformations of the function $f(x,y)$ are defined, respectively, by

$$\mathcal{J}^{\pm 1}[f(x, y)](v_\xi, v_\eta) = \frac{1}{2\pi}\int_{-\infty}^{\infty}\int_{-\infty}^{\infty} f(x, y)\exp[\mp i2\pi(v_\xi x + v_\eta y)]dxdy \quad (7)$$

In Eq. (7) v_ξ and v_η are the spatial frequencies corresponding to the spatial variables ξ and η in the reconstruction plane and they are related to the reconstruction wavelength λ and to the reconstruction distance d by the following relations

$$v_\xi = \frac{\xi}{\lambda d} \quad v_\eta = \frac{\eta}{\lambda d} \quad (8)$$

With off-axis geometry, the object wave and the reference wave arrive at the hologram plane from separate directions and, according to the above equations, the different terms of the numerically reconstructed wavefront propagate along different directions, owing to their different spatial frequencies. In fact, if Eq. (2) is substituted into Eq. (5), it is clear that the reconstruction of the zero-order term, the virtual, and the real image are essentially governed by the frequency content of the respective spectra at the reconstruction distance d, which ultimately imposes restrictions on the spatial bandwidth of the object and reference beam. If the reference field is given by $R(x, y) = \sqrt{I_R}\exp\left[i(k_x x + k_y y)\right]$ where $I_R = |R(x,y)|^2$ is the intensity of the reference field and $\mathbf{k} = (k_x, k_y, k_z)$ is the corresponding wave vector, the three terms are separated in the Fourier domain corresponding to the reconstruction plane $\xi - \eta$ at distance d. The zero-order is located around the origin while the image and the twin image are symmetrically centered on $(k_x/2\pi,\ k_y/2\pi)$ and $(-k_x/2\pi,\ -k_y/2\pi)$, respectively.

To achieve a good quality reconstruction in DH, the sampling theorem (Nyquist criterion) has to be fulfilled across the whole CCD array.[16] This criterion requires at least two pixels per fringe period and this implies that the maximum interference angle $\alpha_{\max}$ between the spherical wavelet from each point of the object and the reference wave field is determined by the pixel size Δx according to the relation

$$\alpha_{\max} = \frac{\lambda}{2\Delta x} \tag{9}$$

Relationship (9) expresses the fact that for recording a hologram by a CCD array with pixel spacing Δx at least two pixels per fringe are needed. For example, in case of a camera with pixel size $\Delta x = 9\,\mu m$, the maximum interference angle is $\alpha_{\max} \approx 1.7°$ for $\lambda = 532\,nm$.

Mathematically, the 2D spatial sampling $I(n\Delta x,\ m\Delta y)$ of the hologram $I(x,y)$ on a rectangular raster of $N \times M$ points can be described by the following relation

$$I(n\Delta x,\ m\Delta y) = I(x,y)\,rect\left(\frac{x}{N\Delta x},\ \frac{y}{M\Delta y}\right)\sum_{n=1}^{N}\sum_{m=1}^{M}\delta(x - n\Delta x,\ y - m\Delta y) \tag{10}$$

where $\delta(x,\ y)$ is the 2D Dirac-delta function, n and m are integer numbers, $N\Delta x \times M\Delta y$ is the area of the digitized hologram, and $rect(x,y)$ is equal to unity if the coordinate point (x,y) is inside the area of the digitized hologram, and is zero elsewhere. Δx and Δy in Eq.(10) are the distances between the neighboring pixels on the CCD array in the horizontal and vertical directions, respectively. If the whole CCD array has a finite width given by $N\Delta x \times M\Delta y$, where N and M are the pixel numbers in each direction, the discrete representation of the Fresnel reconstruction integral given by Eq. (5) can be written as

$$Q(r\Delta\nu_\xi, s\Delta\nu_\eta) = \frac{1}{i\lambda d} \exp\left(i\frac{2\pi}{\lambda}d\right) \exp\left[i\pi\lambda d(r^2\Delta\nu_\xi^2 + s^2\Delta\nu_\eta^2)\right]$$

$$\times \Delta x \Delta y \sum_{n=-N/2}^{N/2-1} \sum_{m=-M/2}^{M/2-1} I(n\Delta x, m\Delta y) R(n\Delta x, m\Delta y) w(n\Delta x, m\Delta y) \qquad (11)$$

$$\exp\left[-i2\pi\left(\frac{rn}{N} + \frac{sm}{M}\right)\right]$$

Equation (11) allows computation of a matrix of $N \times M$ complex numbers corresponding to the reconstructed field via the discrete 2D fast Fourier transform algorithm. According to the theory of discrete Fourier transform, the sampling frequency intervals are $\Delta\nu_\xi = 1/N\Delta x$ and $\Delta\nu_\eta = 1/M\Delta y$ which, together with relations (8) allow determination of the dimensions $\Delta\xi \times \Delta\eta$ of the reconstruction pixel, namely

$$\Delta\xi = \frac{\lambda d}{N\Delta x}, \quad \Delta\eta = \frac{\lambda d}{M\Delta y} \qquad (12)$$

According to Eq. (12) the pixel width in the reconstructed plane is different from those of the digitized hologram and it is scaled inversely to the aperture of the optical system, i.e. to the side length $S = N\Delta x$ of the hologram (limiting the analysis to the x-direction for the sake of simplicity). This result is in agreement with the theory of diffraction, which predicts that at a distance d from the hologram plane the developed diffraction pattern is characterized by the diameter $\lambda d/S$ of its Airy disk (or speckle diameter). Therefore, the resolution of the reconstructed image (amplitude or phase image) is the diffraction limit of the imaging system through the automatic scaling imposed by the Fresnel transform.

If the spatial frequencies of the hologram $I(x,y)$ are smaller than those in the quadratic phase factor $w(x,y)$, the main problem in calculating Eq. (11) is adequately sampling the exponential function $w(x,y)$ inside the integral and of the global phase factor $\exp\left[i\pi\lambda d(r^2\Delta\nu_\xi^2 + s^2\Delta\nu_\eta^2)\right]$ multiplying the expression in Eq. (11). Assuming the sampling of $w(x,y)$ in the Nyquist limit, it is easy to obtain the approximate condition (limiting our analysis to one dimension only) that determines the range of distances d where the discrete Fresnel reconstruction algorithm (cfr. Eq (11)) gives good results, namely

$$d \geq d_c = \frac{N\Delta x^2}{\lambda} \qquad (13)$$

The same argument can be applied to the global phase factor $\exp\left[i\pi\lambda d(r^2\Delta\nu_\xi^2 + s^2\Delta\nu_\eta^2)\right]$, which may vary too rapidly with increasing spatial distance d, and this gives the condition $d \leq d_c$. A good reconstruction of both the amplitude and phase image is accomplished only for the equality $d = d_c$, while, for only an amplitude reconstruction, the less restrictive condition $d \geq d_c$ must hold. In fact, in this case the global phase factor is unessential for evaluating intensity distribution and the intensity profiles have less variation than the corresponding phase. Note that the size $\Delta\xi_c$ of the reconstruction pixel at distance $d = d_c$ in the Nyquist limit coincides with pixel size of the sampled hologram (i.e., $\Delta\xi_c = \lambda d_c / N\Delta x = \Delta x$). As an example, for $N = 512$ pixel, $\lambda = 632\,\text{nm}$ and pixel size $\Delta x = \Delta y = 11\,\mu\text{m}$, the Fresnel method is valid for distances greater than 98mm, for $N = 1024$ pixel, $\lambda = 532\,\text{nm}$ and $\Delta x = \Delta y = 6.7\,\mu\text{m}$, the distance has to be greater than 86.4 mm.

From Eq. (12) we can easily deduce that the lateral extension $S_I = N\Delta\xi$ of the numerical reconstruction increases linearly with reconstruction distance d according to the scaling law

$$S_1 = \frac{\lambda\, d N}{S} \tag{14}$$

where N and the lateral length $S = N\Delta x$ of the hologram are input parameters in the reconstruction process. Nevertheless, this result is only compatible with condition given by Eq. (13), derived from the appropriate sampling of a reconstructed amplitude image. This means that maintaining the number N as a constant may lead to badly sampled reconstructed image if the reconstruction distance does not satisfy the above requirements, and, after Fresnel diffraction, the external part of the reconstructed hologram may extend beyond the matrix and appear on the opposite side of the matrix due to aliasing. This problem can be avoided by padding the recorded hologram with zeros around its border.

3.2.1.2 Convolution Transformation Method

An alternative numerical reconstruction of holograms is through the calculation of the propagated angular spectrum, the so called "convolution approach" to DH. In this case, the reconstructed field in the paraxial approximation can be written in the following form

$$Q(\nu_\xi,\, \nu_\eta) = \frac{1}{2\pi} \exp\left(i\frac{2\pi}{\lambda}\, d \right)$$
$$\left\{ \mathcal{J}^{-1}\left[\exp\left[i\pi\lambda\, d(\nu_\xi^2 + \nu_\eta^2) \right] \mathcal{J}^{+1}[R(x,\, y) h(x,\, y)] \right\}(\nu_\xi,\, \nu_\eta) \tag{15}$$

where the Fourier transform of the chirp function $w(x,y)$ given by Eq. (4) has been used, namely

$$\mathcal{J}^{+1}[w(x,\, y)](\nu_\xi,\, \nu_\eta) = id\lambda \exp\left[-i\pi\lambda\, d(v_\xi^2 + v_\eta^2) \right] \tag{16}$$

It can be shown that when the angular spectrum is used, the use of two Fourier transforms for computing Eq. (15), once for taking the Fourier transform of the hologram (multiplied by the reference wave) and another time for taking the inverse Fourier transform, leads to a cancellation of the scale factor between the input and output field to obtain that the pixel size of the reconstructed image is equal to that of the sampled hologram (i.e., $\Delta\xi = \Delta x$ and)

$\Delta\eta = \Delta y$ and the actual size of both the input hologram and reconstructed image is identical ($S_I = S$).

We point out that, although Eqs. (6) and (15) are formally equivalent, the different use of the DFT algorithm to perform the calculation of the same diffraction integral, makes the convolution-based algorithm valid for near distances $d \le d_c$. Both methods overlap at distance $d = d_c$. Clearly this method is computationally more expensive than the direct evaluation of the Fresnel integral, since it requires two Fourier transforms (one direct and one inverse) but it is advantageous for keeping constant the length scales of the reconstructed images for all distances satisfying the near-field approximation.

From the discrete complex values of the reconstructed field, the intensity $I(r,s;d)$ and phase distribution $\phi(r,s;d)$ of the reconstructed image can be determined by the discrete version of Eqs. (4a) and (4b), namely

$$I(r,s;d) = |Q(r\Delta\xi, s\Delta\eta)|^2 \tag{17a}$$

$$\phi(r,s;d) = \arctan\frac{\text{Im}[Q(r\Delta\xi, s\Delta\eta)]}{\text{Re}[Q(r\Delta\xi, s\Delta\eta)]} \tag{17b}$$

3.2.2 Performance and Limitations of Digital Holography

In the previous section, it was noted that in order to separate the various diffraction components in the reconstruction plane, an offset angle θ is introduced in the off-axis setup. This offset angle must be greater than the minimum value $\theta_{\min} = \sin^{-1}(3B\lambda)$, where B is the highest spatial frequency of the object. For an object with dimensions $L_\xi \times L_\eta$ located at distance d from the hologram-recording plane, the x-y bandwidth of the object in the hologram plane is confined to a rectangle with dimensions

$$2B_x \times 2B_y = \frac{L_\xi}{\lambda d} \times \frac{L_\eta}{\lambda d} \tag{18}$$

According to Eq. (8), this is the bandwidth of the numerically reconstructed object after application of the Fresnel-based reconstruction algorithm for backward propagation from hologram plane to object plane. If the object is offset along the ξ axis at distance b from the optical axis of the system then the minimum angle $\theta_{\min}$ is determined by the corresponding bandwidth of the object in the hologram plane (i.e. $2B_x = L_\xi/2\lambda d_{\min:off-axis}$) and is approximately given by $\theta_{\min} = 3L_\xi/2d_{\min:off-axis}$, where $d_{\min:off-axis}$ is the minimum recording distance of the object of lateral size L_ξ in the off-axis setup. Therefore, the offset angle θ has to comply with the requirement $3L_\xi/2d_{\min:off-axis} \le \theta \le \alpha_{\max}$ where $\alpha_{\max}$ is the maximum interference angle given by Eq. (9).[17-20] Consequently, the offset distance $b = \theta_{\min} d_{\min:off-axis}$ has to be determined in order to conform to the limitation imposed by the minimum recording angle $\theta_{\min}$. For a given object size L_ξ at an offset distance b, as $\alpha_{\max}$ increases, the minimum allowable recording distance $d_{\min:off-axis}$ must decrease accordingly. In fact, the result for an off-axis system is that $d_{\min:off-axis} = (L_{CCD} + L_\xi + 2b)/2\alpha_{\max}$. For a sensor size $L_{CCD} = N\Delta x$, the above relations lead to the requirement that the minimum allowable recording distance increases linearly with object size L_ξ, namely $d_{\min:off-axis} = (L_{CCD} + 4L_\xi)/2\alpha_{\max}$ or, equivalently, in terms of the pixel number N of the sensor array and of the pixel width Δx

$$d_{\min:off-axis} = \frac{\Delta x}{\lambda}(N\Delta x + 4L_\xi) \tag{19a}$$

The above analysis has to be slightly modified for an in-line setup ($b = 0$). In this case it can be assumed that the center of the object of lateral size L_ξ and CCD array are both located on the optical axis of the system. The minimum allowable recording distance is obtained from the maximum interference angle $\alpha_{\max}$ according to the relation $d_{\min:in-line} = (L_{CCD} + L_\xi)/2\alpha_{\max}$ where $\alpha_{\max}$ still has to fulfill the condition given by Eq. (9). The result in terms of the pixel number N of the sensor array and of the pixel width Δx, is

$$d_{\min:in-line} = \frac{\Delta x}{\lambda}(N\Delta x + L_\xi) \tag{19b}$$

Equations 14 (a) and (b) show that in both the in-line and off-axis cases the minimum recording distance increase linearly with object size and that

$$d_{\min:off-axis} = d_{\min:in-line} + 3\frac{\Delta x}{\lambda}L_\xi \tag{20}$$

Equation (20) indicates a shorter recording distance in the in-line setup compared to $d_{\min:off-axis}$ in the off-axis arrangement, which leads to a more compact setup and more efficient use of the pixel area of the recording sensor array. In general, a shorter recording distance helps to achieve higher resolution in DH, since, according to Eq (12), the width $\lambda d_{\min}/N\Delta x$ of the reconstruction pixel is minimum at the distance $d_{\min}$.

3.2.3 Phase Shifting

An alternative approach to DH employs phase shifting to suppress both the zero-order term and the twin image,[21–23] which makes practical in-line recording where the three numerically reconstructed components are not angularly separated. When the phase shift algorithm is applied to DH, three or more holograms are recorded with a shift in phase between the object field and the reference field, and these may be combined to obtain the object field amplitude $|O(x,y)|$ and phase $\phi_o(x,y)$ at the recording plane. Many phase-shifting algorithms have been proposed, and it is impossible to discuss them all here, but we will give an example of application of the four-step quadrature-phase shifting algorithm in which four holograms are recorded with 90° increments of phase shift between them.

Figure 3.3 shows a reconstruction of a digital hologram recorded with off-axis geometry. The experimental setup was a Mach-Zehnder interferometer. The linearly polarized collimated beam from a diode-pumped, frequency-doubled NdYVO$_4$ laser (wavelength $\lambda = 0.532\,\mu$m) was divided by the beam splitter into the object beam and a reference beam. The object consisted in a photographic transparency of the two words "CNR INOA" set at distance $d = 175$ mm from a CCD camera with pixel size $\Delta x = \Delta y = 6.7\,\mu$m. The object beam illuminated the photographic plate, and the reference wave interfered with the object at small angle ($\leq 0.5^{\circ}$), as required by the sampling theorem. The digital hologram was recorded as an array of $N \times M = 1024 \times 1024$ 8-bit encoded numbers, and it is shown in Fig. 3.3a. The contrast of the interference fringes has been maximized by changing the transmittance of a neutral density filter inserted in the reference arm. The numerical intensity of the reconstruction at a distance d $= -175\ mm$ from the hologram plane and is shown in Fig. 3.3b, where the entire area of the reconstructed image is presented. The reconstructed image of Fig. 3.3b shows clearly the zero order diffraction term corresponding to the bright, square component at the center of the reconstructed image. The square occupies $N^2\Delta\xi^2/d\lambda$ of the N pixels in the ξ direction. This term is disturbing because it covers part of the reconstructed image, and several methods have been developed to suppress it.[10,24,25]

To apply a phase shifting algorithm, the reference beam is reflected at a piezoelectric transducer mirror and the reference phase is shifted between recordings. In the case of the four-step algorithm incremental steps of $\pi/2$ are introduced, and the complex amplitude $O(x,y)$ of the object at the hologram plane is determined from the intensity distribution values of four holograms by the following formula

$$O(x,\,y) = \frac{1}{4R^*(x,\,y)}$$
$$\left\{ I(x,\,y;\alpha = 0) - I(x,\,y;\alpha = \pi) + i\left[I\left(x,\,y;\alpha = \frac{\pi}{2}\right) - I\left(x,\,y;\alpha = \frac{3\pi}{2}\right)\right]\right\} \tag{21}$$

where $\alpha = 0, \pi/2, \pi$ and $3\pi/2$ are the phase shifts. The reconstruction at distance d from the hologram plane is performed by the Fresnel transform of the derived complex amplitude $O(x,y)$, namely

$$Q(\xi,\,\eta) = \frac{1}{i\lambda d}\exp\left(i\frac{2\pi}{\lambda}d\right)\exp\left[i\frac{\pi}{\lambda d}(\xi^2 + \eta^2)\right]$$
$$\times \int_{-\infty}^{\infty}\int_{-\infty}^{\infty} O(x,\,y)\exp\left[i\frac{\pi}{\lambda d}(x^2 + y^2)\right]\exp\left[-i\frac{2\pi}{\lambda d}(\xi x + y\eta)\right]dxdy \tag{22}$$

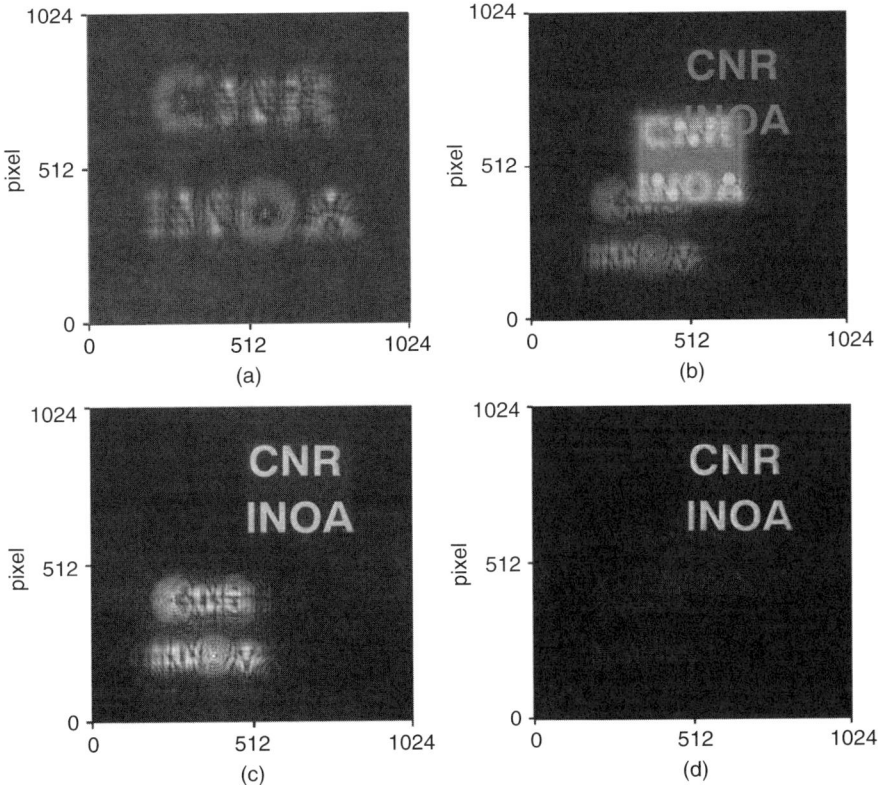

Figure 3.3. Numerical reconstruction of the photographic target in the off-axis digital holography: (a) original recorded off-axis hologram; (b) reconstructed amplitude-contrast at a distance $d = 175\,mm$; (c) suppression of the zero-order diffraction term; (d) reconstruction of the photographic target by the four-step algorithm.[23]

Figure 3.3d shows the amplitude of the numerically reconstructed image after application of the four-step algorithm. The central bright square and the twin image have been substantially suppressed leaving the image of the object visible in the upper right.

In this example the reference wave has been modeled by $R(x,y) = 1$ which represents a collimated unit amplitude wave impinging normally at the CCD recording plane. For phase imaging, the digital reference in the reconstruction algorithm should match as closely as possible the experimental reference wave; otherwise a set of fringes will be superposed on the reconstructed phase image. Alternately, the reconstruction integral can be postmultiplied by a phase correcting factor, which does not alter the amplitude image but can compensate for spurious interference fringes.

3.2.4 Digital Holographic Interferometry

The fringe pattern that results from the interference of the reference beam and an object beam carries phase information of the object under test and any change in its state gives rise to a corresponding modification of that phase information. If the complex fields $Q(\xi,\eta;s_1)$ and $Q(\xi,\eta;s_2)$ are the numerical reconstructions of two holograms recorded at different states s_1 and s_2 of the object, the corresponding phase change $\Delta\phi(\xi,\eta;s_2 - s_1)$ is given by

$$\Delta\phi(\xi,\eta;s_2 - s_1) = Arg[Q(\xi,\eta;s_2)] - Arg[Q(\xi,\eta;s_1)] \tag{23}$$

or an alternate form,

$$\Delta\phi(\xi,\eta;s_2 - s_1) = \frac{Re[Q(\xi,\eta;s_2)]Im[Q(\xi,\eta;s_1)] - Re[Q(\xi,\eta;s_2)]Im[Q(\xi,\eta;s_1)]}{Re[Q(\xi,\eta;s_2)]Re[Q(\xi,\eta;s_1)] + Im[Q(\xi,\eta;s_2)]Im[Q(\xi,\eta;s_1)]} \tag{24}$$

In the case of deformation measurement, s_1 and s_2 are states of deformation of the object under investigation and the calculated interference phase provides information about the displacement of the surface of an opaque object or the full optical path variation that may occur in a transparent object. By using the above Eq. (24) it is possible to obtain the full-field phase map corresponding to the well known technique of Holographic Interferometry which is well known in classical holography.[3,4] Digital Holographic Interferometry has been used to measure deformations of both large and very small objects, to investigate the refractive index changes, and to compare the shape of objects, etc.[17,18,26–32]

3.2.5 Compensation of Aberrations

The possibility of managing phase in DH method is very attractive because aberrations can be removed by adopting a procedure analogous to classical Holographic Interferometry using Eq. (24). Compensation of aberrations is necessary to remove the defocus aberration introduced by the microscope objective when quantitative phase determination is used in microscopic applications of metrology.[33,34] Moreover, it could be used to correct spherical aberrations introduced by high numerical aperture objective lenses employed in digital holographic microscopy where the paraxial approximation implicit in the Fresnel treatment often fails. In this case, the aberrations introduced by a lens can be corrected numerically by introducing a compensating phase factor into the phase of the reconstructed object beam.[23,33–39] This is equivalent to introducing an ad hoc extra phase factor multiplying the Fresnel integral which accounts for the wavefront curvature and which is sensitive to the geometrical parameters of the recording geometry. Alternately, a suitable modification of the quadratic phase factor $w(x, y)$ inside the Fresnel integral can be exploited. Given that the chirp factor $w(x, y)$ itself can be regarded as a compensating factor to obtain the refocused image of the object, modifications of $w(x, y)$ can accommodate and compensate for aberrations that might alter the refocusing process.

Obviously, the main difficulty of these approaches is the need for accurate knowledge of the focal lengths of the optical components used for imaging the object and of the lens-to-object and lens-to-CCD distances as well as knowledge of the relevant aberrations introduced by the recording setup. An alternative approach that does not require fine digital adjusting of the compensating parameters makes use of two recordings. The first hologram is made of the object under investigation and the second hologram is made of a flat reference surface in proximity of the object. This procedure works like double exposure holography in Holographic Interferometry, as discussed previously. The phase change between the two exposures can be numerically calculated by Eq. (23). In fact, the second hologram allows the recording of just the wavefront containing all the aberrations introduced by the optical components, including the defocus aberration introduced by the microscope objective.

3.3 Digital Holographic Microscope

DH is ideal for retrieving the phase distribution of the object wave field for quantitative phase imaging in microscopy, which means that the reconstructed phase distribution can be directly used for metrological applications and, in particular, for surface profilometry. The reconstruction process, in fact, is uniquely flexible because focusing can be adjusted while other aberrations can be removed. Moreover, phase distributions that cannot be observed in optical reconstruction of photographic holography are easily computed and displayed quantitatively. Actually, there are other 3D imaging methods based on interferometry that allow the measurement of minute displacements and surface profiles. Methods like holographic interferometry, fringe projection, and speckle metrology can provide full-field noncontact information about coordinates, deformations, strains, stresses, and vibrations. However, an important advantage of DH, in comparison with interference microscopy, is that the curvature introduced on the object beam by the microscope objective lens need not be compensated by the very same curvature introduced on the reference beam. In fact, in interference microscopy this problem is solved experimentally by inserting the same microscope objective in the reference arm, at an equal distance from the exit of the interferometer. For example, the Linnick interferometer requires that if any change has to be made in the object arm, then the same change must be precisely reproduced in the reference arm in such a way that the interference occurs between equally deformed wavefronts. As a consequence, the experimental configuration requires a very high degree of precision. By contrast, DH allows the direct calculation of the full-field map of the object through the calculation of the complex wavefront from a single exposure. As a consequence, both the acquisition time and the sensitivity to thermal and mechanical stability are reduced. Several applications have been demonstrated by using DHM.[17,18,26,40–44]

In a DHM configuration, high magnification ratios are obtained by inserting an imaging system that magnifies the size of the object, however, the image of the object need not be imaged directly on the CCD array. The image of the object may lie in a plane behind or in front of the sensitive array. Refocusing by a digital holographic microscope relies on the possibility of obtaining the complex optical field at any plane along the propagation of the object beam, and in case of a 3D object, different parts of the object at different distances can be focused separately. The DHM can be used to obtain the profile map of an opaque object, and in this case a DHM a reflection configuration has to be adopted, as shown in Fig. 3.4. The height distribution $h(\xi, \eta; d)$ of the object at distance d from the hologram plane is the information to be retrieved. The height distribution is related to the reconstructed phase distribution $\phi(\xi, \eta; d)$ by the simple relationship

$$h(\xi, \eta; d) = \frac{\lambda}{4\pi} \phi(\xi, \eta; d) \tag{25}$$

As previously described, the values of the measured phase are restricted in the interval $[-\pi, \pi]$ and ambiguities arising from height differences greater than $\lambda/2$ can be resolved by phase-unwrapping methods.

One important factor in DHM is the accuracy of the focal resolution of the image plane, δd. As digital holography refocuses in a way that is similar to that of a conventional optical microscope, δd is the geometrical depth of focus of the

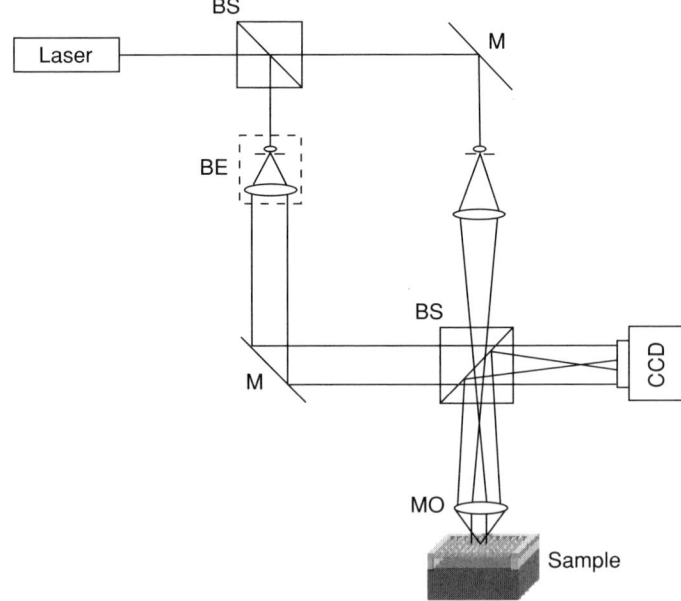

Figure 3.4. Experimental set-up for recording digital holograms; BS – beam splitter; M – mirror; MO – microscope objective; BE – Beam expander.

imaging system defined as the defocus distance that enlarges the point image to a confusion disk with a diameter equal to sampling distance across the object. The sampling distance, or equivalently the spatial resolution RS imposed by the sampling process across the CCD array is $RS = \Delta x/M$ where M is the magnification of the lens interposed between the sample and the recording plane. In terms of the spatial resolution and of the numerical aperture $N.A.$ of the imaging lens, the depth of focus δd is given by

$$\delta d = \frac{\Delta x}{M^2 N.A.} \tag{26}$$

Focal resolution better than a micrometer can be obtained for magnification $M \geq 10$ and $N.A. \approx 0.25$.

Figure 3.4 shows a possible setup for a digital holographic microscope. The setup consists basically of a Mach-Zehnder interferometer for reflection imaging; however, for transparent objects an alternate transmitting configuration can be arranged. In the reference arm, a beam expander is introduced in order to produce a plane wave. In the object arm, in order to illuminate the sample with a collimated beam, a combination of a beam expander, a lens with a long focal length, and a microscope objective is used. This imaging system obtains a magnified image of the object that is used for the hologram creation. In order to control the intensities in both the arms, a combination of a neutral density filter, a half-wave plate, and a polarizing beam splitter is used. The advantage of a Mach-Zehnder configuration is that it allows the recording of off-axis holograms with very small angles between the directions of propagation of the object and reference waves. This feature is important when low-resolution media are used as image acquisition systems. Finally, a CCD camera acquires an image of the hologram. Resolution better than 10 nm has been demonstrated for step-height measurement with a DHM in direct measurements of profiles of minute steps.

In the following section, two examples are described of DH applications as a tool for topographic characterization of microstructures.

3.3.1 Inspecting Microstructure by DHM

When a DHM is used to investigate small objects having mirror-like surfaces, the digital hologram recorded by the camera consists of an interferometric pattern made of circular fringes. Such fringes are due to the interference of the parabolic phase factor (the object beam) superimposed onto the characteristic phase distribution of the object wavefront and plane wavefront of the reference beam. The parabolic phase factor accounts for the wavefront curvature introduced by the imaging lens, the microscope objective. When the phase is retrieved by the numerical reconstruction of the digital hologram, the phase information about the object under investigation is hindered by the parabolic phase factor. Different approaches can be adopted to remove the disturbing parabolic phase factor in the reconstructed image plane.[34,40]

For example, Fig. 3.5a shows one digital hologram acquired for an etched periodically poled Lithium Niobate structure.[41] Figure 3.5b presents the density plot of the corresponding phase map reconstructed at a distance $d = 100$ mm from the hologram. In this case the correction of the curvature phase introduced by the microscope objective is accomplished by exploiting the mirror-like surface of the Lithium Niobate sample.

In the hologram reported in Fig. 3.5(a) it is possible to recognize an area that contains only circular fringes because the surface of the Lithium Niobate substrate acts as a plane mirror surface and the fringe pattern is only due to the curvature introduced by the microscope objective lens. The remaining area in the hologram contains information about the object under investigation superimposed on the circular fringes due to the imaging lens. Assuming that the

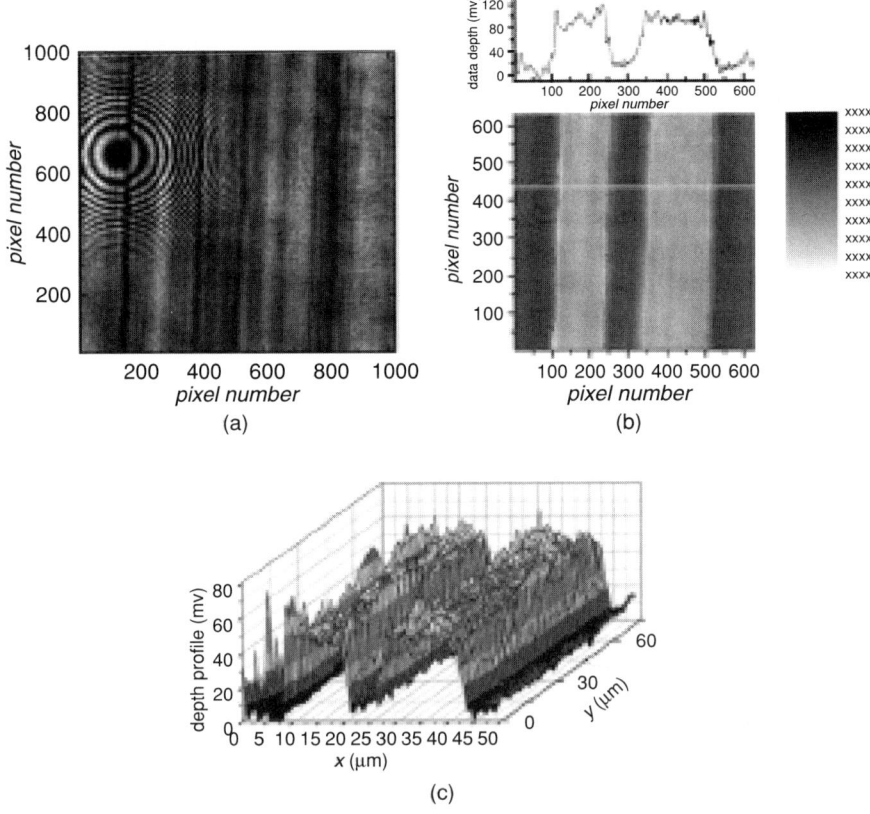

Figure 3.5. (a) Digital hologram acquired for a etched PPLN structure;(b) density plot of the corresponding digitally reconstructed phase map at a distance d = 100 mm and corresponding linear profile along a line; (c) 3D plot of the depth profile obtained from the phase map by converting the data into units of length.[41]

phase calculated for the first area is only due to wavefront curvature introduced by the microscope objective, its distribution can be recovered by performing a nonlinear fit of the unwrapped values. The correction wavefront phase can be written as

$$\boldsymbol{\Phi}_{corr}(x,\ y) = \frac{2\pi}{\lambda}\left(\frac{x^2 + y^2}{2R}\right) \tag{27}$$

where R is the radius of curvature of the correction wavefront (i.e., its defocus radius). In Fig. 3.5(b), it is evident that this approach allows complete removal of the curvature introduced by the microscope objective. Figure 3.5(c) shows the 3D plot of the numerically reconstructed phase-map and provides a full field 3D image of the microstructure. The data have been converted into units of length on the z-axis by the formula $h = \lambda\varphi(x,y)/4\pi$ given by Eq. (25). Another example of noncontacting, whole-field measurement of surface shape with high resolution is shown in Fig. 3.6. In this case, a MicroElectro–Mechanical-System (MEMS) silicon structure is inspected. Figure 3.6a shows a recorded hologram of two polysilicon cantilevers realized on a silicon substrate by means of a silicon oxide sacrificial layer. The two micromachined beams are 130 μm and 60 μm long, respectively, and 20 μm wide.

In order to correct the curvature phase introduced by the microscope objective, the wafer is translated slightly in a transverse direction. An additional hologram (Fig. 3.6b) of the substrate surface is recorded and used as reference. This approach is possible because the micromachined parts have been realized on a flat silicon substrate and the area around the micromachined structure offers a very good mirror-like surface. Thus, the reference hologram is obtained through the interference of the object field scattered by the reference flat surface near the MEMS and the reference beam. In this way, the reference hologram only comprises the effect of the wavefront curvature introduced by the imaging microscope objective. Both holograms are numerically reconstructed at distance $d = 100$ mm. Therefore, in the reconstructed image plane, it is possible to calculate the phase difference and, as consequence, to calculate the wrapped phase map (Fig. 3.6c) of the object cleared of the wavefront curvature introduced by the imaging microscope objective. Finally, as shown in Fig. 3.6d, the quantitative surface profile of the MEMS can be obtained applying a standard unwrapping procedure.

The possibility to obtain accurate quantitative information about the surface profiles of MEMS structures without contact appears crucial to the study of the shapes of microobjects. Figure 3.7 shows an example of the profile of an array of silicon MEMS structures obtained by DHM.

From the previous examples, it is evident that DH is a suitable method for inspection and quantitative evaluation of microstructure surface morphology and hence for influencing microstructure fabrication, functionality, and reliability.

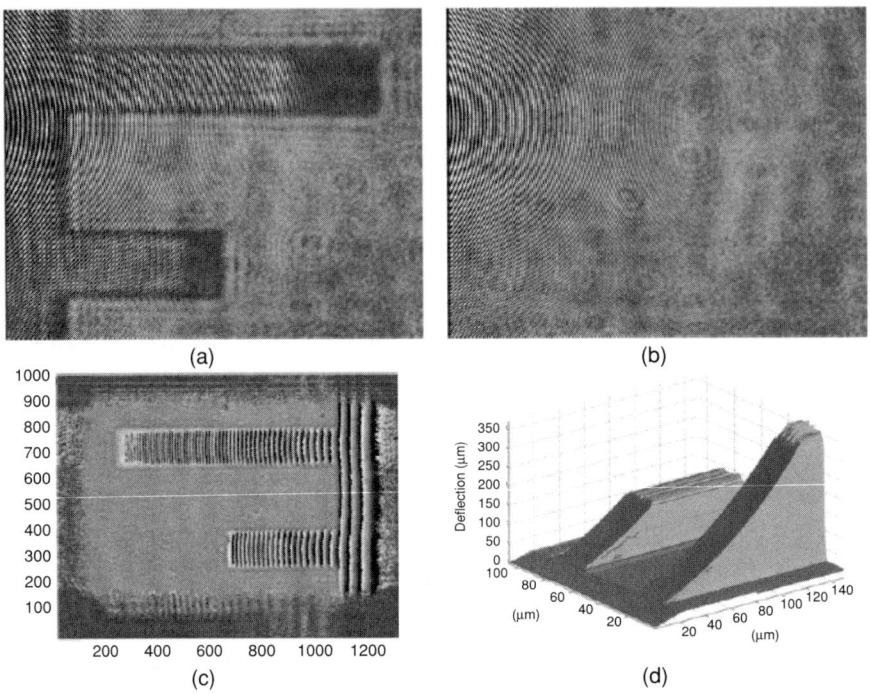

Figure 3.6. (a) Hologram of the polysilicon micro cantilever beams; (b) hologram recorded on a reference surface in proximity of the micro-machined beams; (c) Phase image, wrapped mod. 2π, of the polysilicon beams reconstructed at distance d = 100 mm; (d) unwrapped phase with out-of-plane deformation and dimensions expressed in microns.[34]

3.3.2 Digital Holographic Microscopy for Real-time Acquisition

With the experimental setup reported in Fig. 3.4, a sequence of holograms of the same object can be digitized and numerically reconstructed to investigate an evolving phenomenon.[45] DH has also been demonstrated for performing interferometric measurements of moving objects.[42] The corresponding sequence of numerically reconstructed amplitude and/or phase images can be used to build a video animation to show the behavior of the sample at different instants while it is experiencing deformations, movements, and/or any modification related to optical path changes. In conventional interference microscopy, it is necessary to hold constant all the different reconstructed parameters such as distance between the microscope objective and both the object and the CCD array plane to avoid longitudinal movements that cause focal change. By contrast, with DHM it is possible to refocus reconstructed images if a change has occurred in the above parameters. In fact, a sequence of images can be recovered by performing reconstructions at different distances analogous to

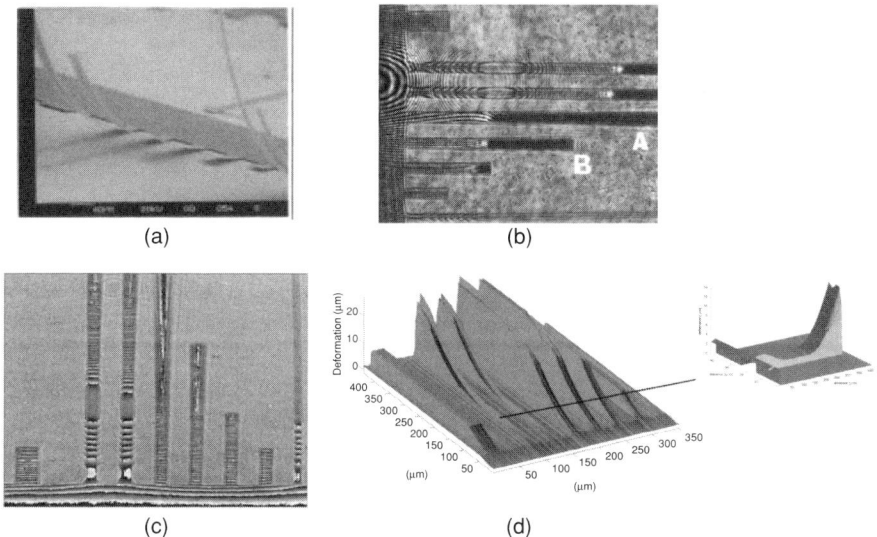

Figure 3.7. Profiles of MEMS: dimensions expressed in microns.[44]

mechanical translation of the microscope objective in conventional microscopy. However, when a long sequence of holograms has been recorded with unpredictable variation in the image plane, such numerical methods can be very time-consuming.

It might be expected that a change in the reconstructed distance would make it impossible to compare two reconstructed images directly since they will have different sizes owing to the different width of their reconstruction pixels (RP) when a Fresnel transform method is adopted. In fact, observing the relationship between the RP and the reconstructed distance d as given by Eq. (12) it is evident that the RP increases with the reconstruction distance so that the size of the image, in terms of number of pixels, is reduced for longer distances. Nevertheless, direct subtraction of unwrapped phase-maps from two holograms at two different distances can give quantitative information on the deformation of the sample.[32,46] The next sections will show how it is possible to use DH to control parameters in the reconstructed images to correct focusing and image size. Moreover, the same approaches can also be used to control the image resolution. The proposed methods add much more flexibility for DH and increase applications of the technique.

3.3.3 Focus Tracking During Dynamic Recording of Digital Holograms in DHM

Refocusing by DH relies on use of the Fresnel equation previously described to calculate the reconstruction at arbitrary focal planes. In DHM, high

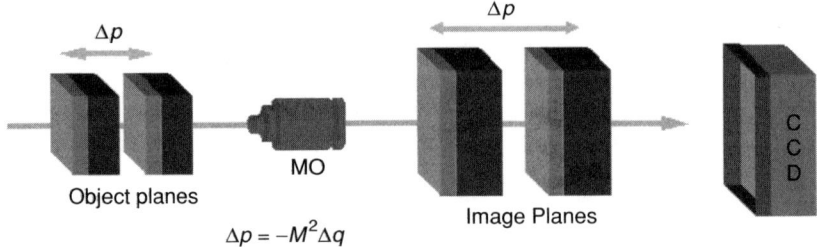

Figure 3.8. Out-of-focus caused by a longitudinal displacement of the imaged object.

magnifications are obtained by using a very short focal length microscope objective lens. If the sample experiences even a very small displacement along the optical axis, a very large change occurs in the distance to the imaging plane and the focus can be lost as depicted in Fig. 3.8. If the distance from the lens to the object is q, p is the distance of the image plane from the lens, and f is the focal length, the following relationships can be written:

$$p = Mq\frac{1}{q} + \frac{1}{p} = \frac{1}{f} \Rightarrow \frac{\partial p}{\partial q} = -\frac{f^2}{(q-f)^2} = -\frac{f^2}{\left(f\left(1+\frac{1}{M}\right)-f\right)^2} = -M^2 \qquad (28)$$

where M is magnification of the imaging system. Any axial displacement Δq of the sample results in a shift of the phase detected at time t given by

$$\Delta\varphi(t) = 4\pi\frac{\Delta q}{\lambda} \qquad (29)$$

and in a translation of the imaging plane in front of the CCD given by

$$\Delta p = -M^2\Delta q \qquad (30)$$

For example, with an imaging system of a magnification $M = 40$, a sample displacement of 10 nm, translates the image plane by $\Delta p = 16\,\mu m$.

Displacement of the object may occur for different reasons and may be dealt with in some cases; however, it is unavoidable in thermal characterization of objects. Temperature changes can cause unpredictable expansion or contraction of the object under study and/or its mechanical support, and the tedious search for new focal planes can be intolerable, especially if there is the need to visualize the phenomena in anything approximating real-time. To overcome this problem it is possible to detect the axial displacement of the object by measuring the phase-shift of the hologram fringes. In fact, with reference to the configuration reported in Fig. 3.8, axial displacement of the object caused a shift in the fringe pattern of the hologram. By recording the phase-shift in a small flat portion of the object, it was possible to determine the displacement and the incremental change of the distance to be used in the numerical reconstruction for each recorded hologram.

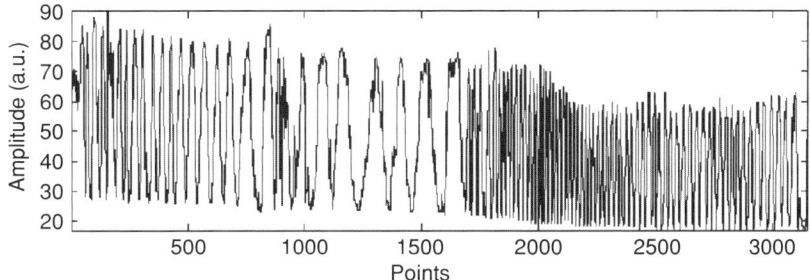

Figure 3.9. Phase shift of hologram fringes recorded in real time.[42]

The effectiveness of the method was demonstrated in a quasireal-time inspection of silicon MEMS structures subjected to a thermal load. Different types of silicon MEMS were investigated that had out-of-plane deformation due to the residual stress induced by the micrifabrication process by the fabrication process; in particular the results about a cantilever ($50 \times 50\,\mu$m) are described in the following. The silicon wafer was mounted on a metal plate and held by a vacuum chuck. The metal plate was mounted on a translation stage in close proximity to a microscope objective with focal length $f = 15.36$ mm and $N.A. = 0.16$. The sample was heated in the range of 23–120°C, by a remote-controlled heating element. Axial displacement was due to the overall thermal expansion of the metallic plate and the translation stage. An initial hologram was recorded before raising the temperature, and numerical reconstruction for a well-focused image was found at an initial distance of 100 mm with estimated magnification $M = 40$. While heating the sample, the phase-shift of the fringes was determined in quasireal-time by measuring the average intensity change in a group of 4×4 pixels. Figure. 3.9 shows the recorded intensity signal.

This signal was analyzed by applying a Fourier-transform method, and for each point recorded the wrapped and unwrapped phase was calculated. The displacement Δp was calculated from Eqs. (29) and (30) and is shown in Fig. 3.10 as a function of the sampled point. The numerical reconstruction distance for each hologram was continuously updated. As shown, the final hologram reconstruction distance differs from the initial distance by about 40 mm. Figures 3.11 (a, b, and c) show the amplitude and phase reconstructions for the cantilever from three different holograms of the recorded sequence, corresponding to three different temperatures. The reconstructions were performed automatically by applying the focus tracking procedure. Figure 3.11a shows the reconstruction of the first hologram at $d = 100$ mm; Figure 3.11b shows that of the second hologram at $d = 117.3$ mm, while Fig. 3.11c shows that of the third hologram at $d = 140.8$ mm. In the phase-map image, the wrapped phase observed on the cantilever indicates that an intrinsic out-of-plane deformation is present. As expected, the reconstructions in Fig. 3.11 are all in focus.

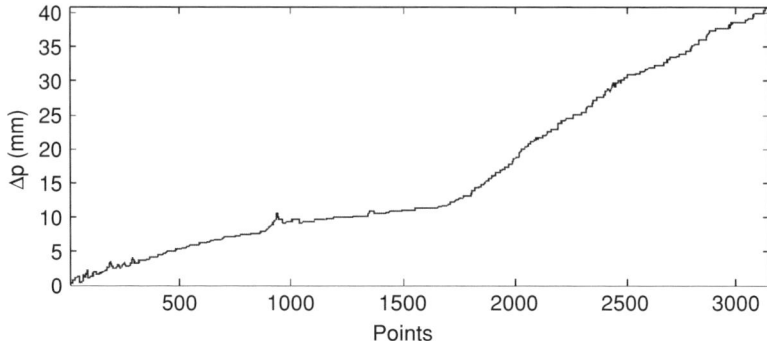

Figure 3.10. Displacement of the sample measured in real time by analyzing the phase shift of hologram fringes.[42]

Thus, by applying the focus-tracking method, the corrected reconstruction distance can be evaluated for each acquired hologram and well-focused amplitude and phase images can be obtained.

3.3.4 Controlling Image Size

As shown in previous section in Fig. 3.11, the size of the reconstructed object decreases with larger reconstruction distance. Thus, although well-focused images were obtained, it was not possible to compare two of them directly since they had different sizes owing to the different width of the Reconstruction Pixel (RP). Similar difficulties arise in MultiWavelength–DH (MWDH) used for color display and for applications in metrology. For each wavelength in MWDH, the width of the *RP* increases with the reconstruction wavelength for a fixed reconstruction distance. Consequently, holograms recorded with different wavelengths produce images with different sizes when numerically reconstructed by means of the Fresnel Transform Method. Color DH display requires simultaneous reconstruction of images recorded with different wavelengths (colors) and the resulting reconstructed images must be perfectly superimposed to get a correct color display.[47,48] This is prevented by the differing image sizes, and this also prevents phase comparison required for holographic interferometry.[46,49]

To avoid the above-mentioned problems, the convolution approach could be employed where the RP remains constant and equal to the size of the pixel of the CCD array, however, for a large reconstruction distance this approach does not work properly. Consequently, it is necessary to use a resizing operation on the reconstructed images[50] at end of the reconstruction process or a scaling operation on the hologram. Recently, a cascaded algorithm for reconstruction of the digital holograms with a variable zooming factor has been proposed.[51] Nevertheless, it is possible to control the image size of the reconstructed images by exploiting the FTM itself,[52] since the size is controlled through enlargement of the number of the pixels of the recorded digital holograms. From Eq. (12) it is clear that the RP size also depends on the lateral number of the pixels N and M. So, the image size can be controlled changing the RP by using a larger number of pixels in the reconstruction process. N and M can be augmented by padding the matrix of the hologram with zeros in both the horizontal and vertical directions such that

$$N_2 = N_1(d_2/d_1)$$
$$M_2 = M_1(d_2/d_1)$$

(31)

getting

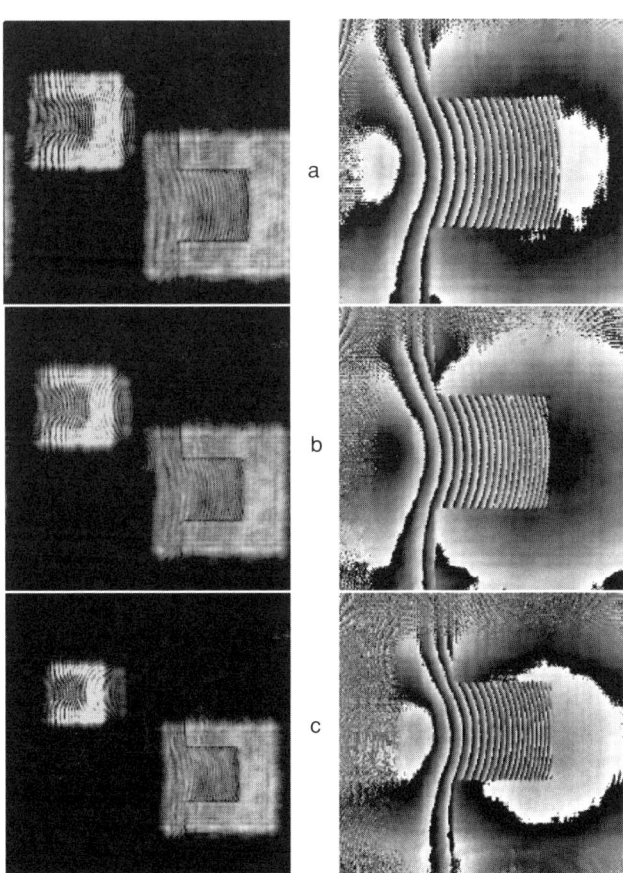

Figure 3.11. In focus amplitude and phase-map for the cantilever beam from three holograms recorded at three different distances of the same sequence, obtained applying the focus-tracking procedure.[42]

$$\Delta\xi_1 = \Delta\xi_2 = \frac{d_1\lambda}{N_1\Delta x} = \frac{d_2\lambda}{N_2\Delta x}$$

$$\Delta\eta_1 = \Delta\eta_2 = \frac{d_1\lambda}{M_1\Delta x} = \frac{d_2\lambda}{M_2\Delta x}$$

$$(32)$$

where $N_1(N_2)$ and $M_1(M_2)$ are the number of pixels of the hologram recorded at the distance $d_1(d_2)$ with $d_1 < d_2$. Thus, in order to obtain two reconstructed images with the same size, the $N_1 x M_1$ matrix of the hologram recorded at distance d_2 has to be padded with $(N_2 - N_1)$ zeros along the "x" direction and with $(M_2 - M_1)$ zeros along the "y" direction. In a similar way in MWDH, if one hologram has been recorded with wavelength λ_1 and a second with λ_2, where $\lambda_1 < \lambda_2$, at the same distance, then the number of pixels of that hologram may be changed such that

$$N_2 = N_1(\lambda_2/\lambda_1)$$

$$M_2 = M_1(\lambda_2/\lambda_1)$$

$$(33)$$

in order to obtain the same width for the RP:

$$\Delta\xi_1 = \Delta\xi_2 = \frac{d\lambda_1}{N_1\Delta x} = \frac{d\lambda_2}{N_2\Delta x}$$

$$\Delta\eta_1 = \Delta\eta_2 = \frac{d\lambda_1}{M_1\Delta x} = \frac{d\lambda_2}{M_2\Delta x}$$

$$(34)$$

The effectiveness of this method was demonstrated both on the cantilevers illustrated in previous section and on a Ronchi grating.[53]

In Fig. 3.12, the three reconstructed phase images illustrated in Fig. 3.11 are reported with a different view, and, as expected, it is evident that the image

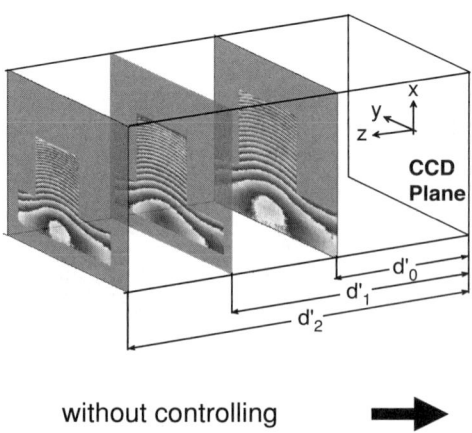

without controlling ➡

Figure 3.12. Wrapped image phases reconstructed at different distances without application of padding operation.[52]

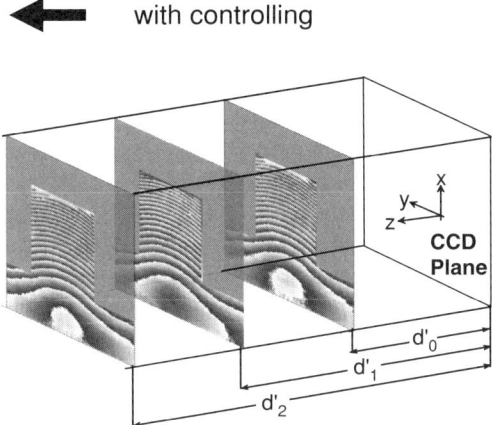

Figure 3.13. Wrapped image phases reconstructed at different distances with application of padding operation.[52]

size, in terms of pixels, is reduced since the three images were reconstructed at different distances from the hologram.

The hologram reconstructed at distance $d_0 = 100$mm had 512×512 pixels. In order to obtain reconstructed images with the same size the holograms reconstructed at distances $d_1 = 117.3$ mm and $d_2 = 140.8$ mm have been padded with zeros up to 614×614 pixels and 718×718 pixels, respectively. The reconstructed phase images obtained applying the padding procedure are shown in Fig. 3.13, where it is clear that the image size is independent of the reconstruction distance so that on these reconstructed images direct phase subtraction can be performed and phase difference can be obtained.

In particular, an example of direct phase subtraction is shown in Fig. 3.14. Where Fig. 3.14a is the unwrapped phase image of the hologram recorded at distance d_1; Fig. 3.14b is the unwrapped phase image of the MEMS at $d_2 = 139.5$ mm without zero padding whereas Fig. 3.14c is the unwrapped phase of the MEMS at d_2 obtained by the padding operation. Finally, Fig. 3.14d shows the difference between the unwrapped phase maps with equal size indicating the small deformation caused by thermal load.

An example of amplitude reconstruction size control is reported in Fig. 3.15. Figure 3.15a shows portion of the reconstruction along the longitudinal axis (Z-axis) for a Ronchi grating without the padding operation. The initial distance of reconstruction was $d_1 = 295$ mm while the final was $d_2 = 695$ mm and the initial number of pixels for the first reconstruction was $N_1 = M_1 = 1024$. In the figure, the Talbot effect noticeable along the Z-axis is due to the dependence of the RP on the reconstruction distance d, since the period of the grating decreases for longer distances. Figure 3.15b shows the results of the same reconstruction when the padding operation is applied, and

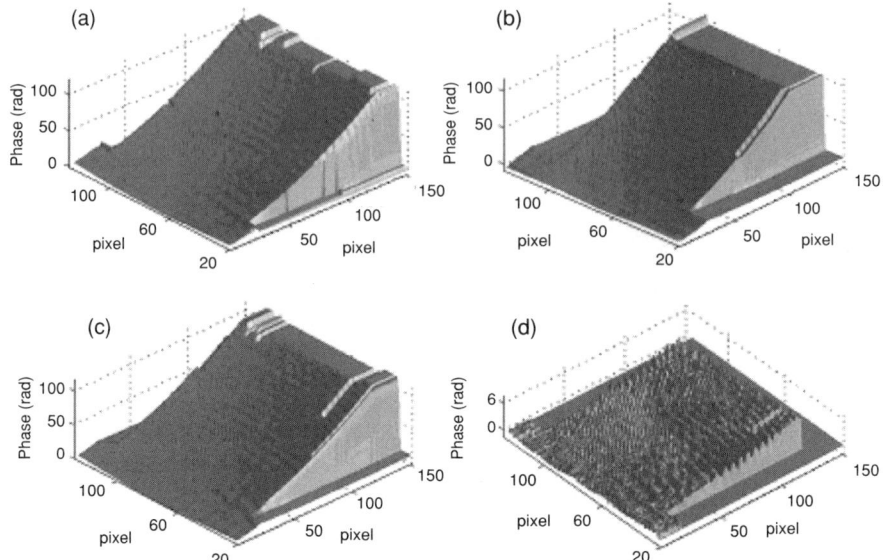

Figure 3.14. Unwrapped phase of MEMS at different distance with padding operation and phase map subtraction between reconstructions at different distances.[52]

the final reconstruction was performed with padding of zeros up to $N_2 = M_2 = 2412$. From Fig. 3.15b it can be noted that the size of the grating has been kept constant.

In order to demonstrate that size can also be controlled in MWDH applications, holograms of a double Ronchi grating, with periods of 5.0 lines/mm and 3.5 lines/mm, were recorded with the two different wavelengths of

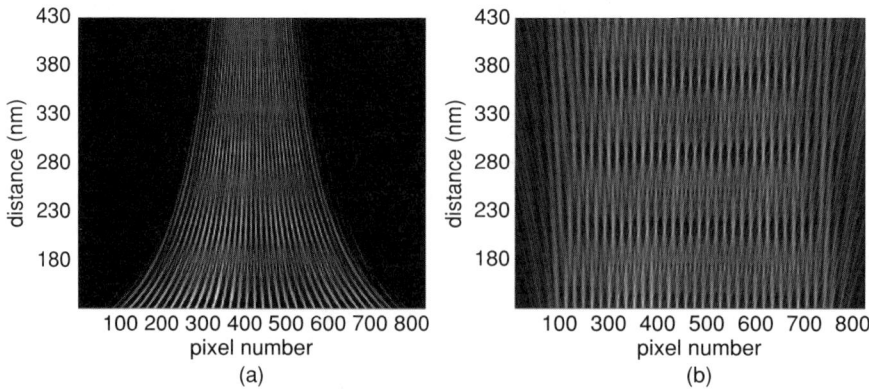

Figure 3.15. Ronchi grating reconstructed at different distances: (a) pitch of the grating decreases for longer distances; (b) size is kept unchanged with padding operation.[53]

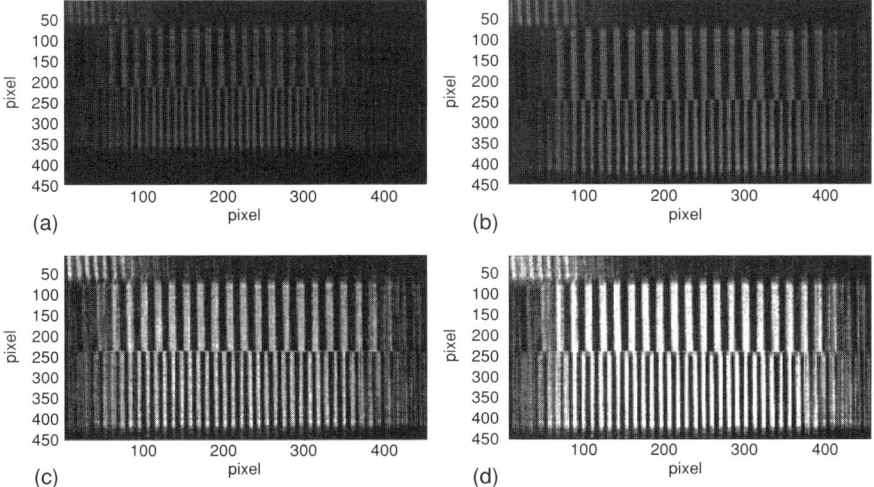

Figure 3.16. MWDH, reconstruction of the red hologram (a) without padding and (b) with padding; (c) reconstruction of the green hologram; (d) superimposition of red and green images (b) and (c).[52]

$\lambda_1 = 532$ nm and $\lambda_2 = 632.8$ nm, respectively. All holograms were initially recorded with $N_1 = M_1 = 1024$. Figures 3.16a and b show the reconstructed amplitude of the grating at λ_2, respectively, without and with padding operation applied to the hologram. Figure 3.16c shows the amplitude image reconstruction at λ_1 (green). The red hologram (that at λ_2) was reconstructed (Fig. 3.16b) according to Eq. (37) after adding a number of zeros around the hologram such that $N_2 = M_2 = 1218$ to obtain an image having equal size in respect to that of Fig. 3.16c.

The RGB combination of red and green images gives perfect superimposition and the new color (yellow) in Fig. 3.16d. From these examples it is clear that by means of a simple padding of the recorded digital holograms with zeros, it is possible to control the size of the reconstructed images independent of distance and wavelength.

3.3.5 Controlling Image Resolution

In DHM it is usual to put the object in the Fresnel region of the image sensor, because use of a single Fourier transform for numerical reconstruction reduces the calculation time by half. This limits resolution, however, because light passing through an object that contains high-spatial frequency components is diffracted at large angles. This limitation is particularly severe when high resolution with visible wavelength light is desired. In fact, elementary pixel width of CCD sensors is at least of several microns, leading to spatial cut-off frequencies lower than 100 mm^{-1}. Consequently, standard CCD sensors limit

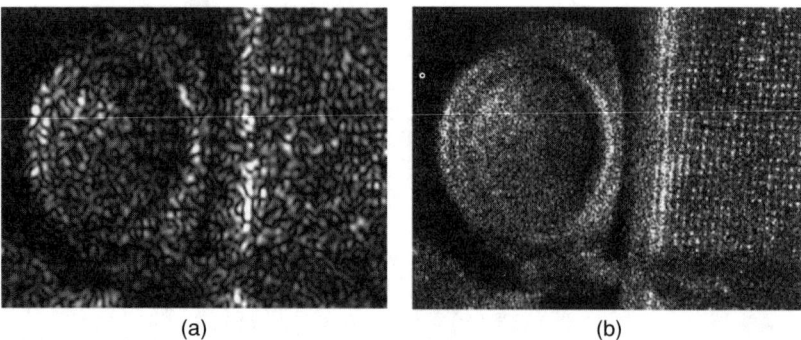

(a)   (b)

Figure 3.17. Reconstruction of the object section from (a) a single hologram and (b) from nine holograms.[56]

incident beams to being quasi-parallel so that only small objects placed at a large distance from the sensor can be recorded with digital holography. So, in most cases, the specimen is magnified during the hologram recording to obtain adequate resolution, but this operation simultaneously magnifies the coherent speckle noise.[54] In order to improve the resolution of digital holography, a kind of space–time digital holography method was proposed.[55] By this method it is possible to produce an image free of speckle noise, but its scanning system makes the related setup and experimental procedure complicated. For improving the resolution, a combination of multiple holograms recorded at different camera positions to give a large digital hologram has been proposed.[56] In this case the CCD camera was mounted upon a translation stage and a series of hologram exposures were recorded as the camera was moved to different positions in a rectangular raster.

In Fig. 3.17 the texture of the radiator of the model car is resolved. In particular, an increase of the resolution by a factor of 2.5 with respect to the size of the CCD sensor can be expected.

Another approach to improve resolution using a single acquisition has been proposed in which a diffraction grating is used to record digital holograms with a wider solid angle. In this approach more object waves reach the CCD camera.[57] In fact, a grating is placed in front of the specimen, so that the light incident upon the grating is split into three beams. One of these beams will propagate along the original direction, and the other two beams will diverge from the original direction. The ray diagrams of the object waves are schematically shown in Fig. 3.18 and, for simplicity, only a point object is described.

Due to the diffraction of the grating placed between the object and the CCD camera, three beams (O0, O1, and O2) reach the area of the hologram and can be digitally recorded. The presence of the grating increases the numerical aperture of the holographic system and consequently, the resolving power is higher. Recovery of the resolution lost intrinsically by the reconstruction FTM has been recently proposed.[58] In fact, if paraxial approximation can be applied,

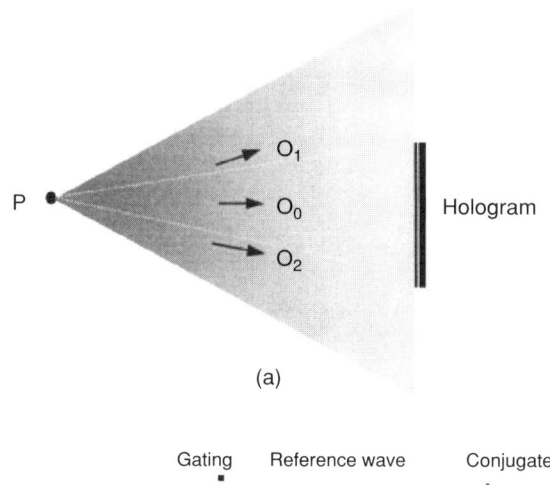

(a)

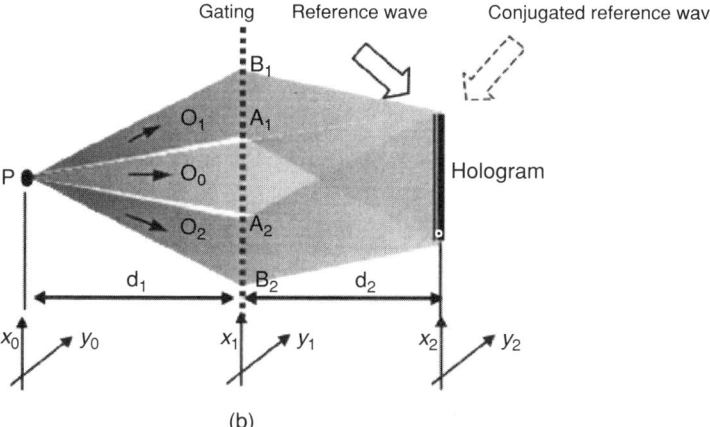

(b)

Figure 3.18. Ray diagram of the object waves when a grating is placed between the object and the CCD camera: of the object wave (a) without grating in set-up and (b) with a grating in set-up.[57]

the FTM can be used in the reconstruction process. However, when the image reconstruction is performed by FTM, the spatial frequencies displayable in reconstructed images are band limited by the size of the reconstruction pixel, which represents the sampling gauge in the image plane. From Eq. (12) it is clear that the reconstruction pixel, and consequently resolution, depends on the wavelength, the distance, number of the pixels N of the sensor array, and their physical size. In other words, spatial frequencies higher then the Nyquist limit, in the image reconstruction plane, are under-sampled and reconstructed incorrectly. Depending on the objects, under-sampling can affect the correctness of reconstructed phase map. Thus, the resolution can be improved through artificial enlargement of the number of the pixels in recorded digital holograms.

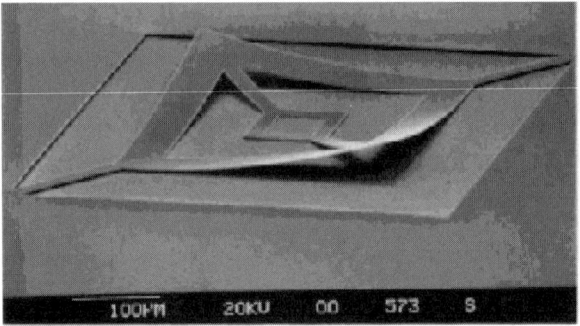

Figure 3.19. SEM picture of MEMS with a large amount of deformation.[58]

The effectiveness of the method has been demonstrated for the characterization of MEMS structure that grows too rapidly as shown in Fig. 3.19.

In fact, in this case the shape of the MEMS induces an undersampling for a fixed distance, wavelength, number, and size of pixels of the CCD, and the resulting profile can be incorrectly reconstructed. Thus, to reconstruct correctly the profile of a structure deformed by a large amount, it is necessary to improve resolution in the imaging plane. Figure 3.20 shows the phase map reconstructed by the digital hologram from a 1024×1024 pixels hologram recorded at distance $d = 200$ mm. In this picture some small under-sampling occurs at the extremity

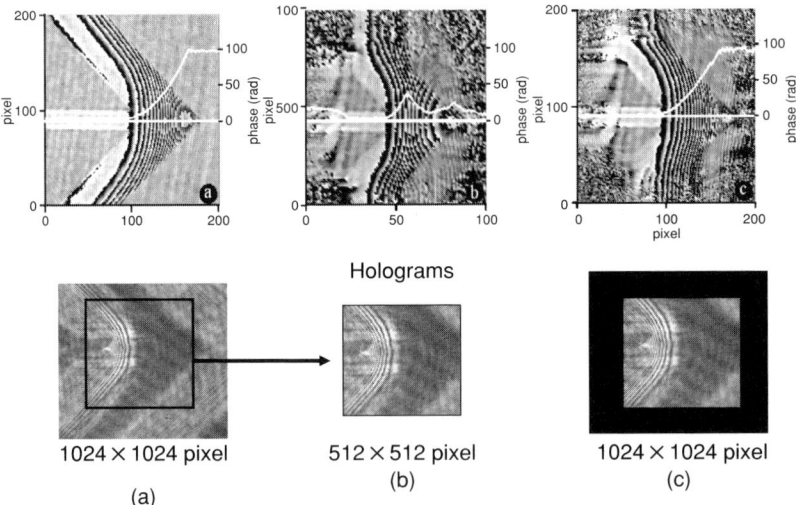

Figure 3.20. Wrapped phase maps of the MEMS: (a) from the original hologram with 1024×1024 pixels shown in the bottom; (b) from a selected central portion of 512×512 pixels; (c) from the previous hologram with 512×512 pixel but padded with zeros up to 1024×1024

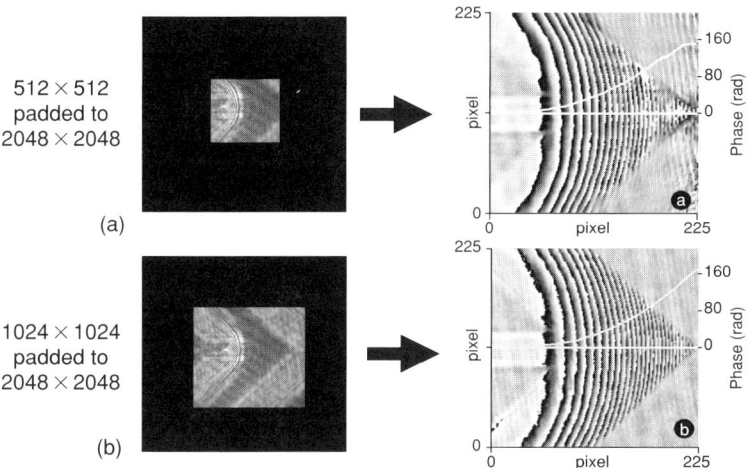

Figure 3.21. Close-up of the wrapped phase map of: (a) the hologram of Fig. 3.7a but padded to 2048 × 2048 pixels (padded digital hologram reconstructed shown on the left); (b) the hologram of Fig. 3.7c but padded to 2048 × 2048 pixels (padded hologram shown on the left).

of the corner. The plot of the unwrapped phase along the diagonal direction of the square is shown superimposed in the same Fig. 3.20a. This profile is correctly retrieved up to about 100 radians, and we may safely assume that the profile up to this value is the actual true profile of the MEMS along that line.

Figure 3.20b shows that cropping from the original hologram a central portion of 512 × 512 pixels causes substantial under-sampling of the wrapped phase in the reconstruction. The phase map up to the original value of 100 radians is completely recovered if zero padding up to 1024 × 1024 pixel is preformed, even if the field of view is reduced. In order to obtain the complete correct profile of the MEMS, the 1024 × 1024 pixel hologram has been padded with zeros up to 2048 × 2048, and the resulting phase map is shown in Fig. 3.21. From the plot of the unwrapped phase along the central line of the MEMS in Fig. 3.21 it is clear that the padding operation allows the recovery of the correct phase map. In fact, in this case a correct profile of the MEMS is obtained along the whole length of the structure. Fig. 3.22(a) shows the 3D profile of the MEMS obtained by the central portion of 512 × 512 pixel hologram; Fig. 3.22 (b) is obtained by the same hologram of Fig. 3.22 (a) but padded with zeros up to 1024 × 1024 pixels; Fig. 3.22 (c) is obtained by the same hologram of Fig. 3.22 (a) but padded with zeros up to 2048 × 2048 pixels.

It is important to note that real content of information in Fig. 3.21a is exactly the same as that producing the phase map of Fig. 3.22b with only a padding operation making the difference. Nevertheless, the profile of the MEMS is almost completely correctly recovered. That means the required information about phase map can be extracted even from a reduced hologram.

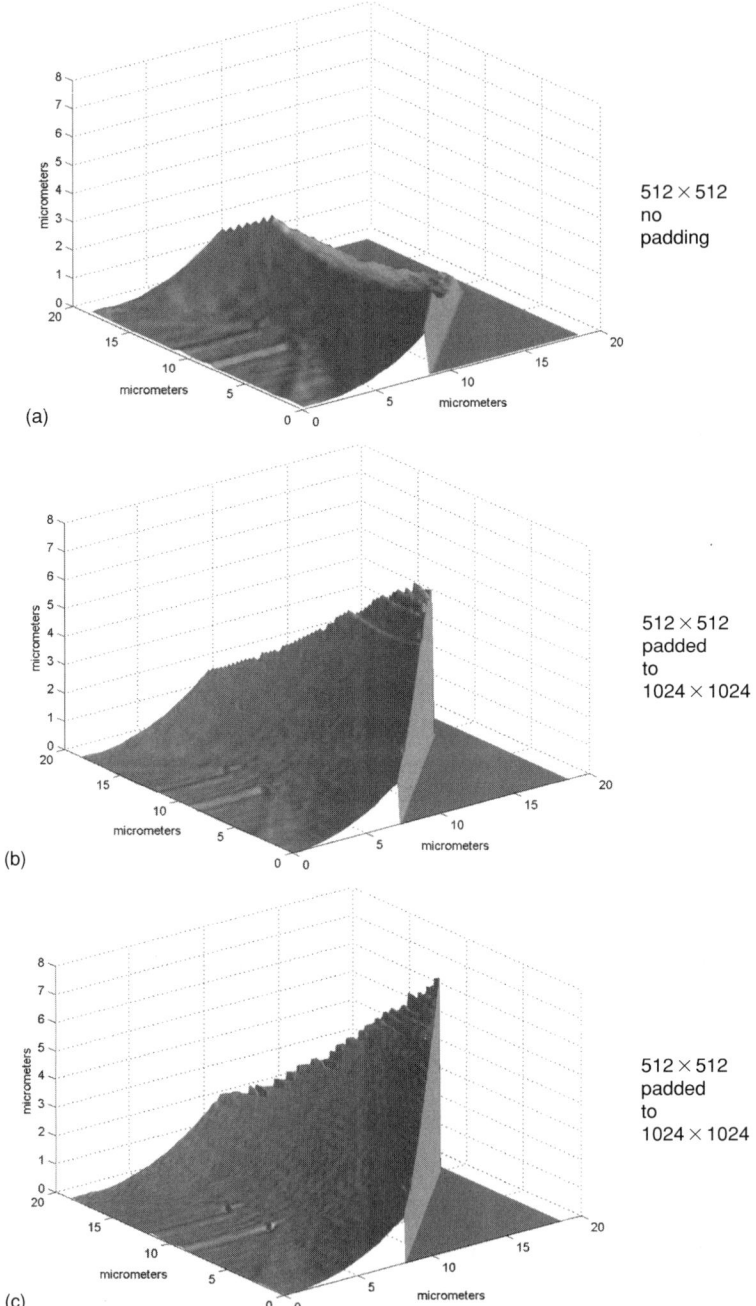

512×512
no
padding

512×512
padded
to
1024×1024

512×512
padded
to
1024×1024

Figure 3.22. Profile of the MEMS obtained by: (a) central portion of 512×512 pixel hologram; (b) same hologram of (a) but padded with zeros up to 1024×1024pixels; (c) padded up to 2048×2048 pixels.

3.4. Conclusions

This chapter has reported a detailed description and discussion of the recent advances and improvements in the novel interferometric technique of Digital Holography. Numerous examples have been shown of applications in microscopy for inspection, characterization, and investigation of different materials and processes. It is believed that the progress achieved in the reconstruction methods will find useful applications in different areas of homeland security, and we hope they can provide inspiration for further investigations for conceptual developments of new methods and systems useful in this field.

Acknowledgments

The authors gratefully acknowledge the invaluable contribution of Andrea Finizio and Giovanni Pierattini.

References

[1] Gabor D. (1948). "A new microscopic principle." *Nature*, 161:777–778.
[2] Stetson KA and Powell RL. (1966). "Hologram interferometry." *.J. Opt., Soc., Am.*, 54:1161.
[3] Vest CM. (1979). *Holographic Interferometry*. John Wiley, New York.
[4] Rastogi PK. (1994). *Holographic Interferometry*. Springer-Verlag, Berlin.
[5] Stetson KA and Brohinsky WR. (1985). "Electrooptic holography, its application to hologram interferometry." *App. Opt.*, 24:3631.
[6] Goodman JW and Lawrence RW. (1967). "Digital image formation from electronically detected holograms." *Appl. Phy. Lett.*, 11:77–79.
[7] Kronrod RW, Merzlyakov NS, and Yaroslavskii LP. (1972). "Reconstruction of a hologram with a computer." *Sov. Phys. Tech. Phys.*, 17:333–334.
[8] Schnars U. (1994). "Direct phase determination in hologram interferometry with use of digitally recorded holograms." *J. Opt. Soc. Am. A.*, 11:2011–2015.
[9] Schanrs U and Juptner W. (1994). "Direct recording of holograms by a CCD target and numerical reconstruction." *Appl. Opt.*, 33:179–181.
[10] Kreis TM and Jüptner WPO. (1997). "Suppression of the dc term in digital holography." *Opt. Eng.*, 36:2357–2360.
[11] Schnars U and Juptner W. (2002). "Digital recording and numerical reconstruction of holograms." *Meas. Sci. Technol.*, 13:R85–R101.
[12] Allaria E, Brugioni S, DeNicola S, Ferraro P, Grilli S, and Meucci R. (2003). "Digital Holography at 10.6 μm." *Opt. Commun.*, 215:257–262.
[13] Leith E and Upatnieks J. (1965). "Microscopy by wavefront reconstruction." *J Opt. Soc. Am.*, 55:569–570.
[14] Goodman JW. (1996). *Introduction to Fourier Optics*. 2nd edn. McGraw-Hill, New York.
[15] Kreis TM and Jüptner W. (1997). *Principles of Digital Holography*. In: Jüptner, Osten, ed. Fringe 97, Academic, Verlag, pp. 253–363.

[16] Onural l. (2000). "Sampling of the diffraction field." *Appl. Opt.*, 39:5929–5935.

[17] Lei X, Xiaoyuan P, Asundi AK, and jianmin M. (2001). "Hybrid holographic microscope for interferometric measurement of microstructures." *Opt. Eng.*, 40:2533–2539.

[18] Lei X, Xiaoyuan P, Jianmin M, and Asundi AK. (2001). "Studies of digital microscopic holography with applications to microstructure testing." *Appl. Opt.*, 40:5046–5052.

[19] Kreis TM. (2002). "Frequency analysis of digital holography." *Opt. Eng.*, 41:771–778.

[20] Kreis TM. (2002). "Frequency analysis of digital holography with reconstruction by convolution." *Opt. Eng.*, 41:1829–1839.

[21] Yamaguchi I and Zhang T. (1997). "Phase-shifting digital holography." *Opt. Lett.*, 23:1268–1270.

[22] Lai S, King B, and Neifeld NA. (2000). "Wavefront reconstruction by means of phase-shifting digital in-line holography." *Opt. Comm.*, 173:155–160.

[23] De Nicola S, Ferraro P, Finizio A, and Pierattini G. (2002). "Wavefront reconstruction of Fresnel off-axis holograms with compensation of aberrations by means of phase-shifting digital holography." *Opt. Laser Eng.*, 37:331–340.

[24] Cuche E, Marquet P, and Depeursinge C. (2000). "Spatial filtering for zero-order and twin-image elimination in digital off-axis holography." *Appl. Opt.*, 39:4070–4075.

[25] Liu C, Li Y, Cheng X, Liu Z, Bo F, and Zhu J. (2002). "Elimination of zero-order diffraction in digital holography," *Opt. Eng.*, 41:2434–2437.

[26] Seebacker S, Osten, Baumbach T, and Juptner W. (2001). "The determination of materials parameters of microcomponents using digital holography." *Opt. Laser Eng.*, 36:103–126.

[27] Jueptner WP, Werner P, Kujawinska M, Osten W, Salbut LA, and Seebacher S. (1987). "Combined measurement of silicon microbeams by grating interferometry and digital holography." In: International Conference on Applied Optical Metrology, Pramod K. Rastogi; Ferenc Gyimesi, eds, *Proc. SPIE.* Vol. 3407, pp. 348–357.

[28] Seebacker S, Osten W, Baumbach T, and Juptner W. (2001). "The determination of materials parameters of microcomponents using digital holography." *Opt. Laser Eng.*, 36:103–126.

[29] Dubois F, Joannes L, Dupont O, Dewandel JL, and Legros JC. (1999). "An integrated optical setup for fluid-physics experiments under microgravity conditions." *Meas. Sci. Technol.*, 10:934–945.

[30] Ferraro P, De Nicola S, Finizio A, Grilli S, and Pierattini G. (2001). "Digital holographic interferometry for characterization of transparent materials." In: *Optical Measurement Systems for Industrial Inspection II: Applications in Production Engineering*, R. Hoefling; WP Jueptner; M Kujawinska, eds., Proc. SPIE Vol. 4399, pp. 9–16.

[31] Nilsson B and Carlsson T. (2000). "Simultaneous measurement of shape and deformation using digital light-in-flight recording by holography." *Opt. Eng.*, 39:244–253.

[32] Osten W, Baumbach T, and Juptner W. (2002). "Comparative digital holography." *Opt. Lett.*, 27:1764–1766.

[33] Cuche E, Marquet P, and Depeursinge C. (1999). "Simultaneous amplitude-contrast and quantitative phase-contrast microscopy by numerical reconstruction of Fresnel off-axis holograms." *Appl. Opt.*, 38:6994–7001.

[34] Ferraro P, DeNicola S, Finizio A, Coppola G, Grilli S, Magro C, and Pierattini G. (2003). "Compensation of the inherent wavefront curvature in digital holographic coherent microscopy for quantitative phase contrast imaging." *Appl. Opt.*, 42(11):1936–1946.

[35] Grilli S, Ferraro P, De Nicola S, Finizio A, Pierattini G, and Meucci R. (2001). "Whole optical wave fields reconstruction by digital holography." *Opt. Exp.*, 9:294–302.

[36] De Nicola S, Ferraro P, Finizio A, and Pierattini G. (2001). "Correct-image reconstruction in the presence of severe anamorphism by means of digital holography." *Opt. Lett.*, 26:974–977.

[37] Grilli S, De Nicola S, Ferraro P, and Pierattini G. (2002). "Experimental demonstration of the longitudinal phase-shift in digital holography." Submitted to *Opt. Eng.*, (2002).

[38] Stadelmaier A and Massig JH. (2000). "Compensation of lens aberrations in digital holography." *Opt. Lett.*, 25:1630–1633.

[39] Pedrini G, Schedin S, and Tiziani HJ. (2001). "Aberration compensation in digital holographic reconstruction of microscopic objects." *J. Mod. Opt.*, 48:1035–1041.

[40] Cuche E, Marquet P, and Depeursinge C. (1999). "Simultaneous amplitude-contrast and quantitative phase-contrast microscopy by numerical reconstruction of Fresnel off-axis holograms." *Appl. Opt.*, 38:6994–7001.

[41] De Nicola S, Ferraro P, Finizio A, Grilli S, Coppola G, Iodice M, De Natale P, and Chiarini M. (2004). "Surface topography of microstructures in lithium niobate by digital holographic microscopy." *Meas. Sci. Technol.*, 15(5):961–968.

[42] Ferraro P, Coppola G, DeNicola S, Finizio A, and Pierattini G. (2003). "Digital holographic microscope with automatic focus tracking by detecting sample displacement in real time." *Opt. Lett.*, 28:1257–1259.

[43] Coppola G, Ferraro P, Iodice M, De Nicola S, Finizio A, and Grilli S. (2004). "A digital holographic microscope for complete characterization of microelectromechanical systems." *Meas. Sci. Technol.*, 15:529–539.

[44] Ferraro P, Coppola G, DeNicola S, Finizio A, Grilli S, Iodice M, Magro C, and Pierattini G. (2002). "Digital holography for characterization and testing of MEMS structures." In: *Proceedings of IEEE/LEOS International Conference on Optical MEMS 2002* (IEEE, New York), pp. 125–126.

[45] Grilli S, Ferraro P, Paturzo M, Alfieri D, De Natale P, De Angelis M, De Nicola S, Finizio A, and Pierattini G. (2004). "In situ visualization, monitoring and analysis of electric field domain reversal process in ferroelectric crystals by digital holography." *Opt. Expr.*, 12(9):1832–1842.

[46] Demoli N, Vukicevic D, and Torzynski M. (2003). "Dynamic digital holographic interferometry with three wavelengths." *Opt. Expr.*, 11:767–774.

[47] Kato J, Yamaguchi I, and Matsumura T. (2003). "Multicolor digital holography with an achromatic phase shifter." *Opt. Lett.*, 27:1403.

[48] Yamaguchi I, Matsumura T, and Kato J. (2002). "Phase-shifting colour digital holography." *Opt. Lett.*, 27:1108–1110.

[49] Gass J, Dakoff A, and Kim MK. (2003). "Phase imaging without 2π ambiguity by multiwavelength digital holography." *Opt. Lett.*, 28:1141–1143.

[50] Kim M. (2000). "Tomographic three-dimensional imaging of a biological specimen using wavelength-scanning digital interference holography." *Opt. Expr.*, 7:305–310.

[51] Zhang F, Yamaguchi I, and Yaroslavsky LP. (2004). "Algorithm for reconstruction of digital holograms with adjustable magnification." *Opt. Lett.*, 29:1668–1670.

[52] Ferraro P, Coppola G, DeNicola S, Finizio A, Pierattini G, and Alfieri D. (2004). "Controlling image size as a function of distance and wavelength in Fresnel transform reconstruction of digital holograms." *Opt. Lett.*, 29(8):854–856.

[53] DeNicola S, Ferraro P, Coppola G, Finizio A, Pierattini G, and Grilli S. (2004). "Talbot self-image effect in digital holography and its application to spectrometry." *Opt. Lett.*, 29(1):104–106.

[54] Pedrini G, Tiziani HJ, and Zoa Y. (1996). "Speckle size of digitally reconstructed wavefronts of diffusely scattering objects." *J. Mod. Opt.*, 43:395–407.

[55] Indebetouw G and Klysubun P. (1999). "Space–time digital holography: A three-dimensional microscopic imaging scheme with an arbitrary degree of spatial coherence." *Appl. Phys. Lett.*, 75:2017.

[56] Massig JH. (2002). "Digital off-axis holography with a synthetic aperture." *Opt. Lett.*, 27:2179.

[57] Liu C, Liu Z, Bo F, Wang Y, and Zhu J. (2002). "Superresolution digital holographic imaging method." *Appl. Phys. Lett.*, 81:3143–3145.

[58] Ferraro P, DeNicola S, Finizio A, Pierattini G, and Coppola G. (2004). "Recovering image resolution in reconstructing digital off-axis holograms by Fresnel-transform method." *Appl. Phys. Lett.*, 85:2709–2711.

4

Hybrid Optical Encryption of a 3D Object by Use of a Digital Holographic Technique

Takanori Nomura

Department of Opto-Mechatronics, Wakayama University, 930 Sakaedani, Wakayama 640–8510, JAPAN nom@sys.wakayama-u.ac.jp

An encryption method of a three dimensional (3D) object based on phase modulation of an object wave is proposed. The phase of the object wave is modulated by a virtual optical random phase mask using a digital holographic technique. The keys of an encryption step are both a phase distribution and a position of the virtual optical phase mask. If either of them is not correct, the three dimensional (3D) object cannot be decrypted. Owing to a characteristic of a hologram, some parallax of a reconstructed three dimensional (3D) object can be seen. Experimental results are presented to confirm the proposed method.

4.0 Introduction

Optical information technologies for security and encryption systems,[1–9] have been studied. Most of them are aimed at encryption of 2D information such as images. An encryption of a three dimensional (3D) object has been proposed by using digital holography.[5] An object is encrypted by use of random phase distribution of the reference wave. In the system, only phase distribution of the reference wave is needed to decrypt. If the phase distribution comes out, the encryption will end in failure. Here, more secure encryption method of a three dimensional (3D) object is proposed by use of phase modulation of an object wave. The encryption is accomplished by a combination of a real optical system and a virtual optical system. Therefore, the method is called hybrid optical encryption. In Section 4.1, the principle of encryption and decryption method of the proposed system is described. In Section 4.2, optical experimental results of three dimensional (3D) object encryption are shown to confirm the proposed system.

4.1 Hybrid Encryption System of a 3D Object

An encryption step and a decryption step of a proposed hybrid encryption system are described here.

4.1.1 Encryption

The scheme of the proposed encryption system is shown in Fig. 4.1. Referring to Fig. 4.1, the hybrid encryption system is described in detail. In the figure, nonencrypted wave and encrypted wave are indicated by solid arrows and dashed arrows, respectively. Let $U(x_o)$ denote a wavefront of a three dimensional (3D) object to be encrypted. One-dimensional notation is used for simplicity. Let $U(x_c)$ denote a wavefront of $U(x_o)$ at a CCD after propagation of a distance z_c. The relation of $U(x_o)$ and $U(x_c)$ may be written as

$$U(x_c) = Pr[U(x_o); z_c], \tag{1}$$

where $Pr[U,z]$ denotes the operation to make a wavefront U propagate at a distance of z. The object wave which comes from the 3D object is recorded electrically as a digital hologram using a phase-shifting digital holography.[10]

Then the hologram is encrypted using a virtual optical system. The wavefront is propagated in opposite direction from the CCD to a plane $(z = z_m)$ where a virtual phase mask (VPM) is placed. The wavefront $U(x_m)$ at the VPM may be written using the same notation as

$$U(x_m) = Pr[U(x_c); z_m - z_c]. \tag{2}$$

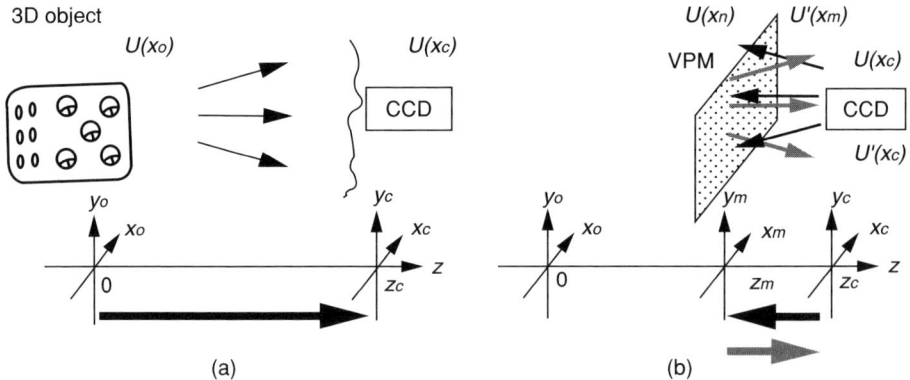

Figure 4.1. Scheme of an encryption step of a hybrid optical encryption of a 3D object using a digital holographic technique: (a) recording a digital hologram of 3D object and (b) encrypting a 3D object as an encrypted digital hologram using a virtual optical system.

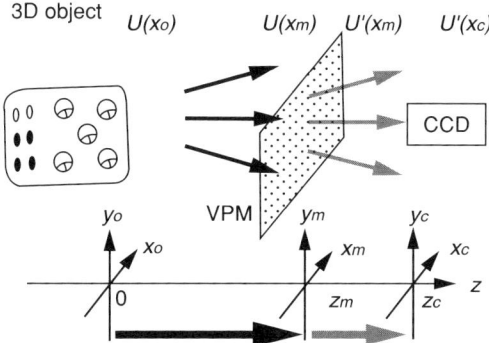

Figure 4.2. Equivalent optical system of an encryption step.

The calculation of a diffraction integral in a virtual optical system is described in Appendix. For encryption, the product of the wavefront and the VPM which has a phase distribution denoted by $\exp[i\alpha(x_m)]$ is calculated. The phase distribution is assumed to be uniformly random. The encrypted wave front $U'(x_m)$ is obtained as follows:

$$U'(x_m) = U(x_m)\exp[i\alpha(x_m)]. \tag{3}$$

Finally, the wavefront is propagated in right direction from the VPM to the CCD. The encrypted wavefront at the CCD is written as

$$U'(x_c) = Pr[U'(x_m); z_c - z_m]. \tag{4}$$

This hologram $U'(x_c)$ may be called an encrypted digital hologram. This hologram is suitable for storage and transmission because of digital data. The data might be compressed by use of digital hologram compression.[11–13] Figure 4.2 shows an equivalent optical system of the above-mentioned encryption step. In the system, the VPM can be placed behind the CCD, because the encryption is performed by a virtual optical system.

4.1.2 Decryption

To decrypt, a virtual optical system shown in Fig. 4.3, is used. The encrypted digital hologram is propagated from the CCD to the VPM placed at $z = z_m$. A wavefront at the VPM written by

$$
\begin{aligned}
V'(x_m) &= Pr[U'(x_c); z_m - z_c] \\
&= Pr[Pr[U'(x_m); z_c - z_m]; z_m - z_c] \\
&= U'(x_m),
\end{aligned}
\tag{5}
$$

is obtained. Next, the product of the wavefront $V'(x_m)$ and a complex conjugate of a phase distribution of the VPM is calculated. The product $V(x_m)$ is written as follows:

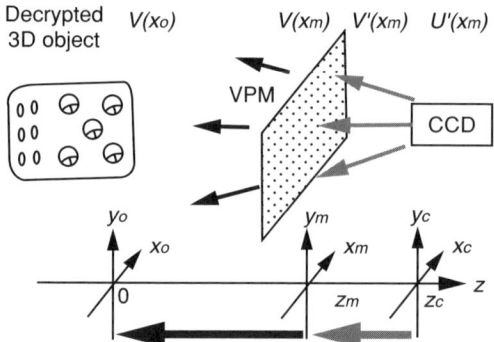

Figure 4.3. Scheme of a decryption step of a hybrid optical encryption of a 3D object using a digital holographic technique.

$$
\begin{aligned}
V(x_m) &= V'(x_m) \exp\left[-i\alpha(x_m)\right] \\
&= U(x_m) \exp\left[i\alpha(z_m)\right] \exp\left[-i\alpha(x_m)\right]. \\
&= U(x_m)
\end{aligned}
\tag{6}
$$

Finally, by applying the diffraction integral the VPM to the original position where the object was placed, the wavefront of the original three dimensional (3D) object can be obtained. The wavefront at the position can be written as

$$
\begin{aligned}
V(x_o) &= Pr[V(x_m); -z_m] \\
&= Pr[U(x_m); -z_m]. \\
&= U(x_o)
\end{aligned}
\tag{7}
$$

From this equation, it is found that the encrypted hologram can be decrypted correctly. Note that both the information of phase distribution and is position of the VPM are needed to decrypt the encrypted hologram.

4.2 Encryption and Decryption Experiments

To confirm the proposed hybrid encryption system, optical experimental results are shown. Figure 4.4 shows the experimental setup to record a digital hologram of 3D objects.

4.2.1 An Experimental System

A He–Ne laser (wavelength, 632.8 nm) is used as a coherent light source. It is collimated by a beam expander consisting of an objective lens, a spatial filter, and lens with a focal length of 250 mm. The reference beam is phase-shifted by a moving mirror driven by a computer-controlled PZT stage. The digital hologram is captured by a CCD camera with 1280 by 960 pixels and 8 bits of

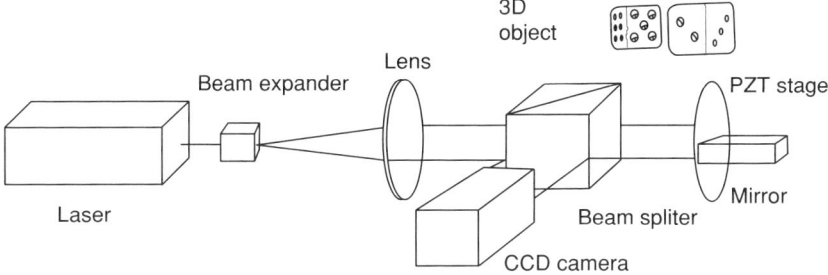

Figure 4.4. Optical setup for recording a digital hologram of 3D objects.

gray levels. The size of a pixel of the CCD is 4.65 μm by 4.65 μm. Two dice are used as 3D objects. They are as large as 10 mm × 10 mm × 10 mm each. The distance from the dice to the CCD are 180 mm and 270 mm, respectively.

4.2.2 Recording Digital Hologram

The digital hologram obtained by using phase-shifting technique is shown in Fig. 4.5. Following each figure of a digital hologram is a part of its original digital hologram. They are the portions extracted 320 by 240 pixels from the center of original 1280 by 960 pixels. In Fig. 4.5(a), the amplitude distribution is normalized by the maximum amplitude. Black and white denote 0 and maximum value, respectively. In Fig. 4.5(b), the phase distribution is normalized by $-\pi$ denoted by black to π denoted by white. The same normalization is applied to the following figures. Figure 4.5 corresponds to $U(x_c)$ given by Eq. (1).

4.2.3 Encryption

For encryption, the wavefront at a VPM is calculated using a computational diffraction integral described in Appendix. In this experiment, it is assumed

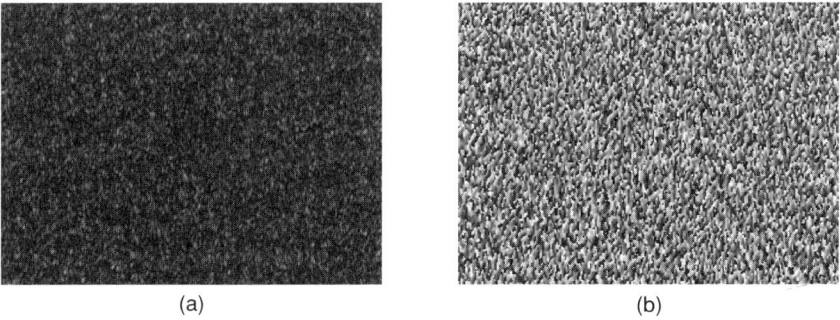

(a)   (b)

Figure 4.5. A nonencrypted digital hologram of 3D objects at the CCD: (a) amplitude distribution and (b) phase distribution.

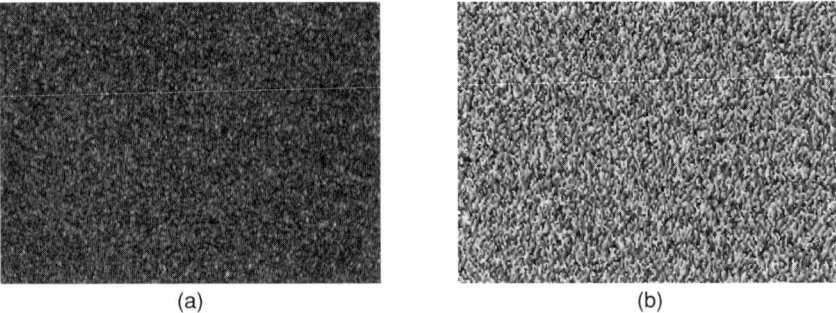

(a)   (b)

Figure 4.6. A nonencrypted wavefront of 3D objects at a virtual phase mask plane: (a) amplitude distribution and (b) phase distribution.

that the distance from the CCD to the VPM is 30 mm. The wavefront shown in Fig. 4.6 corresponds to $U(x_m)$ given by Eq. (2). Then it is multiplied by the VPM shown in Fig. 4.7, which has uniform amplitude distribution and a uniformly random phase distribution corresponding to $\exp[i\alpha(x_m)]$. After multiplying, the encryption wavefront $U'(x_m)$ of 3D objects is obtained. It is given by Eq. (3) at the VPM shown in Fig. 4.8.

By applying Fresnel diffraction integral to the encrypted digital hologram $U'(x_m)$ from the VPM to the CCD, the encrypted digital hologram of a 3D object at the CCD is obtained. It is shown in Fig. 4.9 corresponding to $U'(x_c)$ given by Eq. (4). This encrypted digital hologram is suitable for electrical storage and transmission because of digital data.

4.2.4 Decryption

To decrypt the encrypted digital hologram, the diffraction integral based on the algorithm mentioned in Section 4.1 is applied. With correct position and phase distribution of the VPM, the decrypted 3D objects are shown in

Figure 4.7. A nonencrypted wavefront of 3D objects at a virtual phase mask plane: (a) amplitude distribution and (b) phase distribution.

Figure 4.8. The encrypted wavefront of 3D objects at the virtual phase mask plane: (a) amplitude distribution and (b) phase distribution.

Fig. 4.10(a). Figure 4.10(b) shows the decrypted 3D objects using no information of the VPM. Figures 4.10(c) and (d) are the decrypted 3D objects if either position or phase distribution is wrong. In Fig. 4.10(c), the distance from the CCD to the VPM is set to 31 mm. In Fig. 4.10(d), to decrypt a VPM, which has a phase distribution independent from the VPM, encryption process is used. The reconstructed objects from an original digital hologram are shown in Fig. 4.10(e). From these experimental results, if only both information of position and phase distribution of the VPM are correct, it is found that the encrypted digital hologram can be decrypted.

If a characteristic of holography is used, parallax of the 3D objects can be seen. After the example of Tajahuerce and Javidi,[5] different perspectives of the decrypted dice are shown. Figure 4.11 show three parallax and different focused objects calculated from decrypted hologram $V(x_m)$ at the VPM. Figures 4.11(a), (b), and (c) are obtained from left half, center, and right half region of the decrypted hologram with different distance from the hologram. A different aspect in the figures can be seen.

Figure 4.9. The encrypted digital hologram of 3D objects at the CCD: (a) amplitude distribution and (b) phase distribution.

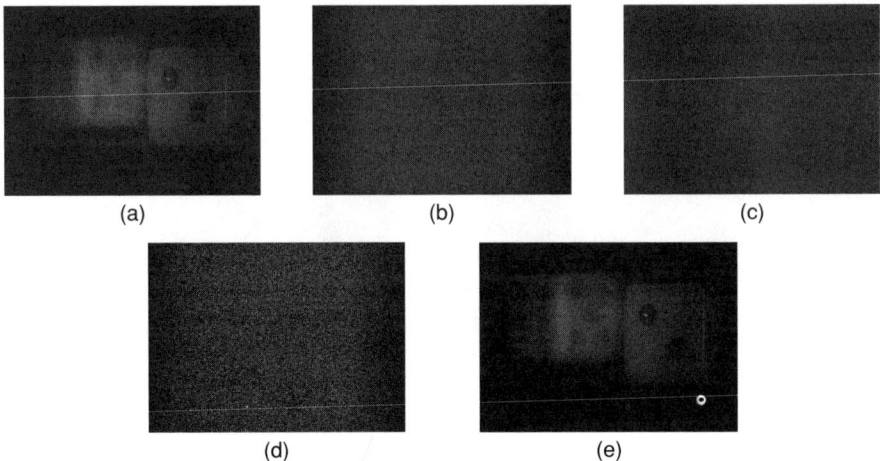

Figure 4.10. The decrypted 3D objects using (a) both correct position and phase distribution, (b) no information, (c) wrong position and correct phase distribution, and (d) correct position and wrong phase distribution, of a virtual phase mask. (e) The reconstruct 3D object from a nonencrypted digital hologram.

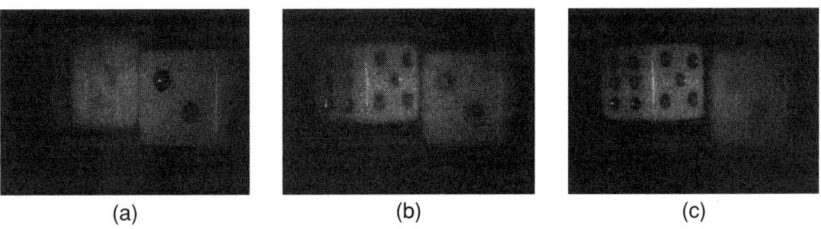

Figure 4.11. The decrypted 3D objects which have a different aspect. (a) Front focused reconstructed objects from left half region, (b) middle focused reconstructed objects from center region, and (c) back focused reconstructed image from right half region, of the decrypted digital holograms, respectively.

4.3 Conclusions

A hybrid optical encryption of a 3D object by using a combination of a real optical system and a virtual optical system has been proposed. The encryption was performed by a phase modulation of the object wave in a virtual optical system. Experimental results including encryption, decryption, and parallax images confirm the proposed system.

The author wishes to thank Dr. Yoshiharu Morimoto, Ms. Kaoru Uota, and Mr. Isao Takahashi for their valuable comments and experimental supports.

4.4 Appendix

The computational calculation methods of wave propagation used in the encryption system are described. Using Fresnel–Kirchhoff diffraction theory [14, 15], the field at (x_2, y_2) at the distance of z from the object $f(x_1, y_1)$ is written as

$$f(x_2, y_2) = \frac{i}{\lambda} \iint_{-\infty}^{\infty} \frac{f(x_1, y_1)}{[z^2 + (x_2 - x_1)^2 + (y_2 - y_1)^2]^{1/2}}$$
$$\times \exp\left\{ -i\frac{2\pi}{\lambda}[z^2 + (x_2 - x_1)^2 + (y_2 - y_1)^2]^{1/2} \right\} dx_1\, dy_1, \quad (8)$$

where λ denotes a wavelength of an illuminating light. In this case the inclination factor is regarded as unity, because it is a paraxial region. If the distance between the object $f(x_1, y_1)$ and the field (x_2, y_2) is in the region of the Fresnel diffraction, the field at (x_2, y_2) at the distance of z from the object $f(x_1, y_1)$ is written as

$$f(x_2, y_2) = \frac{i}{\lambda z} \exp\left(-\frac{i2\pi z}{\lambda} \right)$$
$$\times \iint_{-\infty}^{\infty} f(x_1, y_1) \exp\left\{ -\frac{i\pi}{\lambda z}[(x_2 - x_1)^2 + (y_2 - y_1)^2] \right\} dx_1\, dy_1, \quad (9)$$

using Fresnel diffraction integral.[5, 15–19] Then the both equations for numerical calculation are rewritten. Equation (8) can be written as

$$f(x_2, y_2) = FFT^{-1}\left[FFT[f(x_1, y_1)] \cdot FFT\left[\frac{\exp\left[\text{sign}(z)i\frac{2\pi}{\lambda}(z^2 + x_1^2 + y_1^2)^{1/2}\right]}{(z^2 + x_1^2 + y_1^2)^{1/2}} \right] \right],$$
$$(10)$$

where

$$\text{sign(x)} = \begin{cases} 1 & x \geq 0 \\ -1 & x < 0 \end{cases}. \quad (11)$$

And FFT^{-1} and FFT denote inverse fast Fourier transform and fast Fourier transform operations, respectively. Here the constant phase at (x_2, y_2) plane is neglected. If $f(x_1, y_1)$ has the resolution Δx_1 and Δy_1, the resolution of $f(x_2, y_2)$ is denoted by

$$\Delta x_2 = \Delta x_1, \quad (12)$$
$$\Delta y_2 = \Delta y_1. \quad (13)$$

On the other hand, Eq. (9) can be written as

$$f(x_2, y_2) = \exp\left[\frac{i\pi(x_2^2 + y_2^2)}{\lambda z} \right] FFT\left[f(x_1, y_1) \exp\frac{i\pi}{\lambda z}(x_1^2 + y_1^2) \} \right]. \quad (14)$$

Here the constant phase at (x_2, y_2) plane is also neglected. If $f(x_1, y_1)$ has the resolution Δx_1 and Δy_1, the resolution of $f(x_2, y_2)$ is denoted by

$$\Delta x_2 = \frac{\lambda z}{L_x}, \tag{15}$$

$$\Delta y_2 = \frac{\lambda z}{L_y}. \tag{16}$$

The choice of the above-mentioned Eqs. (10) and (14) can be decided based on the following inequalities shown as

$$z \leq \frac{L_x^2}{\lambda M}, \tag{17}$$

$$z \leq \frac{L_y^2}{\lambda N}, \tag{18}$$

where L_x and L_y denote size of the field of (x_1, y_1), respectively, and M and N denote pixel size of the field of (x_1, y_1). If the inequalities are satisfied, the expression described in Eq. (10) is used; if not, Eq. (14) is used.

As the distance between CCD and VPM in the experiments is 30 mm, the distance satisfies the inequalities. As the distance between a 3D object and VPM is more than 150 mm, the distance does not satisfy. So, Eq. (10) is used for calculating wavefront between VPM and CCD, and Eq. (14) for between VPM to a 3D object.

References

[1] Réfrégier P and Javidi B. (1995). Optical image encryption based on input plane and Fourier plane random encoding. *Opt. Lett.*, **20**:767–769.
[2] Javidi B and Nomura T. (2000). Securing information by use of digital holography. *Opt. Lett.*, **25**:28–30.
[3] Nomura T and Javidi B. (2000). Optical encryption system with a binary key code. *Appl. Opt.*, **39**:4783–4787.
[4] Tajahuerce E, Mataoba, O, Verrall, SC, and Javidi B. (2000). Opto-electronic information encryption with phase-shifting interferometry. *Appl. Opt.*, **39**:2313–2320.
[5] Tajahuerce E and Javidi B. (2000). Encrypting three-dimensional information with digital holography. *Appl. Opt.*, **39**:6595–6601.
[6] Nomura T and Javidi B. (2000). Optical encryption using a joint transform correlator architecture. *Opt. Eng.*, **39**:2031–2035.
[7] Unnikrishnan G, Joseph J, and Singh K. (2000). Optical encryption by double-random phase encoding in the fractional Fourier domain. *Opt. Lett.*, **25**:887–889.
[8] Wang B, Sun C, Su W, and Chiou AET. (2000). Shift-tolerance property of an optical double-random phase-encoding encryption system. *Appl. Opt.*, **39**:4788–4793.

[9] Nomura T, Mikan S, Morimoto Y, and Javidi B. (2003). Secure optical data storage with random phase key codes by use of a configuration of a joint transform correlator. *Appl. Opt.*, **42**:1508–1514.

[10] Yamaguchi I and Zhang T. (1997). Phase-shifting digital holography. *Opt. Lett.*, **22**:1268–1270.

[11] Nomura T, Okazaki A, Kameda M, Morimoto Y, and Javidi B. (2001). Digital holographic data reconstruction with data compression. *Proc. Soc. Photo-Opt. Instrum. Eng.*, **4471**:235–242.

[12] Naughton TJ, Frauel Y, Javidi B, and Tajahuerce E. (2002). Compression of digital holograms for three-dimensional object reconstruction and recognition. *Appl. Opt.*, **41**:4124–4132.

[13] Naughton TJ, McDonald JB, and Javidi, B. (2003). Efficient compression of Fresnel fields for internet transmission of three-dimensional images. *Appl. Opt.*, **42**:4758–4764.

[14] Goodman JW. (1996). *Introduction to Fourier Optics*, 2nd edn. McGraw-Hill, New York.

[15] Milgram JH and Li W. (2002). Computational reconstruction of images from holograms. *Appl. Opt.*, **41**:853–864.

[16] Aoki Y and Ishizuka S. (1974). Numerical two-dimensional Fresnel transform methods. *Trans. IECE B* **57-B**, pp. 511–518. (in Japanese)

[17] Kreis TM. (2002). Frequency analysis of digital holography. *Opt. Eng.*, **41**:771–778.

[18] Kreis TM. (2002). Frequency analysis of digital holography with reconstruction by convolution. *Opt. Eng.*, **41**:1829–1839.

[19] Mas D, Garcia J, Ferreira C, Bernardo LM, and Marinho F. (1999). Fast algorithms for free-space diffraction pattern calculation. *Opt. Commun.*, **164**:233–245.

3D Object Recognition using Gabor Feature Extraction and PCA–FLD Projections of Holographically Sensed Data

Seokwon Yeom and Bahram Javidi

Department of Electrical and Computer Engineering, U-2157, University of Connecticut, Storrs, CT 06269-2157, USA

5.0 Introduction

Object recognition technique plays a significant role in Homeland Security. It has broad applications, such as automated target recognition (ATR) and surveillance system, to identify unknown vehicles or hidden equipment. Growing numbers of three-dimensional (3D) optical and imaging techniques have been researched for this purpose, presenting many challenges and benefits.[1–13]

The acquisition of 3D information is one challenge to be handled. There are several approaches to analyze a 3D scene according to its acquisition methods. The 3D correlation[1] has been presented for 3D object recognition. Two-dimensional (2D) correlation technique combined with Fourier transform profilometry has been proposed.[2] In a hologram, we can sense the 3D information of objects as the complex amplitude.[5–8,13] Images reconstructed at different planes were correlated with the reference image to find the best matching position and angle as well as the object itself.[6,7] Three-dimensional images were reconstructed at various orientations, and neural network technique was applied to recognize the 3D reference object.[8]

Another challenge is increased data size. In general, the 3D image is high dimensional. Each frame of holographic data may consist of millions of pixels. Therefore, 3D object recognition techniques are computationally demanding. The high-dimensional complexity also causes the difficulties in distinguishing one pattern itself from another, which is called "curse of dimensionality".[14]

In this chapter, 3D objects that are sensed by digital interferometry are classified. A 3D object can be reconstructed at different planes from a single hologram. The feature extraction, dimensionality reduction projections, and statistical pattern classification to computer-reconstructed holographic images are applied.

We use Gabor-based wavelets to extract distinguished features from 3D objects while reducing their dimensionality. Gabor-based wavelets are selective in terms of spatial frequency and bandwidth as well as spatial location.[15–17] These have been used for various purposes in pattern recognition and image compression.

We adopt Fisher linear discriminant (FLD) combined with principal component analysis (PCA) to classify the feature vectors extracted from 3D objects.[18–21] Both are optimal projection methods to reduce the dimension of a vector according to some criteria. In the PCA, feature vectors are mapped into a vector space spanned by eigenvector bases. This projection is optimal in terms of mean square error. Fisher linear discriminant projects vectors onto a low-dimensional subspace, thereby maximizing the ratio of the between-class scatter to the within-class scatter.

Another advantage of dimension reduction lies in the flexible selection of decision rules. When large-dimensional data are considered, it is impractical to use decision rules based on statistical analysis. However, in this chapter, we employ statistical distance (Mahalanobis distance) measure[20,21] that is equivalent to maximum likelihood (ML) decision rule if we assume Gaussian probability density function (PDF) with equal covariances between classes.

In the following sections, various components of the 3D classification system have been reviewed. The computational holographic imaging is presented in Section 3. Gabor scheme and feature vector extraction are described in Section 4. Principal component analysis and FLD are presented in Section 5. The statistical decision rule is presented in Section 6. Experimental and simulation results are shown in Section 7, and conclusion follows in Section 8.

5.1 System Description

Figure 5.1 shows a block diagram of the 3D object classification. The overall system consists of several subsystems for sensing and computational processing. First, we sense and reconstruct holographic images of 3D objects. During training procedures, we extract feature vectors from reconstructed holographic images by Gabor-based wavelets. The computations of the PCA and FLD matrices as well as sample means and covariances follow in the next step. During test procedures, PCA and FLD matrices are applied to feature vectors of test images. Statistical decision rule determines the class of the test images. In the following sections, various components of the 3D classification system have been described and experimental results are presented.

5.2 Computational Holographic Imaging

We use phase-shift interferometry to sense the 3D information of an object.[6–8,22] We use a CCD camera in Fresnel diffraction region to measure

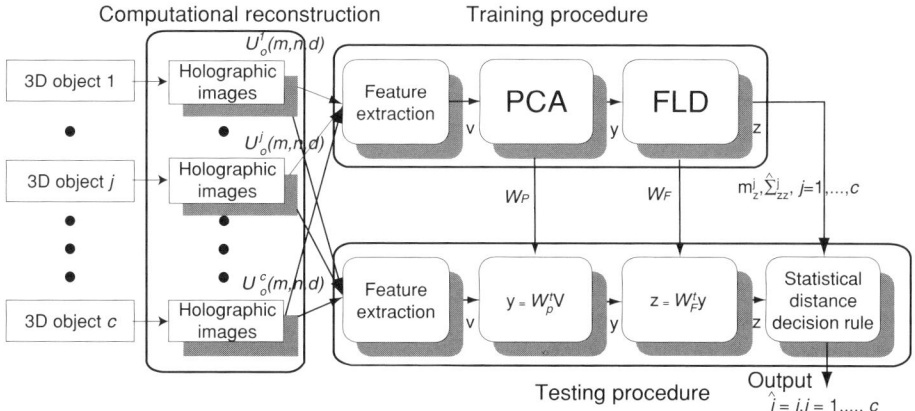

Figure 5.1. Three-dimensional object classification, $U_o^j(m, n, d)$ is the computationally reconstructed holographic image at depth d when the object class is j; (m, n) is 2D discrete coordinates at the image plane; c is the number of classes; $\mathbf{v}$ is feature vector extracted from Gabor-based wavelets; $\mathbf{y}$ is feature vector after PCA projection; $\mathbf{z}$ is feature vector after PCA–FLD projection; W_P and W_F are PCA and FLD matrix, respectively; $\mathbf{m}_z^j$ is the sample mean of the class j; and $\hat{\mathbf{\Sigma}}_{zz}^j$ is the sample covariance of the class j.

the amplitude and the phase information of a 3D object in the scene [see Fig. 5.2(a)]: L is refractive lens; D is diaphragm; BS$_1$ and BS$_2$ are beam splitters; M is plane mirror; and RP$_1$ and RP$_2$ are two wave retardation plates with retardation of $\lambda/2$ and $\lambda/4$, respectively.[5] Computer reconstruction of the 3D object can be performed using discrete inverse Fresnel transformation. Let $H_o(m', n')$ be the discrete complex amplitude of the Fresnel pattern, where m' and n' are discrete coordinates in the hologram plane. Varying the longitudinal depth between the output plane and the reconstruction plane produces different image planes of the 3D object into focus.

We can reconstruct many 2D complex-valued images rather than 3D discrete volume itself by generating slices of the 3D object in 2D planes. On the other hand, we can also reconstruct different perspectives of the 3D object. In this work, we concentrate on images reconstructed at different longitudinal depths only. The discrete Fresnel transformation for the reconstruction becomes:[5]

$$U_o(m,n,d) = \exp\left[\frac{i\pi}{\lambda d}(\Delta x^2 m^2 + \Delta y^2 n^2)\right] \times \sum_{m'=0}^{N_x-1} \sum_{n'=0}^{N_y-1}$$

$$\left\{ \begin{array}{l} H_o(m',n') \exp\left[\frac{i\pi}{\lambda d}(\Delta x'^2 m'^2 + \Delta y'^2 n'^2)\right] \\[2mm] \times \exp\left[-i2\pi\left(\frac{mm'}{N_x} \times \frac{nn'}{N_y}\right)\right] \end{array} \right\}, \tag{1}$$

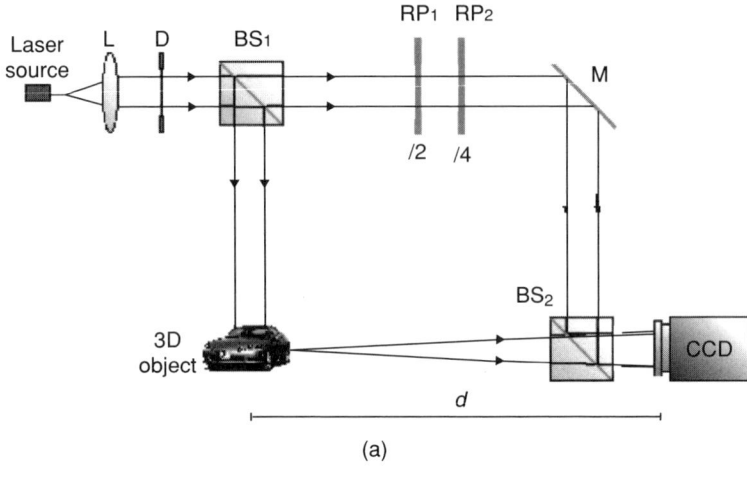

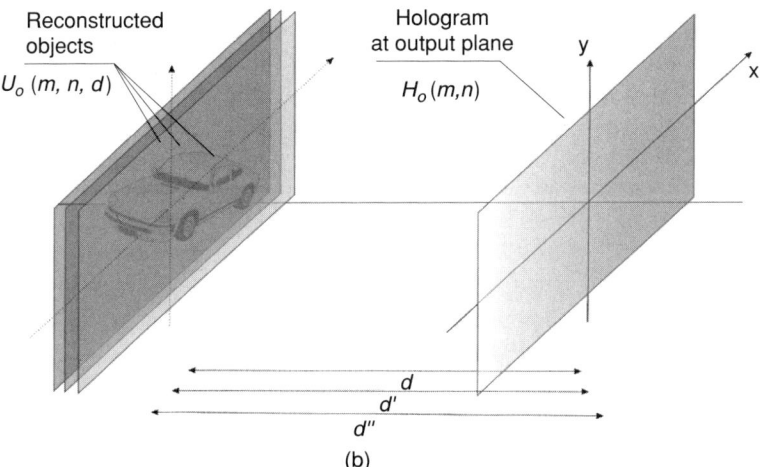

Figure 5.2. Optical sensing system for 3D object: (a) phase-shift interferometry to detect the Fresnel diffraction pattern, (b) hologram reconstruction at specific longitudinal depths.

where $U_o(m, n, d)$ is the 3D reconstructed image with depth d; λ is the wavelength of the incident light; m and n are discrete coordinates at the image plane; N_x and N_y are the numbers of pixels in the hologram window in x and y directions; $\Delta x'$ and $\Delta y'$ are the spatial resolutions of the CCD detector; and Δx and Δy are the resolutions of the object plane. Figure 5.2(b) shows the process of reconstruction at specific longitudinal depths.

In the experiments two toy cars are used to obtain computationally reconstructed holographic images. Each toy car is approximately $2.5\,\text{cm} \times 2.5\,\text{cm} \times 4.5\,\text{cm}$. Each hologram represents one object (car) class. For each hologram,

100 images are reconstructed at different longitudinal depths. The range of the reconstruction depth is from -922.5 to -823.5 mm. One hundred images are reconstructed at every millimeter. Figures 5.3 and 5.4 show examples of reconstructed image intensity.

The proposed technique is substantially different from conventional 2D imaging by a camera. In that case, the imaging is achieved according to the Lens law for an image of a single image plane. To obtain the images at different longitudinal planes, the focal length of the camera has to be changed according to the Lens law and the object distance from the lens. This should be repeated during the training as well as inspecting the input scene that can be impractical for a large depth of 3D object field. In the proposed technique, after the initial exposure, the Fresnel fields of the scene are recorded. The reconstruction is achieved through inverse Fresnel transformation. Thus, there is no need for adjusting the imaging setup or changing the focal length of the lens. Instead, the inverse Fresnel transformation reconstructs the 3D object of different longitudinal depths by selecting the distance from the detector array. Thus, we can detect multiple objects at different longitudinal depth locations from the detector array.

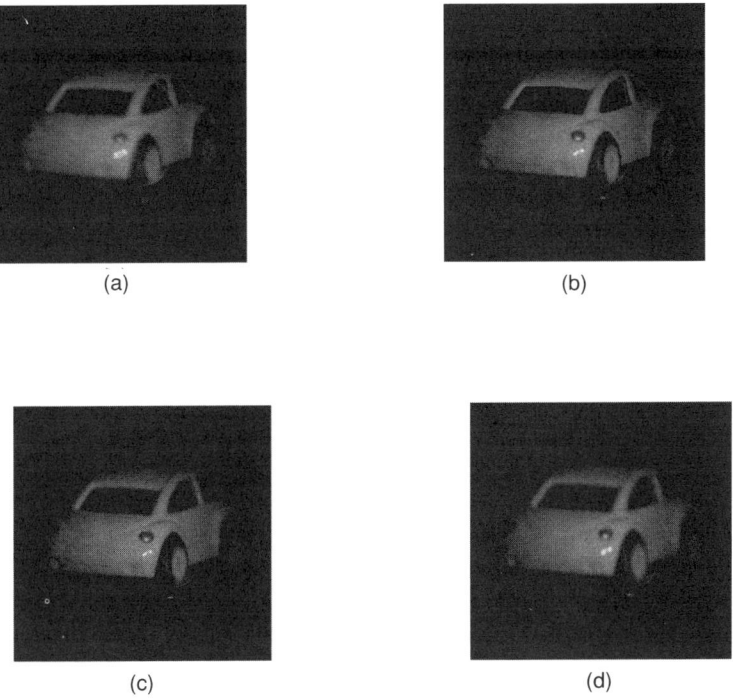

Figure 5.3. Reconstructed images of the class 1 at different longitudinal depths: (a) $d = 922.5$ mm, (b) $d = -882.5$ mm, (c) $d = -842.5$ mm, and (d) $d = -23.5$ mm.

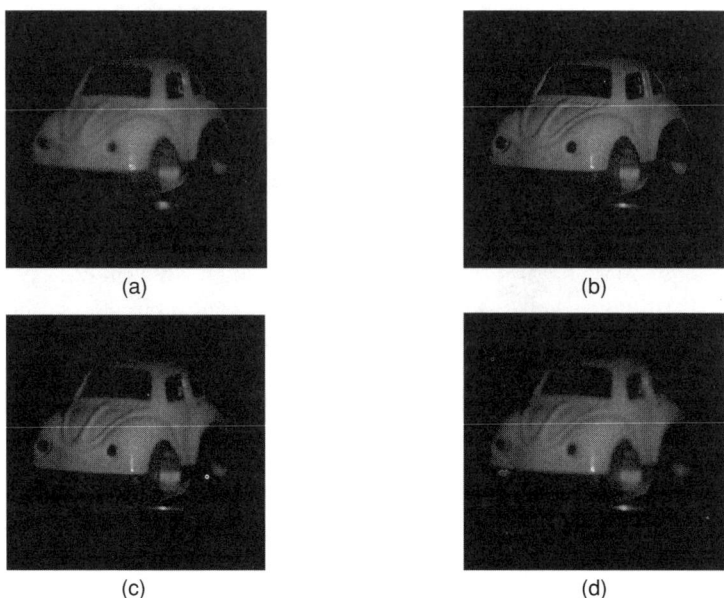

(a)

(b)

(c)

(d)

Figure 5.4. Reconstructed images of the class 2 at different longitudinal depths: (a) $d = -922.5\,\mathrm{mm}$, (b) $d = -882.5\,\mathrm{mm}$, (c) $d = -42.5\,\mathrm{mm}$, and (d) $d = -823.5\,\mathrm{mm}$.

5.3 Gabor-Based Wavelet and Feature Vector Extraction

In this section, feature vector extraction using Gabor-based wavelets and experimental results are discussed. To extract features from training and test data we apply Gabor scheme to the intensity of reconstructed holographic images. The Gabor-based wavelets act as bandpass filters with special selection of passband width according to its Gaussian envelope and carrier frequency of the complex plane wave.[15,16] The 2D impulse response (or kernel) of Gabor-based wavelets is:

$$g(\mathbf{x}) = \frac{|\mathbf{k}|^2}{\sigma^2} \exp\left(-\frac{|\mathbf{k}|^2 |\mathbf{x}|^2}{2\sigma^2}\right) \left[\exp\left(j\mathbf{k} \cdot \mathbf{x}\right) - \exp\left(-\frac{\sigma^2}{2}\right)\right], \qquad (2)$$

where $\mathbf{x}$ is a position vector, $\mathbf{k}$ is a wave number vector, and σ is the standard deviation of Gaussian envelope. By changing the magnitude and direction of the vector $\mathbf{k}$, we can scale and rotate the Gabor kernel to make self-similar forms. The size of the Gaussian envelope is the same in x and y directions, which is proportional to $\sqrt{2}\sigma|\mathbf{k}|$. The second term in the square brackets, $\exp\left(-\sigma^2/2\right)$ subtracts the DC value so it has zero mean response.[17] The frequency response of $g(\mathbf{x})$, $G(\mathbf{k}')$ is given by

$$G(\mathbf{k}') = 2\pi\left\{\exp\left[-\frac{\sigma^2}{2|\mathbf{k}|^2}|\mathbf{k}'-\mathbf{k}|^2\right] - \exp\left[-\frac{\sigma^2}{2|\mathbf{k}|^2}(|\mathbf{k}'|^2+|\mathbf{k}|^2)\right]\right\} \quad (3)$$

We can define a discrete version of the Gabor kernel as $g_{uv}(m,\,n)$ at $\mathbf{k} = \mathbf{k}_{uv}$ and $\mathbf{x} = (m,\,n)$, where m and n are discrete coordinates in 2D space in the x and y directions, respectively. Sampling of $\mathbf{k}$ is done as $\mathbf{k}_{uv} = k_{0u}[\cos\phi_v\sin\phi_v]^t$, $k_{0u} = k_0/\delta^{u-1}$, and $\phi_v = [(v-1)/V]\pi$, $u = 1,\,\ldots,\,U$ and $v = 1,\,\ldots,\,V$, where k_{0u} is the magnitude of the wave number vector; ϕ_v is the azimuth angle of the wave number vector; k_0 is the maximum carrier frequency of the Gabor kernels; δ is the spacing factor in the frequency domain; u and v are the indexes of the Gabor kernels; U and V are the total numbers of decompositions along the tangential and radial axes, respectively; and t stands for the matrix transpose.

The carrier frequency of bandpass filter is determined by $\mathbf{k}$ and spatial and orientation frequency bandwidths depend on the inverse of the Gaussian envelope and the direction of the plane wave in the spatial domain. The symbol σ also determines the ratio of window width to the wavelength. It is noted that the number of plane wave oscillations is approximately $\sigma\sqrt{2}/\pi$ in the Gaussian window. Such complex amplitude enables the Gabor kernel to work as the bandpass filter. Sampling parameters (k_0, δ, U, V) as well as σ should be chosen carefully. If δ is 2, resolution level is half octave; and if it is $\sqrt{2}$ the resolution level becomes one octave.

A feature vector of the image is computed by a set of Gabor kernels. Let $h_{uv}(m,\,n)$ be the output of filtered input image $I_d(m,\,n) = |U_O(m,\,n,\,d)|$ by the Gabor kernel $g_{uv}(m,\,n)$; $h_{uv}(m,\,n)$ is also called "Gabor coefficient" and the magnitude of the Gabor coefficient is called "Gabor jet." One Gabor jet vector is composed of a set of the Gabor jets: $\mathbf{v}(m,\,n) = -|h_{uv}(m,\,n)|$; $u = 1,\,\ldots,\,U,\,v = 1,\,\ldots,\,V\}$. Figures 5.5–5.9 show the selected Gabor jets of the reconstructed image of the hologram in Fig. 5.4(c). The parameters are set up at $\sigma = \pi$, $k_0 = \pi/2$, $\delta = 2$, $U = 5$, and $V = 6$. There is no optimal way to choose these parameters, but several values are widely used heuristically. The Gabor jets illustrate some characteristics of the reconstructed 3D object as shown in Figs. 5.5–5.9. The Gabor-based wavelet has strong response to the

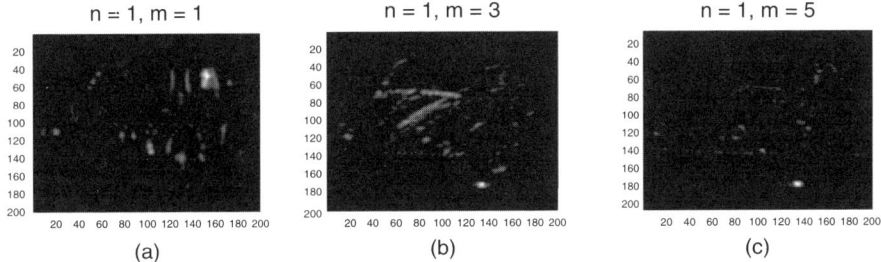

Figure 5.5. Gabor jets for the object in Fig. 5.4(c) when $u = 1$, (a) $v = 1$, (b) $v = 3$, and (c) $v = 5$.

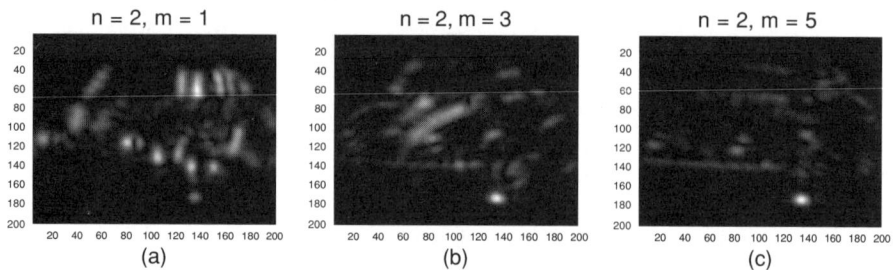

Figure 5.6. Gabor jets for the object in Fig. 5.4(c) when $u = 2$, (a) $v = 1$, (b) $v = 3$, and (c) $v = 5$.

edges if the wave number vector $\mathbf{k}$ is perpendicular to the direction of edges. In addition, local features are well presented in high-frequency bandwidth (small u) with higher precision. Conversely, global features are maintained in low-frequency bandwidth (large u) with lower precision. The speckle noise is always present in the hologram due to the coherent imaging system. The speckle noise produces degradations for the high-frequency bandwidth Gabor kernels as shown in Fig. 5.5. This effect makes the high-frequency bandwidth kernels unsuitable for recognition and classification.

5.4 PCA–FLD Projections

In this section, we discuss the PCA and FLD projections applied to the Gabor jet vectors extracted from computer-reconstructed holographic data and present experimental results. The PCA is a projection method to represent d-dimensional vectors in the subspace of l dimension ($l \leq d$). For a real d-dimension vector $\mathbf{v}$, the mean vector is $\boldsymbol{\mu}_v = \mathrm{E}(\mathbf{v})$, and the covariance matrix is $\Sigma_{vv} = \mathrm{E}(\mathbf{v} - \boldsymbol{\mu}_v)(\mathbf{v} - \boldsymbol{\mu}_v)^t$. The basis vectors in the PCA space are given by orthonormal eigenvectors of its positive definite covariance matrix;

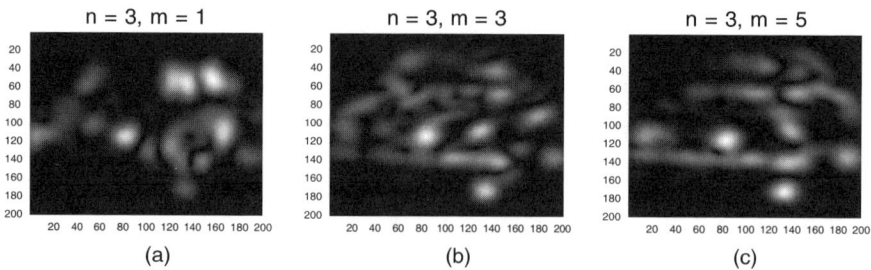

Figure 5.7. Gabor jets for the object in Fig. 5.4(c) when $u = 3$, (a) $v = 1$, (b) $v = 3$, and (c) $v = 5$.

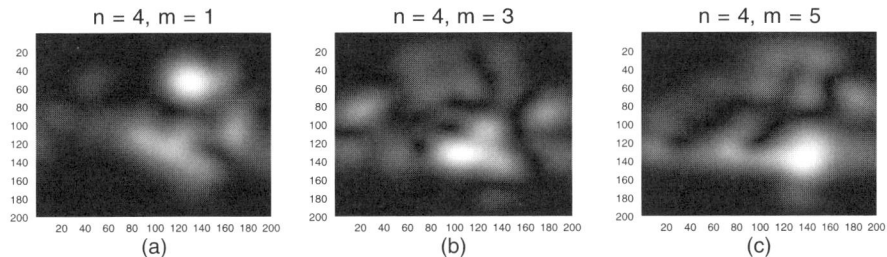

Figure 5.8. Gabor jets for the object in Fig. 5.4(c) when $u = 4$, (a) $v = 1$, (b) $v = 3$, and (c) $v = 5$.

that is, $\Sigma_{vv}E = E\Lambda$ where E is an eigenvector matrix and Λ is the eigenvalue matrix of the covariance matrix. The column vectors of E are normalized eigenvectors $\mathbf{e}_i$'s, i.e., $E = [\mathbf{e}_1, \ldots, \mathbf{e}_d]$, and the diagonal components of Λ are eigenvalues λ_i's, i.e., $\Lambda = \mathrm{diag}(\lambda_1, \ldots, \lambda_d)$. A new vector $\mathbf{y}$ after the PCA projection of $\mathbf{v}$ is $\mathbf{y} = W_P^t \mathbf{v} = E^t \mathbf{v}$ where E is the eigenvector matrix of the covariance matrix Σ_{vv}.

The PCA diagonalizes the covariance matrix of $\mathbf{y}$, i.e., $\Sigma_{yy} = \mathrm{E}(\mathbf{y} - \boldsymbol{\mu}_y)(\mathbf{y} - \boldsymbol{\mu}_y)^t = \Lambda$ where $\boldsymbol{\mu}_y = \mathrm{E}(\mathbf{y})$. If we choose l coefficients from $\mathbf{y} = [y(1), \ldots, y(l)]^t$ the PCA subspace is spanned by corresponding l eigenvectors. It is a well-known property of the PCA that by choosing eigenvectors of the largest l eigenvalues, the projected vector minimizes the mean square errors, $J(\hat{v}) = \mathrm{E} \parallel \mathbf{v} - \hat{\mathbf{v}} \parallel^2 = \sum_{i=l+1}^{d} \lambda_i$ when $\hat{\mathbf{v}} = \boldsymbol{\mu}_x + W_p(\mathbf{y} - \boldsymbol{\mu}_y)$.

The FLD transforms l-dimension vectors onto a subspace of k dimension $(k \leq l)$. The FLD maximizes the ratio of determinant of between-class scatter matrix to determinant of within-class scatter matrix[14]. Total scatter matrix S_T is defined as $S_T = S_B + S_W = \sum_{i=1}^{n_t} (\mathbf{y}_i - \mathbf{m})(\mathbf{y}_i - \mathbf{m})^t$ where n_t is the total number of data; $\mathbf{y}_i$ is the training vector; and $\mathbf{m}$ is the sample mean vector of $\mathbf{y}_i$. We define between-class scatter matrix as S_B and within-class scatter matrix as S_W. Let us denote c as the number of classes and n_j as the number

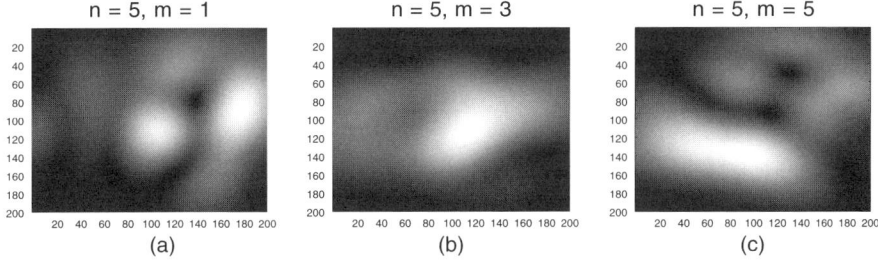

Figure 5.9. Gabor jets for the object in Fig. 5.4(c) when $u = 5$, (a) $v = 1$, (b) $v = 3$, and (c) $v = 5$.

of training data in the class j. Let the FLD transformation matrix be W_F and $\mathbf{z}_i$ be a new vector after transformation of $\mathbf{y}_i$, USA, i.e., $\mathbf{z}_i = W_F^i \mathbf{y}_i$. After applying W_F to each scatter matrix, we have

$$\tilde{S}_B = W_F^t S_B W_F = \sum_{j=1}^{c} n_j (\tilde{\mathbf{m}}_j - \tilde{\mathbf{m}})(\tilde{\mathbf{m}}_j - \tilde{\mathbf{m}})^t, \tag{4}$$

$$\tilde{S}_W = W_F^t S_W W_F = \sum_{j=1}^{c} \sum_{i=1}^{n_j} (\mathbf{z}_i^j - \tilde{\mathbf{m}}_j)(\mathbf{z}_i^j - \tilde{\mathbf{m}}_j)^t, \tag{5}$$

where $\tilde{\mathbf{m}}_j$ is the sample mean of the class j; $\mathbf{z}_i^j$ is the training vector in the class j after projection; $\tilde{\mathbf{m}}$ is the sample mean of all training data after projection; and t denotes matrix transpose. W_F maximizes the cost function, $J(W) = |\tilde{S}_B|/|\tilde{S}_W| = |W^t S_B W|/|W^t S_W W|$ when the column vectors of W_F are the eigenvectors of $S_W^{-1} S_B$ with the largest nonzero k eigenvalues.[14] Note that k is a reduced dimension that is less than c because the rank of S_B is at most $c - 1$. In other words, the maximum number of nonzero eigenvalues of $S_W^{-1} S_B$ is $c - 1$. Therefore, the maximum dimension of the FLD projection for the c-class problem is $c - 1$.

We must also consider the constraint by the decision rule. Any sample covariance or its function for the class j has at most $n_j - 1$ ranks. If we adopt decision rules based on the statistical distance, the maximum number of the FLD subspace is also limited by $\min\{n_j - 1\}$, $j = 1, \ldots, c$. Finally, we get $k \leq \min\{c - 1, \min\{n_j - 1\}\} \leq l$.

In the FLD, usually S_W is singular because the total number of training data n_t is much less than the dimension of the feature vector d. We can overcome this problem by applying the PCA first to reduce the dimensionality of the vector. Reduced dimension of the PCA should be less or equal to $n_t - c$ because the number of independent vectors in S_W is at most $n_t - c$. We know that dimension k in the FLD subspace should satisfy the relation, $k \leq \min\{c - 1, \min\{n_j - 1\}\} \leq l$. So, we can combine them as $k \leq \min\{c - 1, \min\{n_j - 1\}\} \leq l \leq n_t - c$.

The optimal l cannot be decided analytically. Usually, the projected vectors of smaller eigenvalues include more noise than the vectors corresponding to larger eigenvalues. However, lower l may not contain enough energy to properly represent the characteristics of the object. In this research, l values are chosen heuristically when better results are produced. The FLD combined with the PCA has two consecutive projections of W_P and W_F. Final projected vector in k-dimensional space is $\mathbf{z} = W_F^t \mathbf{y} = W_F^t W_P^t \mathbf{v}$ with the cost function, $J(W_F, W_P) = |W_F^t W_P^t S_B W_P W_F|/|W_F^t W_P^t S_W W_P W_F|$.

Figures 5.10–5.14 show basis images of the PCA and FLD subspace at the selected Gabor jets. The feature vectors contain all the pixels of a single Gabor jet for illustration. One basis image represents a column vector in the PCA or PCA–FLD matrix. The dimension of the PCA subspace (l) is set at 3 and the

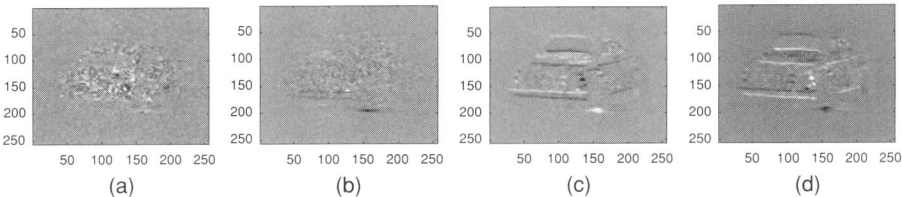

Figure 5.10. Basis images for the subspace for 6 randomly chosen training Gabor jets when $u = 1$ and $v = 1$: (a) PCA1, (b) PCA2, (c) PCA3, and (d) FLD.

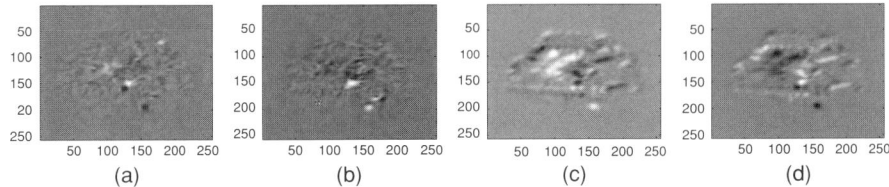

Figure 5.11. Basis images for the subspace for 6 randomly chosen training Gabor jets when $u = 2$, $v = 2$: (a) PCA1, (b) PCA2, (c) PCA3, and (d) FLD.

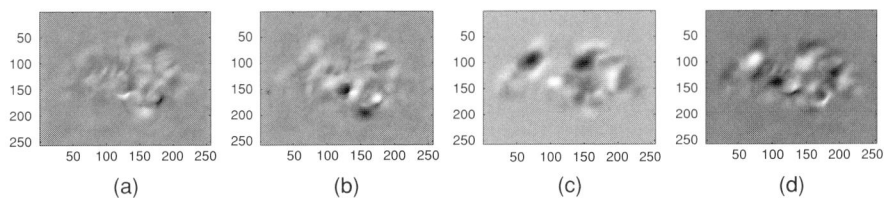

Figure 5.12. Basis images for the subspace for 6 randomly chosen training Gabor jets when $u = 3$, $v = 3$: (a) PCA1, (b) PCA2, (c) PCA3, and (d) FLD.

dimension of the FLD subspace (k) is 1. Three training data were chosen randomly for each class. PCA3 is a basis image of the largest eigenvalue and PCA1 corresponds to a basis image of the smallest eigenvalue.

5.5 Statistical Distance Decision Rule

For the final classification, the discriminant function of the statistical distance is used according to $g_j(\mathbf{z}) = (\mathbf{z} - \mathbf{m}_z^j)^t (\hat{\Sigma}_{zz}^j)^{-1} (\mathbf{z} - \mathbf{m}_z^j)$ where $\mathbf{m}_z^j$ is the sample mean vector of the class j and $\hat{\Sigma}_{zz}^j$ is the unbiased sample covariance matrix of the class j. We classify the object $\mathbf{z}$ as class $\hat{j}$ in the following manner:

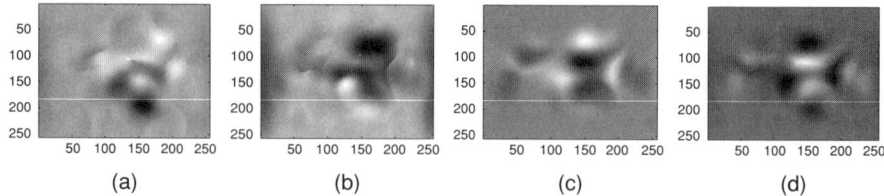

Figure 5.13. Basis images for the subspace for 6 randomly chosen training Gabor jets when $u = 4$, $v = 4$: (a) PCA1, (b) PCA2, (c) PCA3, and (d) FLD.

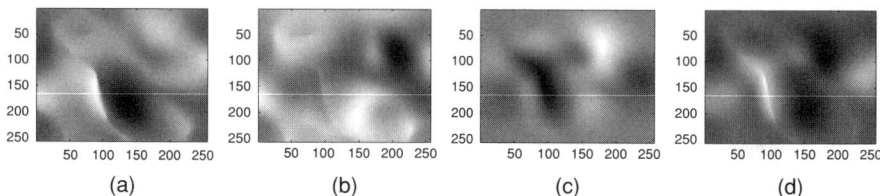

Figure 5.14. Basis images for the subspace for 6 randomly chosen training Gabor jets when $u = 5$, $v = 5$: (a) PCA1, (b) PCA2, (c) PCA3, and (d) FLD.

$$\mathbf{z} \in C_{\hat{j}} \quad \text{if} \quad \hat{j} = \arg j \min g_j(\mathbf{z}), \tag{6}$$

where $C_{\hat{j}}$ is the data set of the class $\hat{j}$. It is equivalent to ML decision when we assume Gaussian distribution with the identical covariance. It is also a powerful similarity measure while being used for the case of unknown distributions.

5.6 Experimental and Simulation Results

We will present experimental results of object classification using two approaches: first when geometric features are previously known or extracted and second when the objects in images are well segmented by the preprocessor.[5] The former classification is based on regional feature vectors and the latter on overall grid feature vectors.

5.6a Regional feature vector

The block diagram of the 3D object classification system is shown in Fig. 5.1. We use the holograms of 3D objects of two toy cars obtained by phase-shift holography as described in Section 3. Regional feature vectors are placed at the positions of two headlights assuming that the positions of headlights are known. Figure 5.15 illustrates two feature vectors of the car: $|h_{uv}|$ is the Gabor jet of the input image with the Gabor kernel indexes, u and v, and $\mathbf{v}(m_{rh}, n_{rh})$ and $\mathbf{v}(m_{lh}, n_{lh})$ are Gabor jet vectors at the positions of

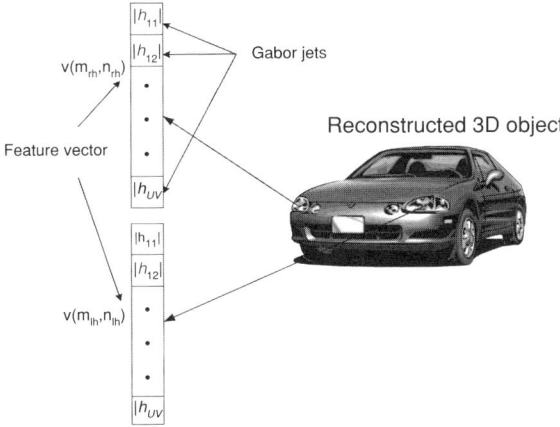

Figure 5.15. Feature vectors from headlights.

the right and left headlight, respectively. The location of each headlight
provides one feature vector that is identical to one Gabor jet vector in
Section 4. The parameters of the Gabor-based wavelet are the same for all
the simulations ($\sigma = \pi$, $k_0 = \pi/2$, $\delta = 2$, $U = 5$, $V = 6$). Figure 5.16 illustrates correct decision rates when the feature vectors of the left headlight
were used.

We perform 1000 runs and average the results. Three training data
were chosen randomly for each class. So, the total number of training data
is 6 at each run, the dimension of the PCA subspace (l) is 2, and the dimension
of the FLD subspace (k) is 1. All image data except for training data were used
for the tests. We adopted the statistical distance decision rule as described in
Section 6. According to Fig. 5.16, correct decision rates of more than 90% are
obtained with a few training data. The overall performance is better for the
second class. It can be interpreted that the car in the second class has more salient
features than the car in the first class. (We can recognize this intuitively from
Figs. 5.3 to 5.4.)

Gabor kernels vary from $u = 1$ to 5 which are the spatial frequency selec-
tions of the Gabor-based wavelet. For each u, Gabor kernels vary from $v = 1$ to
6, having different orientations in the frequency domain. So, the feature vector
has six components for a fixed u. In Fig. 5.16, $u = \text{ALL}$ implies that 30
components of Gabor jets were used. As u is increased, Gabor filtering acts
similar to the low-pass filter of spatial frequencies. When the Gabor filter is
used as a very high spatial frequency filter ($u = 1$), correct decision
rates decrease due to the speckle noise that can lead to the incorrect classifications of the label. The simulation shows better results when the feature vectors
corrupted by the speckle noise were not used; that is, when $u = 3$, 4, or 5 is used
for the classification.

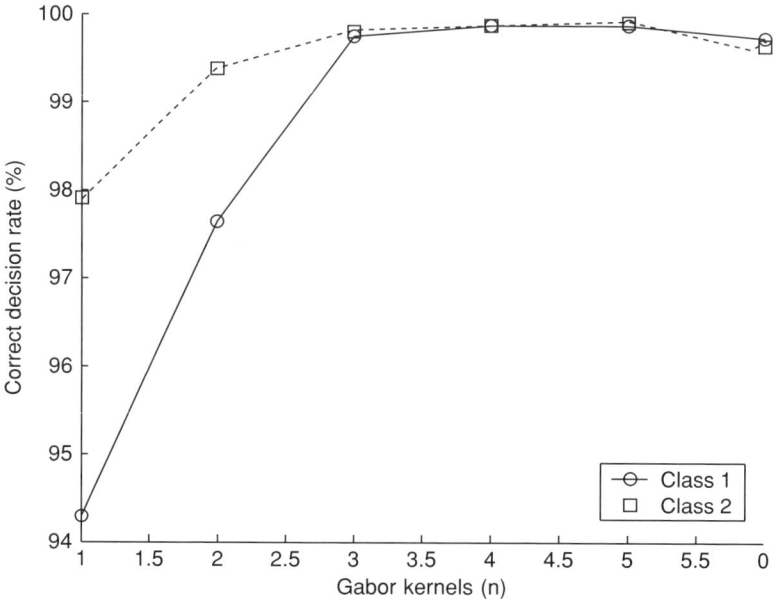

Figure 5.16. Correct decision rate (%) of feature vectors from the left headlight. ALL implies that 30 components of Gabor jets were used.

5.6b. Overall grid feature vector

We experiment with the 3D object classification using an overall feature vector formed at all the nodes of a rectangular grid placed on the object. Assuming that the 3D objects are segmented in the input scene, we can overlay a grid of a certain size on the object. The number of nodes is 45; with 9 nodes in the x direction and 5 nodes in the y direction. One feature vector is composed of 45 Gabor jet vectors as shown in Fig. 5.17. In this figure, $|h_{uv}|$ is the Gabor jet of the input image with the Gabor kernel indexes u and v, and $\mathbf{v}(m_1, n_1)$, $\mathbf{v}(m_2, n_1)$, $\ldots$, $\mathbf{v}(m_9, n_5)$ are the Gabor jet vectors at the positions of nodes in the grid. When the objects are not well segmented or include deformations or occlusions, we can apply a similarity matching technique, such as "dynamic link association."[23,24"]

In this experiment, the Gabor jet vector is obtained by using 30 different Gabor kernels. So, the dimension of the feature vector is 270 when u ranges from 1 to 5 and 1350 when u is denoted as "ALL" in Figs. 5.18 and 5.19. The number of training data is also 3 for each class. We perform 1000 runs and average the results. Figures 5.18 and 5.19 illustrate the correct decision rates when the feature vectors of the overall grid technique were used. The dimensions of the PCA space (l) and the FLD space (k) in Fig. 5.18 are the same as in the simulation of the regional feature vector. In Fig. 19, the PCA dimension is 4.

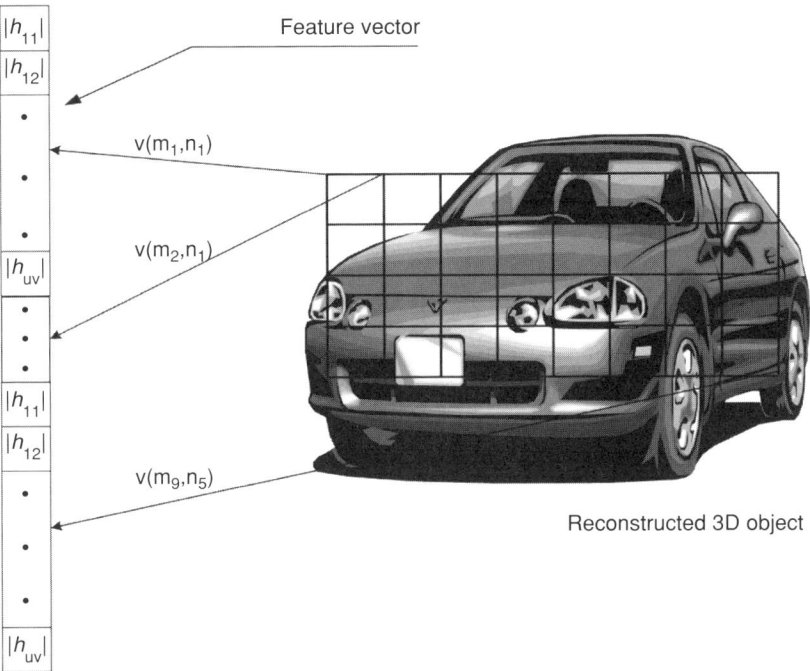

Figure 5.17. Feature vectors at nodes of a grid.

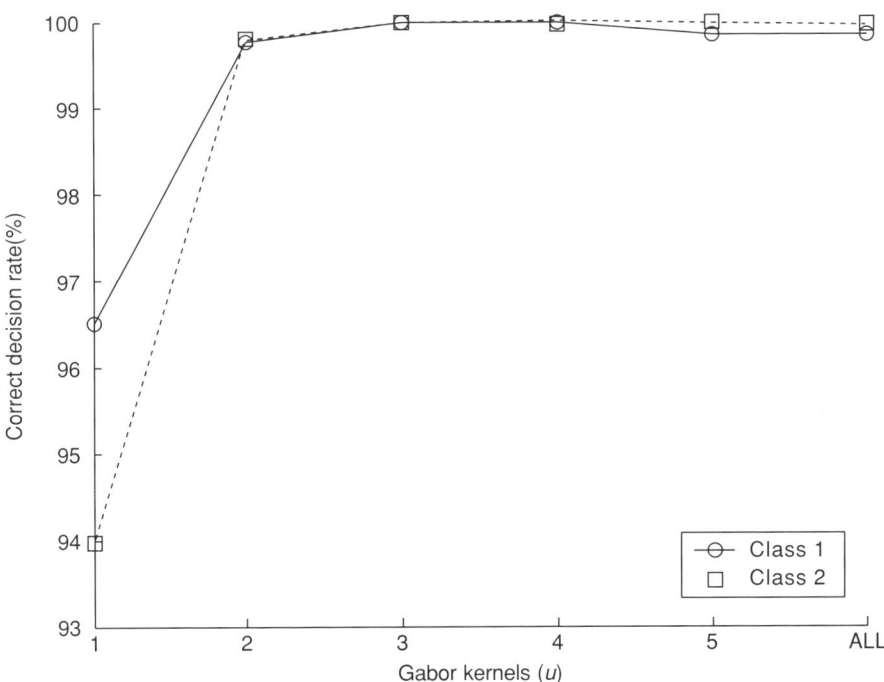

Figure 5.18. Correct decision rate (%) of feature vectors from the overall grid ($l = 2$). ALL implies that 1350 components of Gabor jets were used.

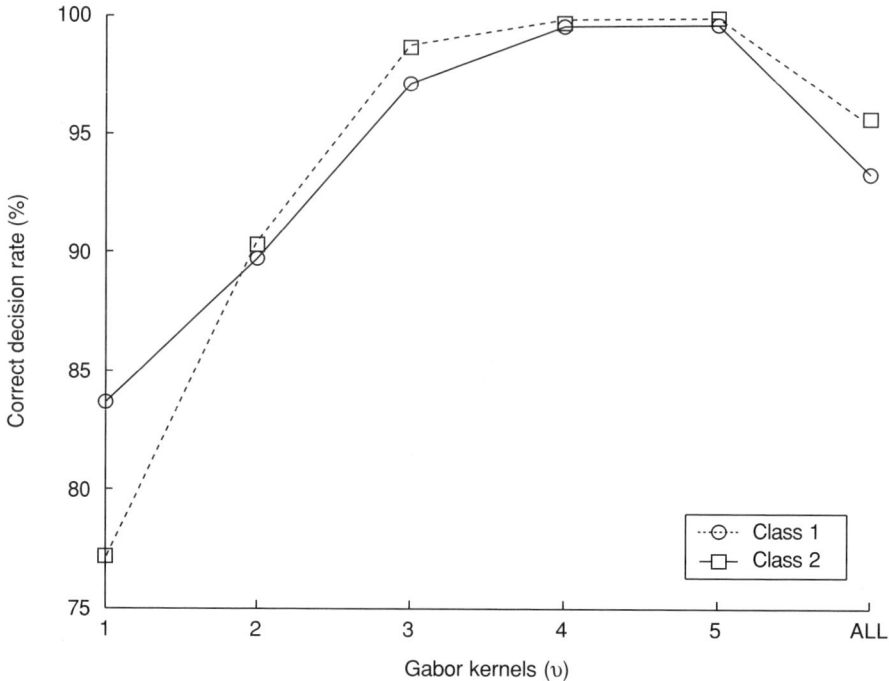

Figure 5.19. Correct decision rate (%) of feature vectors from the overall grid ($l = 4$). ALL implies that 1350 components of Gabor jets were used.

Note that possible values of l range from 2 to 4. In Fig. 5.19, when n equals to 1, correct decision rates are less than 90% due to the speckle effects. The effects of speckle noise for the high frequency Gabor kernels can be observed for small u. The Gabor kernels of $u = 3$–5 provide the best result in the experiments.

5.7 Conclusions

In this research, a 3D object classification technique using a single hologram has been presented. The PCA–FLD classifier with feature vectors based on Gabor wavelets has been utilized for this purpose. Training and test data of the 3D objects were obtained by computational holographic imaging. We were able to classify 3D objects used in the experiments with a few reconstructed planes of the hologram. The Gabor approach appears to be a good feature extractor for hologram-based 3D classification. The FLD combined with the PCA proved to be a very efficient classifier even with a few training data. Substantial dimensionality reduction was achieved by using the proposed technique for

3D classification problem using holographic imaging. As a consequence, we were able to classify different classes of 3D objects using computer-reconstructed holographic images.

References

[1] Rosen J. (1998). "Three-dimensional joint transform correlator." *Appl. Opt.*, 37(32):7538–7544.

[2] Esteve-Taboada JJ, Mas D, and Garcia J. (1999). "Three-dimensional object recognition by Fourier transform profilometry." *Appl. Opt.*, 38(22):4760–4765.

[3] Goudail F and Refregier P. (2001). "Statistical algorithms for target detection in coherent active polarimetric images." *J. Opt. Soc. Am.*, 18(2):3049–3060.

[4] Sadjadi FA. (2002). "New results in the use of polarization diversity for classification of radar targets." in *Automatic Target Recognition XII*, FA. Sadjadi, ed., Proc. SPIE 4726, pp. 26–34.

[5] Javidi B, ed. (2002). *Image Recognition and Classification: Algorithms, Systems, and Applications.* Marcel Dekker, New York.

[6] Javidi B and Tajahuerce E. (2000). "Three-dimensional object recognition by use of digital holography." *Opt. Lett.*, 25(9):610–612.

[7] Frauel Y, Tajahuerce E, Castro M, and Javidi B. (2001). "Distortion-tolerant three-dimensional object recognition with digital holography." *Appl. Opt.*, 40(23): 3887–3893.

[8] Frauel Y and Javidi B. (2001). "Neural network for three-dimensional object recognition based on digital holography." *Opt. Lett.*, 26(9):1478–1480.

[9] Frauel Y and Javidi B. (2002). "Digital three-dimensional image correlation by use of computer-reconstructed integral imaging." *Appl. Opt.*, 41(26):5488–5496.

[10] Perona MT, Mahalanobis A, and Norris-Zachery K. (1999). "Ladar automatic target recognition using correlation filters." in *Automatic Target Recognition IX*, FA. Sadjadi, ed., Proc. SPIE 3718, pp. 388–396.

[11] Perona MT, Mahalanobis A, and Norris-Zachery K. (2000). "System-level evaluation of ladar ATR using correlation filters." in *Automatic Target Recognition X*, FA. Sadjadi, ed., Proc. SPIE 4050, pp. 69–75.

[12] Mahalanobis, (1996). "Review of correlation filters and their application for scene matching." in *Optoelectronic Devices and Systems for Processing*, B Javidi; KM Johnson, eds., Proc. SPIE CR65, pp. 240–260.

[13] Yeom S and Javidi B. (2004). "Three-dimensional object feature extraction and classification with computational holographic imaging." *Appl. Opt.*, 43(2): 442–451.

[14] Duda RO, Hart PE, and Stork DG. (2001). *Pattern Classification* 2nd edn. Wiley, New York.

[15] Daugman JG. (1985). "Uncertainty relation for resolution in space, spatial frequency, and orientation optimized by two-dimensional visual cortical filters." *J. Opt. Soc. Am.*, 2(7):1160–1169.

[16] Daugman JG. (1988). "Complete discrete 2-D Gabor transforms by neural networks for image analysis and compression." *IEEE Trans. ASSP.*, 36(7):1169–1179.

[17] Lee TS. (1996). "Image representation using 2-D Gabor wavelets." *IEEE Trans. PAMI.*, 18(10):959–971.

[18] Belhumer PN, Hespanha JP, and Kriegman DJ. (1997). "Eigenfaces vs. Fisherfaces: Recognition using class specific linear projection." *IEEE Trans. PAMI.*, 19(7):711–720.

[19] Swets DL and Weng J. (1996). "Using discriminant eigenfeatures for image retrieval." *IEEE Trans. PAMI.*, 18(8):831–836.

[20] Lyons MJ, Budynek J, and Akamatsu S. (1999). "Automatic classification of single facial images." *IEEE Trans. PAMI.*, 21(12):1357–1362.

[21] Liu C and Wechsler H. (2002). "Gabor feature based classification using the enhanced Fisher Linear Discriminant model for face recognition." *IEEE Trans. Image Processing*, 11(4):467–476.

[22] Yamaguchi I and Zhang T. (1997). "Phase-shifting digital holography." *Opt. Lett.*, 22(16):1268–1270.

[23] Lades M, Vorbruggen JC, Buhmann J, Lange J, Christoph v.d. Malsburg, Wurtz RP, and Konen W. (1993). "Distortion invariant object recognition in the dynamic link architecture." *IEEE Trans. Comput.*, 42(3):300–311.

[24] Yeom S, Javidi B, Roh YJ, and Cho HS. (2004). "Three-dimensional object recognition using X-ray imaging." Submitted to *Opt. Eng.*

Distortion-tolerant 3D Volume Recognition Using X-ray Imaging

Sekwon Yeom,[1] Bahram Javidi,[1] Young Jun Roh,[2]
and Hyung Suck Cho[2]

[1]Department of Electrical and Computer Engineering, U-2157, University of Connecticut, Storrs, CT USA 06269-2157
[2]Department of Mechanical Engineering, Korea, Advanced Institute of Science and Technology, Daejeon, Korea

6.0 Introduction

There are challenges and benefits in three-dimensional (3D) object recognition. In addition to conventional issues in two-dimensional (2D) object recognition, there have been new challenges facing us with 3D information. One of the challenges is to reconstruct 3D structure itself. Various techniques have been developed to constitute 3D structure according to applications and environments. More accurate acquisition of 3D information on the objects leads to more successful recognition. Another challenge regarding 3D object recognition is that it generally places high demands on the computation and storage of data due to the huge amount of 3D information. Despite these drawbacks of 3D object recognition, a growing amount of research represents potential advantages in studying 3D space.[1–25]

In this chapter, we present distortion-tolerant 3D volume object recognition. Volume information is reconstructed by an advanced X-ray imaging technique, called Uniform Simultaneous Algebraic Reconstruction Technique (USART). It was improved by employing spherical voxel elements for fast implementation and accurate estimation of voxel density.[26,27]

The proposed object recognition system is composed of three stages as shown in Fig. 6.1: feature extraction, feature matching, and decision making. For feature extraction, the conventional 2D Gabor filtering[28–30] is extended to 3D space in order to analyze volume data. The Gabor feature is a multi-resolution representation of object structure and energy in spatial frequency domain. Three-dimensional Gabor filtering extracts salient features according

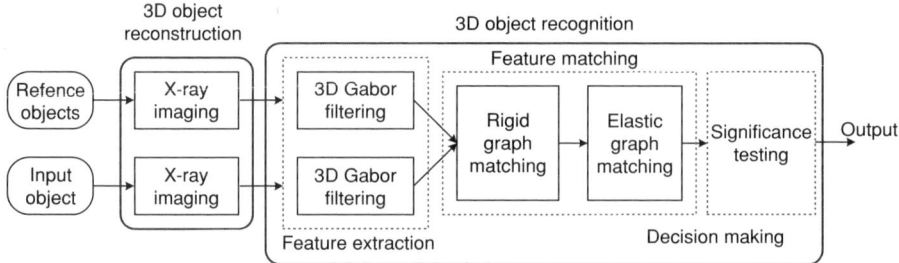

Figure 6.1. Frameworks of 3D volume object reconstruction and recognition.

to 3D location, spatial frequency, and bandwidth. We also achieve dimensionality reduction by sampling Gabor features on 3D volume objects.

Dynamic Link Association (DLA) is a graph-matching technique between a reference and an unknown input object. Theoretically, the DLA scheme is tolerant to any distortion, rotation, and scaling of input objects. Since the main idea of the DLA was proposed in Ref. 31 many efforts have been made to apply the DLA to realistic problems.[32–35] Lades et al.[32] propose Elastic Graph Matching (EGM) to realize the DLA in suboptimal way.

In this chapter, we extend the 2D DLA technique to 3D space and modify it in a simple and straightforward way. The modified DLA scheme is equipped with 3D rotation-tolerant property and efficient realization allowing distortion to some extent. Similar to the conventional model, the modified DLA scheme is composed of two stages: coarse matching and fine matching. However, during the coarse matching stage we search for the best-matched orientation as well as position of the 3D rigid graph, which is placed on the input scene. Rotation-invariant feature vectors are computed from selected Gabor jets. During the fine-matching stage, nodes of the graph are elastically moved by searching for the best-matched positions. We also develop sequential and recursive realization for the fine-matching stage.

As the final step, we employ a statistical testing to classify unknown input objects. It evaluates the statistical significance of each reference after the feature matching.

The main contributions of this chapter can be summarized as: 1) Application of 3D Gabor-based wavelets to volumetric information for feature extraction and dimensionality reduction; 2) Extension of the 2D DLA method to 3D DLA with innovations (i.e., rotation and distortion-tolerant properties and efficient realization); and 3) Proposed volume feature extraction/matching/recognition technique not limited to presented volume data. (we can apply this technique to any volumetric information).

In Section 6.1, we briefly summarize the USART and demonstrate the reconstructed volume data. The 3D Gabor-based wavelets and the 3D DLA are presented in Sections 6.2 and 6.3, respectively. Statistical testing is explained in Section 6.4. In Section 6.5, experimental results and performance analysis are demonstrated. Conclusions follow in Section 6.6.

6.1 Three-dimensional Volume Reconstruction by X-ray Imaging

In this chapter, the 3D volume reconstruction of X-ray imaging is achieved by the USART. The USART estimates voxel density by combining several X-ray images, which are projected from different perspective angles. The USART has been improved by employing spherical voxel elements for fast implementation and accurate estimation.[27] In this section, we briefly revisit the USART technique using spherical voxel elements and present several volume objects.

6.1.1 Overview of the USART

Figure 6.2 shows the X-ray imaging system.[26,27] The system consists of a scanning X-ray source, a stage, and an X-ray digital imaging device; $\mathbf{X}_s$, $s = 1, \ldots, S$ denotes the position which the X-ray source is electromagnetically scanned to; $I_s(u_s, v_s)$ is a gray scale image exposed on the X-ray imaging device corresponding to $\mathbf{X}_s$; u_s and v_s are coordinates in x and y directions from the reference point $\mathbf{P}_s$. We image one stationary object from S different views.

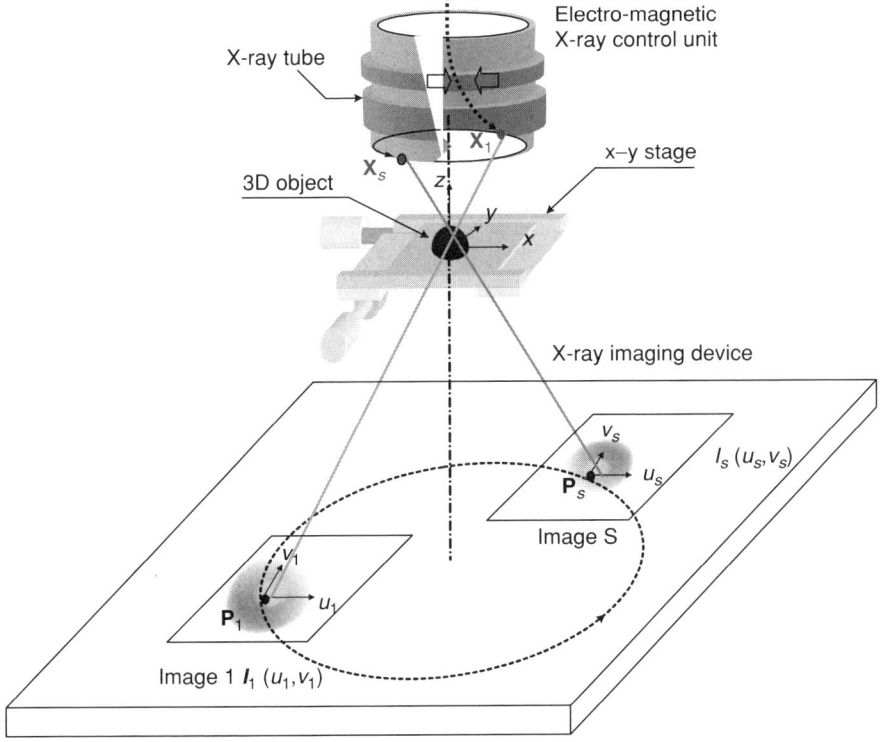

Figure 6.2. X-ray imaging system.

A fast USART has been implemented by means of the spherical voxel model instead of conventional cube voxels. Figure 6.3 shows X-ray projection and reconstruction model of the spherical-voxel USART. Reconstruction process of the spherical-voxel USART is:

$$\hat{f}_i(t+1) = \hat{f}_i(t) + \frac{\lambda_u}{S} \sum_{s=1}^{S} \frac{1}{D_i^s} (g_i^s(t) - h_i^s(t)), \qquad (3)$$

where $\hat{f}_i(t)$ is a density estimate of the voxel i after t iterations; λ_u is a relaxation parameter which controls the convergence of the estimation. We assume an interpolated ray Ω_i^s is emitted from the X-ray source s and passes through the center of the voxel i; $g_i^s(t)$ is a measured value and $h_i^s(t)$ is a modeled value of the Ω_i^s on the image plane; and D_i^s is the total intersection length of the Ω_i^s in the whole reconstruction boundary. The projection of the Ω_i^s is modeled as:

$$h_i^s(t) = d \sum_{p=1}^{P} \tilde{f}_p^{i,s}(t), \qquad (4)$$

where d is the diameter of spherical voxels; $\tilde{f}_p^{i,s}(t)$ is the density of the sphere p on the Ω_i^s after t iterations; $\tilde{f}_p^{i,s}(t)$ can be computed by the interpolation of

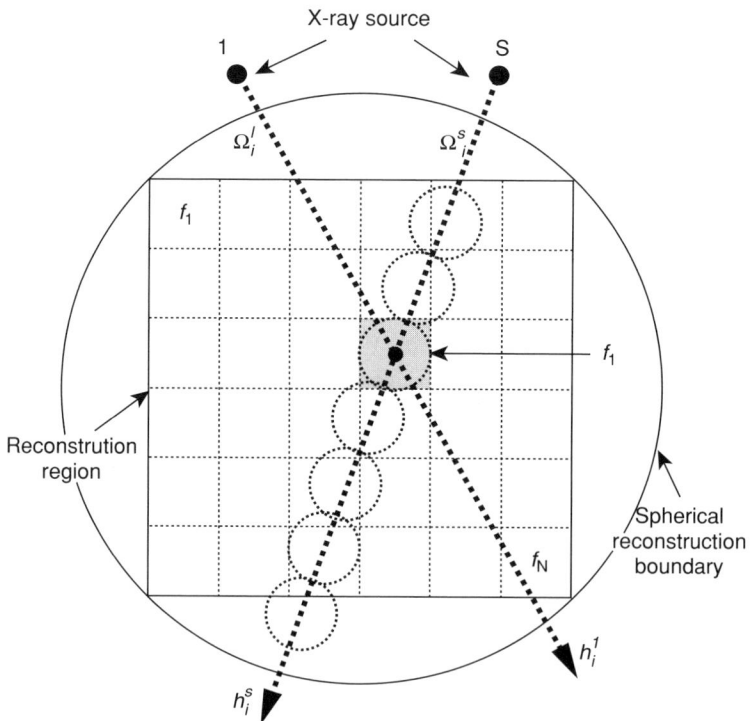

Figure 6.3. X-ray projection and reconstruction model.

density estimates of neighbor voxels of the voxel i; and P is the number of spheres on the Ω_i^s in the spherical reconstruction boundary. It is noted that the computation of geometric parameter is not required for the spherical-voxel USART. It results in great saving of computational time and memory.

6.1.2 Volume Objects Reconstructed by the Spherical Voxel USART

We present five classes of volume objects reconstructed by the spherical-voxel USART: pyramid, hemisphere, cone, short screw (screw number 1), and long screw (screw number 2). X-ray images of pyramid, hemisphere, and cone data are synthesized from their geometric models. For two screws, experimental data are obtained. Their geometric models of three classes of objects and real images of two types of screws are shown in Fig. 6.4.

Eight X-ray images are used for volume reconstruction and the size of volume data is $60 \times 60 \times 60$ voxels. Figure 6.5 shows the outer surface and inner structure of the 3D screw (screw number 1) after one and 30 iterations of the USART reconstruction. The voxel density is normalized so that the maximum value is 100. The outer surface represents the mean value of voxel density in each object. As shown in Fig. 6.5, the artifact errors decrease as the USART iteration reconstruction process is increased. The reconstruction error is defined by the difference between the measured value $g_i^s(t)$ and the modeled value $h_i^s(t)$

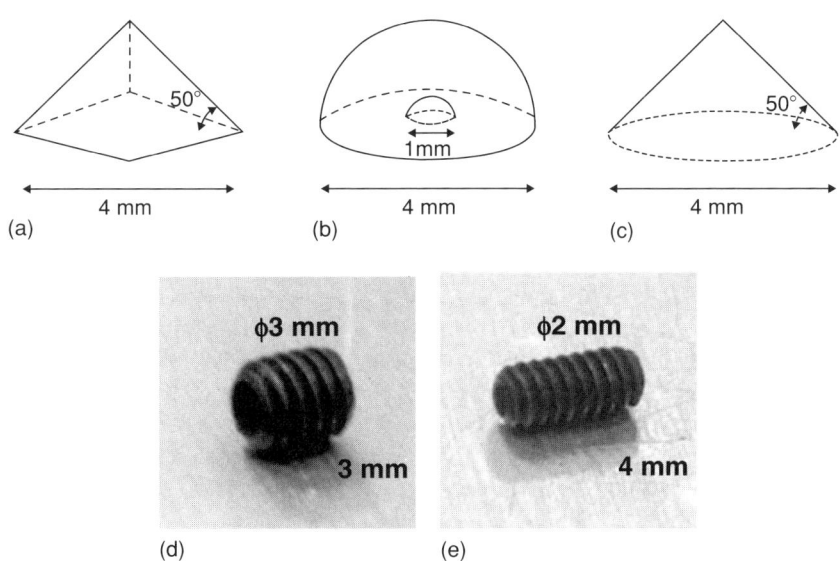

Figure 6.4. Object models for synthetic X-ray data: (a) pyramid, (b) hemisphere, (c) cone and 3D images for experimental X-ray data, (d) screw number 1, and (e) screw number 2.

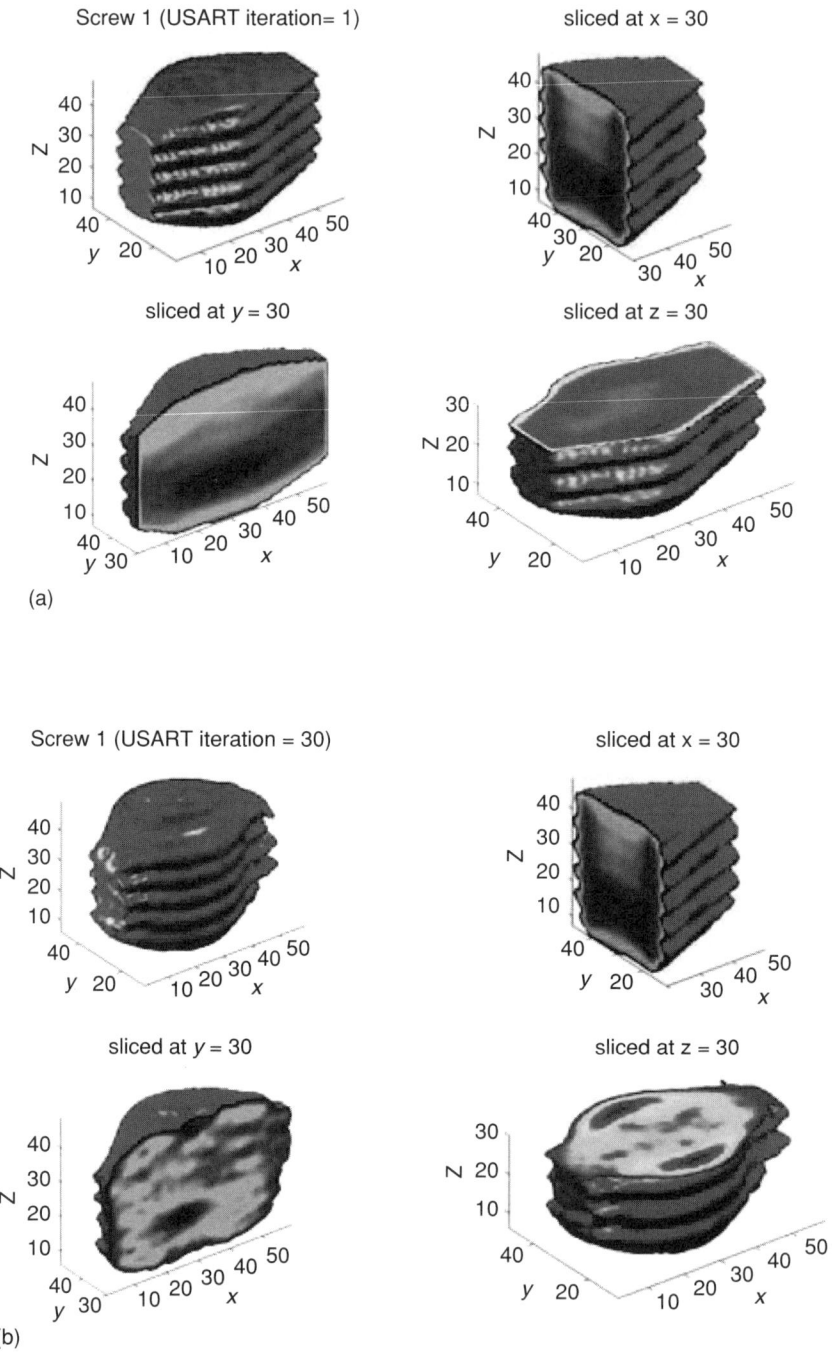

Figure 6.5. Three-dimensional volume object of screw number 1 and sliced views at $x = 30$, $y = 30$, and $z = 30$ after USART (a) 1 iteration and (b) 30 iterations.

of the X-ray projection. It was shown that the reconstruction errors approach steady state levels after 20–30 USART iterations.[27]

6.2 Three-dimensional Gabor Filtering and Feature Vector Extraction

In this section, we discuss 3D Gabor-based wavelets and present Gabor jets of volume objects. We employ 3D Gabor wavelets for feature extraction and data reduction of volume objects.

6.2.1 Three-dimension Gabor-based Wavelets

The 2D Gabor filter acts as a bandpass filter with the special selection of passband according to its Gaussian envelope and the carrier frequency of the complex plane wave.[28-30] The 2D Gabor wavelets can be easily extended to 3D dimensions. The 3D impulse response (or kernel) of the Gabor-based wavelets is:

$$g(\mathbf{x}) = \frac{|\mathbf{k}|^3}{\sigma^3} \exp\left(-\frac{|\mathbf{k}|^2|\mathbf{x}|^2}{2\sigma^2}\right) \left[\exp\left(j\mathbf{k}\cdot\mathbf{x}\right) - \exp\left(-\frac{\sigma^2}{2}\right)\right] \tag{5}$$

where $\mathbf{x}$ is a position vector, $\mathbf{k}$ is a wave number vector, and σ is the standard deviation of 3D Gaussian envelope. The size of the Gaussian envelope is the same in x, y, and z directions, which is proportional to $\sqrt{2}\sigma/|\mathbf{k}|$. The second term in the square brackets, $\exp\left(-\sigma^2/2\right)$ subtracts the DC value so it has zero mean response[30]. The frequency response of $g(\mathbf{x}), G(\mathbf{k}')$ is given by:

$$G(\mathbf{k}') = (2\pi)^{3/2}\left\{\exp\left[-\frac{\sigma^2}{2|\mathbf{k}|^2}|\mathbf{k}'-\mathbf{k}|^2\right] - \exp\left[-\frac{\sigma^2}{2|\mathbf{k}|^2}\left(|\mathbf{k}'|^2+|\mathbf{k}|^2\right)\right]\right\} \tag{6}$$

The sampling of $\mathbf{k}$ is done by $\mathbf{k}_{lmn} = k_{0n}[\sin\theta_l\cos\phi_m \ \sin\theta_l\sin\phi_m \ \cos\theta_l]^t$, and $\theta_l = [(l-1)]/L]\pi$, and $\phi_m = [(m-1)/M]\pi$, and $k_{0n} = k_0/\delta^{n-1}$ where $l = 1,\dots, L$, $m = 1,\dots, M$, and $n = 1,\dots, N$; k_{0n} is the magnitude of the wave number vector; ϕ_m is the azimuth angle; θ_l is the elevation; δ is the spacing factor in the frequency domain; l, m, and n are the indexes of the Gabor kernels; L, M, and N are the total numbers of decompositions along two tangential axes and a radial axis, respectively; and t denotes a matrix transpose throughout this chapter. The carrier frequency of the bandpass filter is determined by $\mathbf{k}$. The Gabor-based wavelets are sensitive to the direction of edges. It has strong response if $\mathbf{k}$ is perpendicular to the direction of edges.

6.2.2 Feature Vector Extraction

We can define $g_{lmn}(\mathbf{x})$ by sampling $\mathbf{k}$ as $\mathbf{k}_{lmn}$. Let $h_{lmn}(\mathbf{x})$ be the output of the filtered input volume $V(\mathbf{x})$ after convolution with $g_{lmn}(\mathbf{x})$:

$$h_{lmn}(x,\,y,\,z) = \sum_{x'=1}^{L_x} \sum_{y'=1}^{L_y} \sum_{z'=1}^{L_z} g_{lmn}(x-x',\,y-y',\,z-z')\,V(x',\,y',\,z'), \quad (7)$$

where L_x, L_y, and L_z is the size of volume data in x, y, and z directions, respectively; $h_{lmn}(\mathbf{x})$ is also called "Gabor coefficient" and the magnitude of the Gabor coefficient is called "Gabor jet". One Gabor jet vector is composed of a set of the Gabor jets: $\mathbf{v}(\mathbf{x}) \equiv \{|h_{lmn(\mathbf{x})}|;\ l=1,\ldots,\,L,\ m=1,\ldots,\,M,\ n=1,\ \cdots,\,N\}$.

6.3 3D Modified Dynamic Link Association (DLA)

In this section, we extend the 2D DLA technique to 3D space for comparison of two-volume objects. The proposed system is composed of two stages: "Rigid Graph Matching (RGM)" and "Elastic Graph Matching (EGM)." The EGM is often referred to as the entire suboptimal system for the DLA in the literature. In this chapter, we use EGM to refer to only the fine matching stage, while we adopt another term, "RGM" for the coarse matching stage.

6.3.1 Rigid Graph Matching (RGM) with Rotation-tolerant Property

We employ 3D graphs of regular hexahedron as shown in Figs. 6.6 and 6.7; however, any arbitrary graph can be used for the 3D DLA technique. Let R and S be two identical and rigid 3D graphs placed on the reference and unknown input data, respectively. During the RGM, we search for the best-matched position and orientation of the graph S with respect to the graph R.

We can describe any rigid motion of the graph S by a translation vector and a rotation matrix. Let $\mathbf{p}$ be a 3D translation vector: $\mathbf{p} = [p_x p_y p_z]^t$ and $\mathbf{e}$ be a vector of three Euler angles: $\mathbf{e} = [\varphi\theta\psi]^t$. Any rigid motion of the graph can be modeled as:

$$\mathbf{x}_i(\mathbf{p},\mathbf{e}) = A(\mathbf{e})(\boldsymbol{x}_i^0 - \boldsymbol{x}_c^0) + \mathbf{p}, \quad i = 1,\ldots,\,K, \quad (8)$$

where $\mathbf{x}_i^0$ and $\mathbf{x}_c^0$ are, respectively, the position of the node i and the center of the graph which is located at the origin without rotation; K is the total number of nodes in the graph; and A is a rotation matrix which is determined by $\mathbf{e}$." Any 3D rotation can be defined by a general rotation matrix $A = BCD$:

$$D = \begin{bmatrix} \cos\varphi & \sin\varphi & 0 \\ -\sin\varphi & \cos\varphi & 0 \\ 0 & 0 & 1 \end{bmatrix}, \quad C = \begin{bmatrix} 1 & 0 & 0 \\ 0 & \cos\theta & \sin\theta \\ 0 & -\sin\theta & \cos\theta \end{bmatrix},$$

$$B = \begin{bmatrix} \cos\psi & \sin\psi & 0 \\ -\sin\psi & \cos\psi & 0 \\ 0 & 0 & 1 \end{bmatrix}, \quad (9)$$

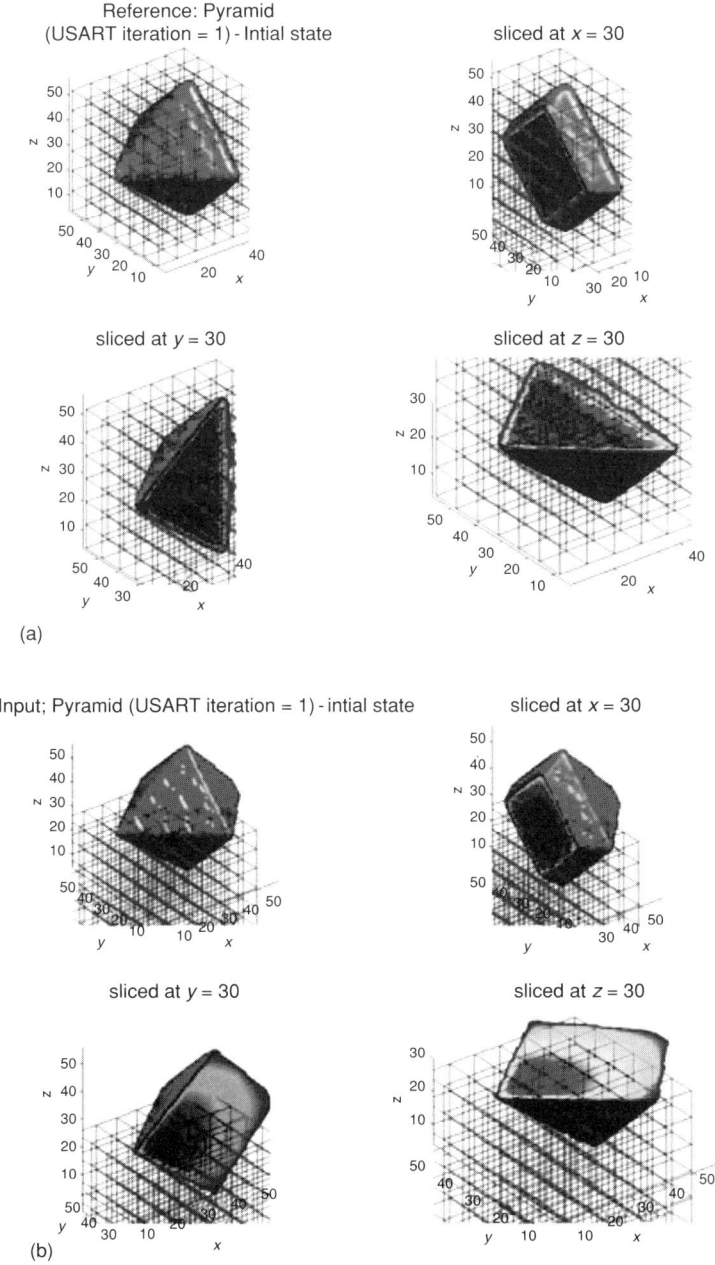

Figure 6.6. Results of the experiment in Section 6.5.1. The reference is the pyramid after 30 USART iterations and the input is the pyramid at the first iteration, which is distorted severely: (a) reference object, (b) input object at the initial state of RGM,

(*Continued*)

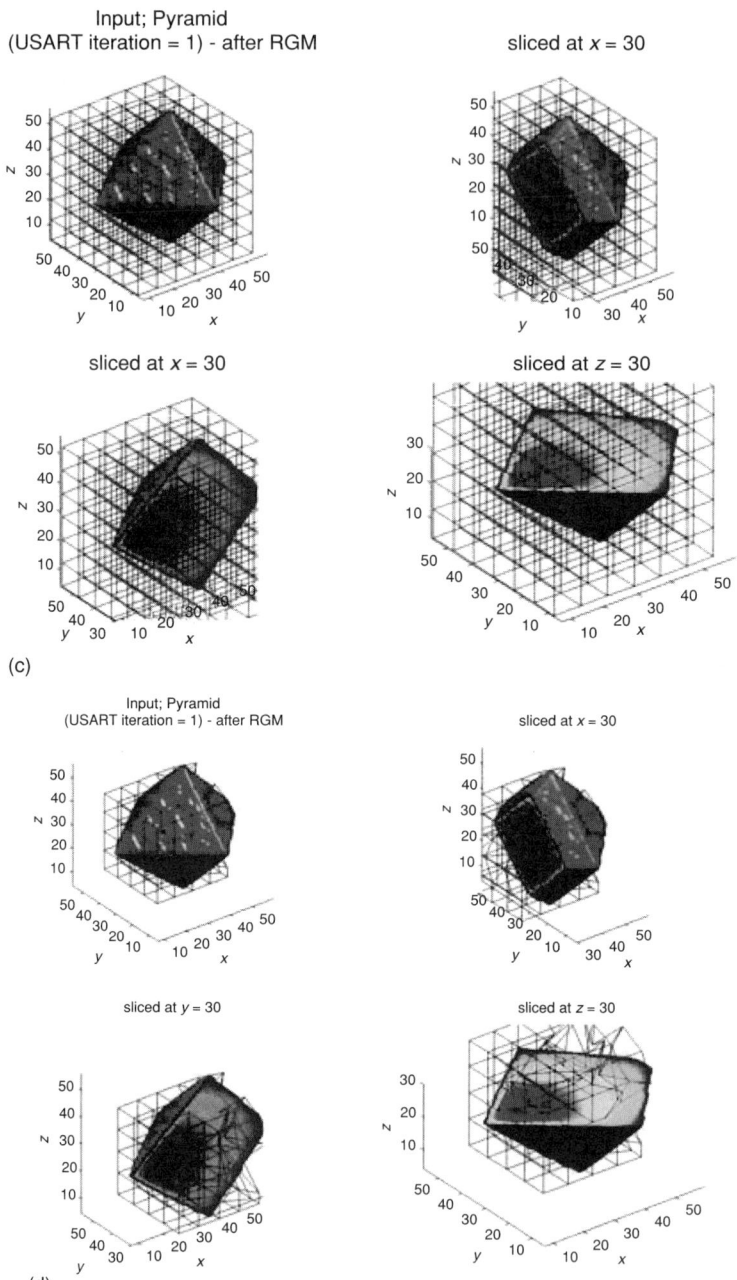

Figure 6.6. (Cont'd) (c) input object after RGM, and (d) input object after EGM, (a)~(d): sliced views of the object at $x = 30$, $y = 30$, and $z = 30$.

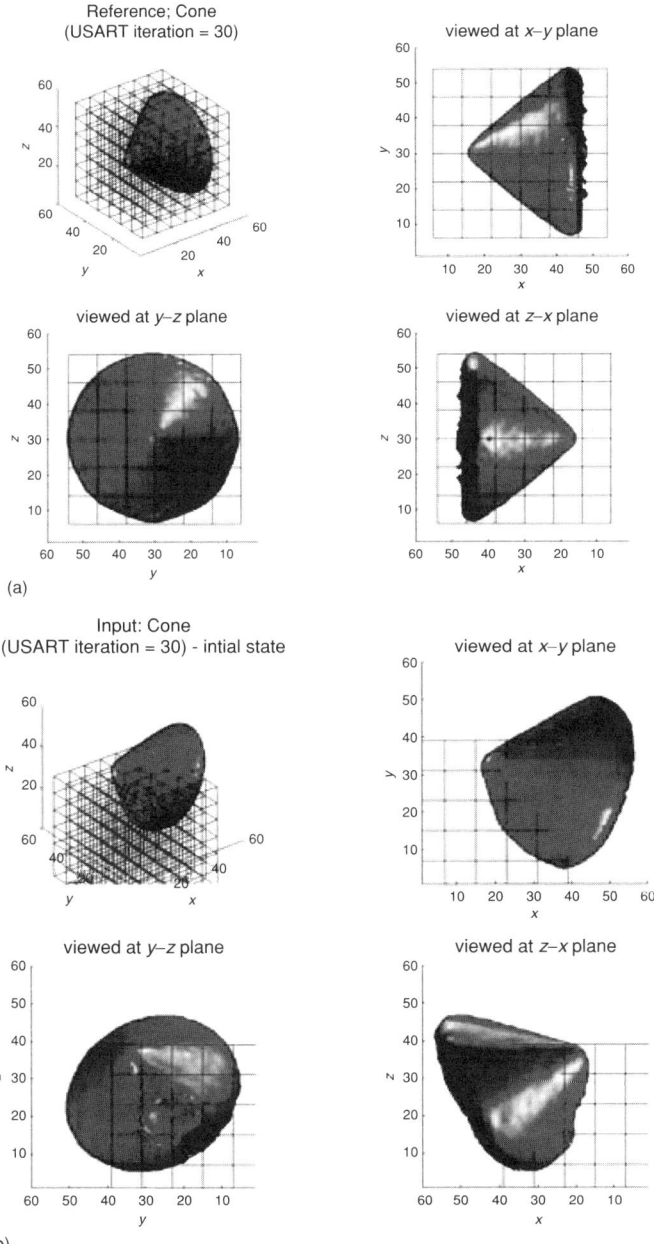

Figure 6.7. Results of the experiment in Section 6.5.2. The reference is a cone after 30 USART iterations and the input is a cone after 30 USART iterations with rotation. The input object is rotated with the rotation angle set 5 ($\varphi = 30°$, $\theta = -60°$, $\psi = 0°$): (a) reference object, (b) input object at the initial state of RGM,

(Continued)

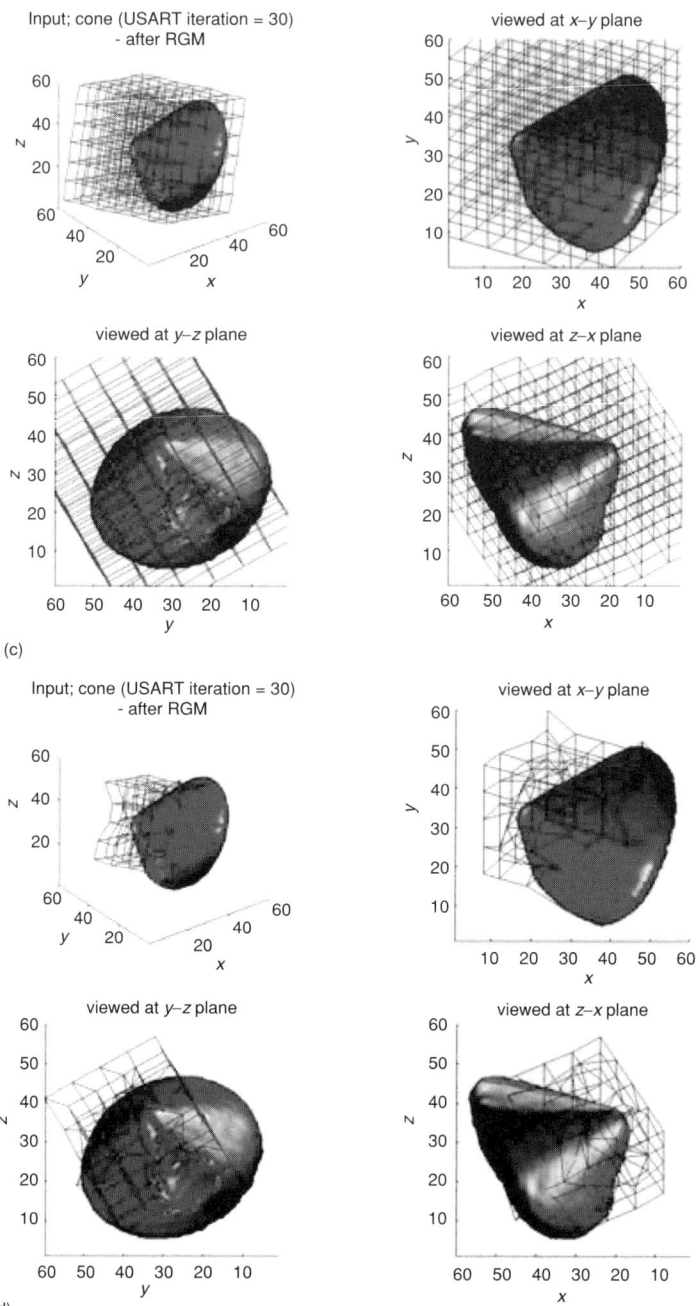

Figure 6.7. (Cont'd) (c) input object after RGM, and (d) input object after EGM, (a)~(d) objects from four different perspectives.

where the first rotation (D) is by an angle φ about the z-axis; the second one (C) is by an angle $\theta \in [0, \pi]$ about the x-axis; and the third one (B) is by an angle ψ about the z-axis again. All axes are rotated in counterclockwise direction. It is noted that $\mathbf{p}$ corresponds to the position vector of the central node in the graph S.

We search for the best-matched position and orientation of the graph S by maximizing a cost function C_{RGM}:

$$\{\hat{\mathbf{p}}, \hat{\mathbf{e}}\} = \arg \max_{\mathbf{p,e}} C_{RGM}(\mathbf{p}, \mathbf{e}), \tag{10}$$

$$C_{RGM}(\mathbf{p}, \mathbf{e}) = \sum_{i=1}^{K} c_i(\mathbf{p}, \mathbf{e}), \tag{11}$$

where $\hat{\mathbf{p}}$ and $\hat{\mathbf{e}}$ are the estimates of the location and orientation of the graph S. A node cost c_i is the cross-correlation coefficient of two function vectors:

$$c_i(\mathbf{p}, \mathbf{e}) \equiv \frac{\langle \mathbf{f}[\mathbf{v}_r(\mathbf{x}_i)], \mathbf{f}[\mathbf{v}_s(\mathbf{x}_i(\mathbf{p}, \mathbf{e}))] \rangle}{\| \mathbf{f}[\mathbf{v}_r(\mathbf{x}_i)] \| \| \mathbf{f}[\mathbf{v}_s(\mathbf{x}_i(\mathbf{p}, \mathbf{e}))] \|}, \quad i = 1, \ldots, K, \tag{12}$$

where $\mathbf{v}_r(\mathbf{x}_i)$ and $\mathbf{v}_s(\mathbf{x}_i)$ are the Gabor jet vectors defined at $\mathbf{x}_i$ in the graph R and $\mathbf{x}_i(\mathbf{p}, \mathbf{e})$ in the graph S, respectively." The rotation-invariant property can be achieved simply by adding up all the Gabor jets along two tangential axes in 3D frequency domain. Therefore, we define a new feature vector $\mathbf{f}$ as:

$$\mathbf{f}(\mathbf{v}) = \begin{bmatrix} f_1(\mathbf{v}) \\ \vdots \\ f_N(\mathbf{v}) \end{bmatrix}, \quad \text{where } f_n(\mathbf{v}) = \sum_{l=1}^{L} \sum_{m=1}^{M} v_{lmn}, \tag{13}$$

where L, M, and N are the total numbers of decompositions for the 3D Gabor wavelets in two tangential axes and one radial axis in 3D frequency domain. We combine total $L \times M$ orientations of Gabor jets to obtain the rotation-invariant feature vector.

Ideally, we can maximize C_{RGM} by searching all possible voxel displacements in 3D integer domain and Euler angles in 3D real domain. In the experiment, we use a predetermined searching interval Δ_p to reduce computational burdens. The searching process is performed by $\mathbf{p} = [p_x p_y p_z]^t \Delta_p$, where $p_x = 1, \ldots, L_x/\Delta_P$, $p_y = 1, \ldots, L_y/\Delta_P$ $p_z = 1, \ldots, L_z/\Delta_P$. The searching process for Euler angles is also restricted by a pre-determined searching angle Δ_e. Allowing large computational cost, we may choose the smallest interval $\Delta_p = 1$ for the searching interval and $\Delta_e = \tan^{-1}(1/L_{\min}) \cong 1/L_{\min}$ for the searching angle; $L_{\min} = \min\{L_x, L_y, L_z\}$, where L_x, L_y, and L_z are the size of volume data in x, y, and z directions, respectively; and $1/L_{\min}$ is approximately the lowest value of the angle when a voxel moves to the nearest neighbor at the end of the volume data by rotation. In the experiments, we set Δ_p and Δ_e arbitrary values with consideration of the computational load.

Another consideration is that the Gabor jet vectors are only defined at integer domain, which is natural for image data. An arbitrary rotation can convert the integer vector into real noninteger values. In that case, we simply apply the 3D Nearest Neighbor interpolation method to overcome this problem in the experiments.

6.3.2 Elastic Graph Matching (EGM) with Sequential and Recursive Realization

The graph S has elastic property during the EGM. We change nodes' positions independently by maximizing a cost function C_{EGM}:

$$\{\hat{\mathbf{x}}_1^s, \ldots, \hat{\mathbf{x}}_K^s\} = \arg \max_{\mathbf{x}_1, \ldots, \mathbf{x}_K} C_{EGM}(\{\mathbf{x}_1^s, \ldots, \mathbf{x}_K^s\}; \hat{\mathbf{p}}, \hat{\mathbf{e}}), \tag{14}$$

$$
\begin{aligned}
C_{EGM} &\equiv \sum_{i=1}^{K} c_i(\hat{\mathbf{p}}, \hat{\mathbf{e}}) - \lambda \sum_{(i,j) \in E^R, E^S} \| \boldsymbol{\Delta}_{ij}^r - \boldsymbol{\Delta}_{ij}^s \|^2 \\
&= C_{RGM}(\hat{\mathbf{p}}, \hat{\mathbf{e}}) - \lambda \sum_{(i,j) \in E^R, E^S} \| \boldsymbol{\Delta}_{ij}^r - \boldsymbol{\Delta}_{ij}^s \|^2,
\end{aligned}
\tag{15}
$$

where $\mathbf{x}_i^r$ is the position vector of the node i in the graph R; $\mathbf{x}_i^s$ is the position vector of the node i in the graph S; K is the total number of nodes in the graph; and $\hat{\mathbf{p}}$ and $\hat{\mathbf{e}}$ are the estimates of position and orientation during the RGM, respectively. We define $\boldsymbol{\Delta}_{ij}^r \equiv \mathbf{x}_i^r - \mathbf{x}_j^r$ and $\boldsymbol{\Delta}_{ij}^s \equiv \mathbf{x}_i^s - \mathbf{x}_j^s$; (i, j) indicates an edge composed of the node i and the node j; the node j can be one of the six nearest neighbors of the node i in the regular hexahedron graph. Let E^R and E^S be sets of node pairs in the graph R and the graph S, respectively. Two sets are composed of one-to-one corresponding nodes and edges. Note that we have already estimated $\hat{\mathbf{p}}$ and $\hat{\mathbf{e}}$ during the RGM. The initial location and orientation of the graph S for the EGM are computed by $\hat{\mathbf{p}}$ and $\hat{\mathbf{e}}$. During the EGM, we relocate all nodes' positions in the elastic graph S.

The parameter λ controls the flexibility of deformation in the graph S. During the RGM, the rigidity of the graph implies infinite penalty for any deformation of the graph. However, during the EGM, we reshape the graph with less constraint. The first part of C_{EGM} is the same as C_{RGM}. The larger the value of λ, the higher the deformation penalty of the graph S. If λ is infinite, C_{EGM} can be maximized when $\boldsymbol{\Delta}_{ij}^r = \boldsymbol{\Delta}_{ij}^s$ for all i and j.

We develop a sequential and recursive method to implement the EGM in fast and effective manner. Table 6.1 demonstrates the overall procedures of the EGM. All steps are presented in one positive dimension; however, they can be easily extended to 2D and 3D space. We simplify the EGM assuming all Euler angle estimates are equal to zero during the RGM. Let $\mathbf{x}_i$ be the initial position of the node i for the EGM. We assume that there exists a global maximum in Eq. (14) and it is located in the small region around the initial value for the EGM.

At the first step, we set d and $l_{\max}$, where d is a fixed displacement for nodes in one dimension and $l_{\max}$ is a maximum iteration number. We then compute the initial C_{EGM} at the next step. At step 3, we re-compute C_{EGM} with a new

Table 6.1. Procedures for the EGM.

Step	Procedures
	Set $0 < d \le d_e$ (e.q. $d = d_e/2$, d_e is edge size)
Step 1	Set maximum iteration number $l_{\max}$
	Let $l = 0$
	Let $l = l + 1$
Step 2	Let node index $i = 0$
	Compute $C_0 = C_{EGM}(\{\mathbf{x}_1^s, \ldots, \mathbf{x}_K^s\}; \hat{\mathbf{p}}, \hat{\mathbf{e}})$
	Let $i = i + 1$ and $x_i^{s+} = x_i^s + d$
Step 3	Re-compute the EGM cost:
	$C_i^+ = C_{EGM}(\{\mathbf{x}_1^s, \ldots, \mathbf{x}_i^{s+}, \ldots, \mathbf{x}_K^s\}; \hat{\mathbf{p}}, \hat{\mathbf{e}})$
	Record i and x_i^{s+} if $C_i^+ > C_0$
	Go to Step 3 until $i = K$
Step 4	Change $x_{s_i} = x_{s_i} + d$ for all the saved nodes "i's."
	Terminate if there is no recorded nodes in Step 4
Step 5	or $l \ge l_{\max}$
	Otherwise, Go to Step 2

node position x_i^{s+} and save the node's index and position if the new position provides a larger C_{EGM}. This task is sequentially performed for all nodes in the graph S. At step 4, we place the recorded nodes into new positions. Finally, we terminate the procedures according to a termination criterion or iterate steps 2 through 5. In this chapter, "sequential" procedures refer to step 3 and "iteration" is used to represent procedures from step 2 to 5.

We described only positive displacement ($d > 0$) in one dimension. It is noted that $x_i^{s-} = x_i^s - d$ should be considered for negative displacement. Thus, in 3D space, one position vector has 6 possible transitions which are $x_i^{s\pm} = x_i^s \pm d$, $y_i^{s\pm} = y_i^s \pm d$ and $z_i^{s\pm} = z_i^s \pm d$.

Recursively, we repeat the whole procedure of Table 6.1 while reducing d gradually. We use the term "recursion" to represent the repetition of the whole procedure in Table 6.1 while reducing d, while "iteration" is used to represent procedures from the steps 2 through step 5 with a fixed d. Different EGM recursions are demonstrated in Section 6.7.

6.4 Statistical Significance Testing

At the final stage, we decide the class of input objects by statistical significance testing used in Refs. 32 and 33. Let C_{sr_i} denote C_{EGM} which is computed with a reference object r_i and an unknown input object s. We order C_{sr_i} in descending sequence, i.e. $C_{sr_i} > C_{sr_{i+1}} \forall i \in \{0, 1, \ldots, N_r - 1\}$ where N_r is the total number of references. We decide the input object s to be the same class as the reference r_0 if $\kappa_{1,s} > \tau_1$ or $\kappa_{2,s} > \tau_2$, where $\kappa_{1,s} = (C_{sr_0} - C_{sr_1})/\sigma$ and $\kappa_{2,s} = (C_{sr_0} - m)/\sigma$; r_0

is the reference of the maximum C_{EGM} among N_r references; m is the mean and σ is the standard deviation of the set $\{\, C_{sr_i} | i = 1, \ldots, N_r - 1 \,\}$; thresholds τ_1 and τ_2 are determined heuristically in the experiment.

We also use two parameters for the performance evaluation:

$$P_D = \frac{\text{Number of correct decisions}(N_D)}{\text{Total number of input objects}(N_O)} \times 100(\%), \tag{16}$$

$$P_F = \frac{\text{Number of wrong decisions}(N_F)}{\text{Total number of input objects}(N_O)} \times 100(\%), \tag{17}$$

where P_D indicates the correct decision rate; N_O is the number of input data tested; N_D is the number of correct decisions accepted by the statistical significance test; P_F is the false alarm probability; and N_F is the number of wrong decisions which are falsely accepted.

6.5 Experimental Result and Performance Analysis

We will present two experiments for the 3D object recognition task: one involves the distortion of input objects and the other their rotation. The former is an experiment based on input objects, which are reconstructed at all USART iterations. The latter is based on input objects reconstructed from the rotated objects at the 30th USART iteration.

The design parameters of the 3D Gabor-based wavelets are the same throughout this chapter ($\sigma = \pi$, $k_0 = \pi/2$, $\delta = 2$, $L = 4$, $M = 3$, $N = 4$). Therefore, one Gabor jet vector at one node is composed of 48 Gabor jets and the dimension of the feature vector is 4 in Eq. (13). The 3D graph of a $7 \times 7 \times 7$ grid is used for the RGM and a $5 \times 5 \times 5$ grid for the EGM, and the edge size (d_e) is set at 8 voxels. We determine the thresholds $\tau_1 = 0.1$ and $\tau_2 = 0.8$ for the first experiment and $\tau_1 = 0.05$ and $\tau_2 = 0.7$ for the second experiment. Those parameters are chosen heuristically when better results are obtained. The control parameter λ is set at 10^{-5}; d is set at 4 voxels for the first EGM recursion, 2 for the second and 1 for the third.

We experiment with five classes of volume objects: pyramid, hemisphere, cone, and screw number 1 and 2. The performance is analyzed in terms of Mean Absolute Error (MAE) and the experiments with different EGM recursions and the control parameter λ are also presented.

6.5.1 Distortion-tolerant Object Recognition

In the first experiment, we perform five different tests according to five different classes of input object sets. Each test has five classes of references and 16 input volume data for each class. For the reference objects, we choose the reconstructed volume objects after 30 iterations of the USART. The input data are composed

Table 6.2. The overall performance of the distortion-tolerant object recognition.

Test Set	Pyramid	Hemisphere	Cone	Screw #1	Screw #2
N_o	16	16	16	16	16
N_D	16	15	14	15	15
N_F	0	1	2	0	1
P_D (%)	100	93.33	86.67	93.33	93.33
P_F (%)	0	6.67	13.33	0	6.67

of volume data after the reconstruction of odd number iterations and the reference itself, that is, volume objects after $1, 3, \ldots, 29$ and 30 USART iterations.

The center of the graph R is placed at a fixed position $[30\ 30\ 30]^t$. The center of the graph S is initially placed at $[15\ 15\ 15]^t$. The searching interval Δ_p is 5 voxels for all three coordinates in 3D space.

Figure 6.6 shows an example of distortion-tolerant object recognition. The RGM process turns out to be a robust detection and aligning process for distorted input objects. We have a finer matching process during the EGM. In Fig. 6.6, the reference is a pyramid after 30 USART iterations and the input is a pyramid at the first iteration. As shown in Fig. 6.7 (b)–(d), the input object is severely distorted because of artifact errors. However, we successfully recognize the input object to be in the same class as the reference object. Table 6.2 shows the overall result of the first experiment.

For the input data set of the screw number 1, N_D is 15; but N_F is zero because the reference of the maximum C_{EGM} is rejected by the statistical significance test.

6.5.2 Rotation-tolerant Object Recognition

In the second experiment, we test rotation-tolerant object recognition. We perform five different tests according to five different classes of input object sets. Each test has five references and five input volume objects. References are the same as in the previous experiment in Section 6.5.1. Input data are composed of reconstructed volume from rotated 3D objects. All input objects are reconstructed with 30 USART iterations. Five different angle sets for rotated input objects are shown in Table 6.3. For the screw number 1 and 2, we rotate reconstructed volume objects computationally using the 3D cubic interpolation according to each rotation angle set.

The graph R and the initial graph S are placed at the same locations as in the previous experiment. We search the best-matched orientation angle as well as the best-matched location during the RGM. We set the searching angle for the rotation Δ_e as $15°$ for all three Euler angles.

Table 6.3. Five rotation angle sets for rotated objects.

Rotation angle	$\varphi(°)$	$\theta(°)$	$\psi(°)$
Set 1	30	0	0
Set 2	45	0	0
Set 3	30	−30	0
Set 4	30	−45	0
Set 5	30	−60	0

Figure 6.7 shows examples of the rotation-tolerant object recognition. In Fig. 6.7, the reference is a cone after 30 USART iterations. The input object is a cone rotated along the rotation angle set 5 ($\varphi = 30°$, $\theta = -60°$, $\psi = 0°$) and reconstructed at the 30-th iteration. The system successfully classifies the input object with correct angle estimates. Table 6.4 shows the overall performance of the rotation-tolerant object recognition. The recognition is performed successfully for most of the input data. All estimated angles are correct except for the rotation angle sets 3 ~ 5 of the hemisphere.

6.5.3 Performance Analysis

There are many factors affecting the performance of the 3D volume object recognition. We analyze the performance according to the similarity between references and input objects. The analysis is only concentrated on distortion-tolerant object recognition. Mean Absolute Error (MAE) is employed for our similarity measure. It is a matching criterion often used in motion-estimation process in image compression.[36] We define the MAE between the reference and the input volume with the location estimate $\hat{\mathbf{p}} = [\hat{p}_x \ \hat{p}_y \ \hat{p}_z]\Delta_p$:

$$MAE = \frac{1}{L_x L_y L_z} \sum_{x=1}^{L_x} \sum_{y=1}^{L_y} \sum_{z=1}^{L_z}$$

$$\times \left| V_S(x - \hat{p}_x\Delta_p, \ y - \hat{p}_y\Delta_p, \ z - \hat{p}_z\Delta_p) - V_R(x, \ y, \ z) \right|, \qquad (18)$$

where V_R is the reference and V_S is the input volume; L_x, L_y, and L_z are the size of volume data in x, y, and z direction, respectively.

Table 6.4. The overall performance of the rotation-tolerant object recognition.

Test Set	Pyramid	Hemisphere	Cone	Screw #1	Screw #2
N_o	5	5	5	5	5
N_D	4	4	5	5	5
N_F	1	1	0	0	0
$P_D(\%)$	80	80	100	100	100
P_F (%)	20	20	0	0	0

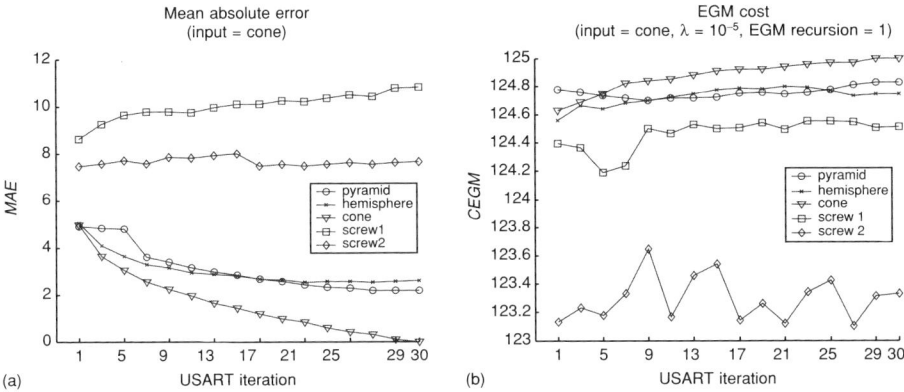

Figure 6.8. Performance analyses from the experiment in Section 6.5.1. The input is the cone after 1, 3, ..., 29 and 30 USART iterations. The reference image is composed of five objects after 30 USART iterations. Horizontal axis shows the USART iteration number of input object: (a) MAE and (b) C_{EGM}.

Figure 6.8(a) shows MAE when the class of input objects is the cone. MAE is computed for five reference objects. Figure 6.8(b) shows C_{EGM}. The x-axis shows the number of the USART iterations of input objects. Figure 6.9(a) and (b) show MAE and C_{EGM} when the input class is the screw number 1. As shown in Figs. 6.8 and 6.9, MAE is smaller when the reference is identified as being of the same class as the input object. It also decreases as the number of the USART iterations increases. When MAE is similar among different reference

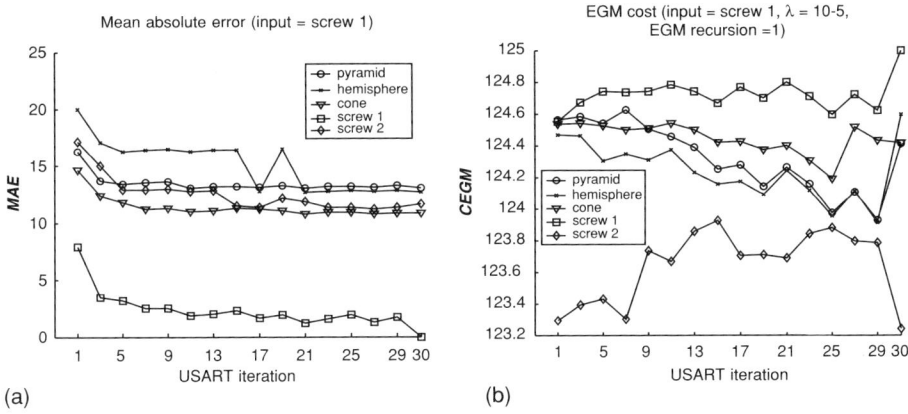

Figure 6.9. Performance analyses from the experiment in Section 6.5.1. The input is the screw number 1 after 1, 3, ..., 29 and 30 USART iterations. The reference image is composed of five objects after USART 30 iterations. Horizontal axis shows the USART iteration number of input object: (a) MAE and (b) C_{EGM}.

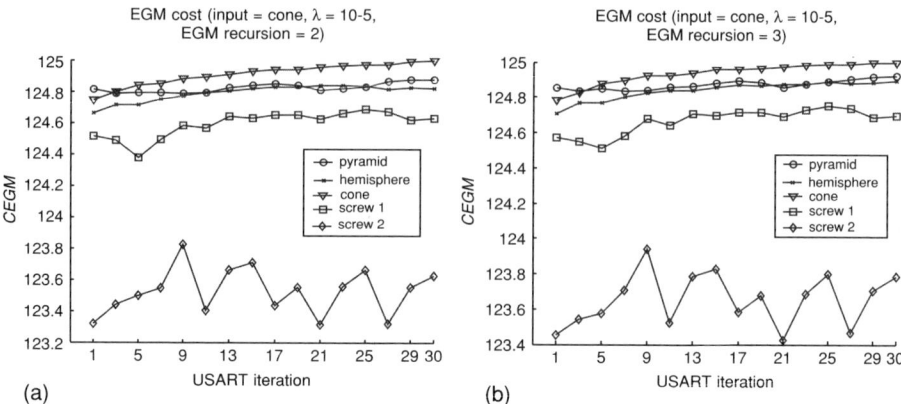

Figure 6.10. Performance analyses of different EGM recursions; the input is the cone after 1, 3, ..., 29 and 30 USART iterations; the reference image is composed of five objects after 30 USART iterations. Horizontal axis shows the USART iteration number of input object; Fig. 6.8(b) corresponds to the EMG with one recursion: (a) EGM recursion is 2 and (b) EGM recursion is 3.

objects, the recognition process has difficulty obtaining correct results Also, when MAE is too large (i.e. similarity is too low) between the input and the true reference, the recognition can fail, providing the highest C_{EGM} from a wrong reference object.

We investigate the effects of the recursive EGM process. The EGM recursion was one ($d = 4$) in Fig. 6.8(b). Figure 6.10(a) shows the results of two EGM recursions ($d = 4$, and 2). Figure 6.10(b) shows the results of three EGM recursions ($d = 4$, 2, and 1). Figure 6.8(b), 11(a), and 11(b) show that C_{EGM} becomes larger with more recursions although the overall shapes of C_{EGM} are similar. Figure 6.11(a) and (b) show the effect of different λ's. Figure 6.8(b) shows the results of $\lambda = 10^{-5}$. Figure 6.11(a) and (b) are the results of $\lambda = 10^{-4}$ and $\lambda = 10^{-6}$, respectively. Large λ implies higher penalty for the graph deformation. A larger λ may provide better results, preserving with less graph deformation but the total cost is proven to become smaller as shown in Fig. 6.11(a).

6.6 Conclusions

In this chapter, we have presented 3D distortion-tolerant volume recognition using 3D modified DLA technique based on 3D Gabor feature vectors. Rotation-invariant features are extracted and constructed by the 3D Gabor wavelets. The 3D Gabor-based wavelets extract localized features of objects according to 3D spatial frequency and bandwidth, as well as location. The modified 3D DLA proves to be a reliable recognition technique, which is tolerant to rotation and distortion. The performance is analyzed in terms of

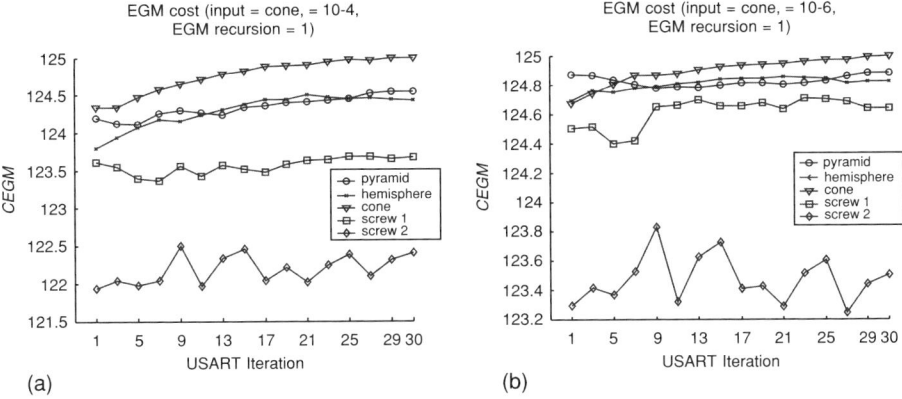

Figure 6.11. Performance analysis of the different control parameter λ the input is the cone after 1, 3, ..., 29 and 30 USART iterations; the USART reference image is composed of five objects after 30 USART iterations. Horizontal axis shows the USART iteration number of input object; Fig. 6.8(b) corresponds to $\lambda = 10^{-5}$, (a) $\lambda = 10^{-4}$, (b) $\lambda = 10^{-6}$.

the MAE. The effects of different EGM recursions and the control parameter λ are also presented.

The scope of applications of presented recognition technique is very broad. It can be applied to any 3D volumetric information for alignment, registration, classification, and identification task.

Some tasks are left for future works. The consideration of noninteger positioning of nodes would be desirable for certain applications such as nonrigid objects rotated or facial expressions. We can utilize advanced statistical classification methods such as linear discriminant analysis with a pool of training data. More testing with various data and different parameters would be helpful for constructing more powerful recognition system.

References

[1] Mahalanobis. (1996). "Review of correlation filters and their application for scene matching." in *Optoelectronic Devices and Systems for Processing*, Proc. SPIE CR65, pp. 240–260.

[2] Yoshikawa N and Yatagai T. (2000). "Fringe pattern correlator for three-dimensional object recognition." *Opt. Lett.*, 25:1424–1426.

[3] Pu A, Denkewalter R, and Psaltis D. (1997). "Real-time vehicle navigation using a holographic memory." *Opt. Eng.*, 36:2737–2746.

[4] Javidi B., ed. (2002). *Image Recognition and Classification: Algorithms, Systems, and Applications*. Marcel Dekker, New York.

[5] Javidi B and Tajahuerce E. (2000). "Three-dimensional object recognition by use of digital holography." *Opt. Lett.*, 25(9):610–612.

[6] Frauel Y, Tajahuerce E, Castro M, and Javidi B. (2001). "Distortion-tolerant three-dimensional object recognition with digital holography." *Appl. Opt.*, 40(23):3887–3893.

[7] Yeom S and Javidi B. (2004). "Three-dimensional object feature extraction and classification with computational holographic imaging." *Appl. Opt.*, 43(2):442–451.

[8] Matoba O, Tajahuerce E, and Javidi B. (2001). "Real-time three-dimensional object recognition with multiple perspective imaging." *Appl. Opt.*, 40(20):3318–3325.

[9] Frauel Y and Javidi B. (2002). "Digital three-dimensional image correlation by use of computer-reconstructed integral imaging." *Appl. Opt.*, 41(26):5488–5496.

[10] Rosen J. (1998). "Three-dimensional joint transform correlator." *Appl. Opt.*, 37(32):7538–7544.

[11] Esteve-Taboada JJ, Mas D, and Garcia J. (1999). "Three-dimensional object recognition by Fourier transform profilometry." *Appl. Opt.*, 38(22):4760–4765.

[12] Johnson AE. (1999). "Using spin images for efficient object recognition in cluttered three-dimensional scenes." *IEEE Trans. on PAMI.*, 21(5):433–449.

[13] Chowdhury AR, Chellappa R, Krishnamurthy S, and Vo T. (2002). "Three-dimensional face econstruction from video using a generic model." *International conference on Multimedia*, Switzerland, I:,pp. 449–452.

[14] Ben-Arie J and Nandy D. (1998). "A volumetric/iconic frequency domain representation for objects with application for pose invariant face recognition." *IEEE Trans. PAMI.*, 20(5):449–457.

[15] Chui H, Win L, Schultz R, Duncan JS, and Rangarajan A. (2003). "A unified nonrigid feature registration method for brain mapping." *Med. Image Anal.*, 7:113–130.

[16] Cass TA. (1998). "Robust affine structure matching for three-dimensional object recognition." *IEEE Trans. PAMI.*, 20(11):1265–1274.

[17] Moshfeghi M, Ranganath S, and Nawyn K. (1994). "Three-dimensional elastic matching of volumes." *IEEE Trans. Image Processing*, 3(2):128–138.

[18] Hemmendorff M, Andersson MT, Kronander T, and Knutsson H. (2002). "Phase-based multidimensional volume registration." *IEEE Trans. Med. Imaging*, 21(12):1536–1543.

[19] Barequet G and Sharir M. (1997). "Partial surface and volume matching in three dimensions." *IEEE Trans. PAMI.*, 19(9):929–948.

[20] Yeom S, Javidi B, Roh YJ, and Cho HS. (2004) "Three-dimensional object recognition using X-ray imaging." Under review for publication in *Opt. Eng.*

[21] Sadjadi FA. (2002). "New results in the use of polarization diversity for classification of radar targets." in *Automatic Target Recognition XII*, FA Sadjadi, ed. Proc. SPIE 4726, pp. 26–34.

[22] Rizvi SA and Nasrabadi NM. (2003). "Automatic target recognition of cluttered FLIR imagery using multistage feature extraction and feature repair." in *Applications of Artificial Neural Networks in Image Processing VIII*, Proc. SPIE 5015, pp. 1–10.

[23] Goudail F and Refregier P. (2001). "Statistical algorithms for target detection in coherent active polarimetric images." *J. Opt. Soc. Am.*, 18(12):3049–3060.

[24] Refregier Ph and Figue J. (1991). "Optimum trade-off filters for pattern recognition and their comparison with Wiener approach." *Opt. Comput. Process*, 1:245–265.

[25] Caulfield HJ. (2004). "Avoiding the accuracy-simplicity trade-off in pattern recognition." in *Applications and Science of Neural Networks, Fuzzy Systems, and Evolutionary Computation VI*, Proc. SPIE 5200, pp. 150–155.

[26] Roh YJ, Cho HS, Kim HC, and Kim JH. (2002). "Three-dimensional volume reconstruction of an object from X-ray images using uniform and simultaneous ART." *J. Control Automation Syst. Eng.*, 8:1.

[27] Roh YJ, Rark WS, Cho HS, and Jeon HJ. (2000). "An implementation of uniform and simultaneous ART for three-dimensional volume reconstruction in X-ray imaging system." in *Optomechatronic Systems III*, T Yoshizawa, ed. Proc. SPIE 4092, pp. 576–587.

[28] Daugman JG. (1985). "Uncertainty relation for resolution in space, spatial frequency, and orientation optimized by two-dimensional visual cortical filters." *J. Opt. Soc. Am.*, 2(7):1160–1169.

[29] Daugman JG. (1988). "Complete discrete two-dimensional Gabor transforms by neural networks for image analysis and compression." *IEEE Trans. ASSP.*, 36(7):1169–1179.

[30] Lee TS. (1996). "Image representation using two-dimensional Gabor wavelets." *IEEE Trans. PAMI.*, 18(10):959–971.

[31] Malsburg C v.d. (1981). *"The correlation theory of brain function."* Internal Report, Max-Planck-Institute for Biophysical Chemistry, Postfach 2841, D-3400 Gottingen, FRG.

[32] Lades M, Vorbruggen JC, Buhmann J, Lange J, Malsburg C v.d, Wurtz RP, and Konen W. (1993). "Distortion invariant object recognition in the dynamic link architecture." *IEEE Trans. Comput.*, 42(3):300–311

[33] Wurtz RP. (1997) "Object recognition robust under translations, deformations, and changes in background." *IEEE Trans. PAMI.*, 19(7):769–775.

[34] Duc B, Fischer S, and Bigun J. (1999). "Face authentification with Gabor information on deformable graphs." *IEEE Trans. Image Proc.*, 8(4):504–516.

[35] Kotropoulos CL, Tefas A, and Pitas I. (2000). "Frontal face authentication using discriminating grids with morphological feature vectors." *IEEE Trans. Multimed.*, 2(1):14–26.

[36] Bhaskaran V and Konstantinids K. (1997). *Image and Video Compression Standards: Algorithms and Architedtures.* Kluwer, New York

3D Imaging and Recognition of Microorganism Using Single-exposure Online (SEOL) Digital Holography

Bahram Javidi, Inkyu Moon, Seokwon Yeom, and Edward Carapezza*

Bahram@engr.uconn.edu
Department of Electrical and Computer Engineering, U-2157, University of Connecticut, Storrs, CT USA 06269-2157
*Defense Advanced Research Projects Agency (DARPA) 3701 N.Fairfax Drive, Arlington, VA 22203-1714

Abstract: We address three-dimensional (3D) visualization and recognition of microorganisms using single-exposure online (SEOL) digital holography. A coherent 3D microscope-based Mach-Zehnder interferometer records a single online Fresnel digital hologram of microorganisms. Three-dimensional microscopic images are reconstructed numerically at different depths by an inverse Fresnel transformation. For recognition, microbiological objects are segmented by processing the background diffraction field. Gabor-based wavelets extract feature vectors with multi-oriented and multi-scaled Gabor kernels. We apply a rigid graph matching (RGM) algorithm to localize predefined shape features of biological samples. Preliminary experimental and simulation results using sphacelaria alga and *Tribonema aequale* alga microorganisms are presented. To the best of our knowledge, this is the first report on 3D visualization and recognition of microorganisms using on-line digital holography with single-exposure.

7.0 Introduction

Optical information systems have proven to be very useful in the design of two-dimensional (2D) pattern recognition systems.[1-6] Recently, interest in 3D optical information systems has increased because of its vast potential in applications such as object recognition, image encryption as well as 3D display.[4,5] Digital holography is attractive for visualization and acquisition of 3D information for these various applications.[7-13]

In this chapter, we address real-time 3D imaging and shape-based recognition of microorganisms.[14] The automatic recognition of living organisms is accompanied by various challenges. Firstly, they are not rigid objects, they vary in size and shape, and they can move, grow, and reproduce themselves depending on growth conditions.[15] In particular, bacteria and algae are very tiny and they have relatively simple morphological traits for image intensity-based recognition and identification. They may occur as a single cell or form an association of various complexities according to the environmental conditions. Therefore, special consideration on the morphological and physiological characteristics of algae and bacteria should be preceded to enhance the recognition system.

The applications of 3D imaging and recognition systems are very broad. It may be used to diagnose an infection caused by specific bacteria or detect biological weapons for security and defense. Identification and quantification of microorganisms are important in wastewater treatment. Monitoring of plankton in the ocean may be another application of the microorganism imaging and recognition system.

Earlier, various researches had been performed to recognize specific 2D shapes of microorganisms based on image intensity. The recognition and identification of tuberculosis bacteria[16] and *Vibrio cholerae*[17] have been studied based on their colors and 2D shapes. In[18], bacteria in a wastewater treatment plant are identified by morphological descriptors. The aggregation of streptomyces is classified into different phases by measuring the aggregation size and reaction time.[19] In[20], plankton recognition is performed using pre-selected geometrical features. More research on image analysis and recognition of microorganism can be found in.[21]

Our research focuses on a new approach to provide real-time 3D visualization, monitoring, and recognition of microorganisms using SEOL digital holography. Off-axis digital holography has been extensively studied in recent years because it requires only a single exposure in separating the original image from the undesired DC and conjugate images. However, off-axis digital holography has a number of drawbacks. Only a fraction of the space-bandwidth product of the photo sensor is used to reconstruct the 3D image, which results in substantially reduced quality of visualization and compromises resolution. As a result, it reduces the accuracy of object recognition. In addition, the angle between the object beam and reference beam during the holographic synthesis is a function of the reconstructed image size, which creates problems in monitoring dynamic scenes containing objects with varying dimensions. Phase-shifting or on-line digital holography has been proposed to avoid these problems. This technique requires multiple interferogram recordings with phase shifts in the reference beam. The multiple exposures are used to remove the DC and the conjugate images in the interferogram. The Fresnel diffraction field of the 3D object is obtained. However, this procedure is not suitable for dynamic events such as moving 3D microorganism and is sensitive to external noise factors such as environmental vibration and fluctuation. Recently, the

SEOL digital holography for 3D object recognition was presented to solve these problems associated with phase-shifting digital holography.[22,23] The SEOL holography can be used for dynamic events because it requires only a single exposure. The additional benefit of our SEOL digital holography for monitoring of a 3D dynamic time varying scene is that various slices of the 3D microorganism and the 3D scene can be digitally reconstructed and numerically focused without mechanical focusing as is required by conventional microscopy. One important benefit of the proposed technique is that microorganism 3D images are recorded in both magnitude and phase, which may provide better classification of algae or bacteria.

In this chapter, we visualize and recognize two filamentous microorganisms (sphacelaria alga and *Tribonema aequale* alga) using SEOL digital holography. Assuming that the microorganisms are individually segmented or they are sparsely aggregated, we identify two different microbiological objects with their morphological traits.

The frameworks of our system are composed of several stages as shown in Fig. 7.1. At the first stage, the SEOL digital holography performs 3D imaging of micro objects. Utilizing a Mach-Zehnder interferometer, the system opto–electronically records the complex amplitude distribution generated by the Fresnel diffraction at a single plane. The 3D information of the wave transmitted from the microorganisms can be reconstructed from the hologram at an arbitrary depth plane. Reconstructed images are resized and objects of interest are segmented at the next stage. We segment foreground objects using the histogram analysis. Gabor-based wavelets extract salient features by decomposing them in the spatial frequency domain.[24,25]

The RGM is a feature matching technique to identify reference shapes. During the RGM, we search similar shapes with that of the reference data by measuring similarity and difference between feature vectors. The feature vectors are defined at the nodes of two identical graphs on the reference and the input images, respectively. The RGM combined with Gabor-based wavelets has proven to be a robust template matching technique that is invariant to shift, rotation, and distortion.[26]

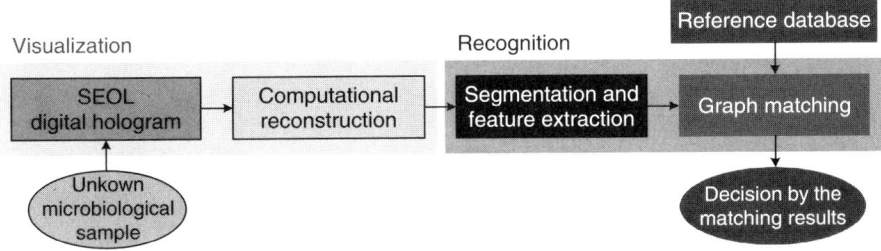

Figure 7.1. Frameworks of the 3D visualization and recognition of microbiological objects.

In our database, two reference graphs are predetermined in order to represent unique shape features of the microorganisms. After the graph matching, the number of detection and the value of feature vectors can be used for further training processes with a pool of training data.[27] In this paper, we present experimental and simulation results as a preliminary step toward a generic and human aided 3D image-based recognition system of microorganisms.

The proposed work is beneficial in a number of ways: 1) the microorganisms are analyzed in 3D topology and coordinates; 2) single-exposure online computational holographic sensor allows optimization of the space bandwidth product for detection as well as robustness to environmental variations during sensing process; 3) multiple exposures are not required and moving bacteria can be sensed within the time constant of the detector; 4) complex amplitude of reconstructed holographic images are decomposed in the spatial frequency domain by Gabor-based wavelets to extract distinguishable features; and 5) a pattern-matching technique measures the similarity of 3D geometrical shapes between a reference microorganism and an unknown sample.

In Section 7.1, we present principles of SEOL digital holography and its advantages. The segmentation and Gabor-based wavelets are presented in Sections 7.2 and 7.3, respectively. The graph matching technique is described in Section 7.5. In Section 7.5, experimental and simulation results are demonstrated. The conclusions follow in Section 7.6.

7.1 Single Exposure Online (SEOL) Digital Holography

In the following, we present the SEOL technique and its advantages over the conventional methods. The 3D optical monitoring system using the SEOL digital holographic recording setup is depicted in Fig. 7.2. Polarized light from an Argon laser with a center wavelength of 514.5 nm, is expanded by use of a spatial filter and a collimating lens to provide spatial coherence. A beam splitter divides the expanded beam into object and reference beams. The object beam illuminates the microorganism sample and the microscope objective

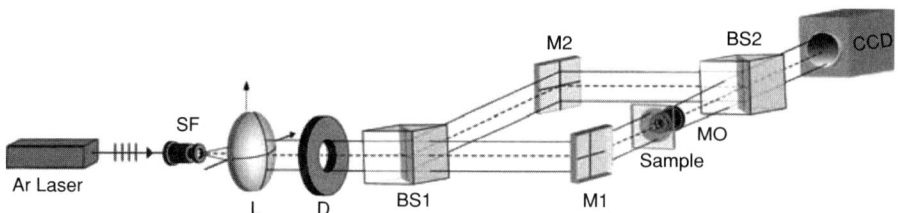

Figure 7.2. Experimental setup for recording an online digital hologram of a microscopic 3D biological object: Ar, Argon laser; SF, Spatial filter; L, lens; D, diaphragm; BS1, BS2, beam splitter; M1, M2, mirror; MO, microscope objective; CCD, charge coupled device array.

produces a magnified image positioned at the image plane of the microscope [see Fig. 7.3]. The reference beam forms an on-axis interference pattern together with the light diffracted by the microorganism sample, which is recorded by the CCD camera. Our system uses no optical components for the phase retardation in the reference beam, which the phase-shifting digital holography technique requires. Also, only a single exposure is recorded in our system. In the following, we describe both on-axis phase-shifting digital holography and SEOL.

We start by describing on-axis phase-shifting digital holography.[12] The hologram recorded on the CCD can be represented as follows:

$$H_p(x,y) = [A_H(x,y)]^2 + A_R^2 + 2A_H(x,y)A_R \times \cos{[\Phi_H(x,y) - \varphi_R - \Delta\varphi_p]} \quad (1)$$

where $A_H(x,y)$ and $\Phi_H(x,y)$ are the amplitude and phase, respectively, of the Fresnel complex-amplitude distribution of the micro objects at the recording plane generated by the object beam; A_R is the amplitude of the reference distribution; φ_R denotes the constant phase of the reference beam; and $\Delta\varphi_p$, where the subscript p is an integer from 1 to 4, denoting the four possible phase shifts required for on-axis phase-shifting digital holography. The desired biological object Fresnel wave function, $A_H(x,y)$ and $\Phi_H(x,y)$ can be obtained by use of the four interference patterns with different phase shifts $\Delta\varphi_p = 0$, $\pi/2$, π and $3\pi/2$.

In this chapter, phase-shifting on-axis digital holography with double exposure, and SEOL digital holography are implemented to obtain experimental results for the visualization and recognition of 3D biological objects. The SEOL results are compared with multiple expose phase-shifting digital holographic results. The double-exposure method requires 1) two interference patterns that have a $\pi/2$ phase difference, 2) the information about a reference

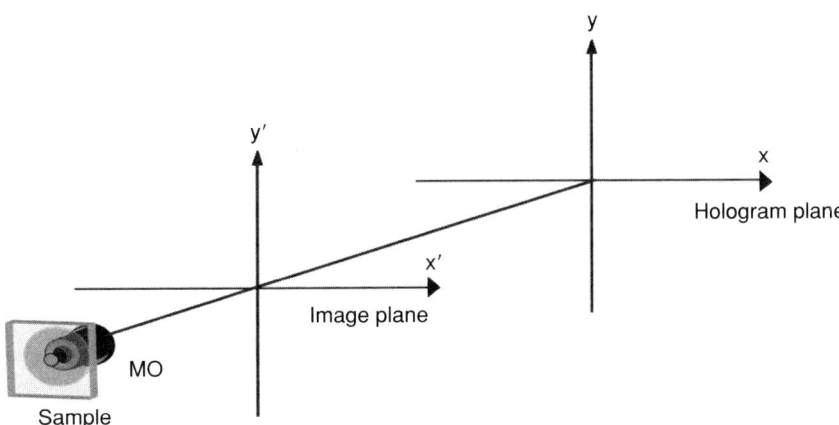

Figure 7.3. Coordinate system for digital hologram and image reconstruction of 3D microorganisms.

beam, and 3) information about the diffracted biological object beam intensity. The complex amplitude of the microscopic 3D biological object wave at the hologram plane from the double-exposure method is represented by:

$$U_h(x,y) = A_H(x,y) \times \cos[\Phi_H(x,y)] + jA_H(x,y) \times \sin[\Phi_H(x,y)]$$

$$= \{H_1(x,y) - A_H(x,y)^2 - A_R^2\}/(2A_R) + j\{H_2(x,y) - A_H(x,y)^2$$

$$- A_R^2\}/(-2A_R) \tag{2}$$

where $H_1(x,y)$ and $H_2(x,y)$ can be obtained from Eq. (1). We assume that the recording between two holograms is uniform and reference beam is plane wave. The former assumption requires stable recording environment and stationary objects.

The SEOL digital holography is suitable for recording dynamic fast events.[21] It needs to record only one hologram to gain information about the complex amplitude of the 3D biological object. The information about the wave front of a 3D biological object contained in the SEOL digital hologram is represented by the following term:

$$U_{h'}(x,y) = 2A_H(x,y)A_R \times \cos(\Phi_H(x,y) - \varphi_R)$$

$$= H_1(x,y) - |A_H(x,y)|^2 - A_R^2 \tag{3}$$

In Eq. (3), $H_1(x,y)$ can be obtained from Eq. (1). To remove DC terms in Eq. (3), the reference beam intensity $|A_R|^2$ is removed by a one-time measurement in the experiment. The object beam intensity $|A_H(x, y)|^2$ can be considerably reduced by use of signal processing (for example, an averaging technique). Even though SEOL digital holography originally contains a conjugate image, we can utilize the conjugate image in the interferogram in recognition experiments since it has information about the biological object. Thus, the 3D biological object wave function $U_{h'}(x, y)$ including a conjugate component in Eq. (3) can be obtained by use of SEOL digital holography. In this paper, we show that the index $U_{h'}(x, y)$ in Eq. (3) obtained by a SEOL hologram can be used for 3D biological object recognition and 3D image formation. The results will be compared with that of index $U_h(x, y)$ in Eq. (2) obtained by on-line phase-shifting holography that requires multiple recordings. The microscopic 3D biological object can be restored by Fresnel propagation of $U_{h'}(x, y)$, which is the biological object wave information in the hologram plane. We can numerically reconstruct 3D section images on any parallel plane perpendicular to the optical axis by computing the following Fresnel transformation with a 2D FFT algorithm:

$$U_{o'}(m', n') = \exp\left[-j\frac{\pi}{\lambda d}(\Delta X^2 m'^2 + \Delta Y^2 n'^2)\right] \times \sum_{m=1}^{N_x}\sum_{n=1}^{N_y} U_{h'}(m, n)\exp$$

$$\left[-j\frac{\pi}{\lambda d}(\Delta x^2 m^2 + \Delta y^2 n^2)\right]\exp\left[j2\pi(\frac{mm'}{N_x} + \frac{nn'}{N_y})\right] \tag{4}$$

where $U_{o'}(m', n')$ and $(\Delta X, \Delta Y)$ are the reconstructed complex amplitude distribution and resolution at the plane in the biological object beam, respectively; $U_{h'}(m, n)$ and $(\Delta x, \Delta y)$ are the object wave function including a conjugate component and resolution at the hologram plane, respectively; and d represents the distance between the image plane and hologram plane.

7.2 Segmentation

In the following, we present the segmentation of digitally reconstructed holographic images. Since the coherent light is scattered by the semi-transparent objects, the intensity in the object region becomes lower than the background diffraction field. Therefore, for recognition, it is more efficient to filter out unnecessary background from computationally reconstructed holographic images.

In this chapter, the threshold for the segmentation is obtained by using histogram analysis. The segmented image (o) is defined as:

$$o(m, n) = \begin{cases} o'(m, n) & \text{if } o'(m, n) < I_s \\ 0 & \text{otherwise} \end{cases} \tag{5}$$

where $o'(m, n)$ is the intensity of the holographic image; and m and n are 2D discrete coordinates in x and y directions, respectively. The threshold I_s is decided from the histogram analysis and the maximum intensity rate:

$$I_s = \min\left[\tau_{\kappa_{\min}}, r_{\max} \cdot \max(o')\right] \tag{6}$$

where $r_{\max}$ is the maximum intensity rate of coherent light after scattering by the microorganisms. The threshold $\tau_{\kappa_{\min}}$ is a minimum value satisfying the following equation:

$$P_s \le \frac{1}{N_T} \sum_{i=1}^{\kappa_{\min}} h(\tau_i) \tag{7}$$

where P_s is a predetermined probability; N_T is the number of pixels; $h(\tau_i)$ is the histogram, i.e., the number of pixels of which intensity is between τ_{i-1} and τ_i; τ_i is the ith quantized intensity level; and $\kappa_{\min}$ is the minimum number of pixels that satisfies Eq. (7). For the experiments, the total number of intensity levels is set at 256. P_s and $r_{\max}$ can be decided according to prior knowledge of the spatial distribution and transmittance of the microorganisms.

7.3 Gabor-based Wavelets and Feature Vector Extraction

In this section, we provide a brief review of Gabor-based wavelets and present feature vectors. Gabor-based wavelets are composed of multi-oriented and multi-scaled Gaussian-form kernels, which are suitable for local spectral analysis.

7.3.1 Gabor-based Wavelets

The Gabor-based wavelets have the form of a Gaussian envelope modulated by the complex sinusoidal functions. The impulse response (or kernel) of the Gabor-based wavelet is:

$$g(\mathbf{x}) = \frac{|\mathbf{k}|^2}{\sigma^2} \exp\left(-\frac{|\mathbf{k}|^2 |\mathbf{x}|^2}{2\sigma^2}\right) \left[\exp\left(j\mathbf{k} \cdot \mathbf{x}\right) - \exp\left(-\frac{\sigma^2}{2}\right)\right] \qquad (8)$$

where $\mathbf{x}$ is a position vector, $\mathbf{k}$ is a wave number vector, and σ is the standard deviation of the Gaussian envelope. By changing the magnitude and direction of the vector $\mathbf{k}$, we can scale and rotate the Gabor kernel to make self-similar forms.

We can define a discrete version of the Gabor kernel as $g_{uv}(m, n)$ at $\mathbf{k} = \mathbf{k}_{uv}$ and $\mathbf{x} = (m, n)$, where m and n are discrete coordinates in 2D space in x and y directions, respectively. Sampling of $\mathbf{k}$ is done as $\mathbf{k}_{uv} = k_{0u} [\cos \Phi_v \sin \Phi_v]^t$, $k_{0u} = k_0/\delta^{u-1}$, and $\Phi_v = [(v-1)/V]\pi$, $u = 1, \ldots, U$ and $v = 1, \ldots, V$, where k_{0u} is the magnitude of the wave number vector; Φ_v is the azimuth angle of the wave number vector; k_0 is the maximum carrier frequency of the Gabor kernels; δ is the spacing factor in the frequency domain; u and v are the indexes of the Gabor kernels; U and V are the total numbers of decompositions along the radial and tangential axes, respectively; and t stands for the matrix transpose.

The Gaussian-envelope in the Gabor filter achieves the minimum space-bandwidth product.[23] Therefore, it is suitable to extract local features with high frequency bandwidth (small u) kernels and global features with low frequency bandwidth (large u) kernels. It is noted that the Gabor-based wavelet has strong response to the edges if the wave number vector $\mathbf{k}$ is perpendicular to the direction of edges.

7.3.2 Feature Vector Extraction

Let $h_{uv}(m,n)$ be the filtered output of the image $o(m,n)$ after it is convolved with the Gabor kernel $g_{uv}(m,n)$:

$$h_{uv}(m, n) = \sum_{m'=1}^{N_m} \sum_{n'=1}^{N_n} g_{uv}(m - m', n - n') o(m', n') \qquad (9)$$

where $o(m,n)$ is the normalized image between 0 and 1 after the segmentation; and N_m and N_n are the size of reconstructed images in x and y directions, respectively. $h_{uv}(m,n)$ is also called the "Gabor coefficient."

A feature vector defined at a pixel (m, n) is composed of a set of the Gabor coefficients and the segmented image. The rotation-invariant property can be achieved simply by adding up all the Gabor coefficients along the tangential axes in the frequency domain. Therefore, we define a rotation-invariant feature vector $\mathbf{v}$ as:

$$\mathbf{v}(m,\ n) = \left[o(m,n) \sum_{v=1}^{V} h_{1v}(m,n) \cdots \sum_{v=1}^{V} h_{Uv}(m,\ n) \right]^{t} \tag{10}$$

Therefore, the dimension of a feature vector $\mathbf{v}$ is $U + 1$. In the experiments, we use only real parts of the feature vector since they are more suitable to recognize filamentous structures. There is no optimal way to choose the parameters for the Gabor kernels, but several values are widely used heuristically depending on the applications. The parameters are set up at $\sigma = \pi$, $k_0 = \pi/2$, $\delta = 2\sqrt{2}$, $U = 3$, $V = 6$ in this paper.

7.4 Rigid Graph Matching (RGM)

In this section, we present the RGM technique. Originally, the RGM is part of a dynamic link association (DLA) to allow elastic deformation of the graph.[25] However, we only adopt the RGM part for our microscopic analysis. The RGM realizes a robust template-matching between two graphs which is tolerant to translation, rotation, and distortion caused by noisy data.

The graph is defined as a set of nodes associated in the local area. Let R and S be two identical and rigid graphs placed on the reference (o_r) object and unknown input image (o_s), respectively. The location of the reference graph R is pre-determined by the translation vector $\mathbf{p}_r$ and the clockwise rotation angle θ_r. A position vector of the node k in the graph R is:

$$\mathbf{x}_k^r = A(\boldsymbol{\theta}_r)(\mathbf{x}_k^o - \mathbf{x}_c^o) + \mathbf{p}_r, \ k = 1, \ldots, K \tag{11}$$

where $\mathbf{x}_k^o$ and $\mathbf{x}_c^o$ are, respectively, the position of the node k and the center of the graph which is located at the origin without rotation; K is the total number of nodes in the graph and A is a rotation matrix.

Assuming the graph R covers a designated shape of the representing characteristic in the reference microorganism, we search the similar local shape by translating and rotating the graph S on unknown input images. We describe any rigid motion of the graph S by translation vector $\mathbf{p}$ and clock wise rotation angle θ:

$$\mathbf{x}_k^s(\theta,\mathbf{p}) = A(\theta)(\mathbf{x}_k^o - \mathbf{x}_c^o) + \mathbf{p}, k = 1, \ldots, K \tag{13}$$

Where $\mathbf{x}_k^s$ is a position vector of the node k in the graphs S. The transformation in Eq. (13) allows robustness in detection of rotated and shifted reference objects.

A similarity function between the graph R and S is defined as:

$$\Gamma_{rs} = \frac{1}{K} \sum_{k=1}^{K} \gamma_k(\theta,\ \mathbf{p}), \tag{14}$$

where the similarity at one node is the normalized inner product of two feature vectors:

$$\gamma_k(\theta,\ \mathbf{p}) \equiv \frac{\langle \mathbf{v}[\mathbf{x}_k^r],\ \mathbf{v}[\mathbf{x}_k^s(\theta,\mathbf{p})] \rangle}{\|\mathbf{v}[\mathbf{x}_k^r]\| \|\mathbf{v}[\mathbf{x}_k^s(\theta,\ \mathbf{p})]\|},\ k = 1, \ldots, K. \tag{15}$$

In Eq. (15), $\langle \cdot \rangle$ stands for the inner product of two vectors; and $\mathbf{v}[\mathbf{x}_k^r]$ and $\mathbf{v}[\mathbf{x}_k^s(\theta,\ \mathbf{p})]$ are feature vectors defined at $\mathbf{x}_i^r$ and $\mathbf{x}_k^s(\theta,\ \mathbf{p})$, respectively.

We define a difference cost function to improve discrimination capability of two graphs R and S as:

$$C_{rs} = \frac{1}{K} \sum_{k=1}^{K} c_k(\theta,\ \mathbf{p}), \tag{16}$$

where the cost at one node is the norm of difference of two feature vectors:

$$c_k(\theta,\ \mathbf{p}) = \|\mathbf{v}[\mathbf{x}_k^r] - \mathbf{v}[\mathbf{x}_k^s(\theta,\ \mathbf{p})]\|,\ k = 1, \ldots,\ K. \tag{17}$$

To utilize the depth information of the SEOL hologram, we simultaneously use multiple references. The similarity function $\Gamma_{r_j s}(\theta_j; \mathbf{p})$ and the difference cost $C_{r_j s}(\theta_j;\ \mathbf{p})$ are measured by the feature vectors between the graph R on the image o_{r_j} and the graph S on the image o_s. The graph R covers the fixed region in the reference images, "o_{r_j}", $j = 1, \ldots, J$; J is the total number of reference images reconstructed at different depths.

The graph S is identified with the reference shape that is covered by the graph R if two conditions are satisfied as follows:

Accept detection at $\mathbf{p}$ if $\Gamma_{r_j s}(\hat{\theta}_j;\ \mathbf{p}) > \alpha_\Gamma$ and $C_{r_j s}(\hat{\theta}_j; \mathbf{p}) < \alpha_C$ \qquad (18)

$$\hat{j} = \max_{j}\ [\Gamma_{r_1 s}(\hat{\theta}_1; \mathbf{p}), \ldots, \Gamma_{r_J s}(\hat{\theta}_J; \mathbf{p})] \tag{19}$$

where $\hat{j}$ is the index of the reference image which produces the maximum similarity between the graph R and the graph S with the translation vector $\mathbf{p}$ and the rotation angle $\hat{\theta}_j$; α_Γ and α_C are thresholds for the similarity function and the difference cost, respectively; and $\hat{\theta}_j$ is obtained by searching the best matching angle to maximize the similarity function:

$$\hat{\theta}_j = \arg\ \max_{\theta}\ \Gamma_{r_j s}(\theta, \mathbf{p}) \tag{20}$$

7.5 Experiments and Simulation Result

We will present experimental results of visualization and recognition of two filamentous algae (sphacelaria alga and *Tribonema aequale* alga). First, we present our 3D imaging of algae using SEOL holography compared with phase-shifting on-line digital holography. Second, recognition process using feature extraction and graph matching are presented to localize the predefined shapes of two different microorganisms.

7.5.1 3D Imaging with SEOL Digital Holography

In this subsection, we experimentally compare the 3D algae visualization of the SEOL digital holography with that of the multiple-exposure phase-shifting on-line digital holography by experiments. In the experiments presented in this paper, the images are reconstructed from digital holograms with 2048×2048 pixels and a pixel size of $9\,\mu$m $\times\,9\,\mu$m. The microorganisms are sandwiched between two transparent cover slips. The diameter of the sample is around $10 \sim 50\,\mu$m. We generate two holograms for the alga samples. The microscopic 3D biological object was placed at a distance 500 mm from the CCD array as shown in Fig. 7.2. The results of the reconstructed images from the hologram of the alga samples are shown in Fig. 7.4. Figure 7.4(a) and (b) show sphacelaria's 2D image and the digital hologram by SEOL digital holography technique, respectively. Figure 7.4(c) and (d) are sphacelaria's reconstructed images from the blurred digital holograms at distance of d = 180 mm and 190 mm, respectively, using the SEOL digital holography. Figure 7.4(e) shows the sphacelaria's reconstructed image at distance d = 180 mm using phase-shifting online digital holography with two interferograms, and Fig. 7.4(f) is *Tribonema aequale*'s reconstructed image at distance d = 180 mm using SEOL digital holography. In the experiments, we use a weak reference beam for the conjugate image, which overlaps the original image. As shown in Fig. 7.4, we obtained the sharpest reconstruction at distance d that is between 180 mm and 190 mm for both holographic methods. The reconstruction results indicate that we obtain the focused image by use of SEOL digital holography as well as from the phase-shifting digital holography. We will show that SEOL digital holography may be a useful method for 3D biological object recognition. That is because the conjugate image in the hologram contains information about the 3D biological object. In addition, SEOL digital holography can be performed without stringent environmental stability requirements.

7.5.2 3D Microorganism Reconstruction and Feature Extraction

To test the recognition performance, we generate eight hologram samples from sphacelaria and tribonema aequale, respectively. We denote eight sphacelaria samples as A1, ..., A8 and eight tribonema aequale samples as B1, ..., B8. To test the robustness of the proposed algorithm, we have changed the position of the CCD during the experiments resulting in different depths for the sharpest reconstruction image. The samples A1–A3 are reconstructed at 180 mm, A4–A6 are reconstructed at 200 mm, and A7 and A8 are reconstructed at 300 mm, and all samples of *Tribonema aequale* (B1–B8) are reconstructed at 180 mm for the sharpest images.

Computationally reconstructed holographic images are cropped and reduced into an image with 256×256 pixels by the reduction ratio 0.25. The probability P_s and the maximum intensity rate $r_{\max}$ for the segmentation are set at 0.25 and 0.45, respectively. We assume that less than 25% of lower

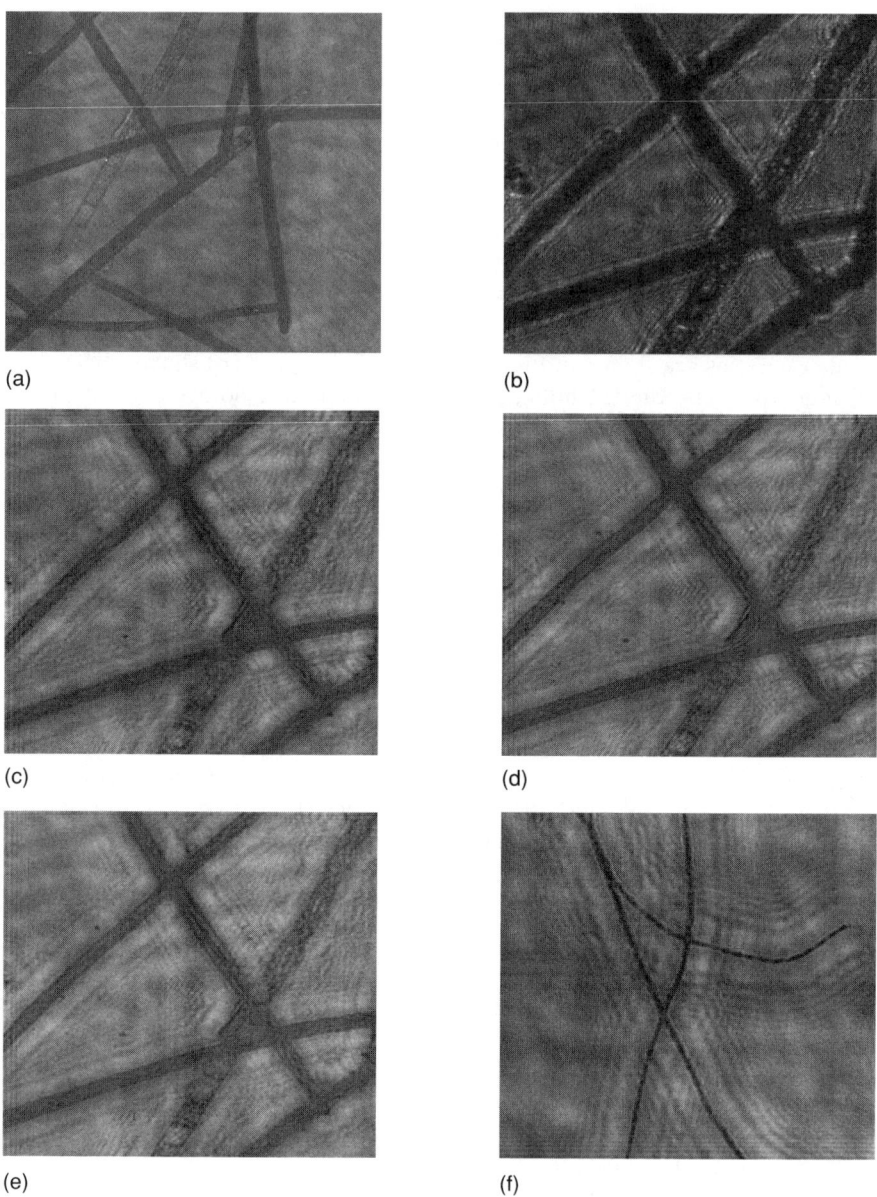

(a)

(b)

(c)

(d)

(e)

(f)

Figure 7.4. Experimental results for biological samples (sphacelaria and *Tribonema aequale*) by use of a 10× microscope objective: (a) sphacelaria's 2D image, (b) sphacelaria's digital hologram by SEOL digital holography, (c) and (d) sphacelaria's reconstructed images by use of SEOL digital holography with only single hologram recording at distance $d = 180$ mm and 190 mm, respectively, (e) sphacelaria's reconstructed image at distance 180 mm using phase-shifting digital holography and (f) tribonema aequale's reconstructed image at distance $d = 180$ mm using SEOL digital holography.

intensity region is occupied by microorganisms and the intensity of microorganisms is less than 45% of the background diffraction field. Figures 7.5(a) and (b) show the reconstructed and segmented image of a sphacelaria sample (A1), respectively. Figures 7.5(c)–(e) show the real parts of Gabor coefficients in Section 7.3.2 when $u = 1$, 2, and 3.

To recognize two filamentous objects that have different thicknesses and distributions, we select two different reference graphs and place them on the sample A1 and B1. The results of the recognition process are followed in the next subsections.

7.5.3 Recognition of Sphacelaria Alga

A rectangular grid is selected as a reference graph for sphacelaria, which shows regular thickness in the reconstructed images. The reference graph is composed of 25×3 nodes and the distance between nodes is 4 pixels in x and y directions. Therefore, the total number of nodes in the graph is 75. The reference graph R is located in the sample A1 with $\mathbf{p}_s = [81, 75]^t$ and as shown in Fig. 7.6(a). To utilize the depth information, four reference images are used. They are reconstructed at $d = 170$, 180, 190, and 200 mm, respectively. The threshold α_Γ and α_C are set at 0.65 and 1, respectively. Thresholds are selected heuristically to produce better results.

Considering the computational load, the graph S is translated by every 3 pixels in x and y directions for measuring its similarity and difference with the graph R. To search the best matching angles, the graph S is rotated by $7.5°$ from 0 to $180°$ at every translated location. When the positions of rotated nodes are not integers, they are replaced with the nearest neighbor nodes.

Figure 7.6(b) shows one sample (A8) of test images with the RGM process. The reference shapes are detected 62 times along the filamentous objects. Figure 7.6(c) shows the number of detections for 16 samples. The detection number for A1–A8 varies from 31 to 251 showing strong similarity between the reference image (A1) and test images (A2–A8) of the same microorganism. There is no detection found in B1–B8. Figure 7.6(d) shows the maximum similarity and the minimum difference cost for all samples.

7.5.4 Recognition of Tribonema aequale Alga

To recognize *Tribonema aequale*, a wider rectangular grid is selected to identify its thin filamentous structure. The reference graph is composed of 20×3 nodes and the distance between nodes is 4 pixels in x direction and 8 pixels in y direction, therefore, the total number of nodes in the graph is 60. The reference graph R is located in the sample B1 with $\mathbf{p}_s = [142, 171]^t$ and $\theta_r = 90°$ as shown in Fig. 7.7(a). Four reference images are used which are reconstructed at $d = 170$, 180, 190, and 200 mm, respectively. The threshold α_Γ and α_C are set at 0.8 and 0.65, respectively.

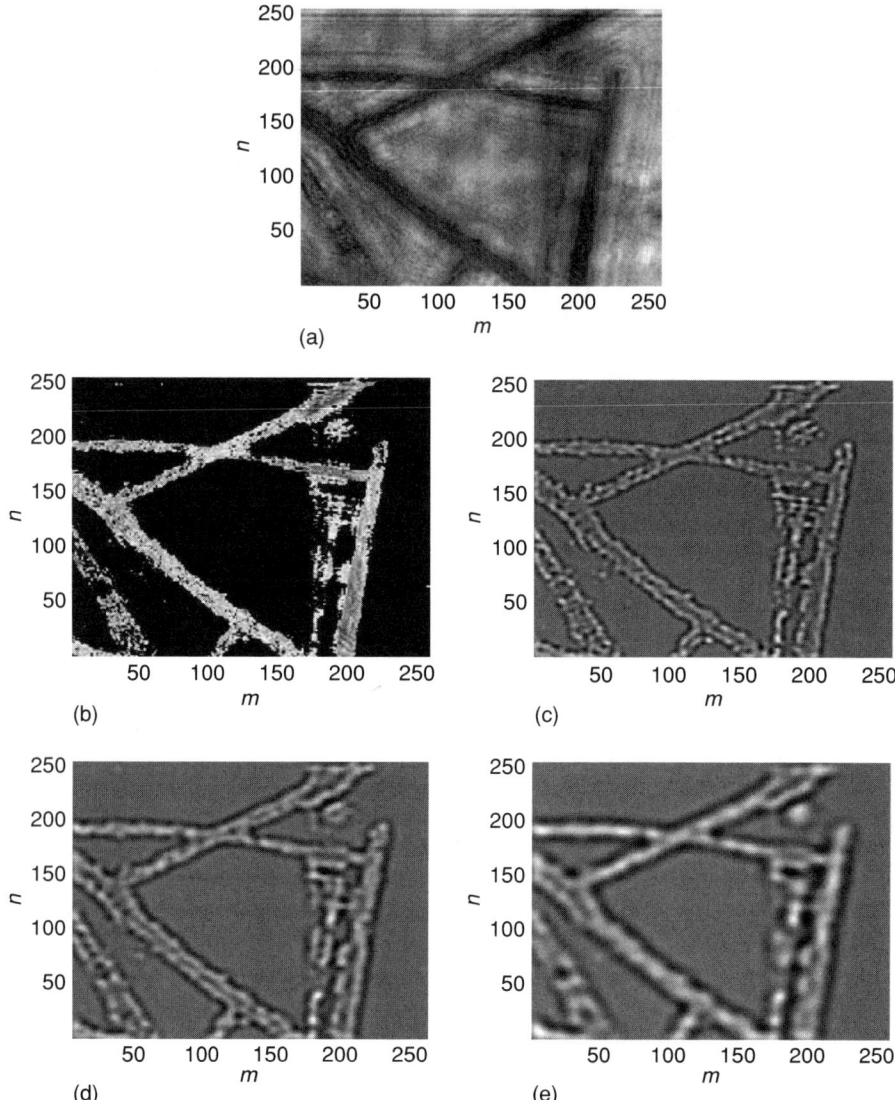

Figure 7.5. Computational reconstruction, segmentation, and feature vector extraction of the sphacelaria sample A1: (a) reconstructed image at $d = 180\,\mathrm{mm}$, (b) segmented image, real parts of Gabor coefficients when (c) $u = 1$, (d) $u = 2$, and (e) $u = 3$.

Figure 7.7(b) shows one sample (B2) of test images with the RGM process. The reference shapes are detected 26 times along the thin filamentous object. Figure 7.7(b) shows the number of detections for 16 samples. The detection number for B1–B8 varies from 5 to 47. One false detection is found in the sample A7. Figure 7.7(d) shows the maximum similarity and the minimum

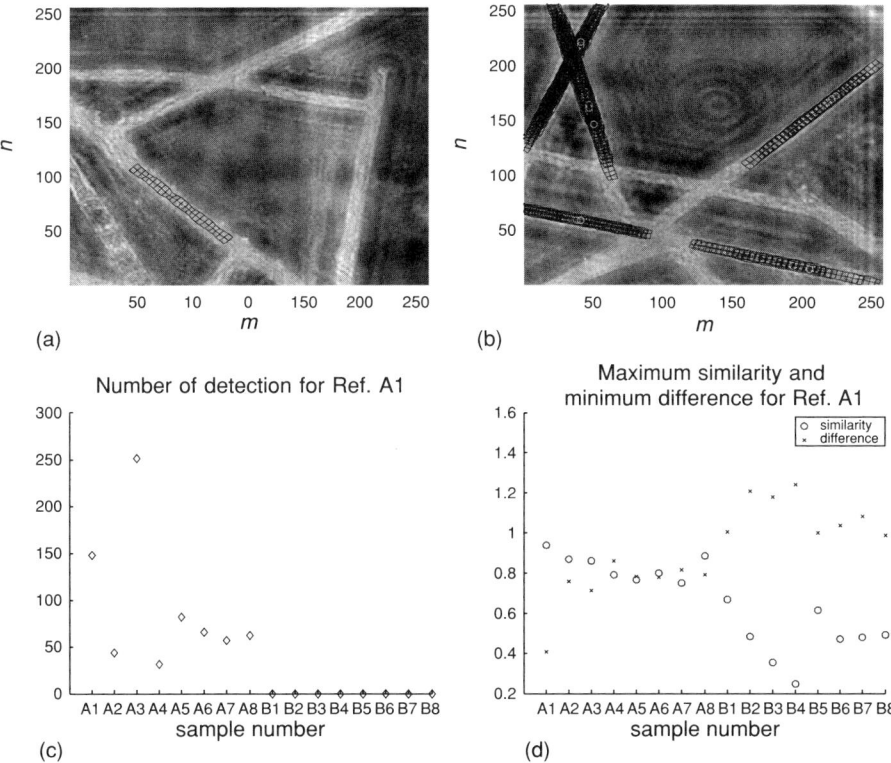

Figure 7.6. Recognition of sphacelaria, (a) reference sample A1 with the graph R, (b) RGM result of one test sample A8, (c) number of detections, (d) maximum similarity and minimum difference cost, (a) and (b) are presented by contrast reversal for better visualization.

difference cost for all samples. As a result, we are able to recognize hologram samples of two different microorganisms by counting the number of detections of each reference shape.

For real-time application, computational complexity should be considered. For numerical reconstruction of the holographic image and Gabor filtering, the computational time of the algorithm is of the same order as the fast Fourier transformation (FFT) which is $O(N) = N \log 2N$, where N is the total number of pixels in the holographic image. For the graph matching, the computational time depends on the shape and size of the graph, the dimension of the feature vector, searching steps for the translation vector, and the rotation angle. Since the largest operation is caused by searching the translation vector, that is $O(N) = N2$, the proposed system requires quadratic computational complexity. Therefore, real-time processing can be achieved by developing parallel processing. Real-time operation is possible because SEOL holography requires a single exposure. Thus, with high-speed electronics, it is possible to have

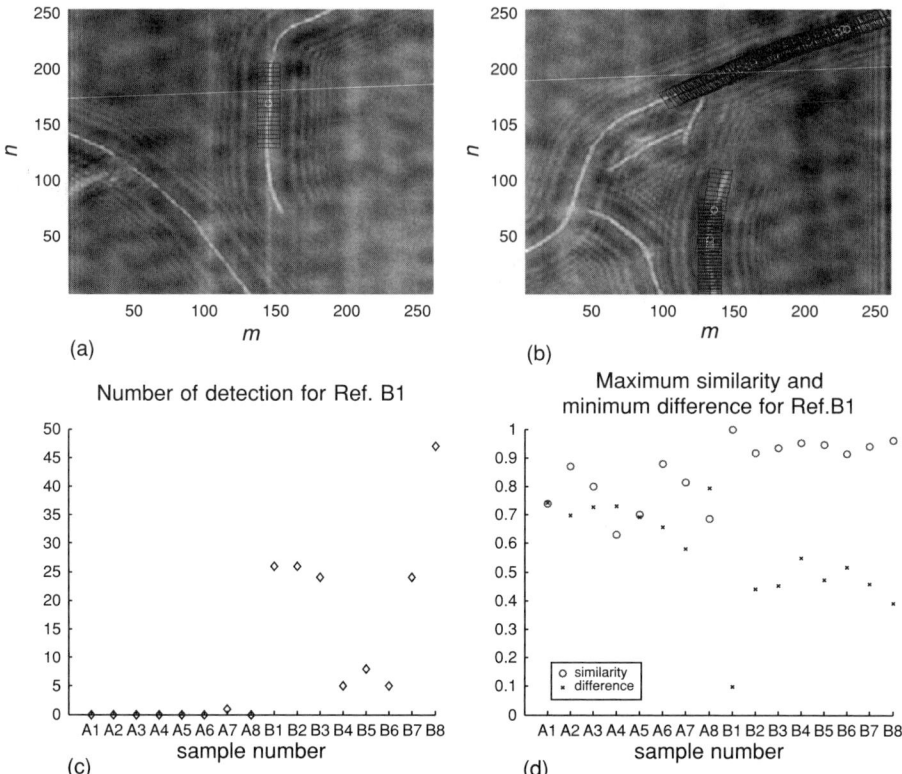

Figure 7.7. Recognition of *Tribonema aequale*, (a) reference sample B1 with the graph R, (b) RGM result of one test sample B2, (c) number of detections, (d) maximum similarity and minimum difference cost; (a) and (b) are presented by contrast reversal for better visualization.

real-time detection. This would not be possible with phase-shift holography, which requires multiple exposures.

7.6 Conclusion

In this chapter, we have presented preliminary results for human aided recognition of microorganisms by examining their simple morphological traits. Three-dimensional visualization and recognition of microbiological objects by SEOL digital holography has been described. 3D imaging and recognition with SEOL digital holography is robust to movement of objects, and to environmental conditions during recording as compared with multiple exposure phase-shifting digital holography. Feature extraction is performed by segmentation and Gabor filtering. They are followed by a feature matching technique to localize specific shape features of two different microorganisms. In this paper, we only detect the reference shapes in the unknown samples. However, they can be used for further training procedures. Indeed, several morphological traits can be combined to recognize different classes of microorganisms more efficiently.

Implementation of a fully automated recognition system of small living organisms presents many challenges due to their spatial and temporal variations. Future work should consider 4D imaging (with consideration of time frames of 3D imaging), which may be a good solution for this task. Also, advanced segmentation techniques using phase distribution can be considered for feature extraction and graph matching. The proposed approach may have great benefits in medicine, environmental monitoring, and defense applications.

References

[1] Mahalanobis A, Muise RR, Stanfill SR, and Nevel AV. (2004). "Design and application of quadratic correlation filters for target detection." *IEEE Trans. AES*, 40:837–850.

[2] Sadjadi FA. (2004). "Infrared target detection with probability density functions of wavelet transform subbands." *Appl. Opt.*, 43:315–323.

[3] Sjoberg H, Goudail F, and Refregier P. (1998). "Optimal algorithms for target location in nonhomogeneous binary images." J. *Opt. Soc. Am. A.*, 15:2976–2985.

[4] Javidi B., ed. (2002). *Image Recognition and Classification: Algorithms, Systems, and Applications*. Marcel Dekker, New York.

[5] Javidi B and Tajahuerce E. (2000). "Three-dimensional object recognition using digital holography." *Opt. Lett.*, 25:610–612.

[6] Matoba O, Naughton TJ, Frauel Y, Bertaux N, and Javidi B. (2002). "Real-time three-dimensional object reconstruction by use of a phase-encoded digital hologram." *Appl. Opt.*, 41:6187–6192.

[7] Sadjadi F. (2002). "Improved target classification using optimum polarimetric SAR signatures." *IEEE Trans. AES.*, 38:38–49.

[8] Javidi B and Okano F., eds. (2002). *Three-dimensional Television Video, and Display Technologies*. Springer, New York.

[9] Goodman JW and Lawrence RW. (1967). "Digital image formation from electronically detected holograms." *Appl. Phy. Lett.*, 11:77–79.

[10] Kreis TM and Juptner WPO. (1997). "Suppression of the dc term in digital holography." *Opt. Eng.*, 36:2357–2360.

[11] Pedrini G and Tiziani HJ. (2002). "Short-coherence digital microscopy by use of a lensless holographic imaging system." *Appl. Opt.*, 41:4489–4496.

[12] Zhang T and Yamaguchi I. (1998). "Three-dimensional microscopy with phase-shifting digital holography." *Opt. Lett.*, 23, 1221.

[13] Stadelmaier A and Massig JH. (2000). "Compensation of lens aberrations in digital holography." *Opt. Lett.*, 25:1630.

[14] Javidi B, Moon I, Yeom S, and Carapezza E. "Three-dimensional imaging and recognition of microorganism using single-exposure on-line (SEOL) digital holography." *Opt. Express*. (to be published).

[15] Lengeler JW, Drews G, and Schlegel HG. (1999). *Biology of the Prokaryotes.* Blackwell Science, New York.

[16] Forero MG, Sroubek F, and Cristobal G. (2004) "Identification of tuberculosis bacteria based on shape and color." *Real-time imaging,* 10:251–262.

[17] Alvarez-Borrego J, Mourino-Perez RR, Cristobal-Perez G, and Pech-Pacheco JL. (2002). "Invariant recognition of polychromatic images of Vibrio cholerae 01." *Opt. Eng.,* 41:872–833.

[18] Amaral AL, da Motta M, Pons MN, Vivier H, Roche N, Moda M, and Ferreira EC. (2004). "Survey of protozoa and metazoa populations in wastewater treatment plants by image analysis and discriminant analysis." *Environmentrics,* 15:381–390.

[19] Treskatis S-K, Orgeldinger V, wolf H, and Gilles ED. (1997). "Morphological characterization of filamentous microorganisms in submerged cultures by on-line digital image analysis and pattern recognition." *Biotechnol. Bioeng.,* 53:191–201.

[20] Luo T, Kramer K, Goldgof DB, Hall LO, Samson S, Remsen A, and Hopkins T. (2004). "Recognizing plankton images from the shadow image particle profiling evaluation recorder." *IEEE Trans. Syst. man, and Cybern., Part B,* 34:1753–1762.

[21] Cabral JMS, Mota M, and Tramper J., eds. (2001). *Multiphase Bioreactor Design,* Chap. 2. "Image analysis and multiphase bioreactor." Taylor & Francis, London.

[22] Javidi B and Kim D. (2005). "Three-dimensional object recognition by use of single exposure on-axis digital holography." *Opt. Lett.,* 30:236–238.

[23] Kim D and Javidi B. (2005), "Distortion-tolerant three-dimensional object recognition by using single exposure on-axis digital holography." *Opt. Expr.,* 12:5539–5548. http://www.opticsinfobase.org/abstract.cfm?id=81653

[24] Daugman JG. (1985). "Uncertainty relation for resolution in space, spatial frequency, and orientation optimized by two-dimensional visual cortical filters." *J. Opt. Soc. Am.,* 2:1160–1169.

[25] Lee TS. (1996). "Image representation using two-dimensional Gabor wavelets." *IEEE Trans. PAMI,* 18:959–971.

[26] Lades M, Vorbruggen JC, Buhmann J, Lange J, Malsburg C v.d., Wurtz RP, and Konen W. (1993). "Distortion invariant object recognition in the dynamic link architecture." *IEEE Trans. Comput.,* 42:300–311.

[27] Yeom S and Javidi B. (2004). "Three-dimensional object feature extraction and classification with computational holographic imaging." *Appl. Opt.,* 43:442–451.

Chapter 8

Integral Imaging Applied to the Digital Reconstruction and Recognition of 3D Scenes

Yann Frauel,[1] Osamu Matoba,[2] Enrique Tajahuerce,[3] and Bahram Javidi[4]

[1]IIMAS, Universidad Nacional Autónoma de México, Mexico
[2]Department of Computer and System Engineering, Kobe University, Japan
[3]Departament de Ciencies Experimentals, Universitat Jaume I, Spain
[4]Department of Electrical and Computer Engineering, University of Connecticut, USA

8.0 Introduction

The technique of integral imaging consists in simultaneously capturing in a single image multiple views of a three-dimensional (3D) scene. The multiple views are provided by a microlens array: each microlens generates an elemental image of the scene obtained from a particular point of view depending on its location with respect to the scene. These views permit the use of the principle of triangulation to find information about the depth of the scene. The distance of a point is obtained by measuring the displacement of the point in the various elemental images.

This technique of integral imaging is derived from the integral photography scheme proposed by G. Lippmann as early as 1908 to capture and render 3D images.[1] This technique has recently regained popularity because of its capability to provide 3D auto-stereoscopic displays[2–12] and also thanks to improvements in microlens arrays manufacturing techniques. The original and most common way of reconstructing the 3D scene is by use of an optical setup similar to the one used for the acquisition of the integral image.[1–3] However, integral images can also be recorded by a high-definition digital camera. In that case, the 3D information can be retrieved by an appropriate digital processing.[13–16] A digital 3D model of the scene can then be reconstructed in a computer using this retrieved 3D information. Once the digital reconstruction is done, the 3D model can easily be visualized and processed as a digital object.

An extremely useful application of the 3D information obtained from integral images is 3D object recognition. The recognition problem has long been focused on two-dimensional (2D) techniques.[17–22] Only recently have technological advances permitted to take into account the three dimensions of the objects. Several techniques have been investigated such as holography,[23–25] moving camera,[26–27] integral imaging,[15,16,25] range images[28] or tomography.[29] The interest of integral imaging is not to need a laser or other type of special illumination. Neither does it require any movement of the object or of the sensor since an integral image captures several 2D perspectives of a 3D scene in a single step. Since this collection of perspectives contains information about the scene depth, it is possible to realize a simple 3D recognition by computing usual 2D correlations between two integral images. This technique has the advantage of offering the possibility of an all-optical implementation.[30] However, it only provides a global recognition and cannot deal with precise longitudinal segmentation of the 3D objects. Another technique consists in using two steps: first a digital reconstruction of the 3D scene from an integral image, and second a digital 3D correlation between reconstructed scenes.[15]

The first section of this contribution explains the basic principles of integral imaging and the properties of an integral image. The second section describes the extraction of the 3D information from an integral image and the reconstruction of a digital 3D model of the scene. The third section presents the two previously mentioned techniques of recognition from integral images and discusses some properties of the resulting 3D recognitions.

8.1 Principle of Integral Imaging

An integral image is a collection of views of a 3D scene obtained through a microlens array (Figure 8.1). In our experiments, we used a hexagonal microlens array with microlenses of diameter $\varphi = 200\,\mu$m and focal length around 2.3 mm. Each of these microlenses generates an elemental image of the scene taken from a different point of view. We make the following assumptions:

(1) The depth of focus of the microlenses is sufficient to assume that the images of all the objects are obtained in the same plane P, independently of their longitudinal position in the 3D scene.
(2) The elemental images generated by neighbor microlenses do not overlap each other.

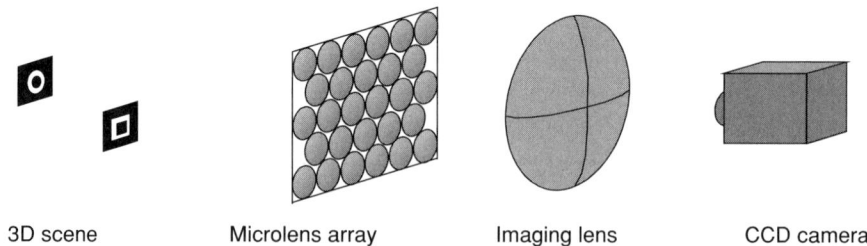

3D scene   Microlens array   Imaging lens   CCD camera

Figure 8.1. Acquisition of an integral image.

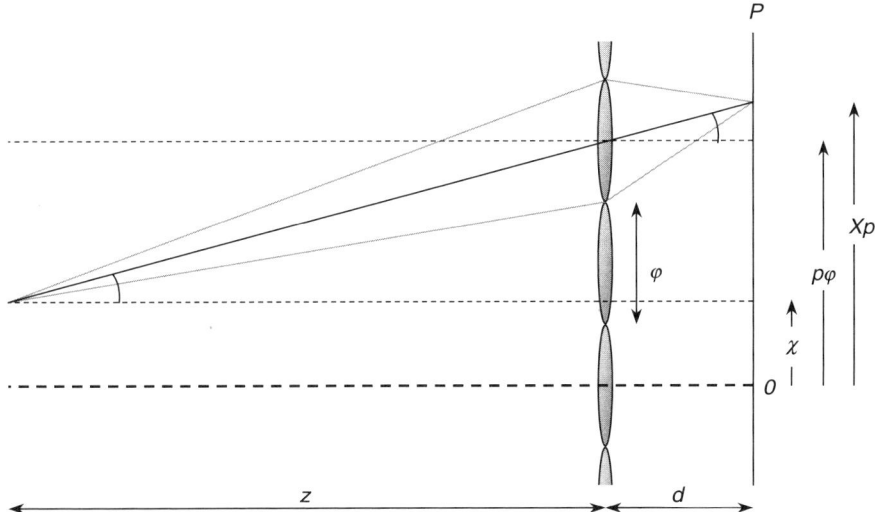

Figure 8.2. Formation of an elemental image by a microlens – Naming convention.

These conditions can always be obtained by placing the objects sufficiently far away from the microlenses. In this case, all the images are obtained in a plane P at distance $d \approx f$ from the array (Figure **8.2**). This plane is then imaged by an additional lens onto a high-definition CCD camera (Figure 8.1). Except for a magnification coefficient – and some amount of aberrations that we will neglect – the image obtained at the camera is identical to the one obtained in the imaging plane P of the microlens array. We will therefore neglect this last imaging step and we will conduct all the calculations in plane P.

As can be seen in Figure **8.2**, the coordinate X_p of an object point projected in plane P by the microlens number p depends on the original lateral coordinate x of the object point as well as on its depth z, according to the following formula:

$$\frac{p\varphi - x}{z} = \frac{X_p - p\varphi}{d}, \tag{1}$$

which yields

$$X_p = p\varphi\left(1 + \frac{d}{z}\right) - \frac{d}{z}x. \tag{2}$$

A similar formula can be obtained with the coordinate y. This relation is illustrated in Figure **8.3** for $X_0(p = 0)$. The distance between the projections of a same object point given by two microlenses p and q is:

$$X_q - X_p = (q - p)\varphi\left(1 + \frac{d}{z}\right). \tag{3}$$

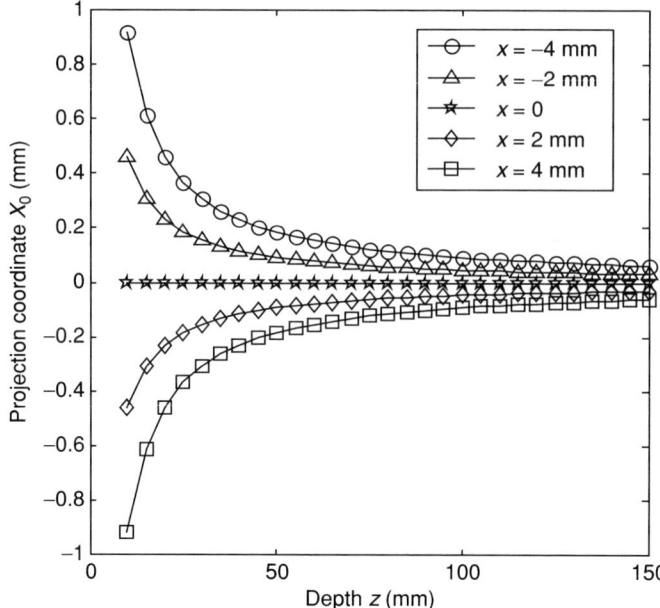

Figure 8.3. Displacement of the projected object point versus distance of the object point for the central microlens.

Thus, the depth of a given object point can be recovered by comparing the projections through different microlenses. This is the principle of triangulation. The first step is therefore to acquire an image of plane P, which contains several elemental images provided by every microlens. In order to improve the quality of this image, we digitally enhance its contrast. Moreover, as we do not know the magnification ratio between plane P and the camera, we also need to calibrate our images. We do this calibration by illuminating the microlens array with a uniform plane wave produced by a He–Ne laser. We thus obtain an image with focused spots that provides the locations – in pixels – of the centers of the microlenses.

8.2 Digital Reconstruction of the Three-dimensional Scene

In this section, we describe the extraction of the depth information from the collection of 2D elemental images that compose the integral image. We then use this information to generate a reconstructed 3D model of the scene by computer.

8.2.1 Retrieval of the 3D Scene Depth

It can be seen on Eq. (3) that two elemental images would be in principle sufficient to determine the depth z of every point. That is what stereoscopical techniques attempt to do. However, the main problem is to find the correspondence between the points of different elemental images. Namely, in order to measure the projected coordinates X_p, we need to know which point in each elemental image corresponds to the original object point. The correspondences of the projected points can be determined by comparing several elemental images. Compared to mere stereoscopic pairs, our integral images have the advantage of providing not only two but several perspectives of the 3D object. The feature matching between only two elemental images can be ambiguous. On the contrary, the redundancy provided by considering several elemental images helps to determine which points of the different views actually correspond to each other. According to Eq. (3), one particular feature has to appear at regular locations among the various elemental images. A distinct object feature, even if it accidentally appears similar to the first elemental image, will not appear in consistent locations [in terms of Eq. (3)] in all the elemental images. Thus, the use of several views permits a more accurate identification of the features, and therefore, a better determination of the depth. Moreover some elemental images may present noise due to defective pixels of the detector or noise in the optical system. In this case, using many images provides redundancy that lowers the effect of noise. On the other hand, the camera has a fixed number of pixels and increasing the number of elemental images obviously reduces the resolution of each of them, hence reducing the lateral resolution of the reconstructed 3D object. Moreover increasing the number of elemental images increases the computational load for determining the depth of each point. In our experiment, the size and number of elemental images were dictated by the available optical components. An example of our integral images is shown in Figure **8.4**. With our particular configuration, we found out heuristically that using only 7×7 of the elemental images (marked in Figure **8.4**) was a good trade-off between computation time and accuracy of the depth estimation. The optimum number of images may vary with the configuration and components used.

In order to find the depth of the object points, we use a stereo matching algorithm that finds the correspondences between points of the various elemental images.[13,15] Let us consider one particular point of the central elemental image – the one corresponding to microlens (0,0). If we fix its depth z arbitrarily, we can determine the corresponding points in the other elemental images according to Eq. (3). We now need to verify that these points are actual projections of the same object point. This cannot be done by comparing only one point to another. We need to compare the surroundings of each of these points. In order to do that, we compute the normalized 2D cross-correlations between pairs of windows centered on the tested points. The size of each window is 9×9 pixels. If I denotes the integral image, the projection of the inspected object point corresponding to the microlens (p,q) is $I(X_p, Y_q)$. The normalized cross-correlation between the window contained in the elemental image (p,q) and the one contained in the elemental image (p', q') is:

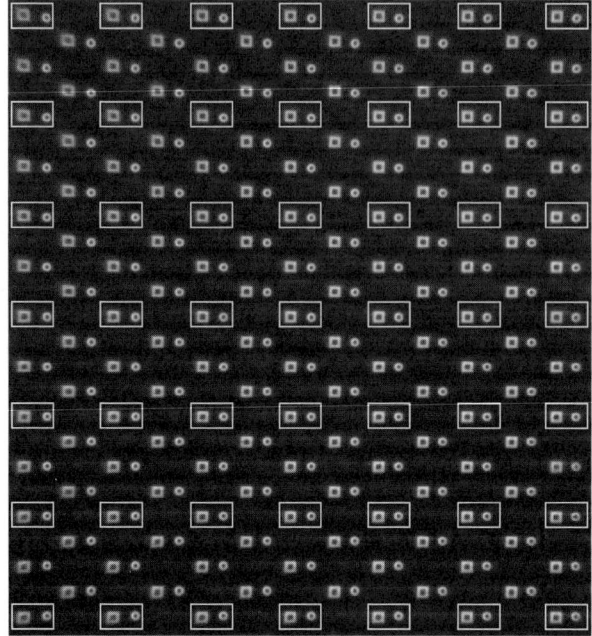

Figure 8.4. Example of integral image of a 3D scene. The elemental images marked are the ones used to determine the scene depth.

$$C[(p, q), (p', q')] = \frac{\sum_{m=-4}^{4} \sum_{n=-4}^{4} I(X_p + m, Y_q + n) I(X_{p'} + m, Y_{q'} + n)}{\left[\sum_{m=-4}^{4} \sum_{n=-4}^{4} I^2(X_p + m, Y_q + n) \sum_{m=-4}^{4} \sum_{n=-4}^{4} I^2(X_{p'} + m, Y_{q'} + n) \right]^{1/2}} \cdot$$

$$(4)$$

This similarity criterion has the advantage of being independent of the intensity variations that can occur between two elemental images. We compare each window with all of its immediate neighbors (horizontally and vertically) and we add together all the correlation values. This gives us a matching criterion that we need to maximize:

$$M(z) = \sum_{p=-3}^{3} \sum_{q=-2}^{3} C[(p, q-1), (p,q)] + \sum_{q=-3}^{3} \sum_{p=-2}^{3} C[(p-1, q), (p, q)]. \quad (5)$$

We compute the value of this criterion for a range of assumed depths z. The depth which yields the highest value for $M(z)$ is the actual depth of the point under consideration. This procedure is repeated for every point of the central elementary image in order to obtain the depth of every point in the 3D scene. Due to the principle of triangulation, for a given setup the longitudinal resolution is lower for distant objects than for closer ones (see Figure **8.3**). In addition, longitudinal resolution of the depth estimation is limited by the resolution of the integral image.[16] In our experiments, the accuracy of the z measurement is about 10 mm. In other words, we are able to reconstruct the object volume with a voxel depth no smaller than 10 mm. This value of 10 mm is due to our particular components and distances of the objects.

8.2.2 Correction of the Depth-dependent Magnification Ratio

In the previous subsection, we have described how to compute the z coordinate of each point of the central elemental image. However, for each of these points, we only know the projected coordinates X_0 and Y_0 and we need to find their actual coordinates x and y in the object space in order to reconstruct a model of the 3D scene. Knowing their depth z, this can be done using Eq. (2), which yields:

$$x = -\frac{z}{d}X_0 \text{ and } y = -\frac{z}{d}Y_0. \tag{6}$$

These equations compensate for the well-known fact that distant objects look smaller than close ones. By using Eq. (6), we can reconstruct their size independently of their distances from the microlens array. At this stage, we obtain a 3D reconstruction of the object space and this 3D model can be used to perform 3D image processing such as correlations.

8.2.3 Example of 3D Reconstruction

In the experiments, we use three planar objects representing three different geometrical shapes, namely a square, a circle, and a triangle. These shapes are about 2 mm large and are located between 90 and 120 mm from the microlens array. The three objects are shown in Figure **8.5**(a)–(c). These images are the central elemental images generated by the microlens array. We create two 3D scenes by placing the square and the circle at various distances from the array. These scenes can be seen in Figure **8.6**(a)–(b). We call Scene 1 the scene in Figure **8.6**(a) and Scene 2 the scene in Figure **8.6**(b). The elemental images

(a)   (b)   (c)

Figure 8.5. Isolated planar objects used in the experiments. These views are the central elemental image of the microlens array.

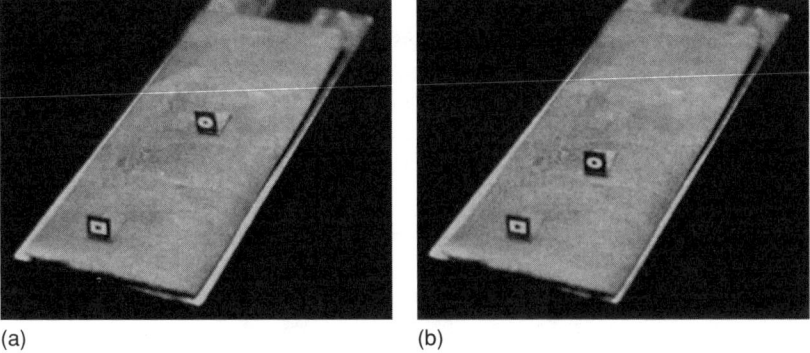

(a)   (b)

Figure 8.6. Three-dimensional input scenes with the planar objects at various distances from the detector – (a) Scene 1 – (b) Scene 2.

obtained with the central microlens of the array are given in Figure **8.7**. Although they look similar, the size of the circle is slightly smaller in Scene 1 [Figure **8.7**(a)] than in Scene 2 [Figure **8.7**(b)] because in Scene 1 the circle is located farther away from the square. Figure **8.8** provides a map of the distances obtained by the matching algorithm for Scene 1 (see Section 8.2.1). The brighter points correspond to smaller distances of z.

Figure **8.9** illustrates the 3D reconstructions of Scene 1 and Scene 2. The contrast has been inverted for a better visualization. It can be noticed that, due to the use of Eq. (6), the circle is now the same size as the square, which was not the case in Figure **8.7**. As mentioned previously, the absolute depth of the objects is estimated with an accuracy close to 10 mm. The errors can be due to a wrong estimation value of the distance d between the microlens array and the projection plane P, or to a poor estimation of the location of the centers of the microlenses.

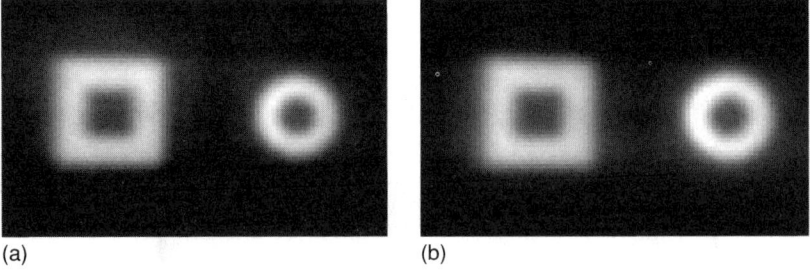

(a)   (b)

Figure 8.7. Elemental images of the 3D input scenes obtained with the central microlens of the array – (a) Scene 1 – (b) Scene 2.

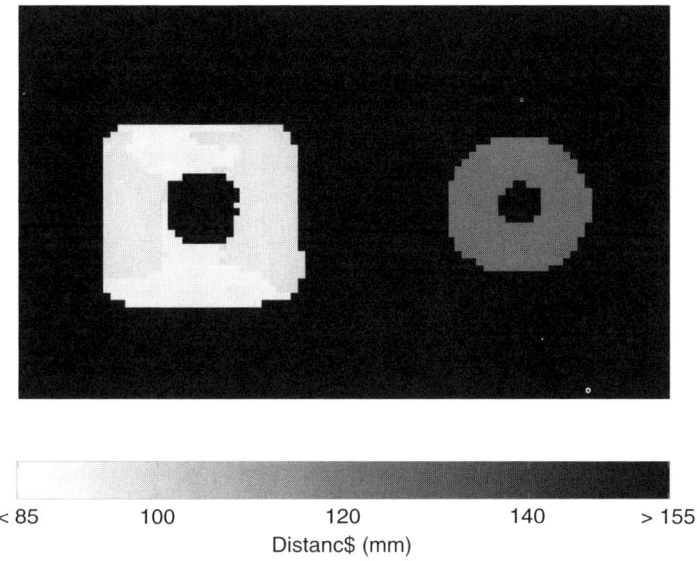

Figure 8.8. Depth map of Scene 1 estimated by the stereo matching algorithm.

8.2.4 Digital Visualization of the 3D Scene

The reconstructed 3D model contains the three coordinates of every visible point of the scene. This data can be used to compute artificial views from any standpoint, simply by computing projections onto an observation plane. Figure **8.10** illustrates the possibilities of visualization by presenting three artificial

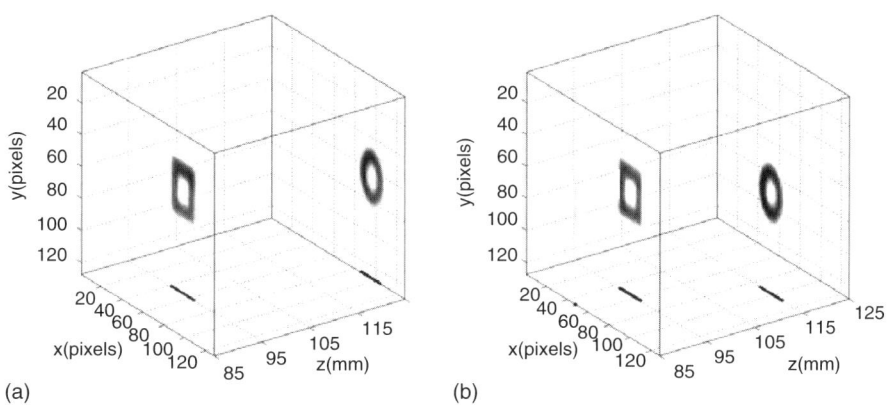

Figure 8.9. Three-dimensional models of the reconstructed scenes – (a) Scene 1 – (b) Scene 2.

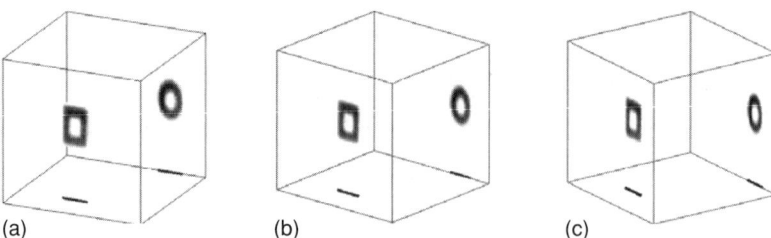

Figure 8.10. Three artificial views of the 3D scene computed from the reconstructed 3D model. (a) 25° azimuth – (b) 40° Azimuth – (c) 55° Azimuth.

views calculated from different standpoints. Obviously a series of such views could be combined to generate a movie that would help a viewer to visualize and understand the original scene.

8.3 Three-dimensional Object Recognition

Integral images are not only useful for 3D visualization. They can also be used in 3D object recognition problems. The recognition process can be performed either directly on the raw integral image or on the reconstructed 3D scenes obtained as described in the previous section. In this section we present both techniques and we discuss their properties.

8.3.1 Direct Correlation of Integral Images

Since the 3D information is contained in the integral image in the form of multiple perspectives of the 3D scene, it is possible – to a certain extent – to compare two 3D objects by computing the 2D correlation between their respective integral images. The advantage of this technique is that it only requires a classical 2D correlation and can therefore be optically performed in real-time.[30] Figure **8.11** presents parts of two integral images of a die with two different orientations. Figure **8.11**(a) is the reference object and Figure **8.11**(b) is the tested object. Note that one face is kept in common and only the other face differs. This choice of objects is intended to prove the discrimination of the system. We use a nonlinear correlation where the joint power spectrum is binarized.

Figure **8.12**(a) shows the autocorrelation of the integral image of the reference object and Figure **8.12**(b) shows the cross-correlation between both integral images. Both correlations are normalized with the value of the auto-correlation peak. The cross-correlation peak is 14 times smaller than the autocorrelation peak, which denotes a successful discrimination of the objects. The correlation works by comparing every elemental image of the reference object to every elemental image of the input object. If the input object is similar

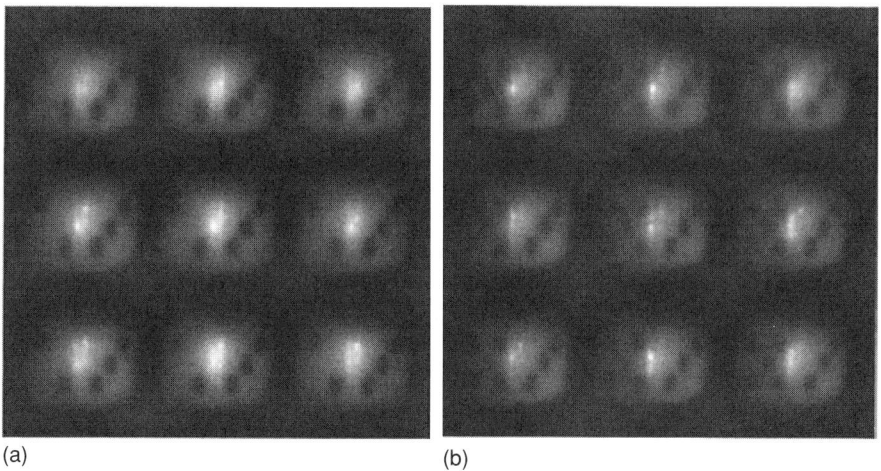

(a)   (b)

Figure 8.11. Parts of integral images of a die with different orientations – (a) is the reference object – (b) is the input object.

to the reference object, all the correlation peaks, obtained for every pair of identical elemental images, superimpose to form the global recognition peak.

The effect of a small out-of-plane rotation of the object is to shift laterally the elemental images. The angle of view previously obtained by one microlens is now obtained by another microlens. Therefore, the global correlation will find a shifted correlation peak. The position of this correlation peak therefore gives information about the rotation angle. The measurable angle depends on both the size of the microlens array and the distance of the object. It is usually small (of the order of 1°). Note that laterally shifting the object also results in shifting the elemental images and therefore the correlation peak. A shift of the correlation peak can thus mean either a lateral shift or a rotation of the object. Another problem of this technique is that it cannot deal with a longitudinal shift of the object, which results in a radial shift of the elemental images.

8.3.2 Correlation of the Reconstructed 3D Scenes

Another possible 3D recognition technique, that is fully shift-invariant in every direction, is to base the recognition on the reconstructed 3D scenes obtained as described in Section 8.2. In this case, we can perform a true 3D correlation between the reconstructed reference scene and a reconstructed input scene.[15]

8.3.2.1 Principle

If $A(x,y,z)$ and $B(x,y,z)$ are two 3D scenes, we define their similarity as the square modulus of their 3D correlation. We compute it through the Fourier domain:

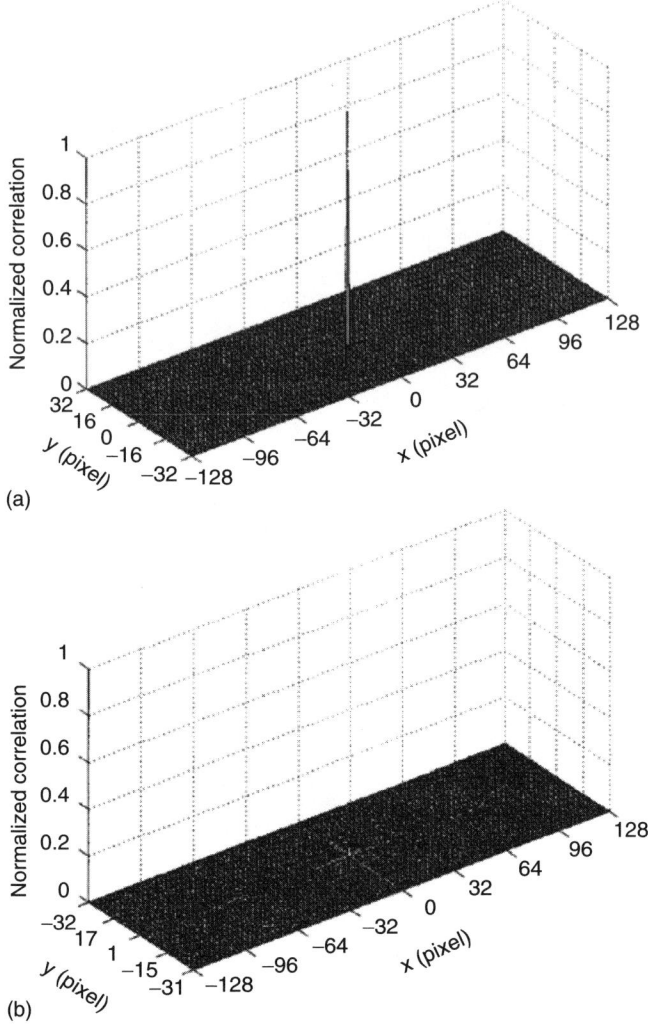

Figure 8.12. Direct correlation of the integral images – (a) Autocorrelation of the reference object – (b) Cross-correlation between the reference and the input objects.

$$S_{AB} = |A \otimes B|^2 = \left| FT^{-1}\{\tilde{A}.\tilde{B}^*\} \right|^2, \tag{7}$$

where the symbol $\otimes$ stands for the 3D correlation, $\tilde{A}$ and $\tilde{B}$ are the Fourier transforms of A and B respectively and FT^{-1} is the inverse Fourier transform. Moreover, in order to improve the recognition performance, we can use the kth-law nonlinear correlation,[21] which provides us with the following degree of similarity:

$$S_{AB}^k = |A \otimes_k B|^2 = \left| = FT^{-1}\left\{ |\tilde{A}|^k \exp(i\varphi_{\tilde{A}}).|\tilde{B}|^k \exp(-i\varphi_{\tilde{B}}) \right\} \right|^2, \tag{8}$$

where $|\tilde{A}|$ and $|\tilde{B}|$ are the modulus of $\tilde{A}$ and $\tilde{B}$, respectively, and $\varphi_{\tilde{A}}$ and $\varphi_{\tilde{B}}$ are their arguments. The value of the nonlinear factor k is usually chosen between 0 and 1. The linear similarity described in Eq. (7) is obtained for $k = 1$. Using a strong nonlinearity – which means k close to 0 – improves the discrimination between similar objects. However, in this case, the recognition also becomes more sensitive to distortions of the objects. A balance has therefore to be found by adjusting the parameter k. In the following, we will use the term "correlation" to designate the similarity criteria defined in Eqs. (7) and (8).

8.3.2.2 Recognition of a 3D Object Using Nonlinear Correlation

In order to study the recognition and discrimination capability of the proposed system, we first consider the two composite scenes of Section 8.2.3 (Scene 1 and Scene 2) as the 3D inputs to be tested. The three single geometrical objects (square, triangle, and circle) are used as the 3D reference objects. The 3D correlations between each input scene and each reference object is computed, which provides $2 \times 3 = 6$ correlation volumes. For each of these correlations, we obtain two peaks that correspond to the two objects present in both input scenes. Thus $2 \times 6 = 12$ peaks are generated of which only four are considered as detection peaks: the ones corresponding to the square in both scenes when using the square as a reference, and the ones corresponding to the circle in both scenes when using the circle as a reference. All the other peaks are undesirable cross-correlation peaks.

We first illustrate the effect of kth-law nonlinear correlation[21] on the values of the correlation peaks as a function of the nonlinear factor k. We determine the relative values of the different peaks for each particular value of k. A different normalization factor is applied for every k, so that one of the four detection peaks is always unity. Figure **8.13** illustrates the normalized peak values versus k. It is clear from this graph that it is possible to separate detection peaks from undesirable peaks if $k \leq 0.5$. A linear correlation is thus excluded. It can be seen that the best discrimination is obtained for $k = 0.2$. For this value of the nonlinear factor, it is easy to find a threshold that will allow us to discriminate between detection peaks and undesirable peaks.

In order to confirm these results, we consider the correlations of Scene 1 or Scene 2 with a reference object (square or circle) that is present in the scene. For each of these correlations, we obtain two peaks: a detection peak P_D and an undesirable peak P_U. We define the discrimination ratio as DR $= P_D/P_U$. The correct recognition can only take place if DR > 1. The plot of this discrimination ratio versus k is shown in Figure **8.14** for each of the correlations. It is evident that nonlinear correlation is required to properly detect the 3D objects.

8.3.2.3 Three-dimensional Object Localization

In the experiment described in the previous subsection, a correlation peak was obtained for each object in the 3D input scene. This peak is obviously three-dimensional and indicates the 3D location of the object in the input scene

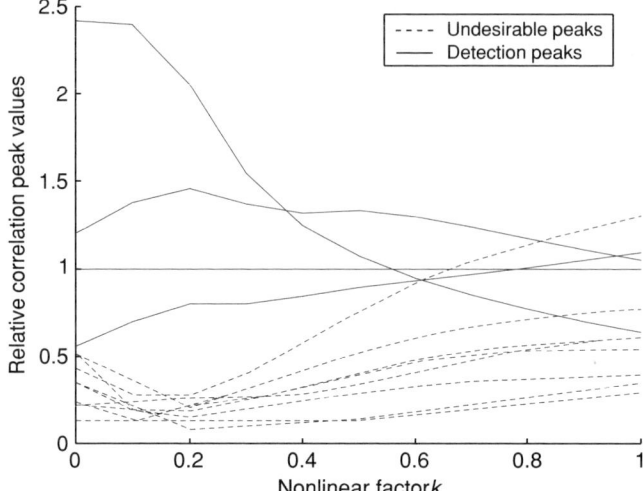

Figure 8.13. Normalized values of the correlation peaks versus kth-law nonlinearity. The detection peaks are the ones corresponding to the presented reference object. The other peaks are undesirable cross-correlations (false alarms).

relative to the location of the object in the reference scene. For instance, Figure **8.15** presents the maximum correlation values at every depth when correlating Scene 1 with the three reference objects. A nonlinear correlation with $k = 0.2$ is used. As mentioned in the previous subsection, the detection peaks are selected

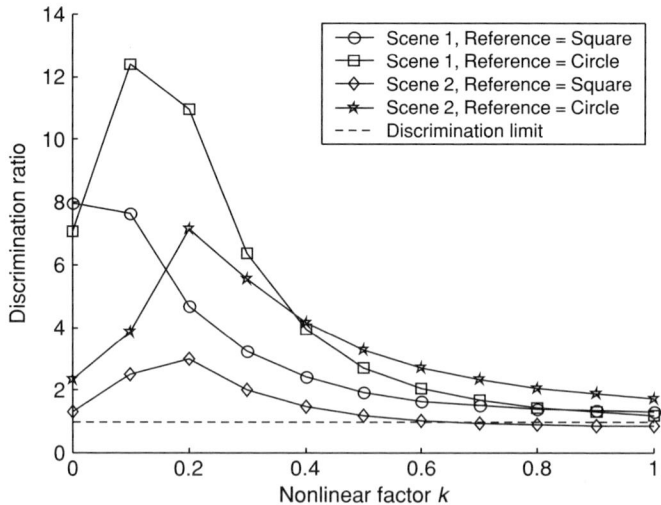

Figure 8.14. Discrimination of 3D correlation versus kth-law nonlinearity.

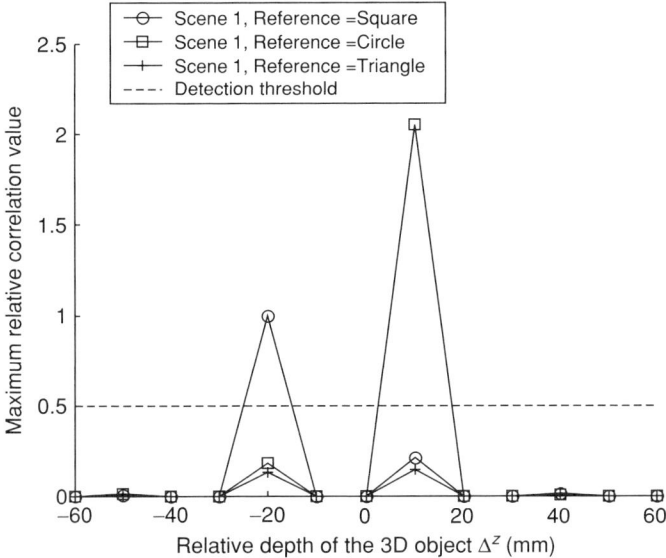

Figure 8.15. Longitudinal locations of the correlation peaks.

by applying a threshold to the output at 0.5. The relative locations Δz of the remaining peaks indicate the longitudinal depths of the corresponding objects.

Figure **8.16** illustrates the correlation planes with fixed z where the maximum peaks for the square reference are generated. This graph demonstrates that the relative (x, y, z) locations of the objects can be found by the positions of the peaks. In this example, the peak in Figure **8.16**(b) would not be taken into consideration because it is below the threshold.

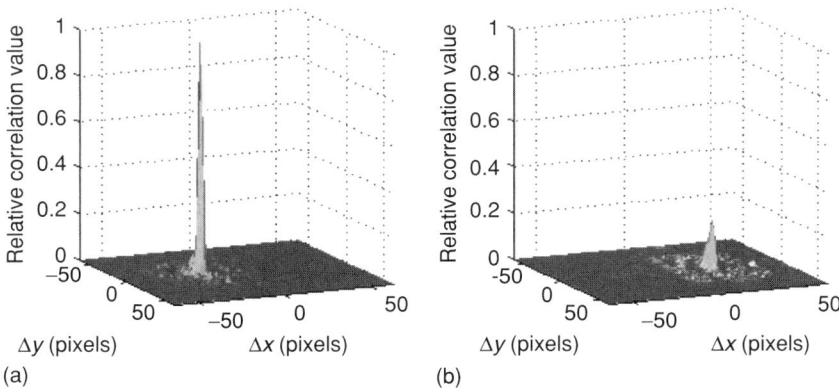

Figure 8.16. Two correlation planes extracted from the 3D correlation between Scene 1 and the square reference object – (a) Correlation plane corresponding to $\Delta z = -20$ mm – (b) Correlation plane corresponding to $\Delta z = +10$ mm.

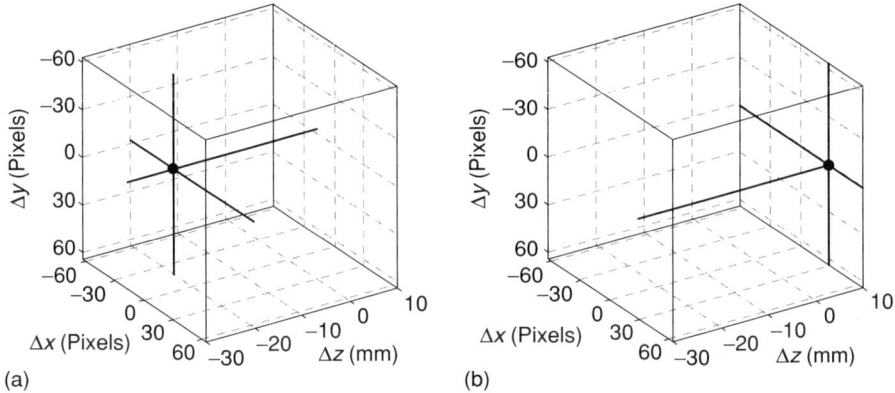

Figure 8.17. Three-dimensional representation of the correlation volumes – (a) Scene 1, square reference – (b) Scene 1, circle reference.

Finally, Figure **8.17** presents the volume representation of the 3D correlation between Scene 1 and both the square reference object [Figure **8.17**(a)] and the circle reference object [Figure **8.17**(b)]. The detection peaks are plotted in 3D. They provide the relative 3D coordinates of the reference objects in the input scene.

8.3.2.4 Comparison Between 2D and 3D Correlation

In the previous subsections, we achieved 3D detection of elemental objects (a square and a circle) in complex 3D scenes (Scene 1 and Scene 2). In this subsection, we compare two complex scenes one to another. Scene 1 and Scene 2 have different 3D structures. They therefore constitute two different 3D objects. The correlation of these two 3D objects is compared by using conventional correlation of 2D images and by the proposed 3D correlation. The 2D correlation is obtained between the images shown in Figure **8.7**(a) and (b), which are the views obtained through the central microlens of the array. The second comparison method consists in digitally reconstructing the 3D scenes and computing their 3D correlation as described previously.

The values C_{1-2} of the cross-correlation peaks for both of these methods is given in Table 8.1. In both cases (2D or 3D), we use a nonlinear correlation with $k = 0.2$. Of course, the correlation values by themselves cannot be compared because they are not normalized. In order to give a comparison scale, we also provide the values of the auto-correlations for Scene 1 (C_{1-1}) and Scene 2 (C_{2-2}) for both the 2D and the 3D methods. Lastly, analogous to what we did in Subsection 8.3.2.2, we define the discrimination ratio as the ratio between the value of the auto-correlation and the value of the cross correlation. Table 8.1 presents the values of this ratio with respect to both scenes. It can be seen that the 3D correlation is roughly three times more discriminant than the 2D

Table 8.1. Comparison between 2D and 3D correlations for discriminating between two 3D objects.

	C_{1-2}	C_{1-1}	C_{2-2}	C_{1-1}/C_{1-2}	C_{2-2}/C_{1-2}
2D	0.157	1.29	1.33	8.2	8.5
3D	1.74×10^9	36.0×10^9	47.5×10^9	21	27

correlation. This is because it takes into account some additional information concerning the depth structure of the 3D objects.

8.4 Conclusion

In this contribution we demonstrated the application of integral imaging to the digital reconstruction and recognition of 3D objects. As we recalled, an integral image is a collection of perspectives of a 3D scene obtained from different points of view. It therefore contains 3D information about the scene in the form of changes in parallax. We explained how to extract the depth information using a triangulation technique. The correspondence between points in various elemental images was found thanks to a stereo-matching algorithm. We then used the retrieved depth information to digitally reconstruct a 3D model of the original scene in a computer. This 3D model can be used to reconstruct synthetic views of the scene and visualize it from new points of view.

Subsequently, we described the use of integral images for 3D object recognition. The first technique we demonstrated is the direct 2D correlation of integral images. This method permits a fast 3D recognition but it is not invariant to longitudinal shifts of the objects. The alternate technique we proposed is to perform numerical 3D correlations on the reconstructed 3D models of the scenes. This technique is fully shift-invariant, that is, it is not affected by a lateral or longitudinal shift of the 3D objects. We presented nonlinear correlation results using various nonlinearities to investigate the discrimination capability. It was demonstrated that the proposed technique might be used to recognize and locate 3D objects in a 3D scene. Although the depth resolution is presently low (about 10 mm for our experiments), it could be increased by improving the resolution of the integral images. Finally, we showed that 3D correlation provides a better discrimination than 2D correlation since it uses the object depth information.

Acknowledgments

The authors are grateful to Dr. MR Taghizadeh at Heriot-Watt University in Edinburgh for providing the microlens array.

References

[1] Lippmann G. (1908). "La photographie intégrale." *Comptes-rendus de l'Académie des Sciences,* 146:446–451.

[2] Okoshi T. (1971). *Three-dimensional Imaging Techniques.* Academic, New York.

[3] Starks M. (1995). "Stereoscopic imaging technology: A review of patents and the literature." *Int. J. Virtual Reality,* 1:2–4.

[4] Okano F, Arai J, Hoshino H, and Yuyama I. (1999). "Three-dimensional video system based on integral photography." *Opt Eng,* 38:1072–1077.

[5] Nakajima S, Masamune K, Sakuma I, and Dohi T. (2000). "Three-dimensional display system for medical imaging with computer-generated integral photography." *Proc SPIE,* 3957, pp. 60–67.

[6] Min SW, Jung S, Park JH, and Lee B. (2001). "Three-dimensional display system based on computer-generated integral photography." *Proc SPIE,* 4297, pp. 187–195.

[7] Zaharia R, Aggoun A, and McCormick M. (2002). "Adaptive 3D–DCT compression algorithm for continuous parallax 3D integral imaging." *Signal Processing: Image Commun.,* 17:231–242.

[8] Stevens RF, Davies N, and Milnethorpe G. (2001). "Lens arrays and optical system for orthoscopic three-dimensional imaging." *Imaging Science J.,* 49:151–164.

[9] Javidi B., ed. (2002). *3D Television, Video, and Digital Technologies.* Springer-Verlag, Berlin.

[10] Park JH, Jung S, Choi H, and Lee B. (2002). "Viewing-angle-enhanced integral imaging by elemental image resizing and elemental lens switching." *Appl. Opt.,* 41:6875–6883.

[11] Shin SH and Javidi B. (2002). "Viewing-angle enhancement of speckle-reduced volume holographic three-dimensional display by use of integral imaging." *Appl. Opt.,* 41:5562–5567.

[12] Jang JS and Javidi B. (2002). "Real-time all-optical three-dimensional integral imaging projector." *Appl. Opt.,* 41:4866–4869.

[13] Park JH, Min SW, Jung S, and Lee B. (2001). "New stereovision scheme using a camera and a lens array." *Proc SPIE,* 4471, pp. 73–80.

[14] Arimoto H and Javidi B. (2001). "Integral three-dimensional imaging with digital reconstruction." *Opt. Lett.,* 26:157–159.

[15] Frauel Y and Javidi B. (2002). "Digital three-dimensional image correlation by use of computer-reconstructed integral imaging." *Appl. Opt.,* 41:5488–5496.

[16] Kishk S and Javidi B. (2003). "Improved resolution 3D object sensing and recognition using time multiplexed computational integral imaging." *Opt. Exp.,* 11:3528–3541.

[17] VanderLugt AB. (1964). "Signal detection by complex spatial filtering." *IEEE Trans. Inf. Theory,* IT-10:139–145.

[18] Goodman JW. (1968). *Introduction to Fourier Optics.* McGraw-Hill, New York.

[19] Casasent D. (1984). "Unified synthetic discriminant function computational formulation." *Appl. Opt.,* 23:1620–1627.

[20] Réfrégier Ph. (1990). "Filter design for optical pattern recognition: Multicriteria optimization approach." *Opt. Lett.,* 15:854–856.

[21] Javidi B. (1989). "Nonlinear joint power spectrum based optical correlation." *Appl. Opt.*, 28:2358–2367.

[22] Javidi B., ed. (2002). *Image Recognition Classification.* Marcel Dekker, New York.

[23] Frauel Y, Tajahuerce E, Castro A, and Javidi B. (2001). "Distortion-tolerant 3D object recognition using digital holography." *Appl. Opt.*, 40:3887–3893.

[24] Frauel Y and Javidi B. (2001). "Neural network for three-dimensional object recognition based on digital holography." *Opt. Lett.*, 26:1478–1480.

[25] Frauel Y, Tajahuerce E, Matoba O, Castro A, and Javidi B. (2004). "Comparison of passive ranging integral imaging and active imaging digital holography for three-dimensional object recognition." *Appl. Opt.*, 43:452–462.

[26] Pu A, Denkewalter R, and Psaltis D. (1997). "Real-time vehicle navigation using a holographic memory." *Opt. Eng.*, 36:2737–2746.

[27] Rosen J. (1998). "Three-dimensional electrooptical correlation." *J. Opt. Soc. Am. A.*, 15:430–436.

[28] Suk M and Bhandarkar SM. (1992). *Three-dimensional Object Recognition from Range Images.* Springer, New York.

[29] Herman GT. (1980). *Image Reconstruction from Projections: The Fundamentals of Computerized Tomography.* Academic, New York.

[30] Matoba O, Tajahuerce E, and Javidi B. (2001). "Real-time three-dimensional object recognition with multiple perspectives imaging." *Appl. Opt.*, 40:3318–3325.

Chapter 9

Real-time Remote Identification and Verification of Objects Using Optical ID Tags

Bahram Javidi

Department of Electrical and Computer Engineering, University of Connecticut, 06269-2157, USA. bahram@engr.uconn.edu

Robust, real-time remote (beyond the range of human vision) identification and verification of objects has potential widespread implications for security, identification, and inventory control. It has become increasingly important to allow only authorized vehicles, into and within a secure distance of installations. Furthermore, a system capable of identifying vehicles from the air could provide additional benefits. We describe a wavelength-hopped laser encoding and decoding system; and a retro-reflective optically phase modulated ID tag based on optical security technologies,[1-8] pattern recognition,[9-18] and optical encoding to yield a secure identification/verification system. The proposed systems is attractive because of the high number of mathematical possibilities optics provides for encoding; and the many degrees of freedom and high bandwidth that optical technologies offer for security and verification. Active imaging systems, such as the wavelength-hopped laser system could be used in tandem with the passive imaging retro-reflective phase encoded tag to increase system flexibility and reliability. The light modulated by the optical tags could be invisible to the human eye. Combining the high level of data storage of optically encoded materials with the free space identification possibilities of active imaging systems offers an attractive combination for remote security, identification, verification, and location of objects.

A laser tagging system based on optical wavelength tuning [Fig. 9.1(a)] can be used to produce optical codes which can identify and authenticate moving or stationary objects at a distance. A commercially available high speed wavelength tunable laser of 1500–1600 nm with sub nm wavelengths resolution and tuning speed of 100 nm/second may be utilized to generate large number of transmitted wavelengths. Tunable laser coupled with a flexible and thin single mode fiber broadcasts the beam. Alternately, the fiber may guide the laser light to a collimating lens such as GRIN attached at the end of the fiber to transmit the wavelength encoded optical beam to uniformly illuminate the receiver. For collimated illumination, tracking may be used for alignment between receiver and transmitter. Also, a slowly diverging beam may be used to have a trade-off between receiver power and alignment between receiver/transmitter. An electronic code assigned to authenticate a particular remote object can be used to produce a specific sequence of output optical waveforms with a unique set of different wavelengths.

A variety of codes such as spread spectrum sequences can be used as the electronic wavelength control of the tunable laser. The tunable laser may generate a large number of wavelengths which enable the system to produce a large number of combinations for a given code length. Multiple fiber optics links can be used to guide the light to the different areas of the vehicle. The output waveforms can be broadcast from the vehicle roof for aircraft inspection. The transmitted wavelengths encoded optical waveforms are inspected by the receiver in Fig. 9.1(a).

The transmitted optical waveforms are focused onto a fiber by a lens such as GRIN. The fiber is connected to a wavelength-sensitive optical component such as a diffraction grating to deflect the received optical beam according to its wavelength sequence. A photo detector array detects the diffracted light to reproduce the wavelength-hopped spread spectrum sequence as a function of time. A correlator will verify the authenticity of the code as a function of its spectral and temporal contents. Since the receiver has a priori information about the code sequence, a wavelength tunable filter or interference filter can

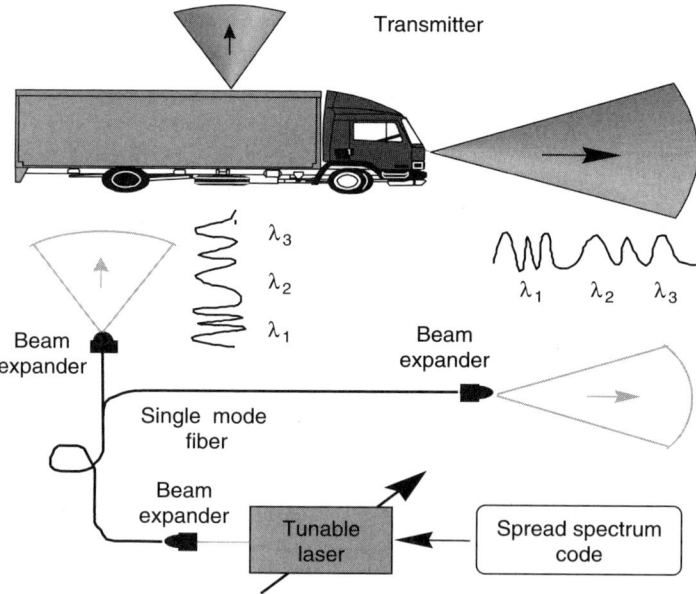

Figure 9.1. (a) Laser tagging based on wavelength division multiplexing. IF is interference filter.

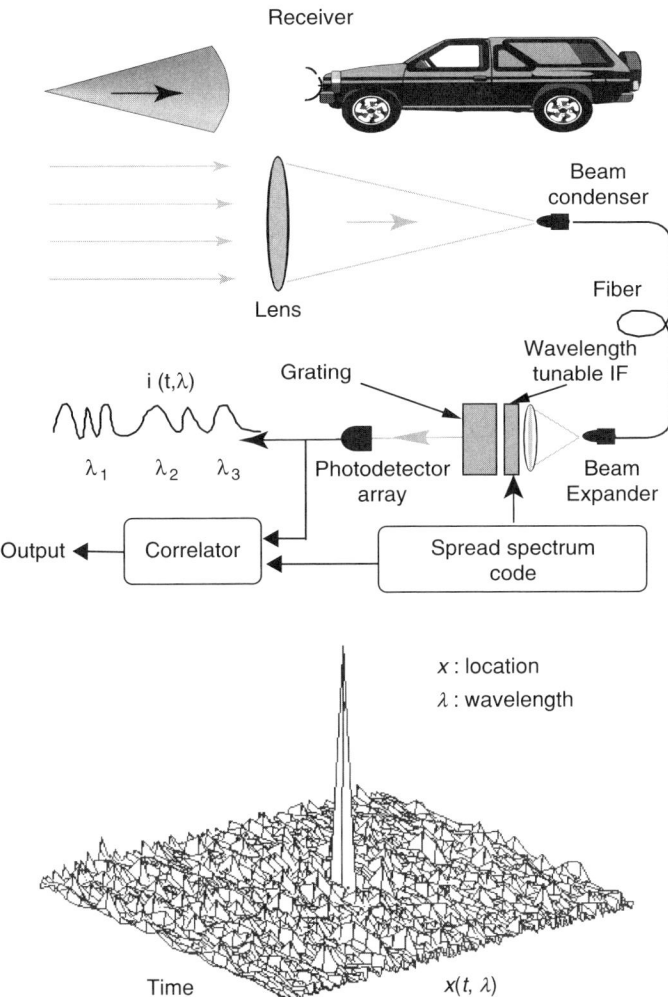

Figure 9.1. (*Continued*) (b) Correlation output of the receiver in Fig. 9.1(a) for a wavelength-hopped sequence of 20 wavelengths in the presence of white noise with mean/standard deviation of 0/0.3.

be used before the detector to remove scattered light or noise that is not part of the authentic code sequence. Figure 9.1 (b) illustrates the receiver output for a sequence of 20 pulses of varying wavelengths spread uniformly between 1400–1600 nm. The code is degraded by zero mean Gaussian noise with variance of 0.3.

Retroreflective or corner cube reflectors can be used to provide self-alignment with respect to illuminating probe beam. An optical identification

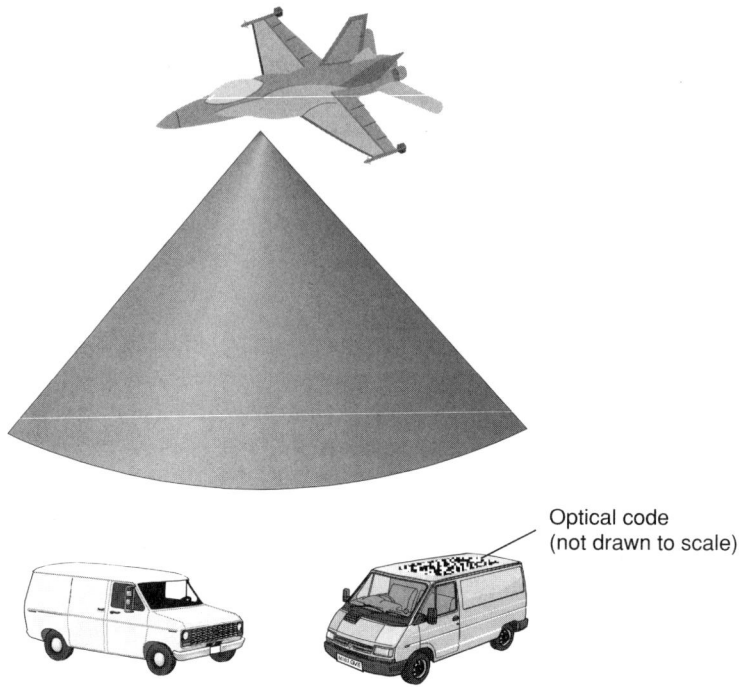

Optical code
(not drawn to scale)

Figure 9.2. Tagging based on retro-reflective optical phase tag.

phase code (tag) manufactured with retro-reflective materials can be inspected with a reader to verify the authenticity of the object (Fig. 9.2). In addition to storing an identification number, a phase-encoded tag could store such information as vehicle image, type, category, model, year, etc. For passive retro-reflective optical ID tags, a power source is not needed for the tag to function. The phase masks can be fabricated by micro-optics, embossing techniques or made of a volume-recording material such as a photopolymer. For high security applications, the volume recording material is preferred because of difficulty of unauthorized duplication due to Bragg effect.

The verification system that reads the phase-encoded identification tag can be a correlator (Fig. 9.3). The optical tag to be verified is imaged onto the input plane. Active imaging optics may be used for compensating environmental degradations and/or tag scale changes. The processor used to verify the input mask can also be used to verify a primary pattern added to the optical tag.

The effects of environmental degradation, variation in the scale and illumination, and noise/clutter suppression can be taken into account in the recognition process.[10] Figure 9.4 illustrates the output for an input optical phase tag code of 16×16 pixels noise with a nonlinear correlator designed for rotationally tolerant recognition of optical codes in the presence of background noise and additive noise.[10] The reference function is displayed in the spatial

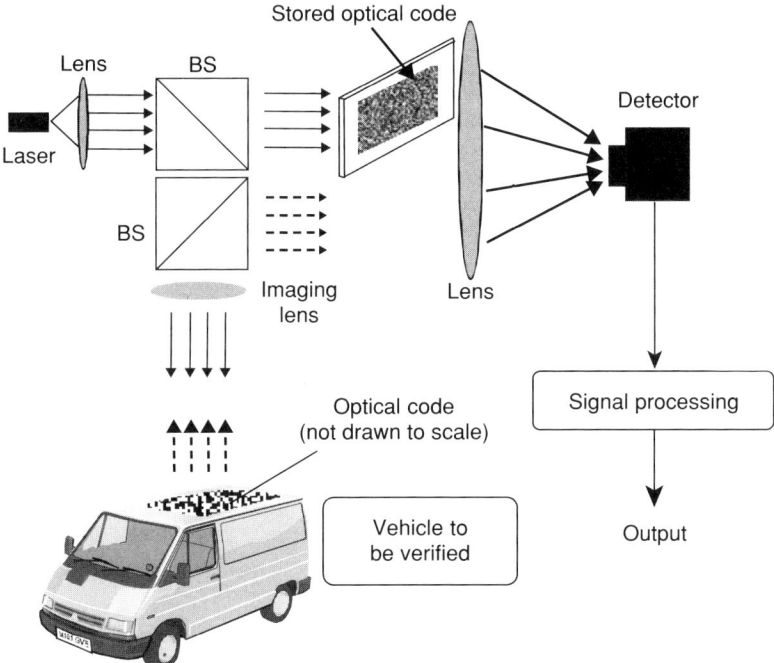

Figure 9.3. Nonlinear joint transform correlator that reads and verifies the optical tags. BS is beam splitter.

domain and can be rapidly updated in real-time using spatial light modulators (SLM).[9]

It is possible to implement 3D security codes for verification and authentication. We could use digital holography for recording a 3D primary pattern of object signature such as the 3D images of a vehicle.[9,14] This provides verification against theft and unauthorized usage of optical tags. The digital holographic pattern can be encoded on the optical tag. Upon a probe beam, the 3D image of the primary pattern of the object is produced at the receiver. Figure 9.5 illustrates the reconstruction of the 3D image of a toy car obtained by phase shift on-line digital holography.[9,14] This 3D optical code combined with the security optical tag can provide a more effective remote authentication/verification system.

Active optical tags (Fig. 9.3), can be written onto an updatable device by a variety of SLMs[9] such as liquid crystal display. One problem to be addressed in this case is light efficiency. With the exception of a few devices including deformable mirror devices and MEMS, light modulators may have low light efficiency,[9] which places additional constraints on the light budget of the system. Alternately, the optical code may be stored electronically to drive a modulating element such as MEMS. Retroreflective based modulating devices will then modulate a probing beam inspected by a receiver.

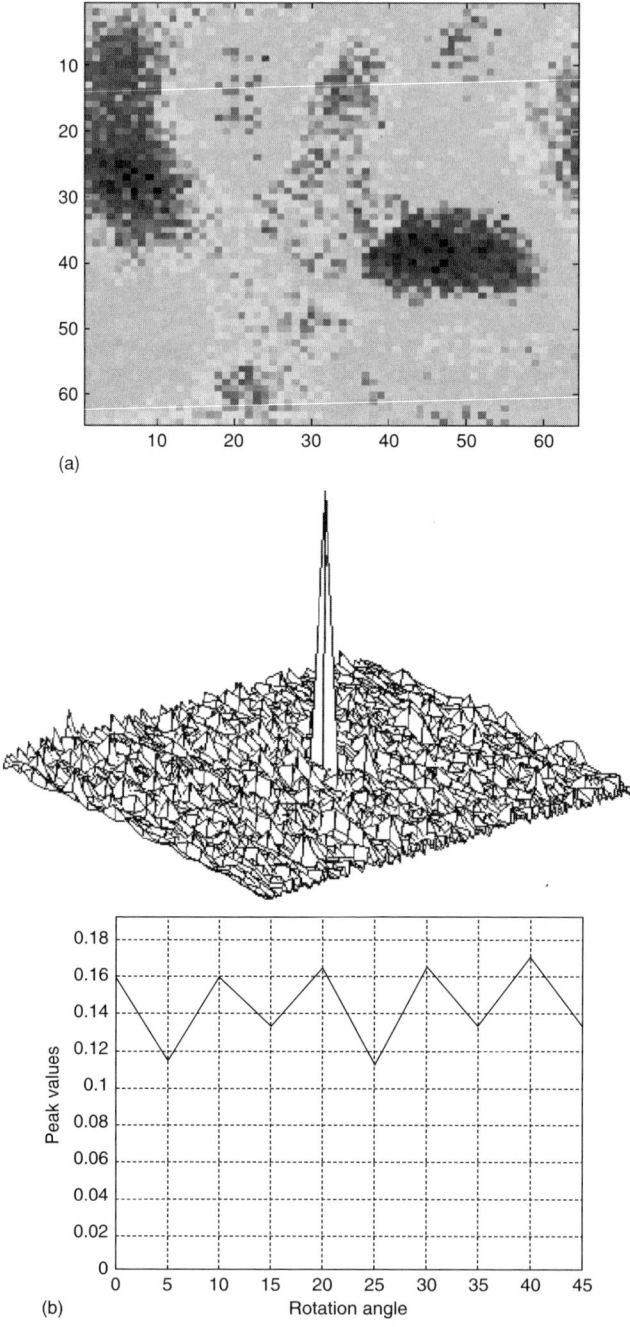

(a)

(b)

Figure 9.4. (a) Input with 45° rotated phase encoded tags. The background noise and additive noise are white with mean/standard deviation of 0.7/0.3 and 0/0.1, respectively. (b) The output of the correlator in Fig. 9.3 for an input optical phase code (Fig. 9.4(a)) with a correlator designed for rotationally tolerant recognition of optical codes in the presence of noise.

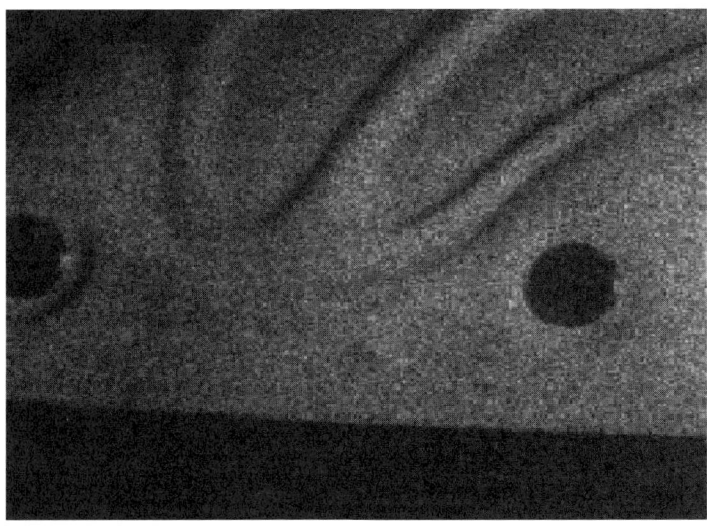

Figure 9.5. Three-dimensional primary image signature reconstruction of a car using online phase shift digital holography.

Acknowledgments

The author wishes to thank S. Hong for his assistance in the preparation of this chapter.

References

[1] Javidi B and Horner J. (1994). "Optical pattern recognition system for security verification." *Opt. Eng.*, 33:6.

[2] Refregier P and Javidi B. (1995). "Optical image encryption using input and Fourier plane random phase encoding." *Opt. Lett.*, 20(7):767–769.

[3] Javidi B. (2003). "Real-time remote identification and verification of objects using optical ID tags." *Opt. Eng.*, 42:8.

[4] Javidi B. (1997). "Encrypting information with optical technologies." *Phys. Today*, 50:3.

[5] Matoba O and Javidi B. (2000). "Encrypted optical memory systems based on multidimensional keys for secure data storage and communications." *IEEE Circuits Devices Magazine*, 16(5):8–15.

[6] Rosen J and Javidi B. (2001). "Optical encryption using embedded images." *Appl. Opt.*, 40(20):3346–3351.

[7] Javidi B and Nomura T. (2000). "Polarization encoding for optical security systems." *Opt. Eng.*, 39:2439–2443.

[8] Goudail F, Bollaro F, Refregier P, and Javidi B. (1998). "Influence of perturbation in a double phase encoding system." *J. Opt. Soc. Am. A.*, (JOSA A), 15:2629–2638.

[9]   Goodman JW. (1996). *Introduction to Fourier Optics*. McGraw-Hill.

[10] Javidi B., ed. (2002). *Image Recognition and Classification: Algorithms, Systems, and Applications*. Marcel-Dekker, New York.

[11] Hong S and Javidi B. (2002). "Optimum nonlinear composite filter for distortion tolerant pattern recognition." *App. Opti.*, 41:2172–2178.

[12] Chan F, Towghi N, Pan L, and Javidi B. (2000). "Distortion tolerant minimum mean squared error filter for detecting noisy targets in environmental degradations." *Opti. Eng.*, 39:2092–2100.

[13] Towghi N, Javidi B, and Li J. (1998). "Generalized optimum receiver for pattern recognition with multiplicative, additive, and nonoverlapping noise." *J. Opt.l Soc. Am. A.*, (JOSA A), 15(6):1557–1565.

[14] Yamaguchi I and Zhang T. (1997). "Phase-shifting digital holography." *Opt. Lett.*, 22(16):1268–1270.

[15] Matoba O and Javidi B. (2002). "Optical retrieval of encrypted digital holograms for secure real-time display." *Opt. Lett.*, 27(5):321–323.

[16] Matoba O, Naughton T, Frauel Y, Bertaux N, and Javidi B. (2002). "Real-time three-dimensional object reconstruction using a phase-encoded digital hologram." *Appl. Opt.*, 41(29):6187–6192.

[17] Frauel Y and Javidi B. (2001). "Neural networks for three dimensional object recognition based on digital holography." *Opt. Lett.*, 26(9):1478–1480.

[18] Javidi B and Tajahuerce E. (2000). "Three-dimensional image recognition using digital holography." *Opt. Lett.*, 25.

Chapter 10

An Adaptive Technique for Minimizing Rate of Sensory Data Transmission in Unmanned Aerial Vehicles

Firooz Sadjadi

Lockheed Martin Corporation
Address: 3400 Highcrest Road, Saint Anthony, MN 55418
email: sadja001@tc.umn.edu

10.0 Introduction

Unmanned Aerial Vehicles (UAV) are intended to perform medium range and/or long endurance surveillance, reconnaissance, relay, targeting, and potentially, attack (both lethal and non-lethal) against a wide variety of possible land and sea-based targets, across the spectrum of conflict[1]. For this purpose, UAVs are equipped with a wide range of high-resolution sensors. Transmitting the acquired data to a command center requires data transmission rate far exceeding what is currently and in the foreseeable future is going to be available.

One way of dealing with this bottleneck is to be optimally selective in what information is transmitted. In this chapter we use a number of techniques for reducing the amount of data that needs to be transmitted (transmission bandwidth)[5]. The techniques include bandwidth reduction by means of mosaicing operation, static regions of interest (ROI) extraction operation, and dynamic regions of interest extraction and tracking operation. Further reductions can be achieved by first classifying both static and dynamic ROIs and by being further selective in transmitting out only regions (or the regions' labels and their attitudes) belonging to a limited number of class or classes. Under different UAV flight patterns, mission scenarios, and sensor parameters different reduction methods spanning different parameters become more appropriate than others. We will report in this chapter the application of a genetic algorithm for selecting an optimum reduction regime and its associated parameters' values.

Figure 10.1 shows the bandwidth management approach and the different methods by means of which the transmission bandwidth from UAV to a command station can be reduced. Depending on the contextual information in the acquired imagery, transmitting bandwidth can be reduced by transmitting

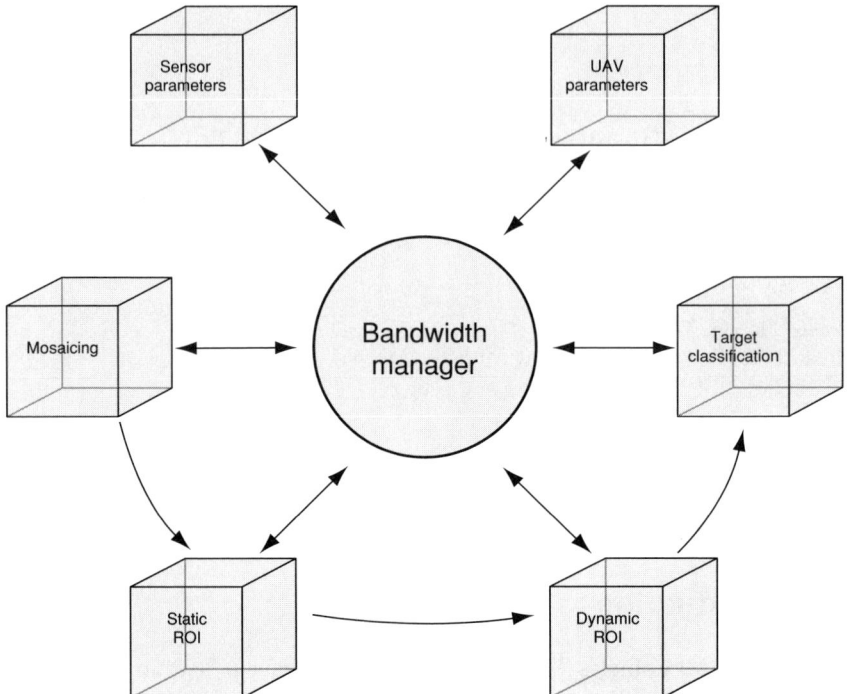

Figure 10.1. Management of transmission bandwidth from a UAV with its command center.

the outputs of the following operations 1) mosaicing, 2) region of interest extraction, 3) object tracking, and 4) automatic object classification.

10.1 Mosaicing

By frame-to-frame image registration and transmission of only non-overlapping pixel data, substantial reduction in transmission rate can be achieved (as shown by Fig. 10.2).[1] Consider an imaging sensor of angular pixel resolution θ radian, array size L by L pixels, with a frame rate of r frames/sec, directly pointing downward on a UAV flying at a constant speed V meters/second at a height of h meters. Assuming flat earth geometry, it can be shown that the ratio of the area of overlap between succeeding image frames to the total areas is:

Percentage-of-Saving by using Mosaicing Operation

$$= 100(1 - \frac{V}{2Lrh\,\boldsymbol{tan}\,\theta})P_{CR} \tag{1}$$

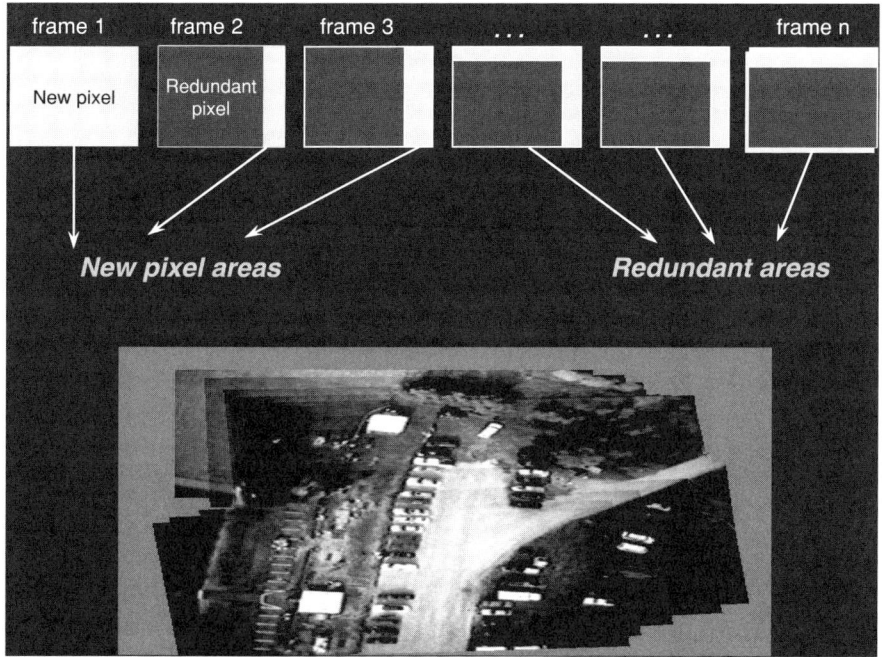

Figure 10.2. Mosaicing operation exploits pixel redundancies.

where P_{CR} is the probability of correct registration, defined as the ratio of total number of pixels that are correctly registered, to the total number of pixels that are being considered. When P_{CR} is very low there would be little overlap areas among succeeding frames and consequently the Percent-of-Saving would be very low.

Figure 10.3 displays plots of variation of this ratio for $P_{CR} = 1$ as functions of v, h, and θ for two values of r. As can be seen, increasing speed of UAV has a more dramatic effect on the amount of the bandwidth saving at lower altitudes than at higher altitudes as can be expected. Moreover, at lower resolutions more saving at lower altitudes can be achieved than at higher latitudes.

10.2 Static Region of Interest Extraction

If the reason for data transmission is viewing the regions containing objects of interest (ROI), further reduction in transmission rates can be achieved. Figure 10.4 shows elements of this concept.

Considering each target size region as a cell, each image of a scene can be viewed to be composed of target cells and clutter cells. Target cells can be characterized by target density (τ) defined as the total area of target cells per unit ground area. Similarly, clutter density (χ) can be defined as the total area of clutter cells per unit ground area.

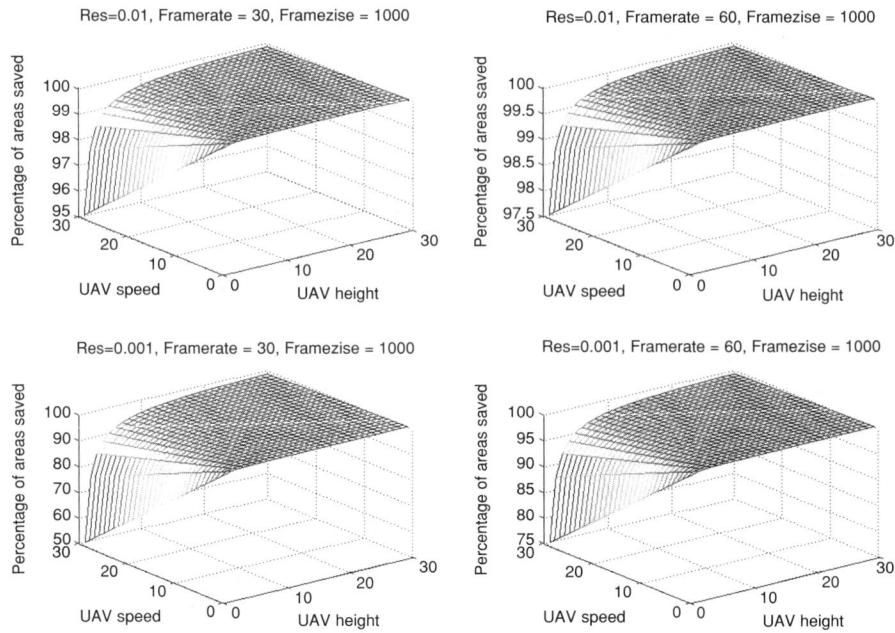

Figure 10.3. Variation of saving ratio for L = 1000.

Based on some rules (and there could be many) every pixel is assigned to be associated with either a target or clutter. False associations generate false alarms (contribute to prob. of false alarm, P_{fa}) and correct associations generate correct detection (contribute to the prob. of detection, P_d). P_d is defined as the ratio of the total number of target cells that satisfied the rules to the total number of target cells in the image. P_{fa} is defined as the ratio of the number of clutter cells that satisfied the rules, to the total number of clutter cells in the image. The relationship between P_d and P_{fa} is best described through the

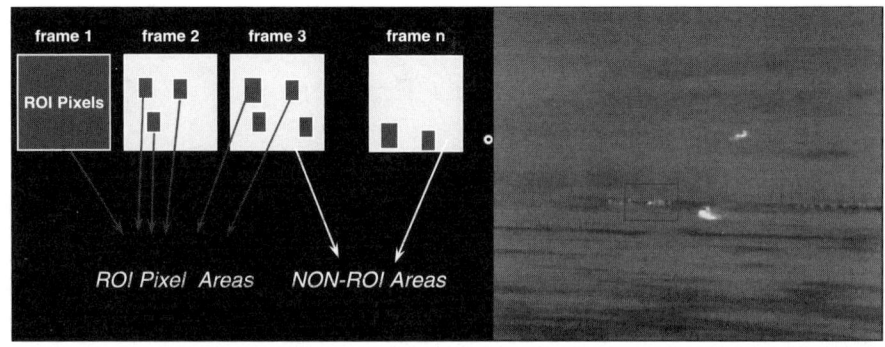

Figure 10.4. Static ROI operation exploits only pixels associated with Target Areas.

receiver operating characteristic (ROC) curve. By selecting the optimum point on this curve where Pd and P_{fa} achieve their optimum values, we can establish the optimum rule quantitatively. The ROI in a single frame is then equal to the total area of target and clutter cells that satisfy this rule quantitatively.

For the first single frame one has:

$$\text{ROI} = (\tau P_d + \chi P_{fa}).\text{Area} = 4L^2h^2(\tau P_d + \chi P_{fa})\tan^2\theta \qquad (2)$$

By associating each ROI with a geographical position in a global coordinate system, once an ROI is extracted, its position in the scene will be fixed and in different frames its retransmission will be avoided.

Then the total saving from ROI operation is:

$$\text{Percent-of-Saving using ROI Operation} = 100(1 - \frac{V(\tau P_d + \chi P_{fa})}{2Lrh\tan\theta})P_{CR} \qquad (3)$$

P_{fa} is related to P_{fa} and scene metrics such as target contrast, h, r, etc, and ROI extraction algorithm internal parameters through the following predictive models[3]:

$$P_D = a_0 + a_1x_1 + a_2x_2 + a_3x_3 + a_{12}x_1x_2 + a_{13}x_1x_3 + a_{23}x_2x_3 + a_{11}x_1^2$$
$$+ a_{22}x_2^2 + a_{33}x_3^2 \qquad (4)$$

where, x_1 = False Alarm Probability (FA), x_2 = Algorithm Internal Parameter, and x_3 = Image/Scene Metric.

For the special case when the probability density functions of observed signature, conditioned on being originated from targets and clutter, are both Gaussian with different means but equal variances one has the following relationship[4]:

$$P_d = Q(Qinv(P_{fa}) - \sqrt{d^2}) \qquad (6)$$

where Q(x) is the complementary cumulative distribution function and Qinv(x) is the inverse complementary cumulative distribution function:

$$Q(x) = \frac{1}{2}erfc(\frac{x}{\sqrt{2}}); \ Qinv(x) = \sqrt{2}erfinv(1 - 2x) \qquad (7)$$

d^2 is defined as:

$$d^2 = \frac{[mean(ROI) - mean(clutter)]^2}{std(ROI)} \qquad (8)$$

Figure 10.5 shows the plots of the variations of bandwidth savings obtained using static ROI extraction operation as functions of UAV height and speed for two different frame rate and resolutions and for some nominal values for $P_{fa} = 0.1$, target area density = 0.25, clutter area density = 0.625. Comparing these results with those shown in Fig. 10.3 it can be seen that, as expected more bandwidth saving can be obtained by transmitting static ROIs than by sending raw data.

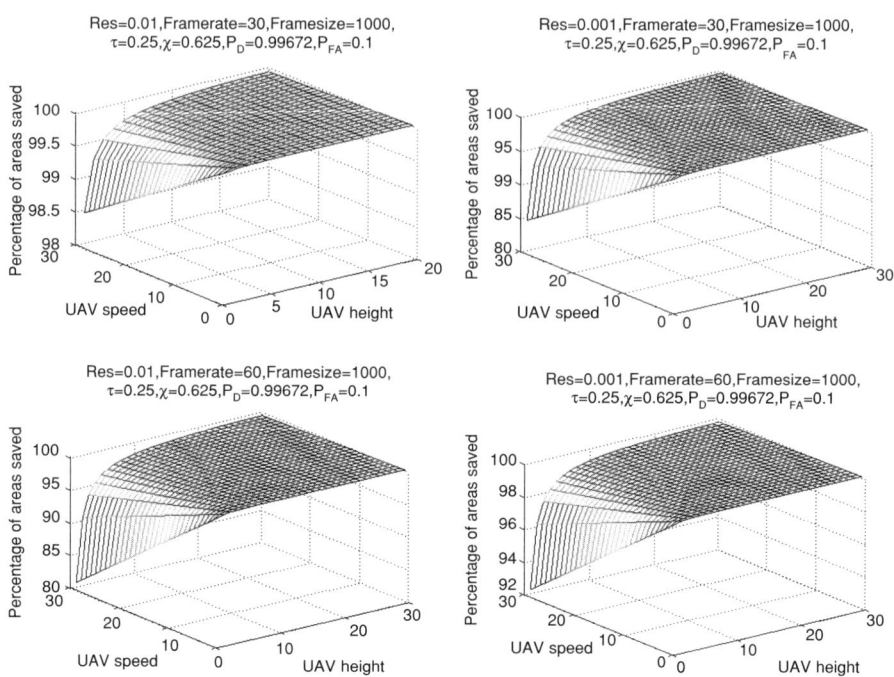

Figure 10.5. Variation of bandwidth saving ratio for the case of static ROI approach for L = 1000.

10.3 Dynamic Regions of Interest Exaction and Tracking

For the cases where the targets of interest are moving (see Fig. 10.6), one can save bandwidth by predicting where the targets of interest will be in the upcoming frames and avoid their retransmission. By assuming that the total density of targets, both moving and stationary, is constant, we can conclude

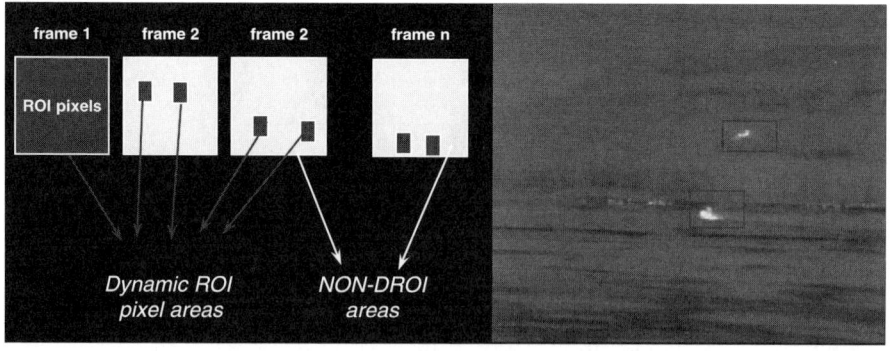

Figure 10.6. DROI exploits those pixels associated with moving targets.

that the total numbers of targets entering and leaving the field of view of an image are identical.

Consider μ to be the moving target density, and σ to be the stationary target density. In the previous section, we considered the total target density as τ, which is total area of target cells per unit ground area. By assuming that $\mu + \sigma = \tau$ is constant, any stationary target that starts moving will increase μ but at the same time will reduce σ so that τ remains unchanged. Moving target set will fit into one or more of the following categories: 1) tracked into next frame, 2) move out of field of vision (FOV), 3) stopped, and 4) lost due to tracking algorithm errors. We will denote the percentage of moving targets that are associated with each of the above categories by P_t, P_o, P_s, and P_l, respectively, where $P_t = 1 - P_o - P_s - P_l$. Then the Saving due to dynamic ROI tracking is:

Percent-of-Saving using Dynamic ROI and Tracking Operation

$$= 100(1 - \frac{V(P_d\mu + P_{fa}\chi)}{2t_{dwell}\,Lh\,\boldsymbol{\tan\theta}})P_tP_{CR} \qquad (9)$$

P_t is related to the projected velocity vector of the UAV and a functional of the moving target velocities (Eigen velocities). $\boldsymbol{t_{dwell}}$ is the tracking dwell time.

Consider the average velocity of the targets as V_t, then if UAV is moving with the same speed in the same direction as the moving target ($V = V_t$) the target will always be in the FOV and maximum $\boldsymbol{t_{dwell}}$ and consequently bandwidth reduction can be achieved. The relative velocity of a target with respect to the UAV can be derived from the following relationship:

$$V_{rel} = V - V_t\,\boldsymbol{\cos\rho} \qquad (10)$$

where, ρ is the angle between the projected velocity vector of the UAV and the average velocity vector of moving ground vehicles. Then we have:

$$t_{dwell} = \frac{VLh\,\boldsymbol{\tan\theta}}{V_{rel}} = \frac{VLh\,\boldsymbol{\tan\theta}}{|V - V_t\,\boldsymbol{\cos\rho}|} \qquad (11)$$

Figure 10.7 shows the Percent-of-Saving as functions of UAV speed and height for two values of sensor resolution and frame rate. In these plots, $P_{fa} = 0.1$, the moving target area density is 0.125, the clutter area density is 0.625, and the static target area density 0.25. As can be seen, saving improves as UAV altitude and speed increase. At lower altitudes, however, increasing the UAV speed diminishes the saving, as is expected.

10.4 Classification of Static and Dynamic Regions of Interest

Further reduction in bandwidth can be achieved by being selective in transmitting static and dynamic regions of interest. This can be achieved by classifying the regions of interest and then transmitting only the labels for each

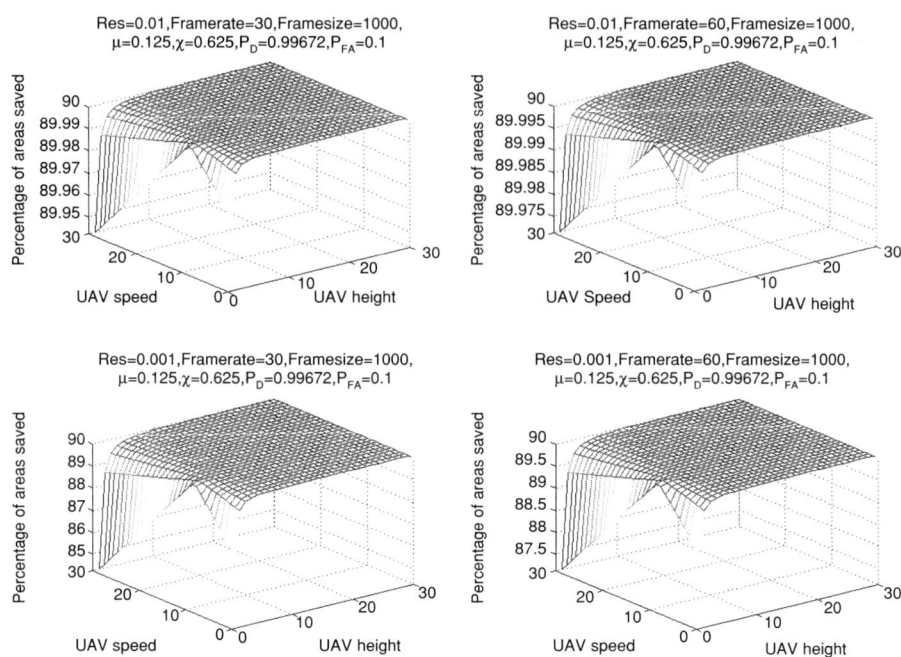

Figure 10.7. Variation of bandwidth saving ratio for the case of dynamic ROI approach for L = 1000.

region and its associated attitude information such as position and velocity (see Fig. 10.8). Furthermore, one can only transmit the labels and attitude information for selective classes of targets, thus reducing the bandwidth requirement even further.

From the previous sections one can show that the total area of dynamic and static ROIs in t_{dwell} seconds is:

$$T_{roi} = \text{Total ROI} = \frac{1}{A_{roi}}[P_t(P_d\mu + P_{fa}\chi)(\frac{2\text{VLh}\tan\theta}{r}) \quad (12)$$
$$+ (\sigma P_d + \chi P_{fa})(2t_{dwell}\,VLh\tan\theta)]P_{CR}$$

Then the saving would be:

Percent-of-Saving By using Target Classification Operation

$$= 100(1 - \frac{T_{roi}N_{bproi}}{4h^2\,t_{dwell}\,L^2N_{bpp}}) \quad (13)$$

N_{bproi} is the number of bits required for transmitting label, and attitude information, for each ROI. N_{bpp} is the number of bits per pixel in the sensor. A_{roi} is the average area of an ROI.

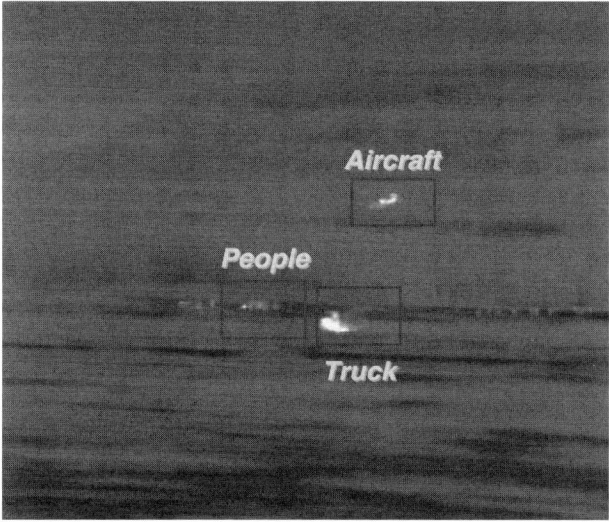

Figure 10.8. Target classification maps the image into a selected set of target labels and their associated state values.

Figure 10.9 shows the Percent-of-Saving as functions of UAV speed and height for two values of sensor resolution and frame rate. In these plots, $P_{fa} = 0.1$, the moving target area density is 0.125, the clutter area density is 0.625, and the static target area density 0.25. As can be seen, saving improves as UAV altitude and speed increase. At lower altitudes, however, increasing the UAV speed diminishes the saving, as is expected.

10.5 Genetic Algorithms

Genetic algorithms are population-based random search strategies that emulate the concepts from reproduction and natural selection of biological systems to produce better solutions from previous solutions. Genetic algorithms[6,7] are useful in a wide variety of applications requiring the optimization of a fitness function. This includes some forms of machine learning. Genetic algorithms (GA) were developed by John Holland[6] and his colleagues at the University of Michigan. Their activities have led to the development of algorithms and software that are robust: can learn from their environment and adapt to the changing conditions. In a biological system the structure that encodes the prescription on how the organism should be constructed is called a chromosome. A set of chromosomes is called a genotype and the resulting organism is called a phenotype. Each chromosome is composed of individual structures called genes. Each gene encodes a particular feature of the organism. Location of the gene in a chromosome or its locus determines what particular

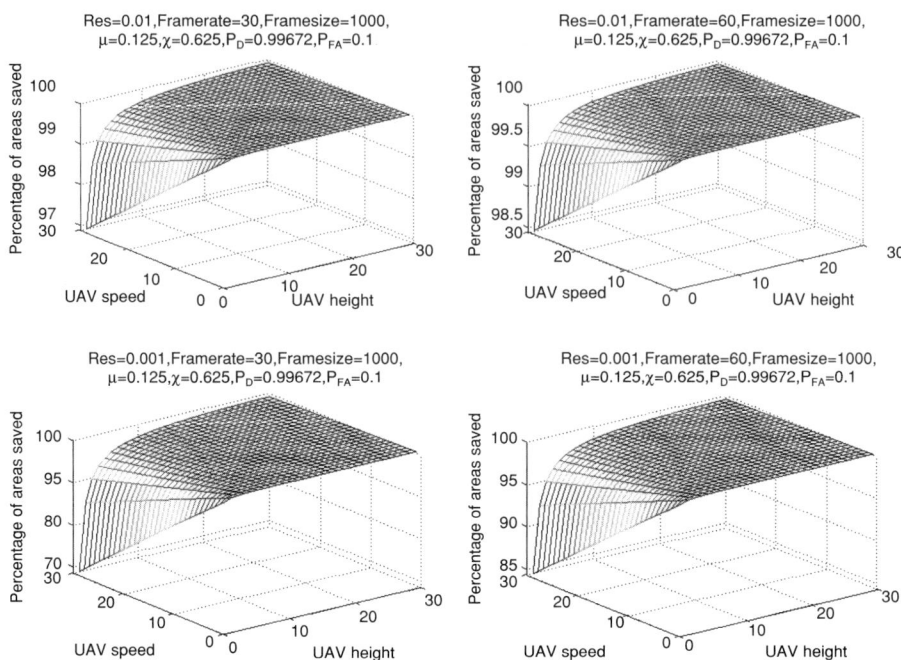

Figure 10.9. Variation of bandwidth saving ratio for the case of static and dynamic ROI classification approach for L = 1000.

characteristics the gene represents. At a particular locus a gene may represent any of several different values of the particular characteristic it represents. These values are called alleles. In the context of genetic algorithm, chromosomes are the strings of data, genes are positions on the string, and the alleles are the values associated with the genes.

In this study, the different UAV bandwidth reduction regimes described in the previous sections are used as fitness functions. The parameters of the UAV state and sensor, are used as variables. The output of the GA is the optimum transmission reduction rate, and the best values for UAV state and sensor parameters. The results of the implementation of this bandwidth management approach are presented in the following section.

10.6 Experimental Results

The reductions in needed bandwidths, gained by four sensor processing techniques of 1) Mosaicing, 2) Static ROI transmission, 3) Dynamic ROI transmission, and 4) Classification output transmission were used as fitness functions in genetic optimization algorithms to obtain the optimum UAV state and sensor parameters.

In the following experiments the maximum number of generation was set at 100; the probability of cross over was 0.98; the probability of mutation was 0.02. The number of genes was 20. And, finally, the number of individuals was 200.

10.6.1 Optimum UAV and Sensor Parameters by use of Mosaicing Operation

Figure 10.10 shows the behavior of the genetic algorithm. The top curve (blue) indicates the maxim reduction values for different generations. The middle curve (green) shows the median of the fitness function, and the bottom curve (red) is indicator of the worst fitness values.

Table 10.1 shows the optimum UAV values for the case of Mosaicing operation. In this exercise the UAV speed was varied between 10 and 100 m/s. The altitude of UAV was set to vary from 1 to 10 km. The frame size of the imaging sensor was varied from 100 by 100 to 1000 by 1000 pixels. The frame rate was between 30 and 100 frames per second. The sensor angular resolution was set to vary from 0.01 to 0.9. The probability of correct registration value was varied from 0.8 to 1.0.

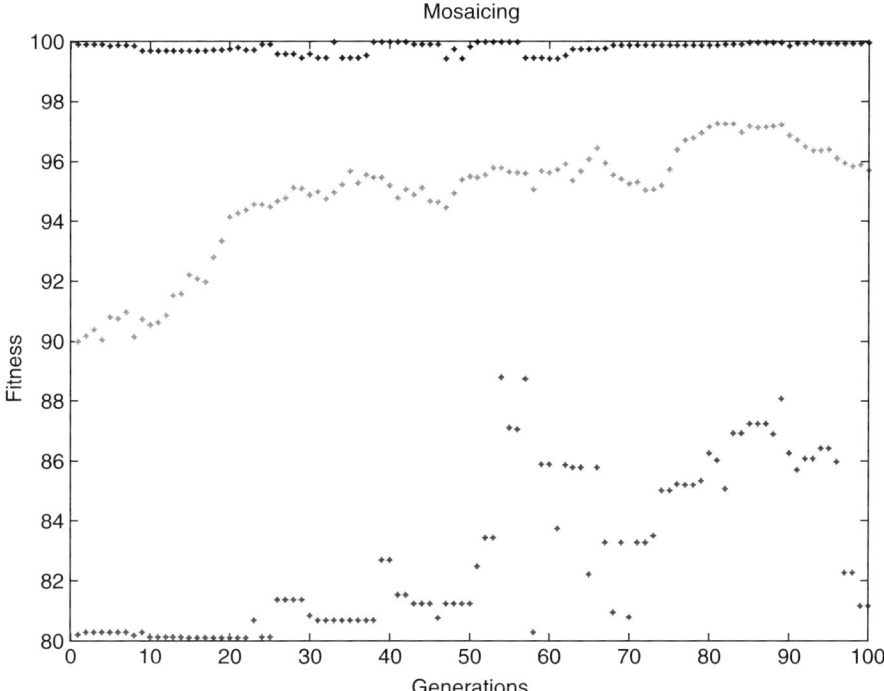

Figure 10.10. The GA output for the Mosaicing operation.

Table 10.1. Optimum UAV and sensor parameters when using Mosaicing Operation.

UAV speed	Altitude	Frame rate	Frame size	Probability of correct registration	Sensor resolution
96.742036	5974.7858	35.577970	889.713470	0.99997711	0.56259403

10.6.2 Optimum UAV and Sensor Parameters When Using Static ROI Operation

Figure 10.11 shows the behavior of the genetic algorithm. The top curve indicates the maxim reduction values for different generations. The middle curve shows the median of the fitness function, and the bottom curve is indicator of the worst fitness values.

Table 10.2 shows the optimum UAV parameters for using Static ROI operation. Additionally, this Table includes optimum values for parameters that are scene related. In situations where scene-related parameters are not controllable but are known, they can be used as input to the Bandwidth Manager module. In this exercise, however, we have assumed that these parameters were controllable and consequently the genetic algorithm generated

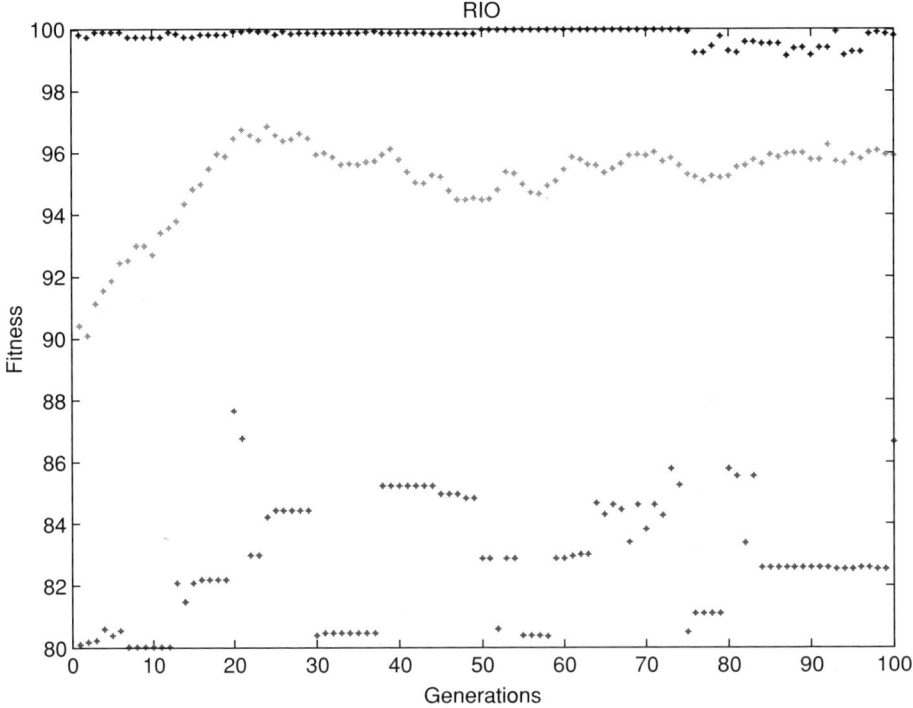

Figure 10.11. The GA output for the ROI Operation.

Table 10.2. Optimum UAV, sensor, and scene parameters when using static ROI operation.

		Standard	
P_{CR}	0.99998283	**deviation**	0.17932136
V	98.069237	χ	0.56081425
τ	0.54462914	**Frame size**	988.21286
P_{fa}	0.63800701	**Frame rate**	98.963260
Background mean intensity value	101.01192	**Altitude**	7687.6504
Target mean intensity value	218.21684	**Sensor resolution**	0.24630208

their optimum values along with the UAV related optimum parameter values. The range of variations for the various UAV related parameters were similar to those of the previous case (Section 10.6.1). However, for the scene parameters we assumed the following ranges: P_{fa} was varied from 0.1 to 1.0; the background and target mean intensity values were varied from 0 to 255; the standard deviation of target intensity was varied from 0.1 to 1.0; the target and clutter densities were varied from 0.1 to 1.0, and, finally the frame rate was varied from 30 to 100 frames per second.

10.6.3 Optimum UAV and Sensor Parameters When Using Dynamic ROI Operation

Figure 10.12 shows the behavior of the genetic algorithm. The top curve indicates the maximum reduction values for different generations. The middle curve shows the median of the fitness function, and the bottom curve is indicator of the worst fitness values.

Table 10.3 shows the optimum UAV parameters for using Dynamic ROI operation. Additionally, this Table includes optimum values for parameters that are scene related. In situations where scene-related parameters are not controllable but are known, they can be used as input to the Bandwidth Manager module. In this exercise, however, we have assumed that these parameters were controllable and consequently the genetic algorithm generated their optimum values along with the UAV related optimum parameter values. The range of variations for the various UAV related parameters were similar to those of the previous case (Section 10.6.2). However, for the scene parameters we assumed the following ranges: Pfa was varied from 0.1 to 1.0; the background and target mean intensity values were varied from 0 to 255; the standard deviation of target intensity was varied from 0.1 to 1.0; the clutter density was varied from 0.1 to 1.0; the frame rate was varied from 30 to 100 frames per second; the moving target density was varied from 0.1 to 1.0; the angle between the projected UAV velocity vector on the ground and a moving target was varied from $\pi/10$ to $20\pi/10$; the range of Pt was varied from 0.8 to 1.0, and finally the average speed of moving targets was varied from 5 to 60 m/s.

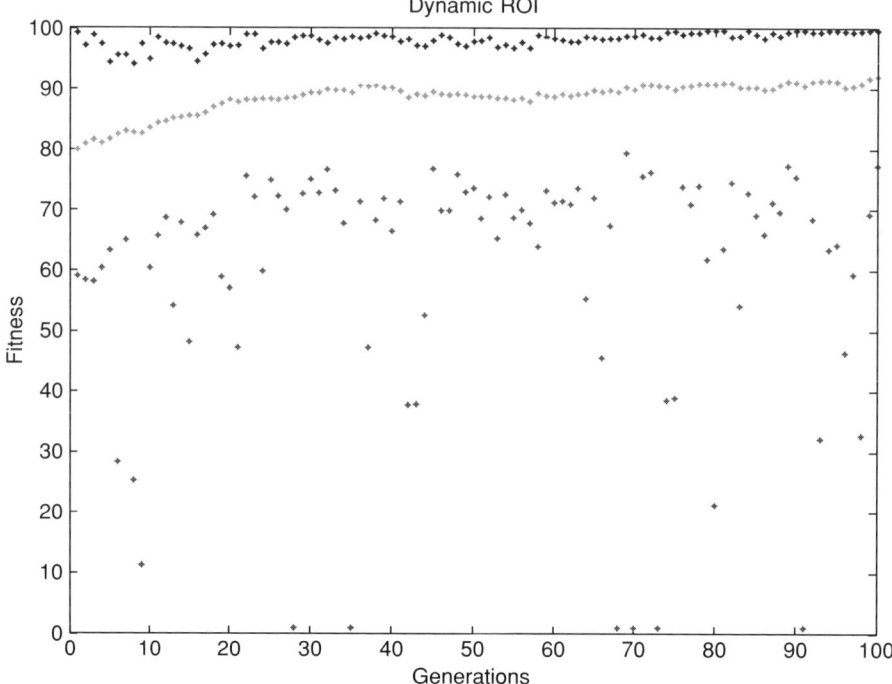

Figure 10.12. The GA output for the dynamic ROI operation.

10.6.4 Optimum UAV and Sensor Parameters When Using Classification of Static and Dynamic ROIs Operation

Figure 10.13 shows the behavior of genetic algorithm. The top curve indicates the maximum reduction values for different generations. The middle curve shows the median of the fitness function, and the bottom curve is indicator of the worst fitness values.

Table 10.3. Optimum UAV, sensor, and scene parameters when using dynamic ROI operation.

		Standard	
$\mathbf{P_{CR}}$	0.99801616	deviation	0.46656005
V	28.315256	χ	0.14743523
μ	0.68511351	**Frame size**	400.44889
$\mathbf{P_{fa}}$	0.38460282	**Frame rate**	80.573702
Background mean intensity value	159.21675	**Altitude**	4035.7743
Target mean intensity value	92.832176	**Sensor resolution**	0.29098477
$\mathbf{P_t}$	0.99984512	ρ	4.4764104
$\mathbf{V_t}$	52.074587		

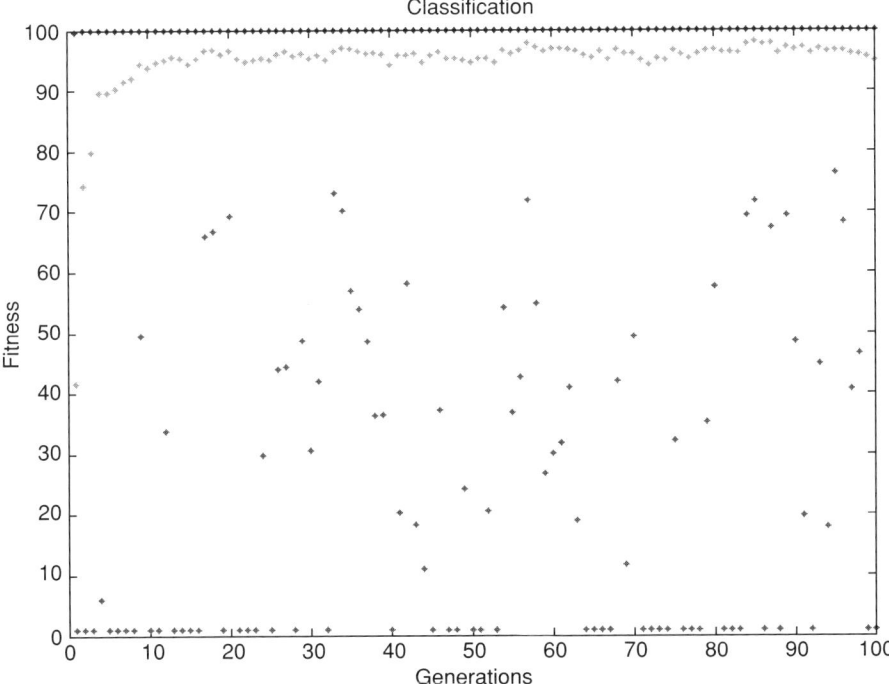

Figure 10.13. The GA output for the classification of static and dynamic ROIs operation.

Table 10.4 shows the optimum UAV parameters for using Classification of Static and Dynamic ROI operation. Additionally, this Table includes optimum values for parameters that are scene related. In situation where scene-related parameters are not controllable but are known, they can be used as input to the

Table 10.4. Optimum UAV, sensor, and scene parameters when using classification of static and dynamic ROI operation.

		Standard	
P_{CR}	0.91026107	**deviation**	0.31232129
V	10.601760	χ	0.17089192
μ	0.85821787	**Frame size**	971.84407
P_{fa}	0.76212774	**Frame rate**	76.467329
Background mean intensity value	23.121491	**Altitude**	2964.2028
Target mean intensity value	20.410171	**Sensor resolution**	0.010234261
P_t	0.87927807	ρ	3.6711248
V_t	45.598531	A_{roi}	90.696827
τ	0.52696946	**NBPROI**	10.006485
NBPP	25.840257		

Bandwidth Manager module. In this exercise, however, we have assumed that these parameters were controllable and consequently the genetic algorithm generated their optimum values along with the UAV related optimum parameter values. The range of variations for the various UAV and scene related parameters were similar to those of the previous case (Section 10.6.3). However, for the additional scene parameters, specific to this operation we assumed that NBPP was varied from 8 to 32; NBPROI was varied from 10 to 210; and finally, A_{roi} was varied form 5 to 100.

A number of observations can be made from Tables 10.1–10.4. The variations of optimum UAV parameters, namely UAV speeds and UAV altitudes, as functions of various bandwidth reduction regimes are shown in Figs. 10.14 and 10.15. As can be seen, lower optimum speed values and lower optimum altitudes correspond to the higher level of processing regimes. Figures 10.16–10.18 show the variations of optimum sensor parameters as functions of various operations used for bandwidth reduction. They indicate that lowest optimum frame rate and coarsest angular resolution are suggested for Mosaicing operation. For the Classification operation, the suggested optimum sensor angular resolution is the smallest value among all of the operations. Finally. Table 10.5 shows a number of statistical information regarding the genetic algorithm behavior. The maximum fitness values for all the operations were larger than 99.5%. The generation numbers at which these maximums reach are tabulated under "Generation Number." As can be seen, for the classification operation this maximum value was reached rather earlier than others. The standard deviations of the fitness function values are shown to be around 90.49 to 96.59. The ratio of maximum fitness value to median fitness value, also called pressure, is an indicator of the

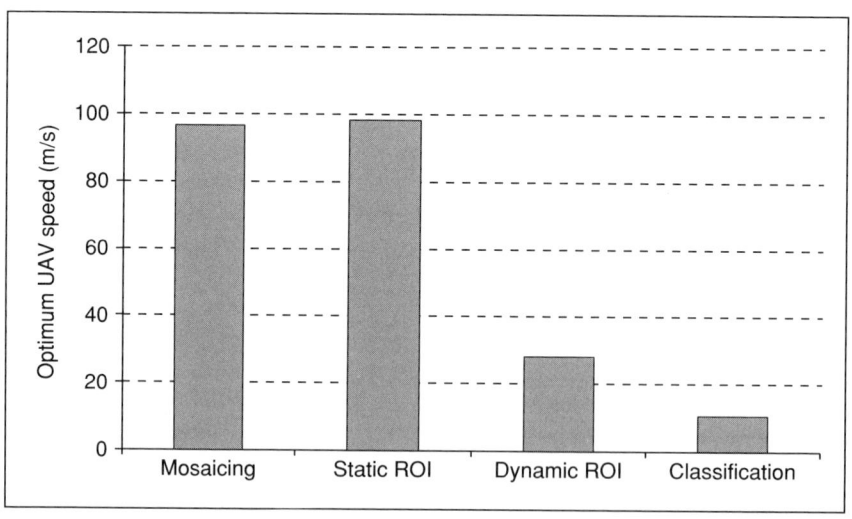

Figure 10.14. Optimum UAV speed obtained for various operations used for bandwidth reduction.

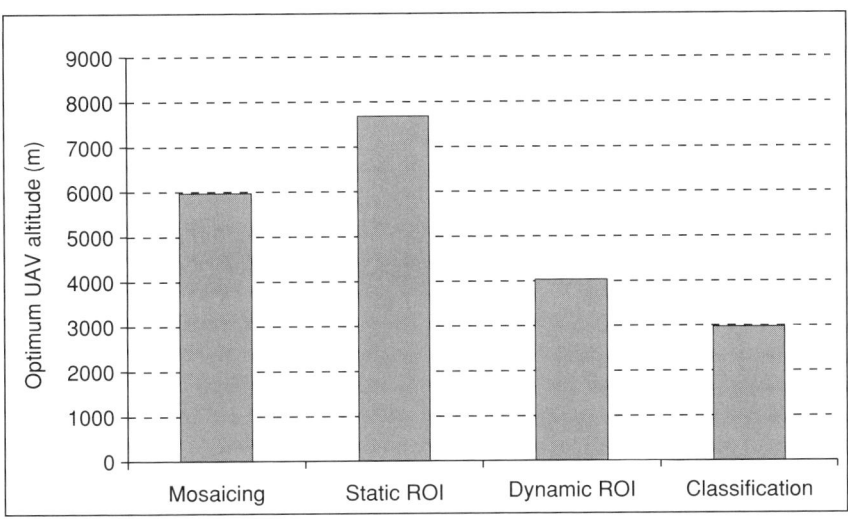

Figure 10.15. Optimum UAV altitude obtained for various operations used for bandwidth reduction.

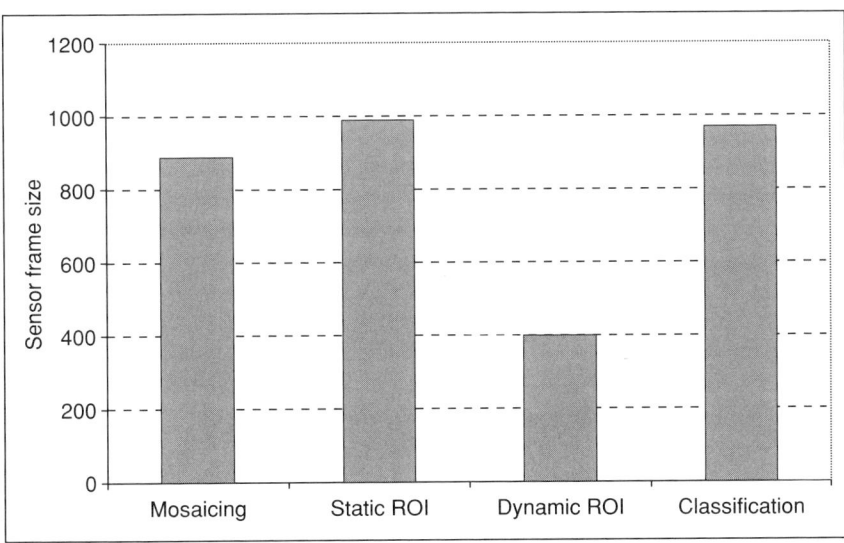

Figure 10.16. Optimum sensor frame size obtained for various operations. used for bandwidth reduction.

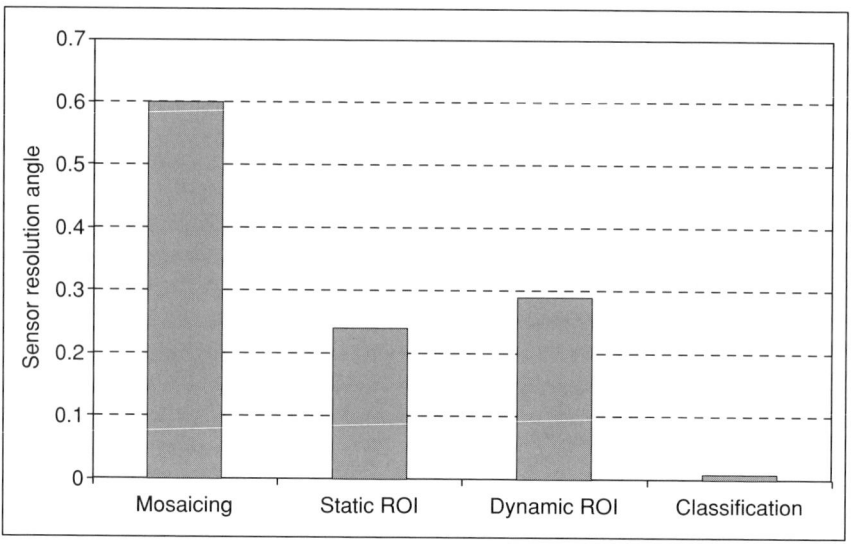

Figure 10.17. Optimum sensor angular resolution angle obtained for various operations used for bandwidth reduction.

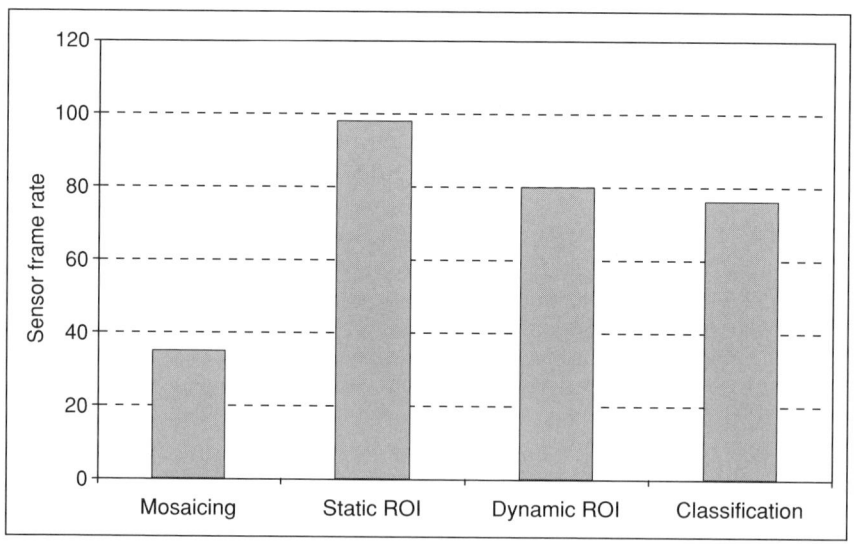

Figure 10.18. Optimum sensor frame rate obtained for various operations used for bandwidth reduction.

Table 10.5. Statistics regarding the genetic algorithm behavior.

BW reduction methods	Maximum fitness value	Generation number	Standard deviation of fitness function	ratio (max. fitness value & median fitness value)	Processing time (s)
Mosaicing	99.999951	91.000000	95.242104	1.0311572	9.6410000
Static ROI	99.998278	70.000000	95.903984	1.0404425	27.625000
Dynamic ROI	99.690316	91.000000	90.492010	1.0842599	34.734000
Classification	99.997571	56.000000	96.594623	1.0088159	55.797000

state of the optimum value. It shows to be closer to 1 for Classification and larger for the other operations. The processing time seems to be largest for the Classification operation and smallest for the Mosaicing.

10.7 Conclusions

In this Chapter, we presented a number of techniques for reducing the transmission bandwidth and explored their effectiveness as functions of UAV and sensor parameters. Then, we maximized these measures of effectiveness by means of a genetic algorithm. An outcome of the approach is an optimum set of UAV state, and sensor parameters, that can be used to automatically control transmission bandwidth. In cases where scene-related parameters could also be controlled, the study shows that another outcome of the approach is an optimum set of these controllable scene related metrics.

References

[1] Stevens R, Sadjadi F, Schantz H, and Esch J. (2002). "A scalable architecture for multirole UAVs utilizing a fiber optic backbone." *Proceedings of the 1st AIAA Unmanned Aerospace Vehicles, Systems, Technologies, and Operations Conference and Workshop*, Portsmouth, VA.

[2] Kumar R., et al. (2001). "Aerial video surveillance and exploitation." *Proceedings of the IEEE*, 89:10.

[3] Sadjadi F., et al. (1991). "Knowledge- and model-based automatic target recognition algorithm adaptation." *Opt. Eng.*, 30:12.

[4] Kay S. (1993). *Fundamentals of Statistical Signal Processing: Detection Theory*. Prentice Hall, Upper Saddle River, NJ.

[5] Sadjadi F, Eekhoff E, and Stevens R. (2003). "On methods for reducing rate of data transmission in unmanned aerial vehicles." *Proceedings of the Association of Unmanned Vehicle Systems International (AUVSI), Unmanned Systems Symposium*, Baltimore, Maryland.

[6] Holland J. (1998). *Adaptation in Natural and Artificial Systems*. MIT.

[7] Goldberg D. (1989). *Genetic Algorithms in Search, Optimization and Machine Learning*. Addison-Wesley.

Information Processing Across Distributed and Netted Systems for Security and Surveillance

Abhijit Mahalanobis,[1] Mubarak Shah,[2] and Alan van Nevel[3]

[1] Lockheed Martin MFC
[2] Dept. of CS, University of Central Florida
[3] NAVAIR

11.0 Introduction

The future of modern security and defense systems lies in managing and exploiting Netted and Distributed systems. The hardware infrastructure for such systems already exists, and is being further revolutionized by new DoD programs. The complexity of managing a network of platforms and sensors is enormous. It requires an organized approach to the uptake, dissemination, and exploitation of information. This task is commonly referred to as Command & Control, Communications, Computers, and Intelligence (C4I). Our objective is not to address the C4I problem, but rather given that a C4I infrastructure exists, we seek to take advantage of the network and its resources to revisit an old problem: that of search and reconnaissance (surveillance), for the detection and recognition of threats.

Even though modern missions are increasingly calling for distributed and netted architectures, the traditional data processing in such systems continues to be typically single sensor and platform centric. In the past, the approaches to target detection, recognition, and tracking have been treated as a stand-alone problem. Thus, the sensing of information has never been guided by the methodology for processing it. There has also been no control over the complexity and/or the quantity of data gathered. Recent efforts seek to integrate sensing and information processing, and represent a fundamental shift in the data processing paradigm that is not only designed for a netted system, but fully exploits the infrastructure to achieve potentially revolutionary performance gains.

The scenario of interest must define the architecture to which the process must be crafted. For instance video cameras abound in civilian life wherever security is of interest (e.g., in commerce, transportation, education, entertainment, and so forth). The viability of distributed security and surveillance capabilities is further enabled by the advent of low-cost cameras, computers, and networking technology (both wired and wireless). Although the component technologies and the infrastructure for such systems already exists today, the challenge is in developing algorithms that work across multiple platforms, and addressing the bandwidth and communication issues. New paradigms must address these issues to ensure the overall system architecture and resources have been taken into account to realize new levels of performance improvements that are critical for mission success. The main questions to address in network-centric security and defense systems are:

1. Where should the data be gathered and how should it be combined (i.e., how should the sensors be tasked and fused)?
2. How much and when should data be gathered (i.e., what is the required resolution of the data, temporally, spatially, and spectrally)?
3. What is the expected gain in performance (in terms of reduced time, increased bandwidth efficiency, increase in detection probability, reduction in false alarm rates etc)?

System concepts can encompass both low-resolution "simple" sensors and high-resolution "detail-oriented" sensors. A large number of relatively low-cost simple sensors can be deployed over a wide area for low-level monitoring purposes. The outputs of these sensors will be processed by appropriate algorithms for estimating target phenomenology and driving the sensor control and scheduling process. Based on the observed "events," detail-oriented sensors may be tasked for detailed analysis of "interesting" regions. It is also feasible that the results of detailed analysis (based on a network of wide-area sensors) will guide the deployment and concentration of simple, near-range ground-based sensors in a region for additional measurements that may be required by the fusion process to confirm the various hypotheses. It is envisioned that the end-to-end process will call for continuous interactions effected between the short-range and wide-area sensors, depending on the specific mission and scenario of interest.

Future netted systems will also allow the possibility of executing missions in a collaborative manner. For instance, in rapid response situations it is essential to reduce the timelines between initial target detection and an effective strike operation. The issue is that targets may move between the time a high-resolution sensor on an airborne platform has detected it, and the decision to strike is made and a weapon is launched. The low-level "disposable" sensors in the vicinity of the designated target can provide useful information about its activity, and update its location to reduce uncertainty associated with weapon delivery and other operational timelines.

Some of the key benefits of netted and distributed systems for Security and Surveillance include:

- **Increased Tolerance to Adverse Conditions** A heterogeneous mix of sensors allows the ability to sense under a wide range of conditions and measure a broad range of phenomenology. The fusion of these sensing modalities is also achieved, and the sensors are scheduled and tasked as needed. This is expected to improve performance under adverse weather, in difficult terrain, and in the presence of time of day variations, obscuration, and camouflage.

- **Improved ATD/R Performance** The overall probabilities of detection and recognition are anticipated to be greater due to the ability to cue sensors to predicted locations of targets, and provide alternate means of detecting targets that are either hidden or not in the field of view of traditional sensors.

- **Efficient Resource Usage:** A balance is achieved between the area of coverage and the bandwidth required to handle the data. Collecting high-resolution data over wide areas not only creates a bandwidth problem, but also imposes heavily on throughput and computational requirements. A mix of sensing resolution allows us to efficiently trade between wide-area surveillance and detailed local area analysis under bandwidth (and throughput) constraints.

- **Improved Strike Capability** The impact of delays in operational time-liness is reduced. Uncertainty is introduced in the position of moving targets between the time it is detected, the decision to strike is made, and a weapon is delivered. In the ISP framework, a target of interest can be "tracked" using low-level sensors to reduce the need to allocate high-level sensors for this task (which may also serve to increase survivability of platforms).

- **Adaptive Self-adjusting Properties** The approach allows a "two-way" control of parameter settings between the sensor network and the algorithms. Parameters internal to the processing algorithms (such as detection thresholds, target clusters and labels, various windows and sample sizes) will be made available to the sensor network to adjust as appropriate. Similarly, the algorithms will influence the choice of sensor and network parameters such as resolution, modality, bit rates, and sensor geometry. It is also anticipated that the network will be able to automatically label potential new targets and include them in its database to achieve rapid target insertion. This ability also reduces the need for exhaustive and time-consuming mission planning as the knowledge base of the system can be updated using "training on the fly" capabilities.

In this chapter, we describe two evolving concepts for information processing on netted systems. The first exemplifies automatic target detection (ATD) and seamless tracking across a network of video cameras. We describe how multiple views are brought together to create a combined view of the world. An overview is given of the KNIGHT human detection and tracking system[1] developed at the University of Central Florida. The integration of the ATD and tracking system and the highlights of the networking and communication process are also discussed. The second example is a collaborative approach to automatic target recognition (ATR). The paradigm requires multiple sensors (platforms) to configure themselves into specific positions relative to each other and the target. We describe a metric that characterizes algorithm performance as a function of sensor position, derive a concept of operation to optimize the performance metric, and discuss the advantages of the process. Both examples illustrate the basic concepts and touch on various aspects of processing information on a netted system.

11.1 Surveillance Using a Network of Video Cameras

In this section, we discuss how it may be possible to take advantage of multiple cameras (with overlapping or nonoverlapping fields of view) in order to monitor activity over a large area. The system must be able to handle both stationary and moving objects. While motion analysis can be used to detect vehicles and humans when they are moving, the ATD/R capability is required for detecting and initiation tracks when they are stationary, and recognizing the detected objects. The system must be able to detect, track, and hand over moving objects between cameras in real time. For seamless operation across platforms, this requires the position of the target in the next field of view (FOV) to be predicted. Based on an analysis of the location of the detections, and registration between the camera views, it becomes possible to depict the positions of the objects and their movements with respect to a site map, thus providing a global composite view of events. This can serve as a powerful monitoring tool by providing situational awareness over the site of engagement.

11.1.1 Target Detection in Multiple Views

We first discuss the approach for detecting stationary vehicular objects (interchangeably referred to as targets). Various target detection and recognition methods may be used depending on the sensor type, range to target, resolution, and other key driving parameters. The objective here is not to build a better ATD/R capability, but to extend the algorithms to work across multiple platforms. For convenience, we use the *maximum average correlation height* (MACH) Correlation Filtering approach[2] for target detection and classification. The basic concept of operation using correlation filters is shown in Fig. 11.1.

Essentially, the input test image is processed by a bank of linear correlation filters that are optimized to respond to the presence of a target by producing a peak at the corresponding location in the output image (also known as correlation plane). Since correlation is a shift-invariant operation, the position of the peak always represents the location of the target, even when it is moving. Each filter is synthesized using representative training images to exhibit distortion tolerance over a limited range of orientations and signature variations. Thus multiple filters are required for every class to accommodate all possible distortions. For example in Fig. 11.1, there are 72 correlation filters for every target

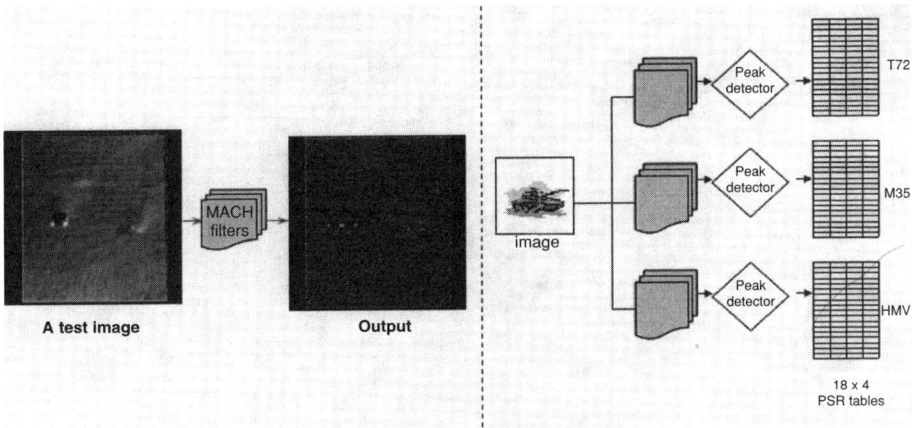

Figure 11.1. An input image is processed by a bank of correlation filters to detect and identify targets. The filter with the highest PSR determines the class of the object, and the position of the correlation peak indicates its location.

to cover 18 aspect bins (each 22.5 degrees wide) and four different signature types (for thermal images these conditions may be hot, cold, day, and night signatures). A metric known as *peak to sidelobe ratio* (PSR) is used to measure the strength of the correlation peaks. The class of the target is declared to be the same as the filter which yields the highest PSR value.

Figure 11.2 illustrates the concept of networking multiple "nodes," each with a camera, processor, and on-board ATD/R capability. The outputs of each node are received at a central "command and control" point where the information is combined. For now, we assume that the sensors and the platforms on which they reside are stationary. Since bandwidth is limited, each node only reports the ATD/R results including pixel position of the detections. Using knowledge of the camera geometry, the location of the target in the

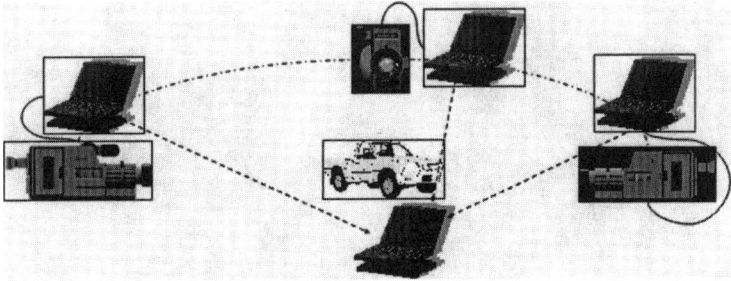

Figure 11.2. Multiple platforms (sensors) are networked to a central computer where ATD/R information is combined into a common reference frame.

sensor view can be converted to a common reference frame, and represented as a point on a site map or an aerial view of the region obtained using an overhead asset. This is further illustrated in Fig. 11.3, where the image in the top window serves as the "site map" and the smaller windows at the bottom represent the three sensor views. The target is detected and recognized in each sensor view, and its pixel position is reported. This data is collected at a central computer where the target coordinates are converted to a common reference frame and fused to depict the location of the target (represented in Fig. 11.3 by the red square) on the site map or overhead image. When the target moves in the sensor views, the site map is updated in real time so that the target position can be seen moving on the site map.

The main advantage of the process illustrated in Fig. 11.3 is that an updated site map that depicts the combined information from multiple sources can be a valuable tool for situational awareness. While individual sensors have only a limited view of the world and may not be able to see around buildings and other obstructions, the combined information can be "dialed-up" by any of the nodes. This allows the local platforms to benefit from the information observed by others in the network. It also allows the command and control center to have a cohesive picture of the battlefield based on multiple observations.

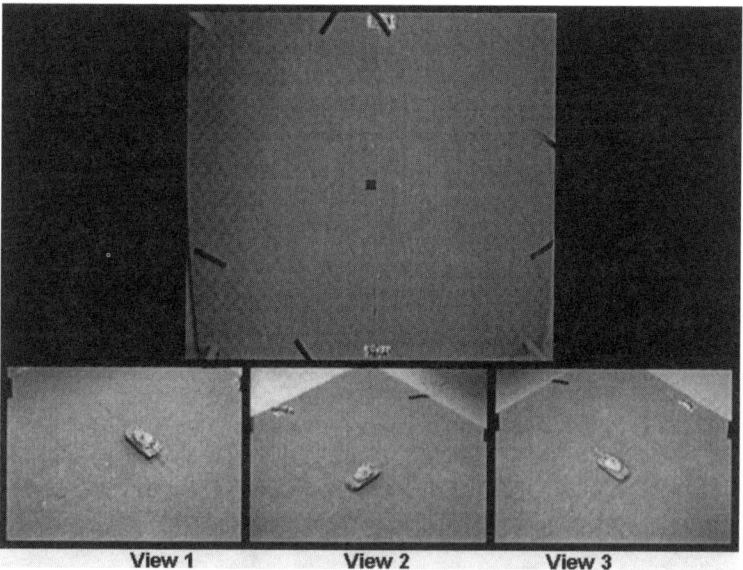

Figure 11.3. The views from the 3 separate nodes (shown at the bottom) are processed locally by an ATD/R, and the position of the target is reported. This data is collected at a central computer where the target coordinates are converted to a common reference frame and fused to depict the location of the target on a site map (or overhead image).

11.1.2 Target Hand-over and Tracking Across Multiple Platforms

Target tracking is an integral and important part of a surveillance system. We first provide a brief overview of the KNIGHT tracking system[1] designed for single camera systems, and then describe its extension to tracking across multiple FOVs. The KNIGHT is a "smart" surveillance system that detects important changes, events, and activities using computer vision techniques, flags significant events, and presents a summary in terms of key frames and textual description of activities to a monitoring officer for final analysis and response decision. The system is robust to illumination changes and weather conditions. The KNIGHT has been installed at four locations in the downtown Orlando area which has Orange Avenue as its primary street, and is currently being field tested. The system employs single camera, and works in real time. KNIGHT consists of four main modules, three of which are shown in Fig. 11.4: object detection and shadow removal, tracking object classification, and activity detection.

Specifically, we view tracking as a region correspondence problem where performance is affected by noisy background subtraction, change in the size of regions, occlusion, and entry/exit of objects. For these reasons traditional approaches cannot be directly applied to tracking humans. To achieve correct correspondence, we have developed a solution based on linear velocity, size, and distance constraints. Furthermore, most of the surveillance systems do not tackle the problems in tracking caused by shadows. To address this issue, we

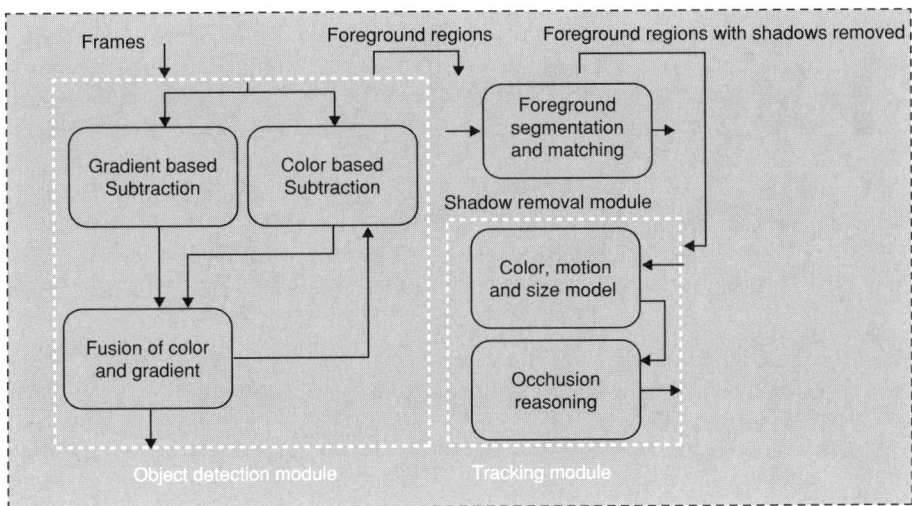

Figure 11.4. An overview of the KNIGHT motion-based detection and tracking system for humans and vehicles. Additional details can be found at http://www.cs.ucf.edu/~vision/projects/Knight/Knight.html

employ a shadow detection approach based on similarity of background and shadow regions.

In addition to tracking moving objects, we believe that motion-based classification helps to reduce the reliance on the spatial primitives of the objects and offers a robust but computationally inexpensive way to perform classification. We have devised a solution to this problem using temporal templates. Temporal templates are used for classification of moving objects. A temporal template is a static vector image in which the value at each point is a function of motion properties at the corresponding spatial location in the image sequence. Motion History and Motion Energy images are examples of temporal templates, proposed by Bobick and Davis.[3] Motion History image is a binary image with a value of one at every pixel where motion occurred. In Motion History image, pixel intensity is a function of temporal history, i.e., pixels where motion occurred recently will, have higher values as compared to other pixels. These images were used for activity detection. We have defined a specific Recurrent Motion template to detect repeated motion. Different types of objects yield very different Recurrent Motion Images (RMIs) and therefore can easily be classified into different categories on the basis of their RMI. We have used the RMIs for object classification and also for detecting carried objects.

11.1.2.1 Tracking Across Multiple Fields of View

To track objects successfully in multiple cameras, one needs to establish correspondence between objects detected and tracked in each camera. Our system is able to discover spatial relationships between the camera FOVs and use this information to correspond between different perspective views of the same person. We employ a novel approach of finding the limits of FOV of a camera as visible in the other cameras that is very fast compared to conventional camera calibration-based approaches. Using this information, when a person is seen in one camera, we are able to predict all the other cameras in which this person will be visible. Moreover, we apply the FOV constraint to disambiguate between possible candidates for correspondence.

When tracking is initiated, there is no information provided about the FOV lines of the cameras. The system can, however, find this information by observing motion in the environment, as illustrated in Fig. 11.5. Whenever there is an object entering or exiting one camera, it actually lies on the projection of the FOV line of this camera in all other ones in which it is visible. Suppose that there is only one target. Then, when it enters the FOV of a new camera, we find one constraint on the associated line. Two such constraints will define the line, and all constraints after that can be used in a least-squares formulation. In an earlier chapter[4], it was demonstrated that the initialization of FOV lines by one person walking in the environment for about 40 s was sufficient to initialize the lines. These lines were then used to resolve the correspondence problem between cameras. However, it is not always possible to have only one target

Figure 11.5. The automatic calibration of three separate cameras with overlapping fields of view (FOVs) is shown. The FOV boundary lines are established by observing where moving objects visible in one camera simultaneously appear at the edge of another camera's view. Places where this occur represent points on the boundaries of FOVs of other cameras that are visible in the current view.

moving in the scene. When multiple targets are in the scene and if one crosses the edge of FOV, all targets in other cameras are picked as being candidates for the projection of FOV line. Since the false candidates are randomly spread on both sides of the line whereas the correct candidates are clustered on a single line, correct correspondences will yield a line in a single orientation, but the wrong correspondences will yield lines in scattered orientations. We can then use Hough transform to find the best line in this case. This method needs more points for a reliable estimate of the lines and therefore takes longer time to set up correctly. Additional constraints derived from categorization of objects and their motion may be used to reduce the number of false correspondences, thus reducing the time it requires to establish the lines.

11.1.3 Network Integration of Tracking and ATD/R

The ability to detect, track, and recognize objects across a network has been demonstrated across both wired and wireless networks. The concept of a wireless peer-to-peer ad hoc network is shown in Fig. 11.6. This system was built and tested using laptop PCs, each equipped with a Synchrotech℠ adapter and a wireless card from MeshLAN℠. We also tested operations on a commercially available 802.11b wireless hub. A socket based communication over TCP/IP was used to network three PCs that acted as "clients" and a fourth one as the "server." Network architecture is traditionally split into layers starting at the top application layer and going progressively down towards the hardware. The Transmission Control Protocol (**TCP**) forms the Transport layer and beneath it the Internet Protocol (**IP**) forms the Network layer. The Transport layer looks after assembling whole messages from individual packets whatever route they may take and the Network layer looks after getting individual packets across the network. If data packets are lost then TCP automatically attempts to retry the operation. It uses a simple acknowledgment interchange to ensure this. Within TCP/IP, the two communicating

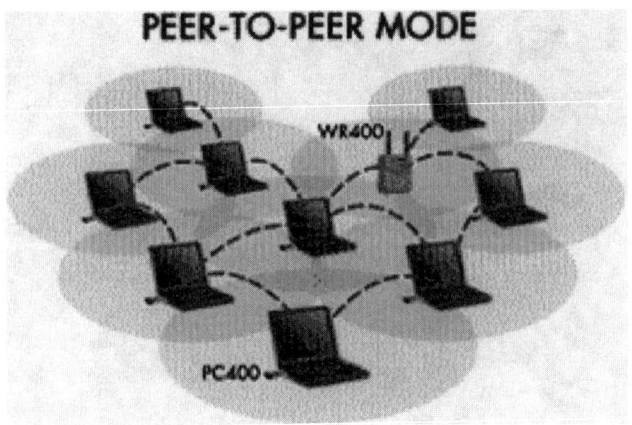

Figure 11.6. Example of an ad hoc peer-to-peer network.

programs (server and client), allocate sockets and then connection is initiated by the client program. The server continually listens for connect requests and then chooses to accept a connection from them. This client–server model is an appropriate scheme for the distributed network as there are many clients making connection requests for information from one place (the server).

The particulars of the interactions are as follows. The KNIGHT tracking system executes locally at each of the clients and the local tracking data is sent to the server. The server ensures that the tracked entities from the clients are deconflicted and properly associated, and assigned global labels as described in Section 3. It is also essential to synchronize the frames processed at the clients so that the proper temporal correspondence can be made. The global labels are then received back at the clients and used for consistent labeling and display purposes. At startup, the server is in a "training mode" to establish the FOV boundaries based on the entry and exit of moving objects across the different FOVs. Thereafter, the main purpose of the server is to generate and return consistent labels for the tracked objects.

The ATR–Tracker interactions occur only at each client, as shown in Fig. 11.7. When a moving object is detected,[a] a 64×128 region of the image containing the tracked object is fed to the ATR for classification, and the result is used for generating the *target call* (class label) and *confidence* associated with that object, which is then sent to the server. When a new object enters the FOV of a client the *target call* sent by the ATR is used as the label. If however the object is already in track (i.e., it corresponds to an existing object) it gets the

---

[a] For now we use the motion-based detection to cue the ATR. The ability to detect stationary targets using the ATD/R and initiate tracks on them will be incorporated in future versions.

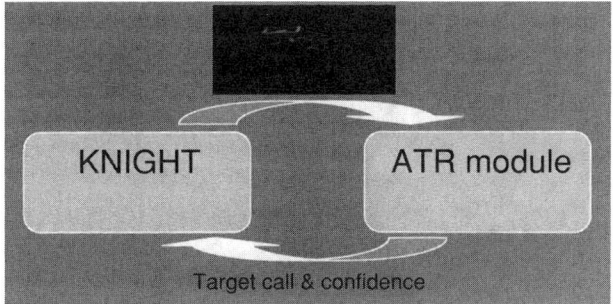

Figure 11.7 ATR–Tracker interactions occur only at each client. The results of the ATR including class label and confidence are sent to the server.

target call with the highest *confidence* (including those from previous classification results) assigned to the object as its label.

Figure 11.8 illustrates the interaction of the ATR and tracker across a network using models for a "Tank" and a "Mini," a relatively smaller vehicle. The pictures represent snapshots of actual events that occurred during a real-time test and demonstration of the algorithms. As these objects move from right to left across the three FOVs, the ATR labels are correctly established and handed over across the clients via the server. The color of the box containing the target is set to green if it is recognized to be the Tank, blue when it is the Mini, and red if it cannot be recognized. In this instance, the Tank is visible and correctly recognized in the left and middle camera views. The Mini is visible in all three views, but is too close to the edge in the left camera view to be recognized. It is however recognized correctly in the middle and right camera views. It should also be noted that the tracking labels P122 and P123 are consistently assigned by the server to the Tank and Mini across all three views. Thus, the KNIGHT video processing system has been extended to work across multiple platforms, achieving ATD/R, tracking, and handover between multiple FOVs in real time.

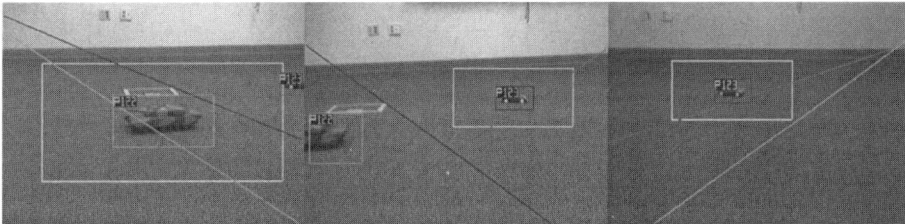

Figure 11.8. Snapshots from real-time demonstration show the detection, tracking, classification, and handover of targets across a network of computers.

11.2 A Collaborative Approach to Object Recognition

In this section, we discuss a novel method where several sensors and ATRs collaborate to recognize objects. Such an approach would be suitable for network-centric application where the sensors and platforms can coordinate to optimize overall ATR performance. We use correlation pattern recognition techniques to facilitate the development of the concept, although other algorithms may be easily substituted. Essentially, a self-configuring network is proposed that positions the sensors optimally with respect to each other depending on the algorithm and the class of the object to be recognized. We show how such a network optimizes overall performance, and illustrate the scheme by means of examples.

11.2.1 Background

Consider a scenario where an object is viewed by multiple sensors from different angles.[b] We note that the concept is applicable to any generalized distributed network of sensors, whether airborne or ground based. The question is how should the sensors be coordinated to provide the optimum views to recognize a class of objects? What methodology should the pattern recognition algorithm exercise on the sensors to optimize overall performance? Some of the fundamental issues involved in answering these questions are addressed in this chapter.

In current approaches, the position of the sensors is not determined by either the processing algorithm or the object to be recognized. The algorithms are designed independently of the sensing process, and therefore, overall system performance is not necessarily optimized. One approach to solving this problem is to define a metric that characterizes performance as a function of a parameter, say viewing geometry, and drive the configuration of the sensors to optimize the metric. Clearly, the metric and performance characteristics will be a function of the processing algorithm as well. In this paper, we develop such a metric for correlation filtering algorithms, and illustrate the concept using SAR images of targets from the public domain MSTAR data set.

The field of Correlation Filters has been reviewed by Kumar.[15] Although many schemes exist that describe the use of multiple correlation filters, of particular interest is the concept of the k-tuple SDF[5] where a bank of filters work together to produce a unique code for each class of objects. In this scheme, the filters are designed to satisfy unique constraints on output produced in response to training images of a particular class. For instance, consider an 8-class problem that must be recognized using k = 3 filters, each producing a binary (1 or 0) output. The possible 3-bit code associated with each class is shown in Table 11.1.

---

[b] For instance, the sensors may be on board multiple UAVs, or the different views may be collected using a sensor on a single platform.

Table 11.1. A simple 3-bit binary code to represent 8 classes.

Class	3-bit Code
1	000
2	100
3	010
4	110
5	001
6	101
7	011
8	111

Each column of the 3-bit code is treated as the output of a particular filter in response to the various classes. Thus, if the vector $\mathbf{x}_i^j$ represents the i-th training image of the j-th class, then the constraints on the first filter $\mathbf{h}_1$ may be expressed as

$$[_1\mathbf{x}_1^1 \wedge_4 \mathbf{x}_N^1 \, _4\mathbf{x}_1^2 \wedge_4 \mathbf{x}_{2N}^2 \wedge_4 \, \mathbf{x}_1^8 \wedge_4 \mathbf{x}_{4N}^8]^T \quad \mathbf{h}_1 = \begin{bmatrix} 0 \\ 1 \\ M \\ 0 \\ 1 \end{bmatrix} \otimes \begin{bmatrix} 1 \\ M \\ 1 \end{bmatrix} \tag{1}$$

where the symbol $\otimes$ denotes a Kronecker product, $\mathbf{u}_1$ is the output constraint vector based on the first bit of the code, and $\mathbf{X}$ is a matrix with the training image vectors as its columns. It is easy to show that the simplest solution to Eq. 1 that minimizes the output variance due to AWGN is simply

$$\mathbf{h}_1 = \mathbf{X}(\mathbf{X}^T\mathbf{X})^{-1}\mathbf{u}_1 \tag{2}$$

Similarly, the second and third filters (i.e., $\mathbf{h}_2$ and $\mathbf{h}_3$) can be designed by using the appropriate columns of the 3-bit code in Table 11.1 as the output constraint vector. In general, the filter outputs do not have to be restricted to binary values. Sudharsanan et al.[6] have discussed methods for selecting optimum values for the output vector to minimize overall probability of error.

The main benefit of using multiple ATRs to form a unique code for each class is 1) a large number of classes can be recognized using relatively few ATRs and 2) the overall robustness is improved if error correction techniques are used. Redundancy in the coding scheme helps to reduce the overall probability of error since several symbols would have to be wrong before one class is confused for another. The selection of good codes for designing ATRs is a subject that is worthy of study on its own.

The solution in Eq. 2 is referred to as a *projection SDF*[5] in the literature. In this chapter we will employ a different filter design technique known as the *maximum average correlation height* (MACH) filter[7] to develop a coded classifier (in fact, any composite filter design technique may be used). The relevant details of the MACH filter design process will be reviewed in Section 11.2.2. Although we will consider an architecture reminiscent of the k-tuple SDFs, our focus is not on the coding scheme, but rather on the development of a concept of operation so that the probability of error for individual bits of the code is minimized. This is done in the context of the relative position of the sensors, the properties of the algorithm used, and the specific targets themselves.

11.2.2 Performance Characteristic Function and Filter Synthesis

For simplicity of discussion, assume that two separate ATRs (that produce a 2-bit code) must be designed to work together to recognize two different classes denoted by ω_x and ω_y. For the purposes of the discussion in this paper, we treat each ATR to be a MACH type correlation filter, and represent them $H_1(k,l)$ and $H_2(k,l)$, respectively. Unless otherwise stated, all quantities are in the frequency domain. We require that if class ω_x is present, $H_1(k,l)$ should produce a large positive output which is treated as a "1" if it exceeds a threshold T_1. Similarly, $H_2(k,l)$ should produce a large negative output which is treated as a "0" if it less than a threshold T_2. Thus, the output code [1 0] should be obtained whenever ω_x is present. Conversely, if ω_y is present, the filters are designed such that $H_1(k,l)$ yields a large negative value while $H_2(k,l)$ yeilds a large positive value which produces the output code [0 1]. The question is, what is the best angular position for the filters relative to the target and to each other?

As noted in Section 11.2.1, we seek a metric that characterizes performance as a function of viewing geometry, and then drive the configuration of the sensors to optimize the performance metric. Towards this end, we define a *distance* or *separation* metric based on the MACH filter algorithm (a similar function can be derived for essentially any ATR algorithm). The formula for the MACH filter is straightforward.[7] Let $X_i(k,l)$ represent the 2D Fourier transforms of N training images of class ω_x, selected to represent viewing angles of the class 1 object around $\theta°$. Similarly, $Y_i(k,l)$ are the 2D Fourier transforms of N training images of class ω_y that represent viewing angles of the class 2 object around $\alpha°$. The *mean* and *spectral variance* for each of the classes are defined as

$$M_x^\theta(k,l) = \frac{1}{N}\sum_{i=1}^{N} X_i(k,l) \quad S_x^\theta(k,l) = \sum_{i=1}^{N} |X_i(k,l) - M_x^\theta(k,l)|^2$$
$$M_y^\alpha(k,l) = \frac{1}{N}\sum_{i=1}^{N} Y_i(k,l) \quad S_y^\alpha(k,l) = \sum_{i=1}^{N} |Y_i(k,l) - M_y^\theta(k,l)|^2 \tag{3}$$

We will first discuss the design of one of the filters, say $H_1(k,l)$, stating that the design of second filter will follow the same paradigm. The expression for the first MACH filter that separates a $\theta°$ view of class 1 from a $\alpha°$ view of class 2 is

$$H_1(k,l) = \frac{M_x^\theta(k,l) - M_y^\alpha(k,l)}{S_x^\theta(k,l) + S_y^\alpha(k,l)} \tag{4}$$

In fact, it is easy to show that *distance* or *separation* produced by this filter as function of the angles α and θ is given by

$$\begin{aligned}
Q(\theta,\alpha) &= \sum_k \sum_l |[M_x^\theta(k,l) - M_y^\alpha(k,l)]^* H_1(k,l)|^2 \\
&= \sum_k \sum_l \frac{|M_x^\theta(k,l) - M_y^\alpha(k,l)|^2}{S_x^\theta(k,l) + S_y^\alpha(k,l)}
\end{aligned} \tag{5}$$

We refer to this function as the *MACH separation metric*. Our strategy is to train the filter at the specific viewing angles of each class that maximize $Q(\theta,\alpha)$.

We now illustrate by means of an example the characteristic behavior of the separation metric as a function of the angles α and θ. Towards this end, we use SAR images of a T72 (class-1 object) and BTR (class-2 object) from the MSTAR public domain data set such as those shown in Fig. 11.9. There were approximately 230 images of each class that cover all aspect views. The function $Q(\theta,\alpha)$ was evaluated for all values of α in increments of $10°$ while fixing $\theta = 0°$. In other words, the zero degree view of the T72 was compared to all possible views of the BTR. The resulting behavior of $Q(0,\alpha)$ is shown in Fig. 11.10. The number of images, N, was determined for both classes by using a $45°$ window centered on angles being compared. Thus, the images of the T72 that fall in the range of angles $\pm 22.5°$ and the images of the BTR that lie in the range $\alpha \pm 22.5°$ were used in the estimation process. The numerical value of the separation is around 315 when both targets are

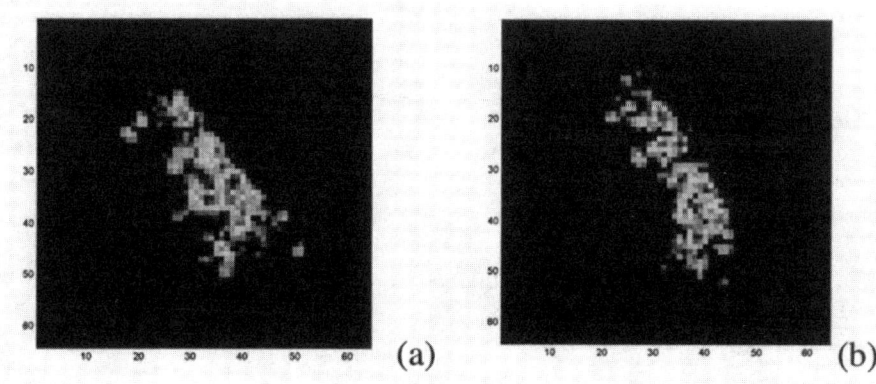

Figure 11.9. Typical SAR images of (a) T72 and (b) BTR from the public domain MSTAR database.

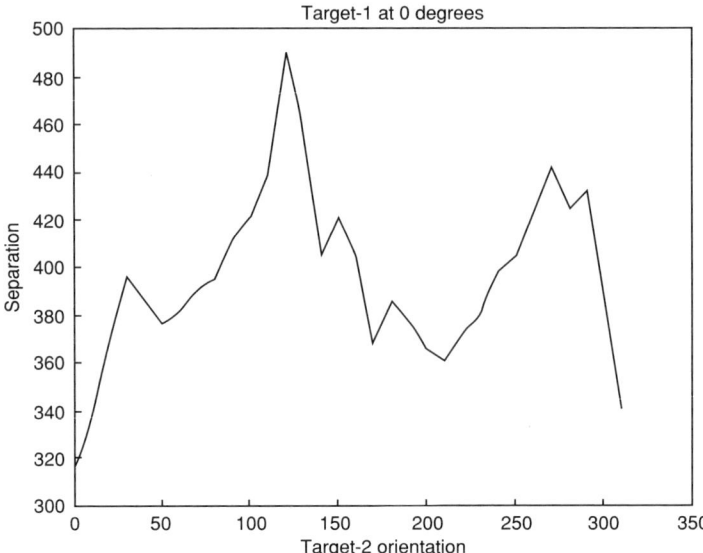

Figure 11.10. The behavior of the MACH separation metric shows that the 0° view of the T72 is most separable from the view of the BTR at about 130°.

compared at 0°, but increases to around 480 when the BTR is at 130°. This shows that the 0° view of the T72 is most separable from the 130° view of the BTR, and that it is not optimum to compare the targets at the same orientation.

The behavior of $Q(\theta, \alpha)$ for all possible values of θ and α is shown as a 2D image in Fig. 11.11. The color bar indicates the magnitude of the separation, red being the greatest and blue being the least. Using a 10° increment and a range of ±22.5° centered on the angles being compared leads to about 32 samples for each angle (i.e., multiplying the sample indices by 10 yields the approximate value of α and θ).

We see immediately that the two targets are most separable if class-1 (T72) is at $\theta_1 \cong 30°$ while class-2 (BTR) is at $\alpha_1 = 130°$. Therefore, these angles are the orientations at which the class means and spectral variances in Eq. 4 are estimated to obtain the best choice for $H_1(k,l)$. Such a filter will produce the largest positive values in response to class-1 when it is viewed at $\theta_1 \cong 30°$ orientation. Similarly, a view of class-2 around $\alpha_1 = 130°$ should induce the filter to produce the most negative values. The question now arises what is the best combination of angles for the second filter? This is easily answered by selecting those values of α and θ for which the second largest value of $Q(\theta, \alpha)$ is obtained. We see that this occurs when class-1 at $\theta_2 \cong 130°$ is compared to class-2 at $\alpha_2 = 40°$. Therefore, images centered on these angles are selected for synthesizing $H_2(k,l)$. To obtain the desired code however, this filter should produce a negative response to the class 1 and a positive response to class 2 which is easily accomplished by using the formula

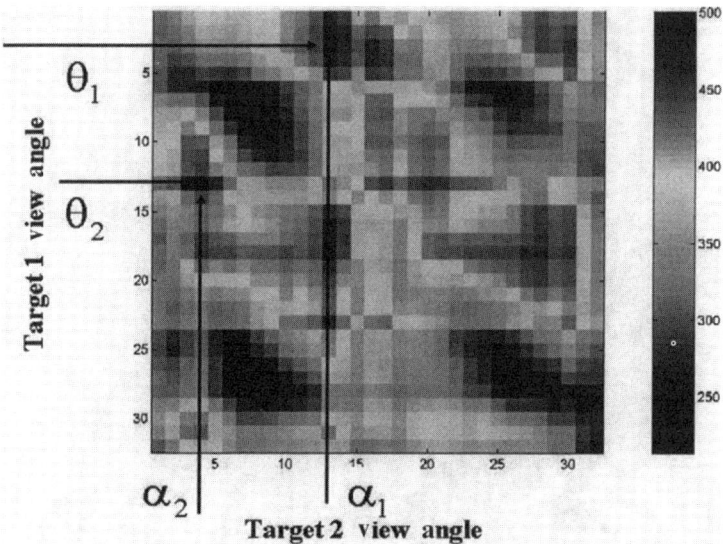

Figure 11.11. The separation metric is depicted here as a function of all possible combinations of the viewing angles of both targets. This 2D array of values characterizes the separability of the two classes as a function of the angular position of the sensors. The best separation between the two classes occurs when class-1 (T72) at $\theta_1 = 30°$ is compared to class-2 (BTR) at $\alpha_1 = 130°$. The second best separation occurs at $\theta_2 = 130°$ and $\alpha_2 = 40°$.

$$H_2(k,l) = \frac{M_y^\alpha(k,l) - M_x^\theta(k,l)}{S_x^\theta(k,l) + S_y^\alpha(k,l)} \qquad\qquad 6$$

Although this is the negative of the expression in Eq. 4, it should be noted that $H_2(k,l)$ is not $-H_1(k,l)$ since each filter is trained at a different combination of viewing angles.

11.2.3 Concept of Operation: A Collaborative Formation of ATRs

The basic concept of operation is shown in Fig. 11.12. Consider a scenario where an object has been detected, and we wish to verify its class. We assume that ATR-1 and ATR-2 ($H_1(k,l)$ and $H_2(k,l)$ in our case) are on separate platforms, each with its own sensor. To drive the relative position of the sensors in an optimum manner consistent with the metric described in Section 11.2.2, the platforms must fly in specific formation. Under the hypothesis that the object belongs to Class-1, ATR-1 should yield a strong positive response (code bit 1) when the object is viewed from angle θ_1. Similarly, ATR-2 should produce a strong negative response (code bit 0) when the object is viewed from the angle θ_2. If the orientation of the object is known, both ATR-1 and ATR-2 can fly to the necessary positions and obtain images at the optimum

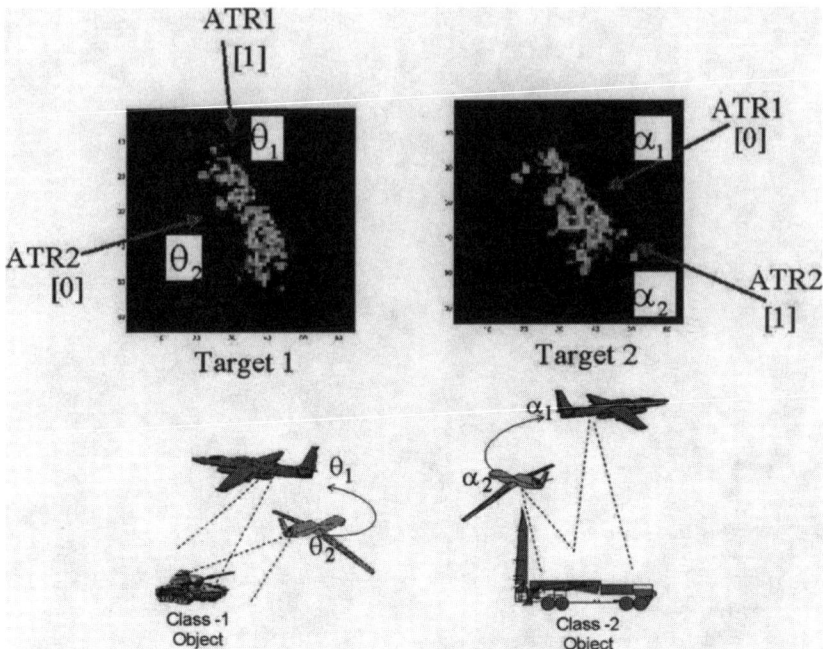

Figure 11.12. The sensors can fly towards the object at optimum angles if its orientation is known. Otherwise the sensors may search for the correct 2-bit code by flying around the object at a relative angular separation of $\theta_1 - \theta_2$ or $\alpha_1 - \alpha_2$, depending on whether it is believed to belong to Class-1 or Class-2.

angles. Otherwise, the two platforms should fly around the object with a relative angular separation of $\theta_1 - \theta_2$, checking to see if a strong [1 0] code is obtained. Similary, to verify the hypothesis that the object belongs to Class-2, the sensors should group into a new formation with a relative angular separation of $\alpha_1 - \alpha_2$ and fly around the object to see if the code [0 1] is obtained.

The method outlined in this paper achieves the process depicted in Fig. 11.13. Multiple ATRs on separate platforms collaborate to produce unique codes for different target classes. The process is driven using a metric that characterizes the dependency of the overall performance on the position of the sensors, relative to each other and to the specific target. *Thus, an optimum configuration of sensors exists for each class of interest.* The overall process treats the relative position of the sensors as a part of the classification algorithm, and configures the formation of the platform to optimize class separation.

11.3 Conclusions

In this chapter we discussed the key issues involved in information processing for security and surveillance application using a distributed and netted system.

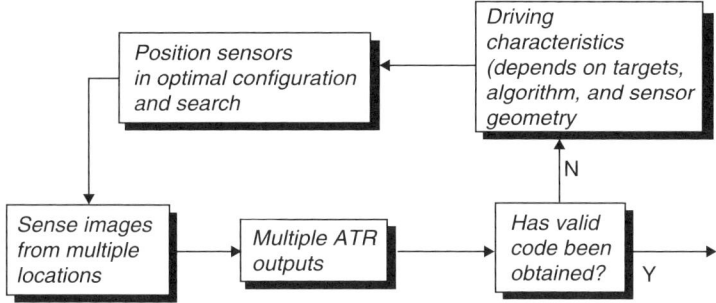

Figure 11.13. The overall performance of multiple ATRs is optimized by first obtaining a metric that characterizes performance as a function of sensor properties (angular position in this case) as well as target types, and then uses this function during operation to drive the sensors into an optimum configuration.

The underlying concepts were illustrated by means of two case studies. In Section 11.1, we discussed target detection, recognition, and tracking across a network of stationary platforms with fixed mounted cameras. The approach illustrated how several existing components such as COTS computers and networking technology, video trackers, and ATD/R algorithms can be brought together to address the need for wide-area surveillance in a distributed processing environment. The tracking system is able to automatically establish where the camera FOVs intersect, and use this information to generate consistent labeling of objects across the network. This process was further augmented using a correlation based ATR algorithm to classify the tracked objects and assign unique labels. The interactions between the ATR and tracking algorithms were defined, and the algorithms were shown to work across a network of three client computers and a server using the TCP/IP protocol. The goal is to eventually evolve the concept of moving and airborne platforms. Here we envision that initially video data will be wirelessly transmitted to receiving computers on the ground where the processing will take place. In the future, it may be advantageous to process imagery aboard the platform and transmit only the salient results across the network. The greater challenge is to solve the FOV registration and the relative calibration between cameras for the moving platform scenario. While we seek a purely image based solution to this problem, it may be advantageous to explore the potential benefits of using GPS and other information about the platforms and their positions relative to one another. In the future, we anticipate that the ability to register the field of views of cameras on moving platforms may potentially lead to novel simplification of the guidance and control required to coordinate the relative behavior of the platforms.

In Section 11.2, we discussed new evolving paradigms for collaborative target recognition that require specific configuration of the platforms around the targets. Such algorithms also heavily enable to automatically calibrate multiple moving FOVs to associate and track objects across them. We introduced a method to design a network of collaborative sensors that reconfigure their position to maximize class separation, and hence maximize overall performance. This was achieved by defining a function $Q(\theta, \alpha)$ that characterizes the separation metric as a function of not only the targets, but also the viewing angles. The proposed approach is general in that such a function can be defined for any ATR algorithm of choice, not just correlation filters. The approach can be extended to parameters other than viewing geometry following the same methodology. Although the approach was described for a Class-2 example, the process can be extended to any number of classes and sensors. In the general case, a swarm of N platforms would operate together and fly in specific formation depending on the class of interest. When only one platform is available, the proposed approach is still helpful to determine how to sample the best views of the targets. Work is currently ongoing to conduct a statistical analysis of the optimality of the proposed approach and to quantify the advantage over conventional methods using numerical simulations.

References

[1] Javed O. and Shah M. (2002). "Tracking and object classification for automated surveillance." in: *The Seventh European Conference on Computer Vision (ECCV 2002)*, Copenhagen.
[2] Javidi B., Ed. (2002). Correlation pattern recognition: An optimum approach. *Image Recognition and Classification: Algorithms, Systems, and Applications.* (BICTL) Marcel Dekker, New York. pp. 295–321.
[3] Bobick A. and Davis J. (2001). "The recognition of human movements using temporal Templates." *Trans. IEEE PAMI*, 23(3).
[4] Khan, S. Javed, O. and Shah M. (2001). "Tracking in uncalibrated cameras with overlapping field of view." *Performance Evaluation of Tracking* and *Surveillance PETS 2001, (with CVPR 2001)*, Kauai, Hawaii, 9th, Dec 2001.
[5] Vijaya Kumar BVK (1992). "Tutorial survey of composite filter designs for optical correlators." *Appl. Opt.*, Vol. 31, pp. 4773–4801, 1992
[6] Sudharshanan, SI Mahalanobis, A. and Sundareshan, MK. (1990). "Selection of optimum output correlation values in synthetic discriminant function design." J. Opt. Soc. Am. A., 7(4): 611–616.
[7] Mahalanobis A, Vijaya Kumar, BVK Sims, SRF and Epperson J. (1994). "Unconstrained correlation filters." *Appl. Opt.* 33: 3751–3759.

Composite Correlation Filters and Neural Networks for Identification and Pose Estimation

Albertina Castro,[1] Yann Frauel,[2] and Bahram Javidi[3]

[1]Instituto Nacional de Astrofísica, Óptica y Electrónica, Apdo. Postal 216, Puebla 72000, Puebla, México, *betina@inaoep.mx* Tel: +52 (222) 266 3100, Fax: +52(222) 247 2940
[2]Instituto de Investigaciones en Matemáticas Aplicadas y en Sistemas, Universidad Nacional Autónoma de Mexico, Apdo. Postal 20-726, Admón. 20 Del. A. Obregón 01000, México DF. México, *yann@leibniz.iimas.unam.mx*
[3]Department of Electrical and Computer Engineering, University of Connecticut, U-1157, Storrs, 06269-2157, Connecticut, USA, *bahram@engr.uconn.edu*

12.0 Introduction

An important problem that arises in three-dimensional (3D) object recognition is due to the changes in orientation of the object under study. These changes in orientation introduce distortions in the two-dimensional (2D) projections, which can impair the recognition task. Consequently, it is essential to consider these changes in orientation in any 3D object recognition problem. Furthermore, there are circumstances where it is desirable to know the orientation of the object under study, that could be the case of some industrial processes where one finished piece should meet another one in a specific orientation; or in surveillance and security issues where it is important to know the place a vehicle is heading to or the direction a person is looking at. The problem of determining the 3D orientation of an object is known as pose estimation.

Pose estimation by itself is one of the oldest computer vision problems. It is important for view-invariant face recognition[1-3] and robot vision tasks.[4] It also has applications in biomedical[5,6] and meteorological imaging[6] or photogrammetry.[7] Many approaches for pose estimation involve the knowledge of a set of reference points[8,9] or the detection of features followed by comparison to a known 3D model.[1,3,10] Here, we focus on a different technique where the reference object is not known through a geometrical model but rather as a collection of 2D projections obtained from various points of view. The pose is then estimated from a single unknown 2D view.

In this chapter we present a technique based on images, linearly weighted synthetic-discriminant function filters and an artificial neural network.[11] The principle of this technique is to feed an artificial neural network with the correlation peak values obtained from the correlation between the input scene and three composite correlation filters. We propose two variations of this technique, using either linear or nonlinear correlations. We present a study of the robustness of the proposed technique when the input scene is affected by noise.

12.1 Pose Estimation Using Linear Correlation

In this section we explain the use of linear correlation filters to estimate the pose of an object. We describe the design of the composite filters and the results of recognition and pose estimation using linear correlation. Although this is the easiest and most obvious way of applying the proposed technique, we show that linear correlation is not satisfactory if object recognition is needed in addition to pose estimation.

12.1.1 Construction of the Composite Correlation Filter

In order to illustrate the proposed technique, we use images of an F15 airplane and consider two pose parameters. These are in-plane rotation and out-of-plane rotation. The out-of-plane rotation is considered as a change of azimuth angle of the object, while in-plane rotation is a rotation around the optical axis of the sensor. Figures 12.1(b) and (c) are images with out-of-plane and in-plane rotation correspondingly with respect to the image of the reference object of Fig. 12.1(a). The method we present here was originally motivated by the work of Monroe and Juday.[12] The starting idea of this technique is to determine the pose of an object by comparing the 2D view under study with several views with known orientations. The comparison can be done through a correlation measurement.[13–15] However, it implies the comparison of the input scene with any stored view of the reference object and it is therefore both storage and time consuming. To lessen the problem, several reference views of the object can be combined into a single synthetic-discriminant function (SDF) filter.[16–22] The filter is constructed by weighting the views in such a way that the value of the correlation peak varies linearly with the pose parameters. The set of views used to construct the filter is called construction set. A basic SDF is given by

$$h = \mathbf{S}(\mathbf{S}^t \, \mathbf{S})^{-1} c, \tag{1}$$

where the upper t stands for a matrix transposition operation, $\mathbf{S}$ represents a matrix whose columns are the images of the construction set rearranged as column vectors. The vector c contains the assigned correlation values for every image of the construction set. We discuss next how to compute these c values.

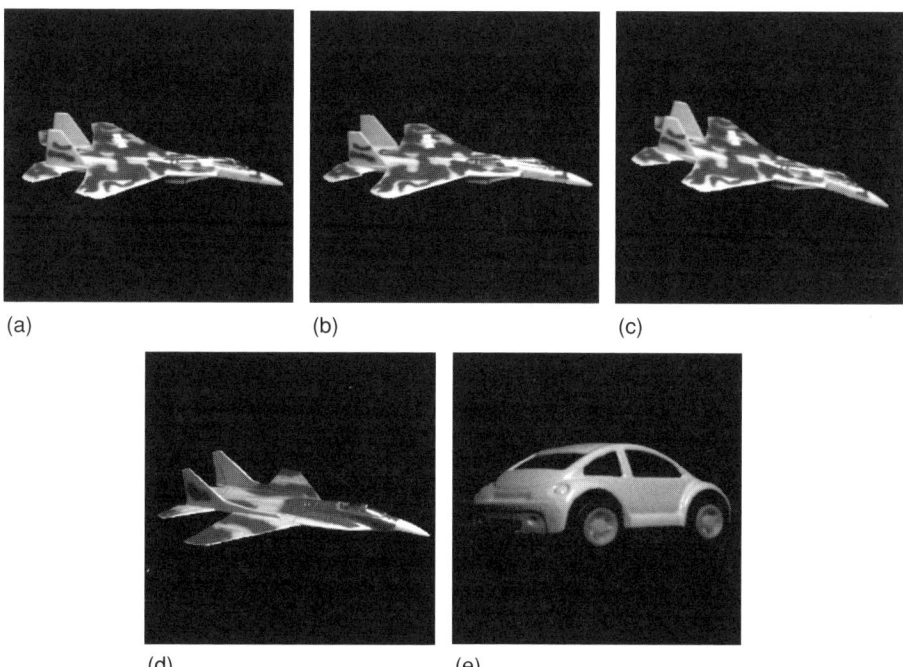

Figure 12.1. Images used for the pose estimation and object recognition: (a) reference object, an F15 airplane, (b) out-of-plane rotated, and (c) in-plane rotated versions of the airplane. (a), (b), and (c) are three of the nine images included in the construction set; (d) and (e) are false targets.

We consider the out-of-plane rotation θ and the in-plane rotation ϕ of the reference airplane, both over a 0 degree to 8 degree angle. The construction set for one linearly weighted composite filter includes nine images corresponding to rotation angles 0, 4, and 8 degrees in both directions. Each image has 256×256 pixels and contains a view of the reference F15 airplane previously extracted from its background. Figures 12.1(a)–(c) show three of the nine images of the construction set. Figures 12.1(b) and (c), respectively, present an out-of-plane and an in-plane rotation with respect to Fig. 12.1(a). Figures 12.1(d) and (e) represent false targets.

In addition to estimating the out-of-plane rotation θ and in-plane rotation ϕ we want to be sure that the presented object is really the reference object. We therefore define the pose vector $p = [\theta, \phi, \mathrm{RF}]^{t}$, where RF is a recognition flag and should be 1 when the tested object is of the class of the reference object. Now, since we want to retrieve three parameters, we need to obtain at least three independent measurements from an image. We therefore construct three different composite filters h_1, h_2, and h_3 with the same construction set but different weights.

Namely, we define the relation between the vector containing the correlation values c and the pose vector p to be linear, so that for a particular image with pose $p = [\theta_0, \phi_0, RF_0]^t$ the correlation values $c = [c_1, c_2, c_3]^t$ for each of the three filters will be given by

$$c = \mathbf{T}p, \tag{2}$$

where $\mathbf{T}$ is a 3×3 transformation matrix. This matrix is fully defined by three independent (p, c) pairs. We use, for instance, three points of our pose space (0, 0), (0, 8), and (8, 8) and as mentioned before, we add a third column with a value of 1 to specify that the images belong to the true class; that is, $p_1 = [0,0,1]^t$, $p_2 = [0,8,1]^t$, and $p_3 = [8,8,1]^t$. We decide that each filter will give a correlation value of 1 for one of these three points and 0,8 for the other two points. Explicitly matrix $\mathbf{T}$ is found as

$$\mathbf{T} = \begin{bmatrix} 1 & 0.8 & 0.8 \\ 0.8 & 1 & 0.8 \\ 0.8 & 0.8 & 1 \end{bmatrix} \begin{bmatrix} 0 & 0 & 8 \\ 0 & 8 & 8 \\ 1 & 1 & 1 \end{bmatrix}^{-1}. \tag{3}$$

Once the matrix $\mathbf{T}$ is obtained as described above, we compute according to Eq. (2) the correlation values that we should get for the other images of the construction set. That is

$$\mathbf{C}_{const} = \mathbf{T}\mathbf{P}_{const}, \tag{4}$$

where $\mathbf{P}_{const}$ is a matrix whose columns are the pose vectors for every image of the construction set. Each row of the matrix $\mathbf{C}_{const}$ contains the correlation values for one of the filters for the nine images of the construction set. As we can observe from Figs. 12.2(a)–(c), each filter has indeed a linear dependency of the correlation peak value with respect to the two parameters of orientation.

Now, to generate the filter h_1 we take the transpose of the first row of $\mathbf{C}_{const}$ as the vector c required in Eq. (1). Then h_1 is converted back from a vector form to an image form. Filters h_2 and h_3 are generated in the same way now using rows 2 and 3 of matrix $\mathbf{C}_{const}$, respectively, as the required vector c in Eq. (1).

12.1.2 Linear Estimation of the Pose

The aforementioned pose estimation approach expects a linear dependency between pose parameters and correlation peak even for in-between images that were not used to construct the filters. Consequently, for our pose domain, the idea is to reverse the problem. Since we know the linear relation – stored in the transformation matrix T– and since we can compute the correlation peak values – we have constructed the SDF filters – we can therefore retrieve the pose parameters of an unknown presented image. That is, we compute the correlation between the presented image with each of the three composite filters h_1, h_2, and h_3. Then for each filter, we just retain the maximum value

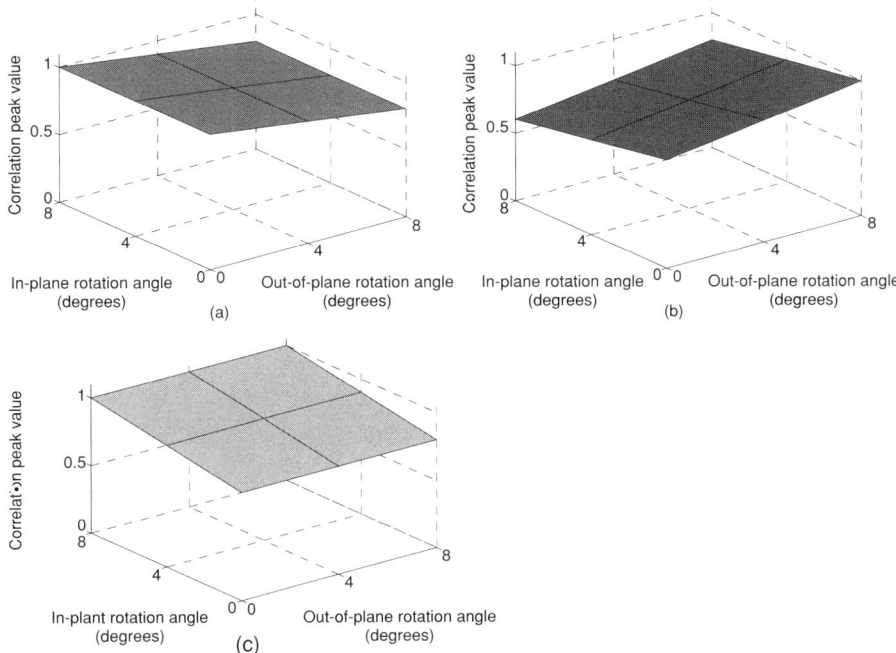

Figure 12.2. Defined relationship between correlation peak values and the orientation parameters: (a) for SDF filter h_1, (b) for SDF filter h_2, and (c) for SDF filter h_3.

of the correlation plane as correlation peak value. This value is independent on the location of the object and thus provides shift invariance. We then combine the three values obtained with the three filters in a correlation vector. We create an evaluation set containing images with rotations from 0 to 8 degrees in steps of one degree in both orientations (see Fig. 12.3). Then we form a matrix C_{eval} whose columns are the correlation vectors of every image of the evaluation set. In this way the pose estimation for the complete set can be retrieved easily as:

$$P_{eval} = T^{-1} C_{eval}, \tag{5}$$

where the matrix P_{eval} contains in its columns the pose vectors for every evaluated image. Namely, each column has the estimated out-of-plane, that is the in-plane rotation and the recognition flag information, respectively. If the recognition flag is close to 1 then the object is accepted as the true class and the estimated pose makes sense.

After computing the correlations we store the correlation peak values into the C_{eval} matrix. Figure 12.4 shows the relationship between the pose parameters (in-plane rotation ϕ and out-of-plane rotation θ) and the correlation value for each of the SDF filter. It can be seen that the variation is not exactly linear in-between the reference views. Therefore, the pose estimation can be slightly

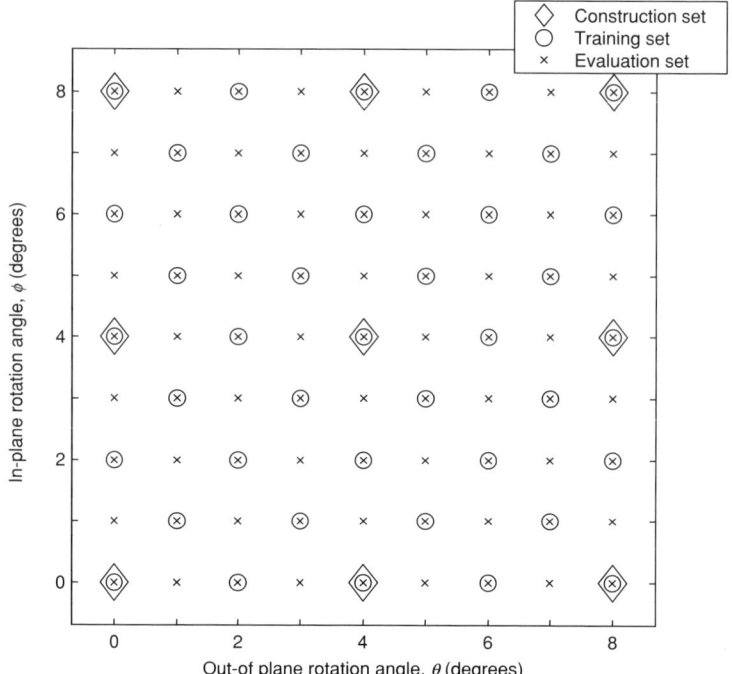

Figure 12.3. Composition of the three sets of images. The construction set is used to design the composite filter. The training set is used for the linear least-square fit and for training the neural network. The evaluation set is used to test the pose estimation.

improved by substituting the original conversion matrix T^{-1} with a matrix F that minimizes the pose estimation errors. In order to estimate these errors we first use a training set composed of 41 images with known orientations (Fig. 12.3). Let us denote by $\mathbf{S}_{train}$ the matrix containing all the training images and by $\mathbf{P}_{train}$ the matrix containing the corresponding known pose parameters. By correlating $\mathbf{S}_{train}$ with each of the filters, we obtain the correlation peaks values for the complete training set $\mathbf{C}_{\mathbf{train}}$.

Now, we compute the conversion matrix F that achieves a least-square fit, that is to say it minimizes the square error $\| P_{train} - FC_{train} \|^2$. It can be shown[23] that the solution is obtained using the Moore–Penrose pseudoinverse of $\mathbf{C}_{train}$ The final result is

$$F = P_{train} C_{train}^{T} \left(C_{train} C_{train}^{T} \right)^{-1} \tag{6}$$

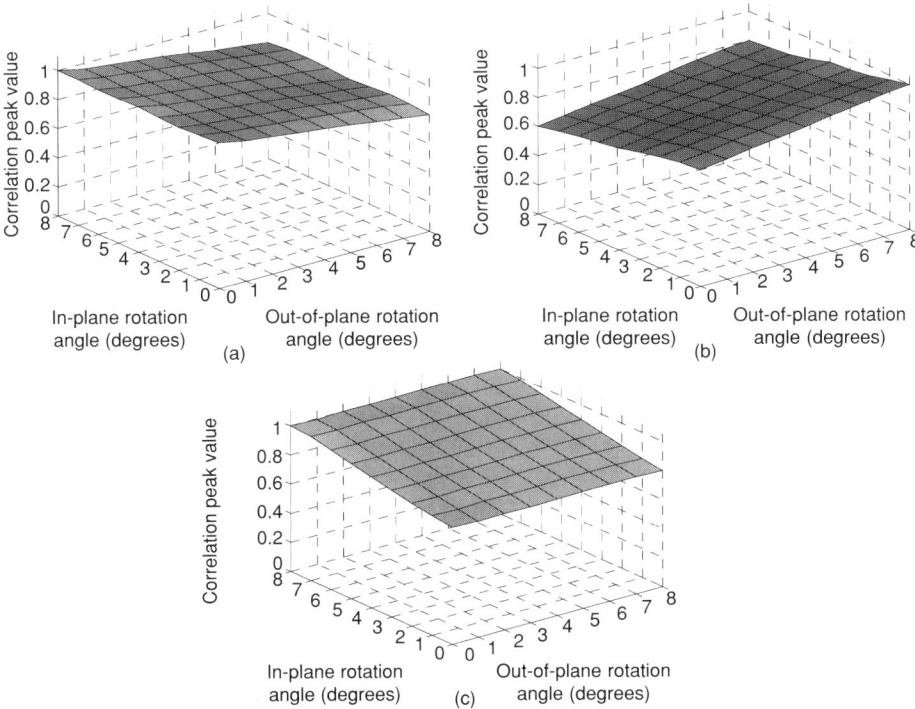

Figure 12.4. Real relationship between correlation peak values and orientation parameters for: (a) SDF filter h_1, (b) SDF filter h_2, and (c) SDF filter h_3.

and, similar to Eq. (5) the estimated pose parameters are

$$\hat{P}_{eval} = FC_{eval}. \tag{7}$$

Figure 12.5 shows the pose estimation results on the evaluation set that includes 81 images with rotations from 0 to 8 degrees in steps of one degree in both θ and ϕ (Fig. 12.3). The estimated poses are located at the nodes of the grid. If the dashed lines were perfectly straight passing all the way through the dots, then the estimation would be exact. Here the estimation error on the out-of-plane rotation θ has a standard deviation of 0.31 degrees and a maximum of 0.87 degrees. The error for the in-plane rotation ϕ has a standard deviation of 0.20 degrees and a maximum of 0.55 degrees. Let us mention that the above technique is slightly different from the one presented in[12], where the least-square fit is done for the transformation matrix T rather than for its inverse. The fitted matrix has then to be inverted to compute the poses. Our technique is therefore slightly more straightforward. Nevertheless, both methods give very similar results. Now our proposal is to introduce a neural network to improve the estimation of the pose.

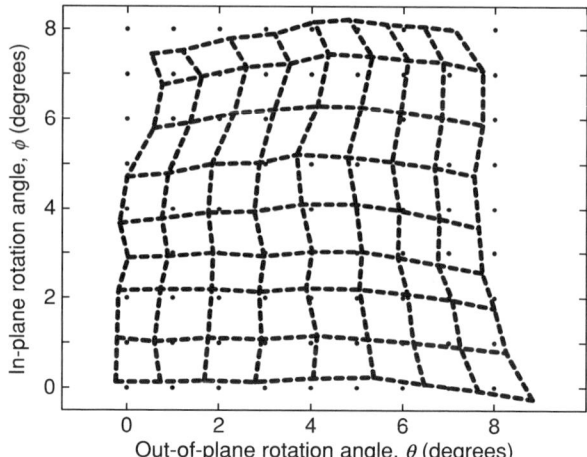

Figure 12.5. Pose estimation with a linear least-square fit of the correlation values – Linear correlation.

12.1.3 Pose Estimation Using a Two-layer Neural Network

The errors in the pose estimation appear because the relationship between the pose parameters and the correlation values is not perfectly linear, as shown in Fig. 12.4. In order to improve the estimation, we now use an artificial neural network (ANN).

In fact, the pose estimation given in Eq. (7) can already be seen as the operation of a single-layer ANN where the matrix F contains the weights of the ANN obtained from training with the correlation values of the training set. To obtain a better estimation, we need a nonlinear function. We therefore use a two-layer back propagation neural network[24] trained with the Levenberg–Marquardt algorithm.[25] The hidden layer contains 20 neurons and the output layer 3, each of which corresponds to one parameter to estimate (Fig. 12.6). To train the network, the inputs are the maximum correlation values obtained for the three SDF filters for each image of the training set. The desired outputs are the corresponding known pose parameters. The learning stage uses 100 epochs. The number of epochs, as well as the number of hidden neurons, was found heuristically. Figure 12.7 shows the pose estimation results. It is evident from this figure that the estimation is significantly improved compared to the previous linear estimation. The results slightly vary from one training to another because of the random initialization of the ANN. However, the error for the out-of-plane rotation θ typically has a standard deviation of 0.07 degrees and a maximum of 0.25 degrees. The error for the in-plane rotation ϕ typically has a standard deviation of 0.05 degrees and a maximum of 0.15 degrees.

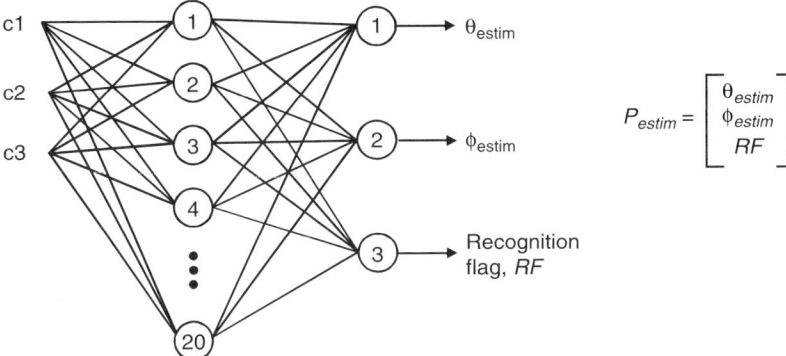

Figure 12.6. The proposed neural network to estimate the pose. A two-layer feed-forward back propagation neural network. The inputs are the correlation values provided by the composite filters. The outputs are the pose parameters.

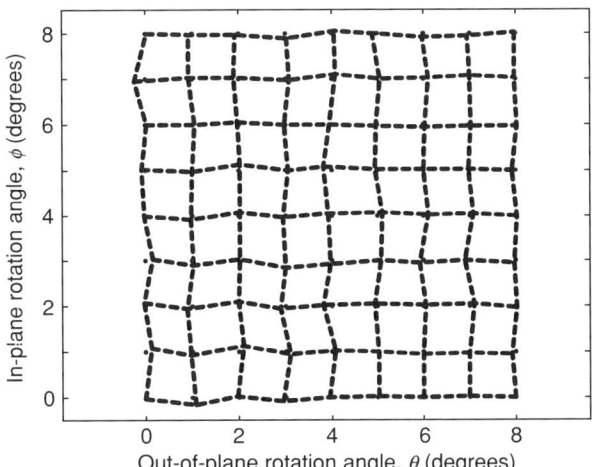

Figure 12.7. Pose estimation results with a two-layer neural network – Linear correlation.

12.1.4 The Recognition Issue

We previously mentioned that the estimated orientation angles only make sense if the presented object is a true target, that is, it belongs to the class of the reference object. In order to know whether this is the case, we included in the pose parameters a recognition flag RF that is supposed to be 1 only for views of the reference object. We thus achieve simultaneously the object recognition and the pose estimation. When retrieving the pose parameters for the evaluation set with the two-layer ANN presented in the previous subsection, the estimated value of the recognition flag is comprised between 0.99 and 1.01. This confirms that the images of the evaluation set correspond to the reference airplane. However, when we present to the system the false target shown in Fig. 12.1(d), the obtained recognition flag is comprised between 0.95 and 1.05 in 82% of the cases (result obtained from 5000 trials with random initializations). For the false target presented in Fig. 12.1(e), the recognition flag is comprised between 0.95 and 1.05 in 67% of the cases. The inability of the system to discriminate between the reference object and these false targets – including a very dissimilar one – shows that the procedure is not satisfactory. This problem is due to the low discrimination capability of linear correlation. Therefore, we explore the use of nonlinear correlation for the pose estimation.

12.2 Pose Estimation Using Nonlinear Correlation

In order to improve the recognition capability of our technique, we replace the previous linear correlation with a kth-law nonlinear correlation.[26,27] As we show next, the problem is then that the relation between the pose parameters and the correlation values becomes strongly nonlinear.

12.2.1 Composite Filter Using Fourier Plane Nonlinear Filters

The optimum nonlinear filter is presented in.[28,29] For simplicity, we use the kth-law nonlinearity, which is easily implemented in the Fourier domain, and is an approximate of the optimum nonlinearity. For every image I, we compute its 2D Fourier transform $\tilde{I}$. The nonlinear operation consists in raising the modulus of this Fourier transform to the power of k, while keeping its original phase. The resulting matrix is

$$\tilde{I}^{(k)} = \left|\tilde{I}\right|^{k} \exp\left(i\varphi_{\tilde{I}}\right), \tag{8}$$

where $\left|\tilde{I}\right|$ is the modulus of $\tilde{I}$ and $\varphi_{\tilde{I}}$ is its phase. The nonlinear factor k is comprised between 0 and 1. The nonlinearity is all the more stronger as k is closer to 0; a value of 1 corresponds to the linear correlation case. If $k = 0$, the comparison is performed only on the phase information. The following is similar to what we explained in the previous section, except that we substitute every image I of an object with its nonlinearly transformed Fourier transform $\tilde{I}^{(k)}$. Thus, the nonlinear SDF filter is obtained with

$$\tilde{h}^{(k)} = \tilde{\mathbf{S}}^{(k)} \left(\tilde{\mathbf{S}}^{(k)^{+}} \tilde{\mathbf{S}}^{(k)}\right)^{-1} c \tag{9}$$

Here $\tilde{\mathbf{S}}^{(k)}$ is matrix whose columns contain the construction set images $\tilde{I}^{(k)}$ rearranged as columns vectors. The symbol $+$ stands for the transpose conjugate. This Fourier-plane SDF filter is then converted back to matrix form and multiplied by the complex conjugate of the nonlinearly transformed Fourier transform of an unknown image. The result of the multiplication is inverse Fourier transformed to provide the correlation plane. Provided that the constraint vector c is real, it can be shown that – because the kth-law nonlinearity preserves Hermiticity – all the obtained correlation values are still real. The correlation result is the value of the maximum correlation peak. The shift invariance is thus preserved.

12.2.2. Pose Estimation and Recognition of the Object

We first evaluate the linearity of the correlation-pose transformation. In the same way as in Section 12.2.2, we construct now the nonlinear SDF filters[26,27] with the images of the construction set. We choose a medium nonlinear factor: $k = 0.5$ and we perform nonlinear correlations between each image of the complete evaluation set and each nonlinear filter. Figure 12.8 shows the real relation between the pose parameters (in-plane rotation ϕ and out-of-plane rotation θ) and the correlation peak value for each of the three nonlinear correlation filters. It is evident that the relation is not linear at all in-between the reference views that compose the filters.

We can therefore expect that a retrieval of the pose from the correlation values through a linear relation like Eq. (5) will be incorrect, even with a least-square fit Eq. (7) that is equivalent to a one-layer neural network. This is confirmed by Fig. 12.9 that presents the result of such an estimation for $k = 0.5$. In the case of nonlinear correlation, it is hence necessary to use a two-layer neural network to retrieve the pose parameters.

We use an ANN similar to the one used in Section 12.2.3 except that it has 40 hidden neurons. The training stage is stopped when the mean-square error reaches 3.10^{-3} (about 100 epochs). Here also the parameters have been found heuristically. The error for the out-of-plane rotation θ typically has a standard deviation of 0.25 degrees and a maximum of 0.9 degree. The error for the in-plane rotation ϕ typically has a standard deviation of 0.1 degrees and a maximum of 0.5 degrees.

Figure 12.10 presents the pose estimation for the entire evaluation set. The results are less accurate than with a linear correlation but are nevertheless acceptable. They are about four times more accurate than the ones obtained with nonlinear correlation by linear fitting.

The reason why we introduced the use of a nonlinear correlation is to get a better discrimination in the recognition of the object. To test this discrimination, we study the values of the recognition flag RF provided by the system when presenting various images. We use a nonlinear factor $k = 0.5$. In this case, for the images of reference airplane contained in the evaluation set, we obtain a

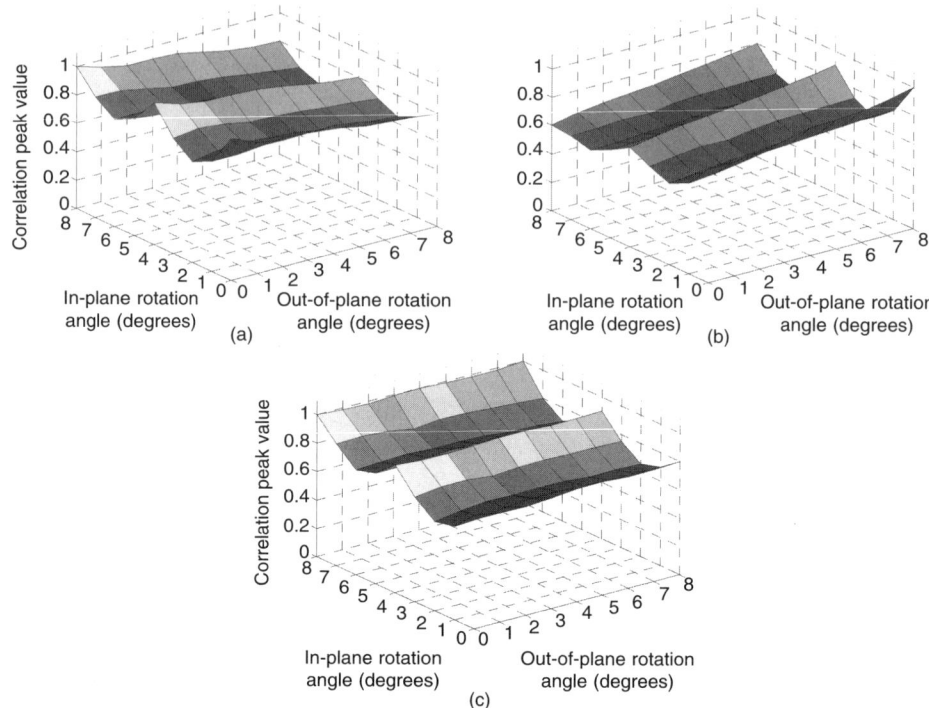

Figure 12.8. Relationship between correlation peak values and orientation parameters for (a) SDF nonlinear correlation filter h_1^k, (b) SDF nonlinear correlation filter h_2^k, and (c) SDF nonlinear correlation filter h_3^k.

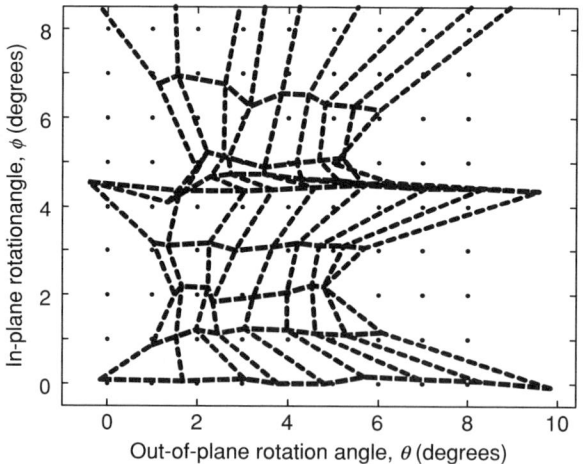

Figure 12.9. Pose estimation results with a one-layer neural network – Nonlinear correlations.

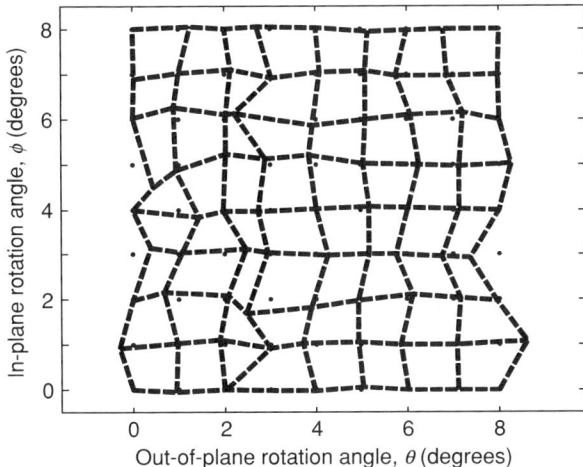

Figure 12.10. Pose estimation results with a two-layer neural network – Nonlinear correlations.

recognition flag usually between 0.95 and 1.05. For the false target of Fig. 12.1(d), the value obtained for the recognition flag varies randomly between -10 and $+10$ approximately, depending on the random initialization of the network. According to a 5000-trial test, the probability that the recognition flag be between 0.95 and 1.05 is only 7%. Thus, using nonlinear correlations, the probability of wrong classification of the false target is substantially lower than with the linear correlation case. Moreover, it can be further reduced to $(0.07)^2 = 0.5\%$ by cross-checking the result with a second – independently trained – network. For the false target of Fig. 12.1(e), which is very similar to the reference object, the probability that the recognition flag be between 0.95 and 1.05 is 21%. It can be reduced to 4.5% by cross checking with a second network.

12.2.3 Alternate Technique

If the shift-invariance property is not needed, the pose estimation results can be improved. This could happen in some industrial processes where the location of the tested pieces is fixed. Rather than using the maximum correlation value, the modified technique uses the value of the center of the correlation plane (corresponding to the inner product between the nonlinearly transformed filter and the image). Indeed, the design of SDF filters only imposes the value of the inner product between the image and the filter. In the case of nonlinear correlation, the maximum correlation peak is not always obtained in the origin and its value may therefore be different from the expected value. Actually, a further improvement can be obtained by taking as correlation value the mean

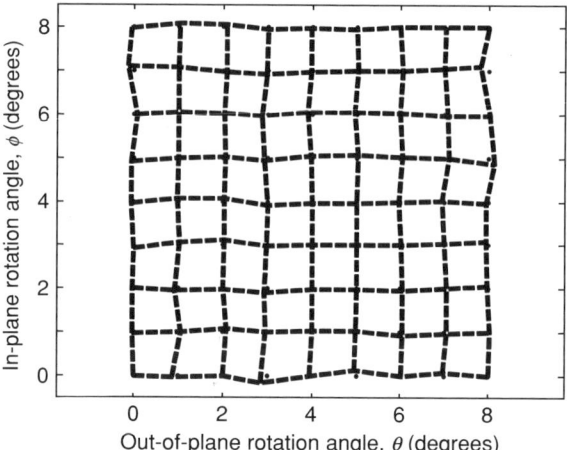

Figure 12.11. Pose estimation results with a two-layer neural network taking as correlation value the mean around the center – not shift invariant nonlinear correlation.

of a small window centered on the center of the correlation plane. We found that a 5 × 5 window gives the best results for our application.

This alternate method supposes that the studied object is exactly located at the same place as the reference objects. This loss of shift invariance is the price to pay for the improved pose estimation. Figure 12.11 presents the results obtained with an ANN having 40 hidden neurons and trained until reaching an error of 10^{-3} (about 200 epochs). The error for the out-of-plane rotation θ typically has a standard deviation of 0.06 degrees and a maximum of 0.2 degrees. The error for the in-plane rotation ϕ typically has a standard deviation of 0.05 degrees and a maximum of 0.15 degrees. Note that this alternate technique can also be applied to the linear correlation. However, the gain in accuracy is less noticeable in that case.

12.3 Discussion

12.3.1 Linear Versus Nonlinear Correlation

As we have shown, target recognition and pose estimation can be performed simultaneously, both with linear and nonlinear correlations. In both cases, the shift invariance property is achievable. However, there is a trade-off between discrimination capability and accuracy of the pose estimation. Discrimination is improved with nonlinear correlation whereas pose estimation is more accurate with linear correlation. We showed that the accuracy of nonlinear correlation could be improved at the cost of losing the shift invariance (see Table 12.1). However, this requires that the position of the object be

Table 12.1. Comparison among the different variations of the proposed technique.

	SI	σ_θ (deg)	Θ_{maxerr} (deg)	σ_ϕ (deg)	$\varphi_{\mathrm{maxerr}}$ (deg)	FP
Linear correlation and one-layer ANN (least-square fit)	yes	0.31	0.87	0.20	0.55	NC
Linear correlation and two-layer ANN	yes	0.07	0.25	0.05	0.15	67%
Nonlinear correlation and one-layer ANN	yes	1.2	3.1	0.30	1.3	NC
Nonlinear correlation and two-layer ANN	yes	0.25	0.9	0.1	0.5	7%
Nonlinear correlation mean value around the correlation peak (two-layer ANN)	No	0.06	0.2	0.05	0.15	NC

SI means shift-invariance; σ_θ and Θ_{maxerr} are standard deviation and maximum error for out-of-plane rotation; σ_ϕ and ϕ_{maxerr} are standard deviation and maximum error for in-plane rotation; FP is the False Positive rate; NC means Not Calculated

known with a one-pixel precision and is therefore of little practical interest in most cases. Actually, the precision achieved with the shift-invariant nonlinear technique is satisfactory (accuracy of 0.9 degree).

If a greater accuracy is needed, one could separate the recognition stage from the pose estimation stage. The recognition alone could be performed with a nonlinear composite filter for high discrimination and the pose estimation of the recognized target could then be performed with the linear correlation technique. Note that even if the recognition is performed separately, the location of the object will not be known with a one-pixel precision so it is still important that the pose estimation be shift invariant.

12.3.2 Choice of the Construction Set

In the case of nonlinear correlation, the nonlinearity of the response of a composite filter decreases when the angle between the construction images is smaller (the filter is better constrained). We used an incremental angle of 4 degrees between our construction images. The linearity, and therefore the pose estimation, might be improved by reducing this angle. On the other hand, the number of reference images that can be included in a single composite filter without degrading its performance is limited. Thus, decreasing the incremental angle between construction images results in reducing the total angular range covered by the filter. Since the full pose space is covered by tiling it with several composite filters, a trade-off has to be made between the total number of tiles and the accuracy of the pose estimation. In order to keep the total number of filters reasonably low, it is necessary to use filters constructed with a sufficient incremental angle. Our choice of an 8-degree range with an incremental angle of 4 degrees between the construction images is not due to theoretical considerations. It is intended to show that the nonlinear response of the filters can be compensated by the use of a neural network. The amount of tolerated nonlinearity can be used to reduce the incremental angle between construction images, and therefore, to reduce the total number of filters. So, the choice of the incremental angles, the number of construction images and the number of filters should be adapted to every particular application. The total number of filters could reach a few hundreds if we want to span the full pose space. However, the advantage of the correlation technique is that it can be performed optically, including the nonlinearity.[29] The computation time is thus strongly reduced.

12.3.3 Robustness and Limitations

Since our pose estimation relies on the values of the correlation peak, it is somewhat sensitive to factors that modify this value such as noise. We tested the influence of several types of distortions applied to the evaluation images. We use 1) an additive Gaussian white noise with mean 0 and standard deviation 20% of the maximum intensity of the image, 2) a multiplicative Gaussian white noise with mean 1 and standard deviation 0.2, and 3) a multiplicative Gaussian noise with mean 1, standard deviation 0.1, and a spatial correlation of 10 pixels to simulate changes in illumination. See Fig. 12.12 for examples of noisy images. Table 12.2 sums up the results for the shift-invariant pose estimation using linear correlation and noisy images. It can be seen that these levels of noise are somewhat acceptable. However, the same experiment for shift-invariant pose estimation using nonlinear correlation gives unacceptable results: errors of more than 1 degree in standard deviation and several degrees as maximum. The nonlinear correlation only tolerates a couple of percents of noise. The pose estimation with linear correlation is thus more robust than the one with nonlinear correlation. This again pleads in favor of a separate recognition stage using nonlinear correlation followed by a pose estimation using linear correlation. For both cases of linear and nonlinear correlation, the pose estimation cannot be performed when there is a uniform change of illumination or when a cluttered background is present. The illumination problem might be solved by normalizing the input images in energy. In addition, the object would have to be extracted from the background using segmentation techniques.[30–32]

12.4 Conclusions

In this chapter, we discussed the problem of pose estimation based on images, correlations and an artificial neural network. We constructed three different composite correlation filters using views of the reference object with known orientations. We constructed them in a way that the correlation peak depends linearly on the pose parameters. However, the interpolation for new

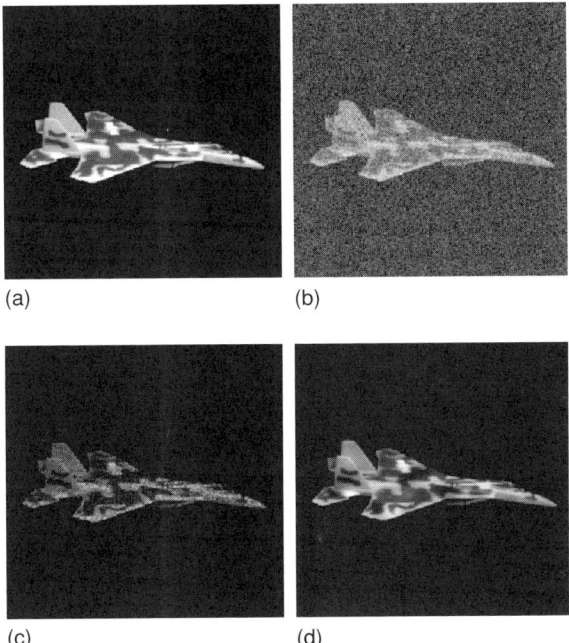

(a)   (b)

(c)   (d)

Figure 12.12. Examples of noisy images: (a) noiseless, (b) additive noise 20%, (c) multiplicative noise 20%, and (d) illumination noise 10%.

orientations within the pose domain was not always straightforward. We showed that the linear-dependency assumption does not hold when a nonlinear correlation is used. Even for a linear correlation, it was only an approximation. We therefore proposed to improve the pose estimation by feeding the correlation peak values of the three linearly weighted composite filters into a neural network. This network was previously trained with a set of reference views. As an illustration, we presented the determination of two rotation angles

Table 12.2. Error (in degrees) for shift-invariant pose estimation with linear correlation and a two-layer ANN.

Noise type	$\sigma_\theta{}^{(\mathrm{deg})}$	$\Theta_{\mathrm{maxerr}}{}^{(\mathrm{deg})}$	$\sigma_\phi{}^{(\mathrm{deg})}$	$\varphi_{\mathrm{maxerr}}{}^{(\mathrm{deg})}$
Noiseless	0.07	0.25	0.05	0.15
Additive 20%	0.25	0.8	0.15	0.5
Multiplicative 20%	0.2	0.5	0.1	0.4
Illumination 10%	0.4	1.0	0.3	0.7

σ_θ and Θ_{maxerr} are standard deviation and maximum error for out-of-plane rotation; σ_ϕ and ϕ_{maxerr} are standard deviation and maximum error for in-plane rotation.

(in- and out-of-plane) from actual pictures of a plane. The best pose estimation was obtained using a two-layer back propagation neural network. This neural network improved the pose estimation results in both cases: linear and non-linear correlations. We showed that nonlinear correlation is needed for a good discrimination but that linear correlation provides a more robust pose estimation. Thus, it might be profitable to separate the recognition stage from the pose estimation stage. Even if only linear correlation is considered, our technique provides a significant enhancement over the simple fitted pose estimation procedure presented in reference[12] As mentioned before, the pose estimation was performed for out-of-plane rotations and in-plane rotations within an 8-degree square range. A larger range can be covered by tiling the desired pose space. It is also possible to estimate more than two distortion parameters by increasing the number of composite filters.

References

[1] Horprasert T, Yacoob Y, and Davis LS. (1997). "Computing three-dimensional head orientation from a monocular image sequence." in: *25th AIPR Workshop: Emerging Application of Computer Vision, Proc SPIE 2962.* Schaefer DH; Williams EF, eds. pp. 244–252.

[2] Wu H, Fukumoto T, Chen Q, and Yachida M. (1996). "Face detection and rotations estimation using color information." in: *Proc RO-MAN,* '96. pp. 341–346.

[3] Shimizu I, Zhang Z, Akamatsu S, and Deguchi K. (1998). "Head pose determination from one image using a generic model." in: *Proc. 3rd IEEE Int Conf on Automatic Face and Gesture Recognition.* IEEE, New York, pp. 100–105.

[4] Wilson W. (1994). "Visual servo control of robots using Kalman filter estimates of robot pose relative to work-pieces." *Visual Serving.* Hashimoto K., ed. Word Scientific Press. pp. 71–104.

[5] Bajura M, Fuchs H, Ohbuchi R. (1992). "Merging virtual objects with the real world: Seeing ultrasound imagery within the patient." *Comp Graph.,* 26:203–210.

[6] Ezquerra N and Mullick R. (1996). "An approach to three-dimensional pose determination." *ACM Trans. Graph.,,* 15:99–120.

[7] Rosenfield GH. (1959). "The problem of exterior orientation in photogrammetry." *Photogrammetric Eng.,* 25:536–553.

[8] Haralick RM, Lee C-N, Ottenberg K, and Nolle M. (1994). "Review and analysis of solutions of the three point perspective pose estimation problem." *Int. J. Comput. Vis.,* 13:331–356.

[9] Lu C-P, Hager GD, and Mjolsness E. (2000). "Fast and globally convergent pose estimation from video images." in: *IEEE Trans. Pattern Anal. Mach Intell., 22.* IEEE, New York. 610–622.

[10] Huttenlocher DP and Ullman S. (1990). "Recognizing solid objects by alignment with an image." *Int. J. Computer Vis.,* 5:195–212.

[11] Castro A, Frauel Y, Tepichín E,and Javidi B. (2003). "Pose estimation from a two-dimensional view by use of composite correlation filters and neural network." *Appl. Opt.,* 42:5882–5890.

[12] Monroe Jr SE and Juday RD. (1990). "Multidimensional synthetic estimation filter." in: *Optical Information-Processing Systems and Architectures II*, *Proc SPIE 1347*, Javidi B., ed. pp. 179–185.

[13] Hassebrook LG, Lhamon ME, Wang M, and Chatterjee JP. (1997). "Postprocessing of correlation for orientation estimation." *Opt. Eng.*, 36:2710–2718.

[14] Dubois F. (1991). "Nonredundant filters for pattern recognition and orientation estimation of rotated images." *Appl. Opt.*, 30:1388–1395.

[15] Vijaya Kumar BVK, Lee a J, and Connelly JM. (1998). "Correlation filters for orientation estimation." in: *Proc SPIE*. 938:190 197.

[16] Caulfield HJ and Maloney WT. (1969). "Improved discrimination in optical character recognition." *Appl. Opt.*, 8:2354–2356.

[17] Caulfield HJ. (1980). "Linear combinations of filters for character recognition: A unified treatment." *Appl. Opt.*, 19:3877–3879.

[18] Hester CF and Casasent D. (1980). "Multivariant technique for multiclass pattern recognition." *Appl. Opt.*, 19:1758–1761.

[19] Casasent D. (1984). "Unified synthetic discriminant function computational formulation." *Appl. Opt.*, 23:1620–1627.

[20] Javidi B and Wang J. (1995). "Optimum distortion invariant filters for detecting a noisy distorted target in background noise." *J. Opt. Soc. Am. A.*, 12:2604–2614.

[21] Mahalanobis A. (1996). "Review of correlation filters and their application for scene matching." *Optoelectronic Devices and Systems for Processing*, Javidi B; Johnson KM. eds. SPIE Press. CR-65: pp. 240–260.

[22] Mahalanobis A. (1997). "Correlation filters for object tracking, target re-acquisition and smart aim-point selection." in: *Optical Pattern Recognition VIII*, Proc SPIE, Casasent DP; Chao TH. eds. 3073:25–32.

[23] Beale R and Jackson T. (1992). *Neural Computing: An introduction*. IOP.

[24] Rumelhart DE, Hinton GE, and Williams RJ. (1986). "Learning internal representations by error propagation." *Parallel Distributed Processing*, Rumelhart DE; McClelland JL. eds. MIT, Cambridge, MA.

[25] Hagan MT and Menhaj M. (1994). "Training feed-forward networks with the Marquardt algorithm." in: *IEEE Transactions on Neural Networks 5*. IEEE, New York. pp. 989–993.

[26] Javidi B and Painchaud D. (1996). "Distortion-invariant pattern recognition with Fourier-plane nonlinear filters." *Appl. Opt.*, 35:318–331.

[27] Hong S and Javidi B. (2002). "Optimum nonlinear composite filter for distortion tolerant pattern recognition." *Appl. Opt.*, 41:2172–2178.

[28] Réfrégier P, Laude V, and Javidi B. (1994). "Nonlinear joint transform correlation: an optimum solution for adaptive image discrimination and input noise robustness." *Opt. Lett.*, 19:405–407.

[29] Javidi B. (1997). "Nonlinear joint power spectrum based optical correlation." *Appl. Opt.*, 28:2358–2367.

[30] Castleman KR. (1996). *Digital Image Processing*. Prentice Hall, Upper Saddle River, NJ.

[31] Haralick RM and Shapiro LM. (1985). "Survey: Image segmentation." *Comp. Vis. Graph. & Im. Proc.*, 29:100–132.

[32] Chesnaud C, Réfrégier P, and Boulet V. (1999). "Statistical region snake-based segmentation adapted to different physical noise model." *IEEE Trans. Patt. Anal. & Mach. Intell.*, 21:1145–1157.

Evolutionary Sensor Fusion for Security

Bir Bhanu and Sohail Nadimi

Center for Research in Intelligent Systems
University of California, Riverside
{bhanu, sohail}@cris.ucr.edu

Abstract: A robust moving object detection system for an outdoor scene must be able to handle adverse illumination conditions such as sudden illumination changes or lack of illumination in a scene. This is of particular importance for scenarios where active illumination cannot be relied upon. Utilizing infrared and video sensors, we develop a novel sensor fusion system that automatically adapts to the environmental changes that affect sensor measurements. The adaptation is done through a cooperative coevolutionary algorithm that fuses the scene contextual and statistical information through a physics-based method. The sensor fusion system maintains high detection rates under a variety of conditions. The results are shown for a full 24-hour diurnal cycle.

13.0 Introduction

Over the past several decades many approaches have been developed for moving object detection for indoor and outdoor scenes. Moving object detection methods fall into two categories: (a) feature-based methods,[21] and (b) featureless methods (e.g., image subtraction, optical flow, statistical modeling).[2,4,6,8,18,24]. Each of these methods offers advantages that are exploited for different applications. For example, temporal differencing is simple and may suffice for indoor type illuminations for slow moving objects, optical flow is useful for a moving camera platform, and statistical modeling can capture the background motion.

Some of the shortcomings of the above approaches for moving detection are:

1) None of these approaches address the problem of low light or no light conditions,
2) No contextual information is used to update the parameters,

3) Generally, a large number of observations are required before a background model can be learned effectively, and
4) The algorithms have been applied to a single sensing modality (usually visible or near infrared) and no results have been shown for extreme conditions, for example, no illumination, sunset, or sunrise condition.

To overcome illumination conditions such as low or no light conditions, other sensing modalities such as cameras operating in near or long-wave IR have been utilized.[3] However, these sensing modalities could still fail due to similar conditions in their respective bandwidth. For example, in a long-wave (thermal infrared) camera, a subject's temperature could reach that of the background, thus having limited contrast, which may cause detection failure.

Multisensor fusion attempts to resolve this problem by incorporating benefits of different sensing modalities. The advantages of multisensor fusion are improved detection, increased accuracy, reduced ambiguity, robust operation, and extended coverage. Sensor fusion can be performed at different levels including signal or pixel level, feature level, and decision level.

This chapter provides a novel sensor fusion system that fuses long-wave (thermal IR) and visible sensors in a unified manner. By utilizing the IR signal, we can overcome some of the limitations of the visible cameras and by combining the visible and IR signal we improve the detection under a variety of conditions. The salient features of our approach are:

a) *Consistent data representation:* At the image level all sensing modalities are represented by mixture of Gaussians in a consistent manner.
b) *Physical models*: Sound physical models are used for each sensing modality (e.g., visible and IR) to provide prediction for each signal.
c) E*volutionary-based approach for fusion*: A cooperative coevolutionary algorithm is developed to systematically fuse and integrate information from both statistical and physical models into a unified structure for detection.
d) C*ontext-based adaptation:* Environmental conditions such as ambient air temperature, wind velocity, surface emissivity, etc., are directly incorporated into the detection algorithm and influence the fusion strategies.

Section 13.1 provides the related work and motivation, Section 13.2 presents the details of the technical approach, Section 13.3 discusses the experimental results, and finally Section 13.4 provides the conclusions of the chapter.

13.1 Related Work and Motivation

Current multisensor fusion and integration approaches use the following paradigms:

(a) **Statistical Paradigm** This paradigm utilizes the statistical properties of signal at pixel, feature, or decision level. It includes statistical methods such as Bayesian, Dempster-Shafer, and Fuzzy approaches. These approaches have been used extensively for fusion due to their well-developed mathematics. In Ref.[1], Bayesian and Dempster-Shafer multi-sensor fusion methods are compared for target identification. In Ref.[9], a Bayesian-based method for lane detection is developed. In Refs.[15,16], statistics-based techniques have been used for fusing video, near infrared (NIR), midwave infrared (MWIR), and long-wave infrared (LWIR) signals for image enhancement. Statistical-based fusion approaches provide a unified framework and methods that can deal with sensor noise; however, they require enormous amounts of data and prior knowledge of statistical properties of the signals.

(b) **Artificial Intelligence (AI) Paradigm** This paradigm attempts to fuse the data through methods such as knowledge-based, rule-based, and information-theoretic methods. Examples of AI-based fusion techniques are[5,20] for image enhancement and target detection, and [7]for robotics. This paradigm has the advantage of incorporating contextual information, heuristics, and domain knowledge by utilizing well-developed algorithms in the AI field; however, once designed, addition of new sensing modalities require a new set of algorithms and/or domain knowledge and heuristics that are generally provided by external experts for expansion of knowledge rules.

(c) **Data Structure Paradigm** This paradigm utilizes various representations such as graphs, trees, tables, and data structure-specific techniques such as graph traversal. Terrian[19] and Waxman[23] provide methods for fusing FLIR and an image intensifier data for image enhancement. In Ref.[10], an approach is introduced to fuse acoustic and video data for underwater vehicle tracking. This paradigm works well when the data can be represented by one of the structures mentioned. The obvious disadvantage of this paradigm is that once the data structure is defined, it may not be possible to extend the method to new sensing modalities; therefore, this paradigm is suitable when all sensing modalities participating in fusion are known in advance.

(d) **Physics Paradigm**: This paradigm utilizes the sensor phenomenology to model the signals, based on the physical aspect of the world. Physical models describe the relation of object parameters (e.g., surface reflectance, orientation, roughness, temperature, material density, etc.) to scene environmental parameters (such as ambient temperature, direction of illumination, wind velocity, etc.) to predict sensor values.

Pavlidis et al.[13] develop an automatic passenger counting system based on subbands below short wave infrared (SWIR). They measure reflectance of many objects including human beings; they note that the human skin reflectance spectral map is very similar to that of distilled water; they relate this phenomenon to the fact that humans are 70% water. In Ref.[12], several physical models have been developed to model the thermal, acoustic, and laser radar signals for various segmentation problems. The fusion is viewed as the problem of relating scene parameters to object parameters. Since IR bands above 3μm increasingly measure thermal fluctuations, they model a surface based on heat conductance and use the conservation of energy to model the interaction of surface and radiation.

Among the four paradigms, AI and data structure-based paradigms are less suited for dynamic conditions whereas the statistics and physics-based paradigms are the methods of choice for integrating sensor information that can change over time. We provide a new sensor fusion technique that combines the statistical and physics-based fusion paradigms through an evolutionary process. We overcome the disadvantage of each of these paradigms by including suitable sensor models that have enormous generalizing power. This generalizing power is then used to complement the limited available sensor data that is required by the statistical methods. The fusion is performed at the pixel level where the information loss is minimal.

13.2 Technical Approach

The sensor fusion architecture for moving object detection is depicted in Fig. 13.1. Observations from the sensors along with the external conditions, which carry the contextual information, are used to build statistical (mixture of Gaussian) background model. The contextual information is also used to update values of internal physical models. Physical models include reflectance models for predicting image intensity values and thermal models for predicting background surface temperature values. Unlike the previous work that updates

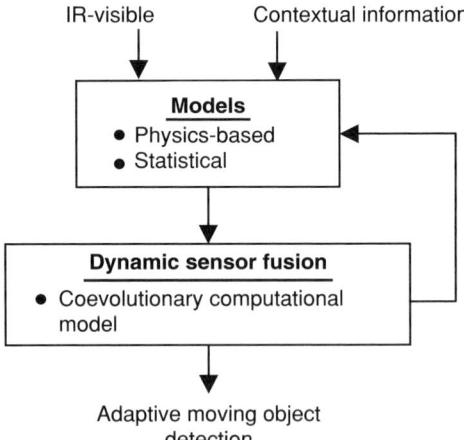

Figure 13.1. Sensor fusion architecture.

Table 13.1. Algorithm for learning background model for a pixel.

Evolutionary Adaptive Background Modeling

S=Training set which includes prediction, observation, and previous classification results per pixel;

Note: An organism represents a solution.

- - - - - - - - - - - -**Cooperative Coevolution Algorithm per Pixel**- - - - - - - - - -

Steps

1. Create and initialize 4 subpopulations for each channel
2. *Loop*
3.   **For each Sub-population**
4.   **For each individual**
5.   Build an organism (*e.g., combine representative individuals from different sub-populations*)
6.   Evaluate the organism using the training set S and $F_{organism}$
7.   Store the new fitness value for the individual
8.   **EndFor**
9.   **EndFor**
10.   Evolve all sub-populations (*Selection, Mutation, Crossover*)
11.   *Until stop Condition*
12.   Return the best organism (*best organism or solution is the best individual from each subpopulation*)

---

the background models solely based on the current observations, we incorporate the physical models into the adaptive loop. The physical models are integrated with the statistical models through a cooperative coveolutionary process.[14] The cooperative coevolutionary process estimates the best representation for the background per pixel. This is done through a genetic evolutionary process that searches for the optimal representation based on the current, and recent past observations and detection results in addition to the predictions given by the physical model.

Our representation of mixture of Gaussians (described in Section 13.2.1) includes Gaussian parameters for the infrared and visible sensors (including RGB channels). A population of this representation is maintained as a pool of individuals for the evolutionary process. Once the evolutionary process is stopped, the best individual represents the background model of that pixel. In this manner, the contextual information plays an active role in contributing to the most ideal sensor for a particular condition.

The detection algorithm in Fig. 13.1 requires a model of the background. This model is estimated by a mixture of Gaussians. Table 13.1 shows this process. The details are explained in the following subsections.

13.2.1 Representation

The probability of a pixel, classified as a background, drawn from a probability distribution can be estimated by a mixture of density functions. Assuming the parametric form of the mixture is Gaussian, probability of observing a background pixel is:

$$P(X) = \sum_{i=1}^{m} W_i \eta(X, \mu_i, \Sigma_i) \tag{1}$$

where X is the pixel value, W_i is the weight of the ith Gaussian, m is the number of Gaussians, and η is the Gaussian form characterized by the mean μ_i and covariance Σ_i. Assuming R (Red), G (Green), B (Blue), and T (Temperature) channels are independent, each pixel is represented by its first order statistics for each respective channel as follow:

$$\mathbf{I_R} = <\mathbf{Fitness_R}, \mathbf{W_{R_1}}, \boldsymbol{\mu_{R_1}}, \boldsymbol{\sigma_{R_1}}, \ldots, \mathbf{W_{R_m}}, \boldsymbol{\mu_{R_m}}, \boldsymbol{\sigma_{R_m}}>,$$

$$\mathbf{I_G} = <\mathbf{Fitness_G}, \mathbf{W_{G_1}}, \boldsymbol{\mu_{G_1}}, \boldsymbol{\sigma_{G_1}}, \ldots, \mathbf{W_{G_m}}, \boldsymbol{\mu_{G_m}}, \boldsymbol{\sigma_{G_m}}>,$$

$$\mathbf{I_B} = <\mathbf{Fitness_B}, \mathbf{W_{B_1}}, \boldsymbol{\mu_{B_1}}, \boldsymbol{\sigma_{B_1}}, \ldots, \mathbf{W_{B_m}}, \boldsymbol{\mu_{B_m}}, \boldsymbol{\sigma_{B_m}}>,$$

$$\mathbf{I_T} = <\mathbf{Fitness_T}, \mathbf{W_{T_1}}, \boldsymbol{\mu_{T_1}}, \boldsymbol{\sigma_{T_1}}, \ldots, \mathbf{W_{T_m}}, \boldsymbol{\mu_{T_m}}, \boldsymbol{\sigma_{T_m}}> \tag{2}$$

where Fitness is an evaluation value assigned to the mixture model for a given channel (see Section 13.2.3). Therefore, background model for a pixel is represented by concatenating the representations of all the channels, which represents a solution instance. An evolutionary-based search algorithm (see Section 13.2.3) is used to search the solution space for an optimal background representation.

13.2.2 Physical Models

The algorithm shown in Table 13.1 uses the physics-based predictions in its evaluation phase. Models of bidirectional reflectance distribution functions (BRDF) and thermal equilibrium based on conservation of energy are used to predict surface color and temperature in the visible and long-wave IR. The models are briefly described here.

13.2.2.1 Physical Models of Reflectance

Several reflectance models including the Lambertian, Phong, dichromatic[17] and Ward[22] models have been developed to describe the reflectance due to normal, forescatter, and backscatter distributions. We utilize the dichromatic model:

$$L(\lambda, \hat{e}) = L_i(\lambda, \hat{e}) + L_b(\lambda, \hat{e}) = m_i(\hat{e})c_i(\lambda) + m_b(\hat{e})c_b(\lambda) \tag{3}$$

where L is the total reflected intensity, L_i and L_b are reflected intensities due to surface and subsurface respectively, m_i and m_b are geometric terms, c_i and c_b are relative spectral power distribution (SPD) of the surface and subsurface respectively, and $\hat{e}$ is a vector representing incident and reflected light angles with respect to the surface normal. The dichromatic model is useful in describing the reflection from inhomogeneous opaque dielectric materials (e.g., plastics). It is also useful in describing material colors since the SPD of the reflected light due to subsurface is decoupled from the geometric terms. To calculate the invariant body color, the image is segmented into regions with uniform reflectivity. For each region, pixel values in the RGB space are formed into a matrix M of size $n \times 3$ where n is the number of rows (pixels) and 3 represents R, G, and B values. Singular value decomposition is then applied to M and the singular vector corresponding to the largest singular value is selected as the body color (c_b), which is the predicted surface color[11].

13.2.2.2 Thermal Physical Model

For predicting surface temperatures in the long-wave IR, the following conservation of energy model is used. $E_{in} = E_{out}$; $E_{out} = E_{rad} + E_{cv} + E_{cd}$, where E_{in} is the input energy, E_{out} the output energy described by three phenomenon E_{rad} (energy radiated), E_{cv} (energy convected), and E_{cd} (energy conducted). Models for each energy flux are described in details in.[11] Briefly the following models are used to describe each of the above fluxes:

$$\boldsymbol{E_{in}} = E_{direct} + E_{skylight} + E_{atm}$$

$$\boldsymbol{E_{direct}} = (1089.5/\mathrm{m_a})\mathrm{e}^{(-0.2819\,\mathrm{m_a})} \tag{4}$$

$$\boldsymbol{E_{atm}} = E(\mathrm{BB,Ta})\{1 - [0.261\mathrm{e}^{-7.77^{*}10-4(273-\mathrm{Ta})2}]\}$$

where E_{direct} = direct irradiation due to sun, $E_{skylight}$ = irradiation due to sky $\approx (40\text{--}70 \text{ W/m}^2)$, E_{atm} = irradiation due to upper atmosphere, m_a = the number of air masses ($m_a \approx \text{secant } (Z)$), T_a = Air temperature, $E(\mathrm{BB}, T_a)$ = radiation of a blackbody at T_a temp, and Z = sun's Zenith angle.

E_{rad} is estimated based on Stephen–Boltzman law:

$E_b = \sigma T^4$, where $\sigma = 5.669 \times 10^{-8} \text{ W/m}^2 \text{ Kelvin}^4$ and the subscript b is for blackbody which is capable of 100% absorption (or emission) of energy.

The convected heat flux is given by:

$$\boldsymbol{E_{cv}} = h_{cv}(Ts - T\infty) \tag{5}$$

where h_{cv} is the convective heat transfer function which is a complex phenomena, Ts and $T\infty$ are surface and fluid temperatures respectively. For laminar flow, h_{cv} can be roughly estimated by the following empirical model:

$$\boldsymbol{h_{cv}} = 1.7|Ts - Ta|^{1/3} + (6\,Va^{0.8})/L^{0.2} \tag{6}$$

where Va = wind speed; L = characteristic lateral dimension of surface, Ts and Ta are surface and air temperature, respectively.

The conducted heat flux is described by:

$$E_{cd} = A(T2 - T1)/(L/k) \tag{7}$$

where A is the area, $T2 - T1$ is the differential temperature and L/k is called the thermal resistance or R-value and is tabulated for many materials. The above equilibrium model is solved for Ts, which is the predicted temperature.

13.3.3 Background Model Estimation

As mentioned in Section 13.2.1, a pixel is represented by concatenating mixture of Gaussian models of all its channels R, G, B, and T. In the mixture model, a single Gaussian is parameterized by W, μ, and σ; therefore, finding the best representation for a pixel with 4 channels represented by m Gaussians in each channel, requires searching in a $4 \times 3 \times m = 12m$ dimensional space. There are several search algorithms including brute force (e.g., depth first, breadth first), gradient methods (e.g., neural networks), heuristic methods (e.g., best first, beam search, A*), and genetic algorithms (GA).

Brute force methods are computationally expensive. Gradient-based techniques are suboptimal and may converge to local maxima. And, heuristic methods suffer from the curse of dimensionality. Genetic algorithms search from a population of individuals, which makes them ideal for parallel architectures. They have the potential to provide the global maximum.

Genetic algorithms are based on evolutionary process, examples of which are abundant in nature. In a typical GA the solution to a problem is encoded in each individual representation. A population of these individuals is randomly created. This population represents the location of individuals in the search space. An evaluation function (fitness function) that plays the role of the environment, rating individuals in terms of their fitness, is defined. The fitness function is used to rank individuals in the population. To continue exploring the search space, new populations are generated where individuals in the new population are selected based on the performance of their predecessors. In other words, solutions that have higher fitness value (e.g., better representations) are given more chance of being propagated in the next generation. In order to explore this search space more effectively, randomization is introduced in the selection of the individuals. There are two main operators for this randomization, referred to as crossover and mutation. Crossover is an operation where two individuals swap portions of their representation in random, effectively creating new offspring (solutions) encoding part of their parents (old solutions). Mutation is an operator that randomly, usually with low probability, changes a representation, for example, flipping a bit in a bit string. By applying the crossover, mutation, and selection operators, the GA effectively explores the search space in a parallel fashion.

The Cooperative Coevolution (CC) algorithm utilized here is a recent evolutionary, GA-like, algorithm[14]. Like the GA algorithm, the CC algorithm explores the solution space in a random fashion. As in GA, the CC algorithm applies the operators crossover, mutation, and selection to generate potential solutions. However, in CC, the representation of a solution is broken down into subparts, each of which encodes part of the solution and is evolved separately. Therefore, subpopulations are generated and maintained in each generation of the CC algorithm. In this manner, the opportunities for searching and exploring different solution subspaces are increased.

By comparing the algorithms in Fig. 13.2, it is clear that the major difference between these two models lies in how the evaluation of individuals is performed. As stated earlier, the evaluation in the GA model is performed on an individual (as a whole) in a population; on the other hand, in the CC model, individuals from separate subpopulations must come together to create an "organism" that is viewed as the solution. Hence, in the CC model an individual cannot provide a meaningful solution to the problem and requires the cooperation of individuals from other subpopulations.

The success of CC depends on four criteria:

1) Problem decomposition,
2) Interdependability,
3) Credit assignment, and
4) Population diversity.

Our sensor fusion algorithm satisfies all four criteria since

a) our problem is naturally decomposed (color video and IR),
b) our representation (mixture of Gaussians for all the four channels (R, G, B, and T)) provides interdependencies between subcomponents,
c) the objective or fitness function minimizes the discrepancy between the physics-based prediction and the actual observations in both IR and video, and
d) population diversity is maintained by roulette wheel selection method.

```
Procedure GA( )
initialize population
loop
    evaluate individuals
    store best individual

    select mating candidates
    recombine parents and use their
    offspring as the next generation
until stopping condition

return best individual
```

```
Procedure CC( )
initialize subpopulations
loop
    evaluate crganisms (solutions)
    store best organism
    for each subpopulation
        select mating candidates
        recombine parents and use their
        offspring as the next generation
    end for
until stopping condition
return best organism
```

Figure 13.2. Comparing a typical GA and CC algorithm.

As mentioned, an important part of the evolutionary algorithm is the evaluation function, referred to as the fitness function. We provide a suitable fitness function that integrates the statistics collected by the system and the physical models that are directed by the contextual information (environmental conditions). The cooperative coevolutionary (CC) algorithm is used to select an optimal representation for a pixel background based on the recent past observations, classification (background versus foreground) results, and physics-based predictions.

13.3.3.1 Fitness Function

For each channel, a population of individuals (see Section 13.2.1) is initially created randomly. These individuals are maintained for both the video channels (R, G, and B) and the thermal channel (T).

Briefly, the CC algorithm (see Table 13.1 and Fig. 13.3), works as follows: Initially, four groups of individuals of type I_R, I_G, I_B, and I_T are randomly initialized. Each group is called a subpopulation and each member of a subpopulation is referred to as an individual, which is also assigned a fitness value. The fitness value (see Fig. 13.3) is a measure of goodness and indicates how well that individual represents the background for its respective channel.

Individuals are rewarded when they perform well together as a team and punished when they perform poorly. This is the key concept in cooperative coevolutionary paradigm. In the example in Fig. 13.3, to evaluate an individual in the red channel, it is combined with representatives from other subpopulations (in this case the representative of a subpopulation is an

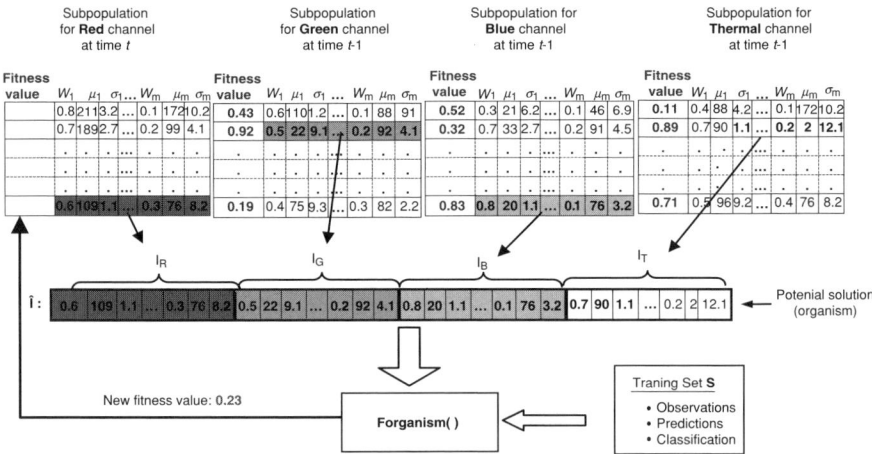

Figure 13.3. Example of evaluating an individual in the red channel – Individuals with highest fitness value in their population from other channels at the previous generation are combined to form an organism (solution). The result is stored back for the individual in the red channel.

individual with the highest fitness value); an organism is then created. The fitness value of the organism, $F_{organism}$, indicates how well the individual (I_R in this example) fits with other channels. In other words, the fitness value indicates the contribution of this individual as part of a whole solution.

The evaluation of the fitness function requires a training set. This training set is a recent past history, which includes observations, predictions, and classifications for each pixel and is kept in a QUEUE. To initialize the algorithm, initial n frames of background from all channels R, G, B, and T are collected and kept in a memory queue. Similarly, a physics-based prediction for each pixel for each frame is kept in the memory queue. Since the initial n frames are assumed to be background, the groundtruth at the initialization stage is known (e.g., all pixels represent background). As a new frame is observed and pixels classified as either background or foreground, this training set is updated in a Last In First Out (LIFO) manner.

Pixel classification is the result of detection where a pixel is classified as background if its value falls within 3σ of any of its Gaussians for all the channels; else, it is considered a foreground.

For each channel, let an individual I_Y in a population be represented as in Section 13.2.1 where Y represents a channel, $Y \in \{R, G, B, T\}$. Let:

Y_{Xob_j} = Observed value of a pixel X at jth frame for channel Y, $j = 1 \ldots n$; n = size of the window in the past.

Y_{Xp_j} = Predicted value of a pixel X by physics for the jth frame for channel Y.

$P(Y_X)$ = The probability distribution function for the pixel X for channel Y.

We keep a moving window of n previous frames for all the channels. This window serves as the groundtruth data, G, for training examples. Unlike most other works that only use the last or current observation (frame) to update the mixture of Gaussians, we elect to keep a window of frames. Let

$$G_{j\{j=1\ldots n\}} = \begin{cases} 1 & Background \\ 0 & Foreground \end{cases} \tag{8}$$

where $\{j = 1 \ldots n\}$ represents the last n frames (e.g., G_1 = current frame, G_2 = previous frame, and so on), and G is used as part of the training set $\boldsymbol{S}$. Initially G for the current frame is obtained by using the mixture of Gaussian parameters that are obtained at time $t - 1$. (After the learning process has been completed for the current frame, G for the current frame is updated based on the learned parameters of mixture of Gaussians). In order to relate statistics-based classification and the physics-based predictions, we introduce the following function, named *credibility function* for each channel:

$$C_Y = e^{-\alpha \left[\frac{1}{n} \sum_{j=1}^{n} G_j \frac{\left| Y_{X_{ob_j}} - Y_{X_{P_j}} \right|}{Y_{X_{ob_j}} + Y_{X_{p_j}}} + (1 - G_j)(1 - \frac{\left| Y_{X_{ob_j}} - Y_{X_{P_j}} \right|}{Y_{X_{ob_j}} + Y_{X_{p_j}}}) \right]} \tag{9}$$

where vectors G, Y_{Xob} and Y_{Xp} are defined as before and α controls the rate of decay of credibility function. As the observed values Y_{Xob} are closer to the predicted values Y_{Xp} for a particular classification G, the value of the credibility approaches 1. For example, it is easy to verify that in the extreme case where a pixel is classified as the background pixel in all the previous n frames, and that the predicted pixel values matched the observed values, the credibility will be close to 1.

The physics-based prediction predicts color and thermal properties of the background, therefore, it will be more credible if the observed pixel value is classified as the background pixel, and the predicted pixel value agrees with the observed value. Similarly, if the physics predicts a very different value than observed value and the system has actually classified the pixel as the foreground, then the physics may still be credible. On the other hand, if the physics-based prediction is very close to that of the observed value but the system has classified the pixel as foreground, then the physics-based prediction may not be reliable and a low credibility must be assigned. This process is depicted in Table 13.2.

The statistical estimation of fitness function based on the recent past observations for an individual, in channel Y, is given by:

$$F(I_Y) = \frac{1}{n} \sum_{j=1}^{n} \left[G_J P(Y_{X_{ob_j}}) + (1 - G_J)(1 - P(Y_{X_{ob_j}})) \right] \qquad (10)$$

The above function is only based on the statistical properties of the current and past observations. Given $F(I_Y)$ and the credibility function C_Y for individuals for all channels R, G, B, and T, then, a fitness function for an organism (solution) made of both video and IR species (e.g., R, G, B, and T channels) can be realized as follows:

$$F_{\text{organism}}(< I_R, I_G, I_B, I_T >) = 1/4[C_R\, F(I_R) + C_G\, F(I_G) \\ + C_B\, F(I_B) + C_T\, F(I_T)] \qquad (11)$$

The above equation is used for evaluating the organisms formed by the video and IR signals, in which the individual being evaluated is part of a complete solution (see Fig. 13.3).

Table 13.2. Credibility table describing the relationship between the predicted and observed values.

	Difference of current observed with predicted by physics	
	High	Low
Classification: Background	LOW Credibility	HIGH Credibility
Foreground	HIGH Credibility	LOW Credibility

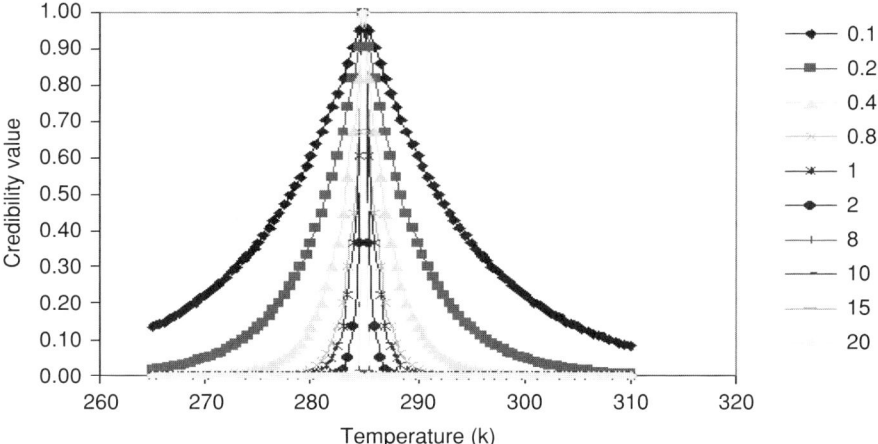

Figure 13.4. Credibility values for various values of alpha (α) for the thermal channel.

The final solution is the organism obtained by selecting the best individual (e.g., individual with highest fitness value) from each subpopulation. This solution is used to classify the current pixel as background or foreground.

The parameter α adjusts the importance of the role the credibility function plays in the fitness function. α can be adjusted depending on how fast the credibility function is desired to be influenced by the agreement between the physics prediction and actual observations.

Figure 13.4 shows how the parameter α affects the rate of change in the credibility function. For the observed temperature of 285K, if the predictions are credible but not as close to the observed values, then lower values of alpha are desired. On the other hand, if tight coupling between physics predictions and observations is required, higher values for α are desired.

13.4 Experiments

The data was gathered at a typical urban location with the latitude $33°50'06''N$ and longitude $117°54'49''W$, from 15:30:00 on January 21, 2003 till 14:24:00 January 22, 2003. Initially, from 15:30:00 till 17:07:04, data was collected at the rate of 1 frame every 2 s, then the temporal resolution was changed to approximately 1 frame per 10 s for the rest of the data collection period. Two cameras, a FLIR system thermal camera operating at 7–13 μm and an Intel Web-cam operating in the visible range were utilized for data acquisition. The thermal camera was fully radiometric, which means that the pixel values obtained by the camera were thermal. The thermal camera included self-calibration that at specified intervals adjusted to internal thermal noise. The radiation-to-temperature conversion was done automatically by the camera for the default

Figure 13.5. Position of the cameras with respect to the scene and the direction of the sun's path.

values of emissivity $= 0.92$, air and ambient temperatures $= 280\,\mathrm{K}$, distance to target $= 100\,\mathrm{m}$, and humidity $= 50\%$.

The video camera was attached to the top of the thermal camera on a tripod (see Fig. 13.5). Both cameras were located 20 ft above the ground looking downward at the scene at an angle of approximately $25°$. In addition to the thermal and the video cameras, a complete weather station was utilized to obtain weather data every minute. The weather station included an anemometer, humidity sensor, wind direction, two temperature sensors, and a barometer sensor. All sensors and the cameras were controlled by a PC. The data from the cameras and the weather station were synchronized through a software control.

To avoid temporal registration, both cameras were triggered simultaneously and in parallel. For spatial registration between the two cameras affine transformation was applied. For predicting correct reflectance and thermal predictions, a split and merge algorithm initially segmented the images for both cameras and a user initially labeled the segments into five regions, asphalt, concrete, grass, bush, and unknown. Only statistical properties were utilized for the unknown surface type.

13.4.1 Physical Model Estimation and Predictions

For surface color estimation, the dichromatic model was utilized. The results for the four different presegmented surfaces are given in terms of unit vectors in the RGB space. Due to lack of illumination during the nighttime, the values were obtained after sunrise and before sunset for various times and are given in Table 13.3 at an hourly illumination condition. The asphalt and concrete had similar vectors due to their neutral color attributes. On the other hand, the chlorophyll in the vegetation such as grass and bush causes the vectors to be shifted toward green. The higher variation in the reflectance of grass and bush are contributed by their surface specularity, which is not modeled by our algorithm.

For surface temperature prediction, the thermal models of Section *13.2.2.2* were used. These predictions were used by the fitness function in Section

Table 13.3. Surface body color estimation (c_b).

Time	Asphalt			Concrete		
	R	G	B	R	G	B
8:30	.5727	.5726	.5867	**.5813**	**.582**	**.5687**
9:30	.5714	.5716	.5889	**.5791**	**.5797**	**.5732**
10:30	.5773	.5714	.5862	**.5824**	**.5824**	**.567**
11:30	.5669	.5676	.5970	**.5737**	**.5745**	**.5838**
12:30	.5695	.5695	.5927	**.5686**	**.5749**	**.5884**
13:30	.5682	.5680	.5954	**.5767**	**.5753**	**.5801**
14:30	.5741	.5720	.5859	**.5681**	**.5752**	**.5886**
15:30	.5635	.5520	.6025	**.5570**	**.5723**	**.6019**
16:30	.5623	.5684	.6006	**.5572**	**.5802**	**.594**
17:30	.5544	.5668	.6095	**.5566**	**.5813**	**.5935**

Time	Grass			Bush		
	R	G	B	R	G	B
8:30	.6336	.7260	.2672	**.5718**	**.6239**	**.5327**
9:30	.6343	.7189	.2844	**.5893**	**.6240**	**.5132**
10:30	.6369	.7128	.2938	**.5662**	**.6368**	**.5234**
11:30	.6320	.7193	.2883	**.5476**	**.6250**	**.5563**
12:30	.6256	.7376	.2543	**.5430**	**.6370**	**.5471**
13:30	.6249	.7364	.2591	**.5749**	**.6404**	**.5093**
14:30	.6210	.7391	.2611	**.5968**	**.6338**	**.4921**
15:30	.6060	.7505	.2636	**.5639**	**.6421**	**.5193**
16:30	.6040	.7572	.2486	**.6567**	**.6369**	**.4039**
17:30	.6231	.7380	.2590	**.6321**	**.6357**	**.4431.**

13.3.3.1. Figure 13.6 shows the result of predictions superimposed on actual measurements by the thermal camera.

As shown, the models were able to track temperature fluctuations for four different surface types closely except for the dots that appear at certain times in these plots. These dots indicate occasional camera self-calibration when the camera shutter is automatically closed. During the self-calibration, the thermal image is ignored. The average difference between the prediction and measurement for all surfaces was about $2°C$ with standard deviation of $1.87°C$.

13.4.2 Detection Results

Moving object detection is performed after an initial background model is built. Once new thermal and video frames are available, they are registered. The registered image then contains red, green, blue, and temperature values at each pixel location. The cooperative coevolutionary algorithm is used to build the

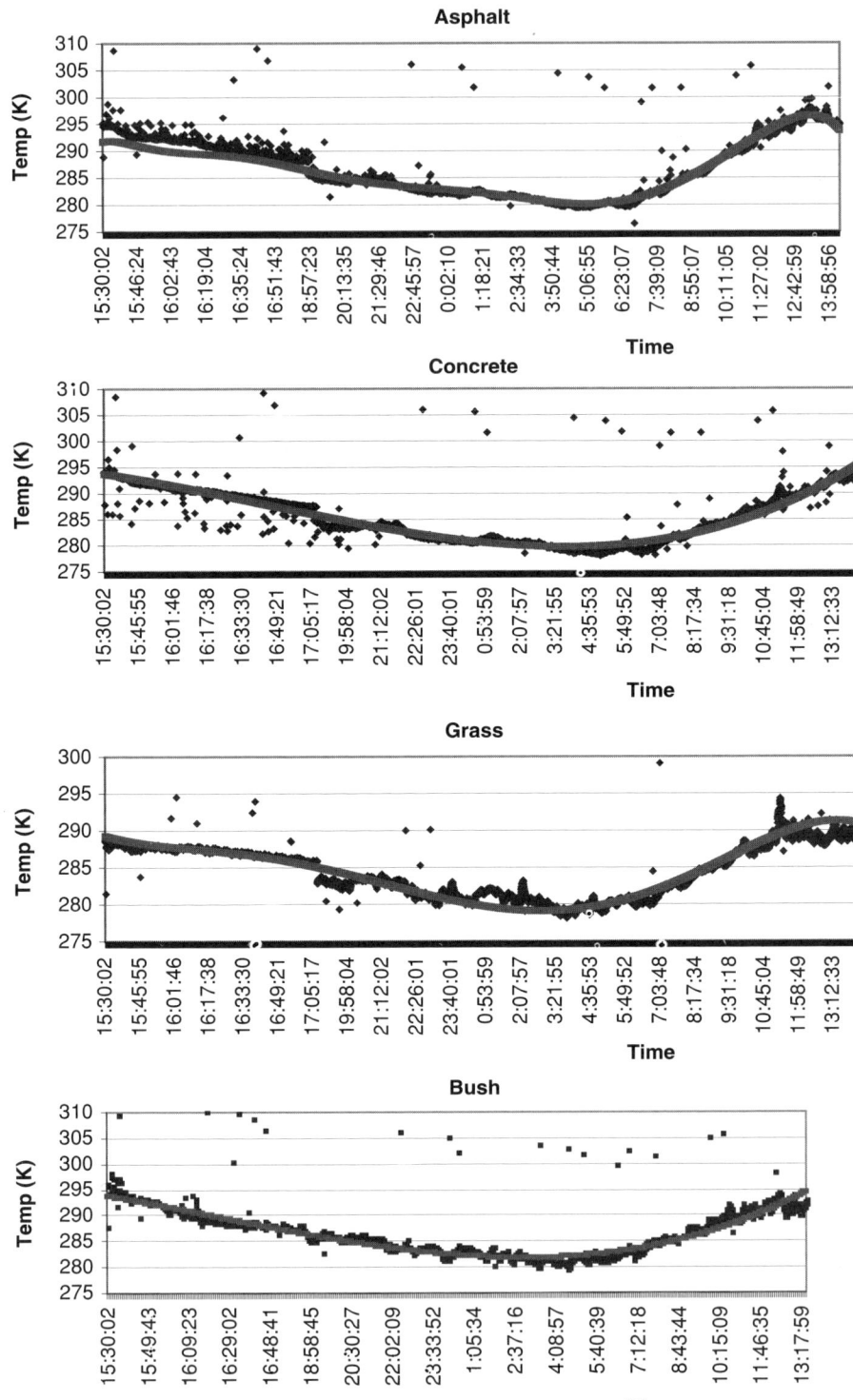

Figure 13.6. Measurement (blue) versus. predicted (red) surface temperature values.

background model. Each pixel is updated independently. The background model is periodically updated to track the environmental changes. The following parameters were used in the cooperative coevolutionary algorithm to update the background models: number of species = 4; population size = 60; crossover = single point; crossover rate = 0.8; mutation rate = 0.01; maximum number of generations = 60; training data = 20 frames; number of Gaussians per sensor = 3; $\alpha = 0.5$.

Once the background model is available for each incoming frame, each pixel is compared to its corresponding model and if its value is within 3 standard deviation of any of its Gaussians, it is classified as a background pixel. This information is kept in a binary image where a detected moving pixel is a binary 1 (white) and a background pixel is 0 (black). These binary frames provide training data for the next background model update. In the following examples, in addition to the thermal IR and video frames, detection for each camera and the fused detection for the registered images are also provided. The following confusion matrix is given for the results for all the moving objects:

% Moving object correctly detected	% Moving object missed
% Background missed	% Background correctly detected

- **Example 1.** Figure 13.7 shows example frames detected in the afternoon and early evening hours. During this period, illumination and heat exchanges are rapid. Depending on the heat stored and reradiated by an object and the background the object may be observed having very similar temperatures as the background (IR frames 2408 and 2685) or very different (IR frames 2422 and 2676). In frame 2408, video signal was much stronger, providing sharp contrast for the moving objects. Despite the lower performance of the IR, the objects were recovered by the video. Similarly, in frame 2422, the detection result of the IR was further enhanced by the registered video as is shown in the fused detected frame. Frames 2676 and 2685 are obtained during early evening hours. The video camera had a 25-lux minimum illumination requirement; therefore, although the scene was not totally dark, the video signal during the night time was very weak. This was compensated by the strong IR signal.
- **Example 2.** Figure 13.8 is an example where the detection algorithm relied heavily on one sensor, IR. Due to lack of illumination and video sensor's low sensitivity, objects could not be detected by video only. A good example is frame 2726 where a car and a person were in the scene. These were not observed in the video; however, they were present in the IR image and were clearly detected in both IR and the fused frame. Frames 2741, 6692, and 6718 indicate that the detection was not influenced by the video. The lights from the vehicles were visible and detected as part of the moving object, and the surface reflection from the lights did not affect the results.

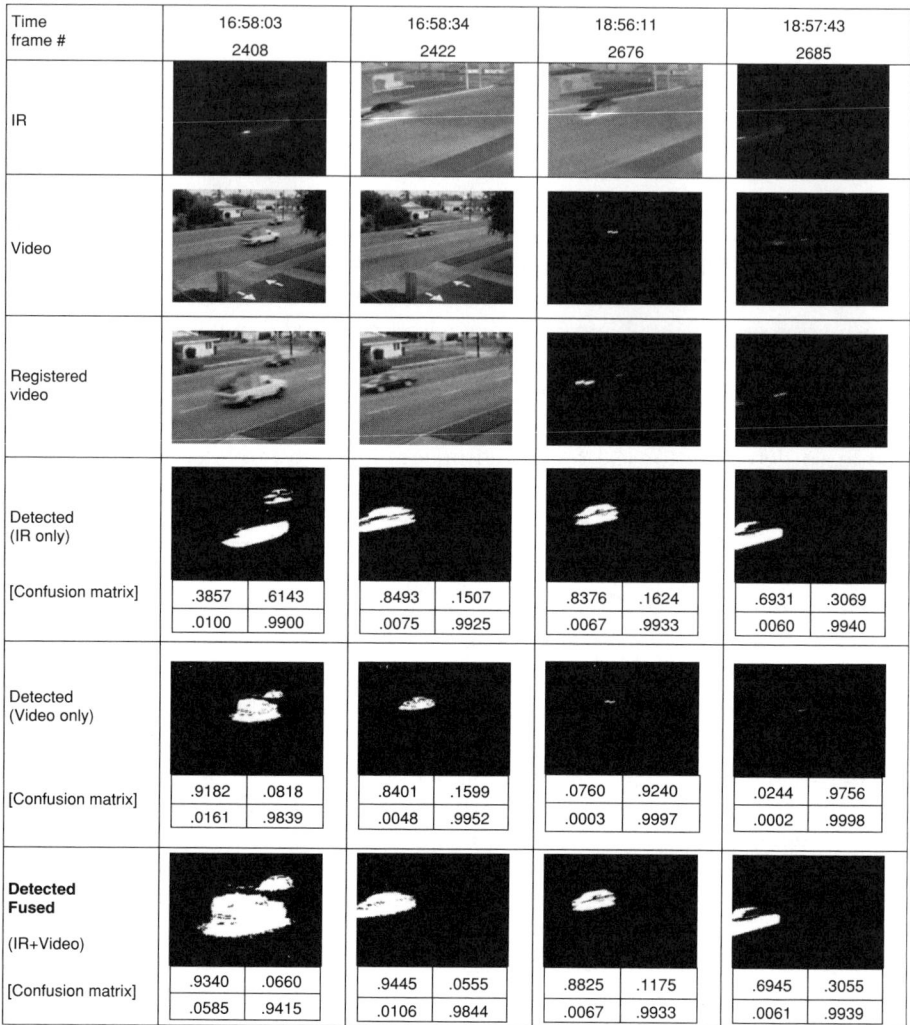

Time frame #	16:58:03 2408		16:58:34 2422		18:56:11 2676		18:57:43 2685	
IR								
Video								
Registered video								
Detected (IR only)								
[Confusion matrix]	.3857	.6143	.8493	.1507	.8376	.1624	.6931	.3069
	.0100	.9900	.0075	.9925	.0067	.9933	.0060	.9940
Detected (Video only)								
[Confusion matrix]	.9182	.0818	.8401	.1599	.0760	.9240	.0244	.9756
	.0161	.9839	.0048	.9952	.0003	.9997	.0002	.9998
Detected Fused (IR+Video)								
[Confusion matrix]	.9340	.0660	.9445	.0555	.8825	.1175	.6945	.3055
	.0585	.9415	.0106	.9844	.0067	.9933	.0061	.9939

Figure 13.7. Example 1: Mixed good and bad IR and video at various times in the afternoon and early evening.

This is due to the fact that the physics-based prediction assigns low credibility to the video signal; hence, low reflections are not detected as foreground. In effect this plays a role in deciding how important a camera's observations are. If a video pixel gets a low credibility, then its values are less meaningful; therefore, in order to observe a change, the signal must be strong (e.g., front head lights of a car). Since the front head lamps of most vehicles are halogen and radiate heat, they are also observed as part of the vehicle in the IR image, thus, they are also being detected as part of the vehicle.

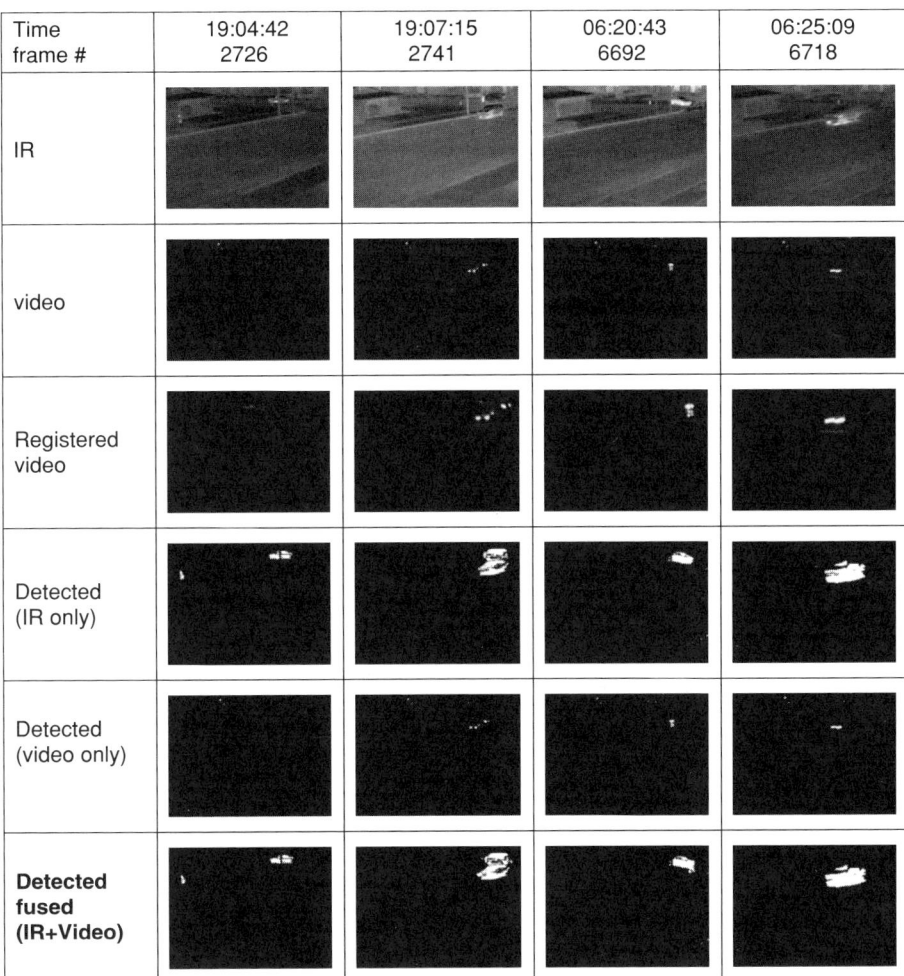

Time frame #	19:04:42 2726	19:07:15 2741	06:20:43 6692	06:25:09 6718
IR				
video				
Registered video				
Detected (IR only)				
Detected (video only)				
Detected fused (IR+Video)				

Figure 13.8. Example 2. Good to excellent IR signal, bad video signal at night. (*Note: Due to lack of video contrast no groundtruth is obtained.*)

- **Example 3.** Figure 13.9 is an example of dramatic illumination changes during the early sunrise and early morning hours. During these periods, the environment changes radically due to the energy of the sun. The sensors must adapt to these rapid changes. Figure 13.6 shows the thermal changes on different surfaces that are tracked by the physics-based models. As shown, the slope of the temperature values changes radically during this period. However, the physics-based models are able to follow these changes and provide high credibility values that affect the background models built by the algorithm. As the illumination reaching the video camera is increased, the detection due to video gets better. This is shown in frames

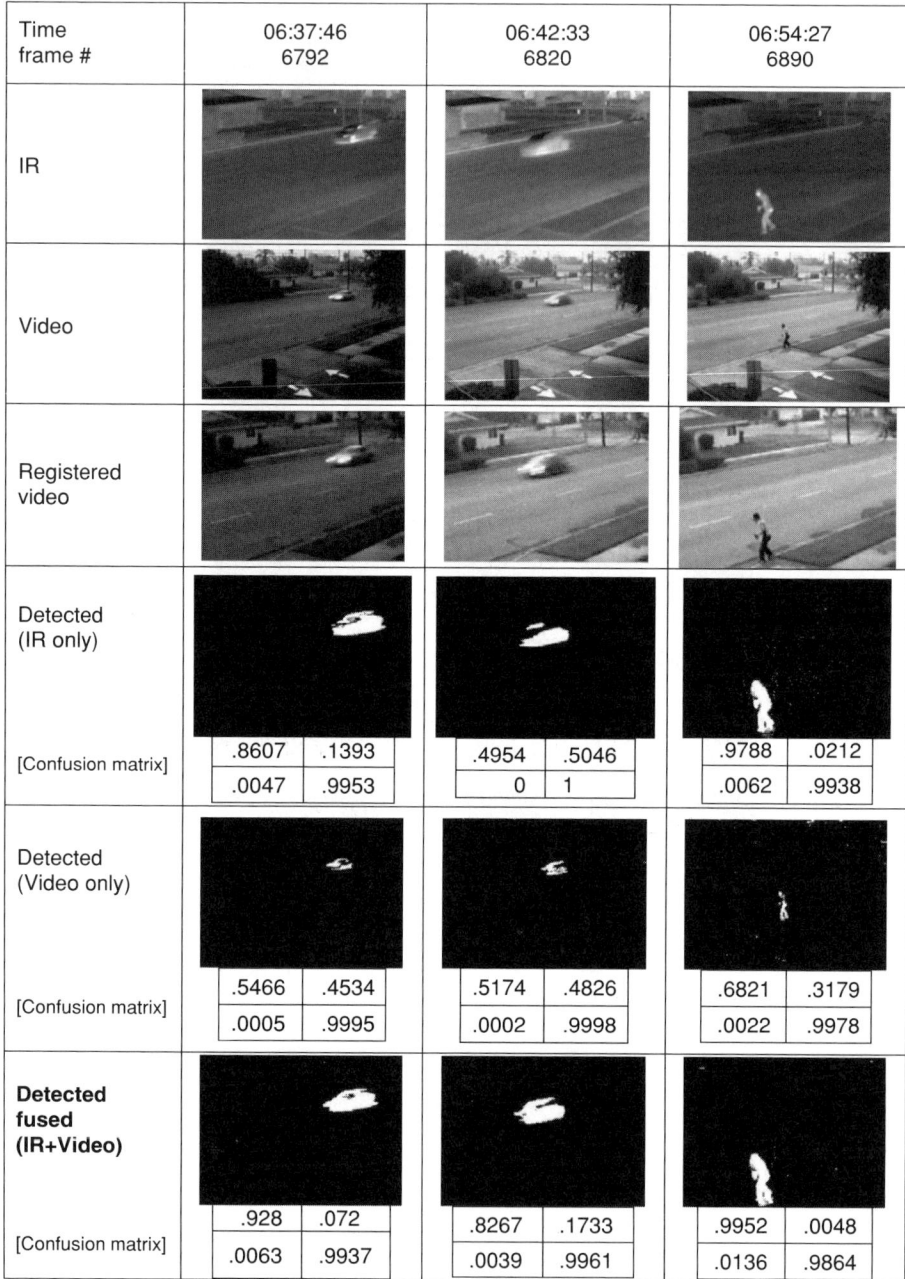

Time frame #	06:37:46 6792	06:42:33 6820	06:54:27 6890
IR			
Video			
Registered video			
Detected (IR only)			
[Confusion matrix]	.8607 .1393 .0047 .9953	.4954 .5046 0 1	.9788 .0212 .0062 .9938
Detected (Video only)			
[Confusion matrix]	.5466 .4534 .0005 .9995	.5174 .4826 .0002 .9998	.6821 .3179 .0022 .9978
Detected fused (IR+Video)			
[Confusion matrix]	.928 .072 .0063 .9937	.8267 .1733 .0039 .9961	.9952 .0048 .0136 .9864

Figure 13.9. Example 3: Fusion while illumination changes at sunrise.

6792 and 6820 where the video camera began participating in the detection process. This is indicated by the increase in the detection performance for the fused image versus the IR or video only images.

- **Example 4.** Figure 13.10 is an example of early morning, noon, and early afternoon hours. As the sun rises, the surfaces are heated up by the incoming energy from the sun, the increase in the surface temperatures approaches closer to the temperatures of some moving object surfaces. Depending on the moving object surface temperatures and emissivities, the contrast in the IR can be radically different. This is obvious between frames 6954 and 8646 for example. Frame 6954 represents an image in the morning with a person in the scene. Surface temperatures are still lower than that of the human body; moreover, human body's emissivity is high (0.98) compared to the background surfaces. The human is clearly visible in the IR image. Although not visible in the video image of frame 6954, the human is also in that image; this is clearer in the registered image. Both sensors provide good contrast in this case and the person is clearly detected.

Frames 8646 and 9350 show moving objects later in the day when surfaces have reached higher temperatures. In this case, it is possible to have a moving object that may have temperature close to the background surface as is indicated by both these frames. On the other hand, video provides excellent signal and contrast. Many pixels are missing from the detected IR only, but the final fused detection recovers most of these missed pixels on moving objects.

13.4.3 Performance Analysis

To compare the performance of the detection algorithm for sensor fusion, we utilize the Receiver Operating Characteristic (ROC) curves and define the probability of detection as percentage of moving object pixels that are correctly detected and probability of false alarm as percent of background pixels that are classified as moving object. We selected frames representing afternoon, early morning, and high noon for this analysis. The nighttime was not selected since no video signal was available at night (6:30 p.m. – 6:30 a.m.) and the detection algorithm relied only on the IR sensor; this was explained in example 2 above. The first ROC curve, Fig. 13.11(a), represents an afternoon time. An example of this is frame 2408 in Fig. 13.7. As is indicated by example 1 frame 2408 and this ROC curve, the video signal provided a higher performance than the IR signal. The fusion method provides a higher level of performance than both the video and the IR.

The ROC curve of Fig. 13.11(b) is an example of early morning hours. This figure is in contrast to that of Fig. 13.11(a) in the afternoon. In this case, the detection rates for both the IR and the fused image were high and the video sensor operated only nominally. This is again due to the fact that a great deal of energy has been dissipated to the environment throughout the night during early morning hours, and a large gradient may exist between natural surface

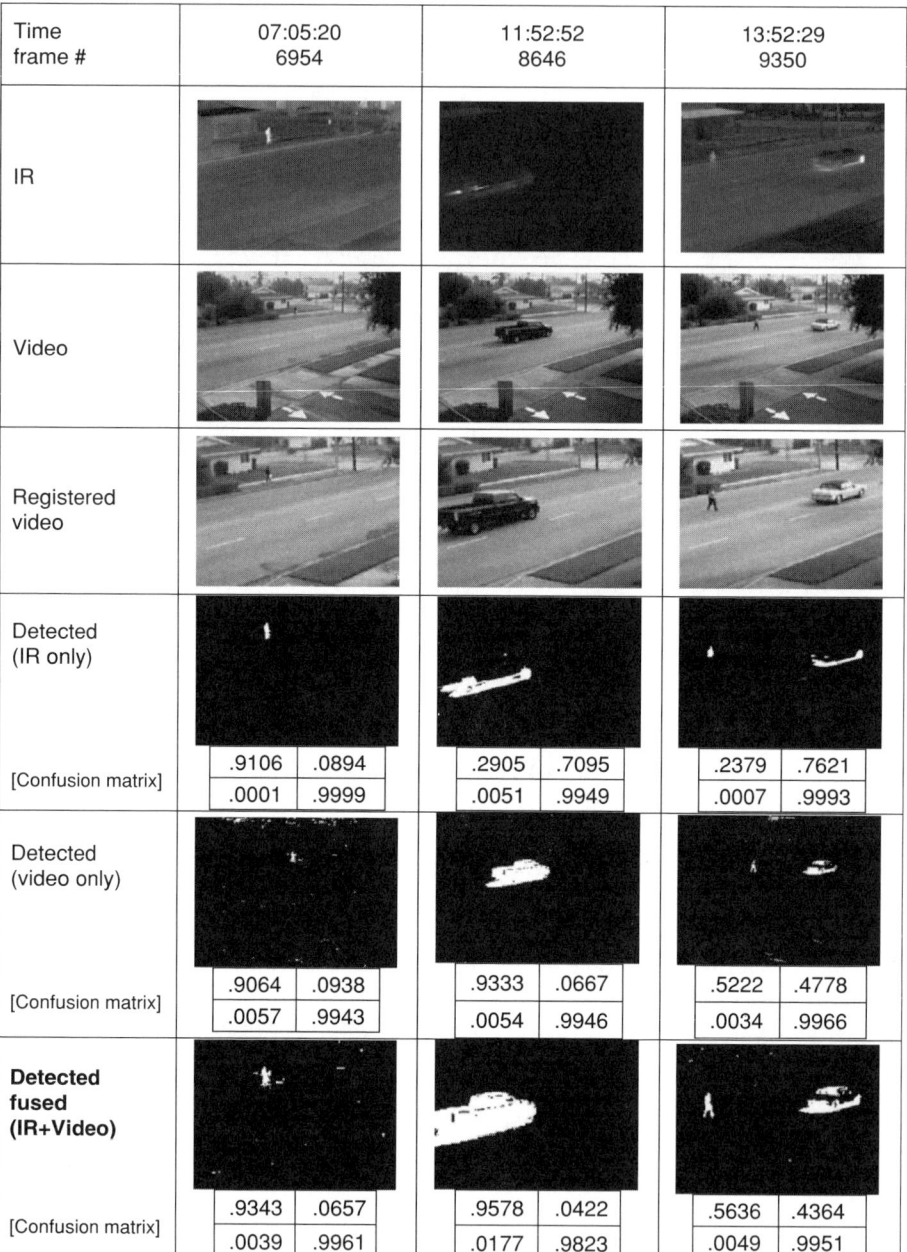

Time frame #	07:05:20 6954		11:52:52 8646		13:52:29 9350	
IR						
Video						
Registered video						
Detected (IR only)						
[Confusion matrix]	.9106	.0894	.2905	.7095	.2379	.7621
	.0001	.9999	.0051	.9949	.0007	.9993
Detected (video only)						
[Confusion matrix]	.9064	.0938	.9333	.0667	.5222	.4778
	.0057	.9943	.0054	.9946	.0034	.9966
Detected fused (IR+Video)						
[Confusion matrix]	.9343	.0657	.9578	.0422	.5636	.4364
	.0039	.9961	.0177	.9823	.0049	.9951

Figure 13.10. Example 4: Mixed IR and good video signal.

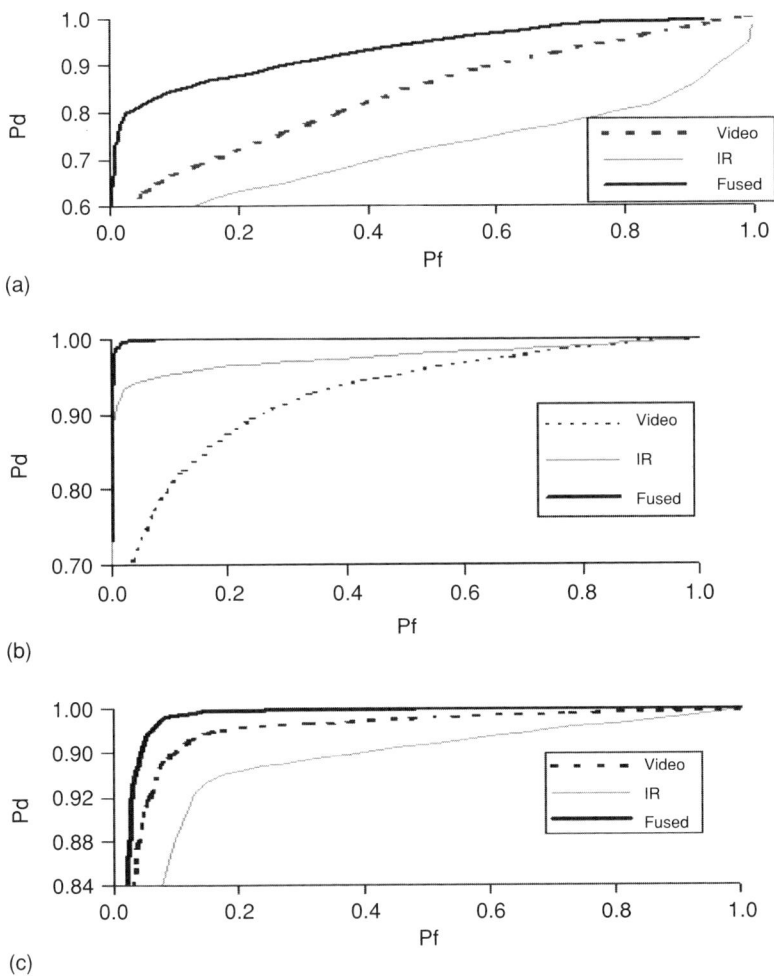

(a)

(b)

(c)

Figure 13.11. ROC curves for various periods of the day: (a) afternoon–evening, (b) early morning, and (c) morning–noon.

temperatures and those of animated objects with internal sources of energy such as vehicles and humans. In addition, the video signal, as indicated in Fig. 13.9, example 3, is rapidly changing due to the illumination changes when sun is rising in the sky.

The third ROC curve, Fig. 13.11(c), is an example of how fusion can enhance the detection when both sensors may be operating at lower rates. This is an example when cooperation between sensors can play a complementary role. This is due to the fact that different sensors may detect different parts of an object. So, one expects sensor fusion to do much better in detecting more pixels on the object than any one of the sensors alone. This is observed from frames 8646 and 9340 of example 4 in Fig. 13.10 when for example, the detected IR and video frames have detected different parts of the same object.

These ROC curves also indicate that as the time of day changes, the dynamic sensor fusion introduced here can automatically adapt to environmental changes. This adaptation is also in the form of adapting to the best sensor at the time. The cooperation among sensors can also take on a complementary role when different cameras are able to detect different parts of an object that may not be visible to the other. This adaptation is done continuously in a cooperative manner.

13.5 Conclusions

In this chapter, a novel physics-based sensor fusion technique for moving object detection was introduced. The sensor fusion architecture integrated the statistical and phenomenology of the sensors in the visible and long-wave IR through an evolutionary computational model. Our representation, mixture of Gaussians, along with the cooperative coevolutionary search algorithm integrated the contextual information through the physics-based and statistical models. We showed that our fusion model adapted to various illumination conditions and is suitable for detection under a variety of environmental conditions.

References

[1] Buede DM and Girardi P. (1997). "A target identification comparison of Bayesian and Dempster–Shafer multisensor fusion." *IEEE Tran. Syst. Man. & Cybern.*, 27(5):569–577.

[2] Cristani M, Bicego M, and Murino V. (2002). "Integrated region and pixel-based approach to background modeling." *IEEE Workshop Motion and Video Comput.* pp. 3–8.

[3] Han J and Bhanu B. (2003). "Detecting moving humans using color and infrared video." in: *IEEE Conference on Multisensor Fusion and Integration for Intelligent Systems.* pp. 228–233.

[4] Haritaoglu I, Harwood D, and Davis L. (2000). "W4: Real-time surveillance of people and their activities." *IEEE Trans. Pattern Anal. Mach. Intell.*, 22(8):809–830.

[5] Heesung K, Sandor ZD, and Nasrabadi NM. (2002). "Adaptive multisensor target detection using feature-based fusion." *Opt. Eng.*, 41(1):69–80.

[6] Horprasert T, Harwood D, and Davis LS. (1999). "A statistical approach for real-time robust background subtraction and shadow detection." in: *Proc. FRAME-RATE Workshop held in conjunction with Intl. Conf. on Computer Vision.* pp. 1–19.

[7]   Joshi R and Sanderson AC. (1999). "Minimal representation multisensor fusion using differential evolution." *IEEE Trans. Syst. Man & Cybernet.*, 29(1):63–76.

[8]   Lipton AJ, Fujiyoshi H, and Patil RS. (1998). "Moving target classification and tracking from real-time video." in: *IEEE Workshop on Applications of Computer Vision*, pp. 8–14.

[9]   Ma B, Lakshmanan S, and Hero III O. (2000). "Simultaneous detection of lane and pavement boundaries using model-based multisensor fusion." *IEEE Trans. Intell. Transportation Syst.*, 1(3):135–147.

[10] Majumder S, Scheding S, and Durrant-Whyte HF. (2001). "Multisensor data fusion for underwater navigation." *Robotics & Autonomous Syst.*, 35(2):97–108.

[11] Nadimi S and Bhanu B. (2003). "Physics-based models of color and IR video for sensor fusion." in: *IEEE Conference on Multisensor Fusion and Integration for Intelligent Systems.* pp. 161–166.

[12] Nandhakumar N and Aggrawal JK. (1997). "Physics-based integration of multiple sensing modalities for scene interpretation." *Proc. of the IEEE*, 85(1):147–163.

[13] Pavlidis I, et al. (1999). "Automatic detection of vehicle passengers through near-infrared fusion." in: *Proc. IEEE Conference on Intelligent Transportation Systems.* pp. 304–309.

[14] Potter MA and DeJong KA. (1994). "A cooperative coevolutionary approach to function optimization." in: *Proc. of the 3rd Conference on Parallel Problem Solving from Nature.* pp. 249–257.

[15] Scribner D, Warren P, Schuler J, Satyshur M, and Kruer M. (1998). "Infrared color vision: An approach to sensor fusion." *Optics and Photonics News.* pp. 27–32.

[16] Scribner D, Warren P, and Schuler J. (1999). "Extending color vision methods to bands beyond the visible." in: *Proc. IEEE Workshop on Computer Vision Beyond the Visible Spectrum: Methods and Applications.* pp. 33–44.

[17] Shafer SA. (1985). "Using color to separate reflection components." *Color Research Appl.*, 10(4):210–218.

[18] Stauffer C and Grimson WEL. (2000). "Learning patterns of activity using real-time tracking." *IEEE Trans. Pattern Anal. Mach. Intell.*, 22(8):747–757.

[19] Therrien CW, Scrofani JW, and Krebs WK. (1997). "An adaptive technique for the enhanced fusion of low light visible with uncooled thermal infrared imagery." in: *Proc. Intl. Conference on Image Processing.* 3:405–409.

[20] Ulug ME and McCullough C. (1999). "Feature and data level fusion of infrared and visual images." *Proc. SPIE: Intl. Soc. Opt. Eng.*, 37(19):312–318.

[21] Viva MMD and Morrone C. (1998). "Motion analysis by feature tracking." *Vis. Research*, 38:3633–3653.

[22] Ward G. (1994). "The RADIANCE lighting simulation and rendering system." *Computer Graphics (SIGGRAPH' 94).* pp. 459–472.

[23] Waxman AM, et al. (1998). "Solid state color night vision: Fusion of low-light visible and thermal infrared imagery." *MIT Lincoln Lab J.*, 11(1):41–57.

[24] Wren CR, Azarbayejani A, Darrell T, and Pentland AP. (1997). "Pfinder: Real-time tracking of the human body." *IEEE Trans. Pattern Anal. Mach. Intell.*, 19(7):780–785.

The Use of Synthetic Data in Eye/Face Recognition

Behrooz Kamgar-Parsi,[1] Behzad Kamgar-Parsi,[2] and Benjamin N. Waber[1,3]

[1]Information Technology Division, Naval Research Laboratory, Washington, DC 20375
[2]Office of Naval Research, Arlington, VA 22217
[3]Department of Computer Science, Boston University, MA 02215

Information is in data. Classifiers cannot make up for inadequate data when performing recognition tasks. The approach presented here is an attempt aiming at increasing data for cases where adequate facial image data is not available, i.e., when only one or a few images of a subject is at hand. Due to significant natural facial variations, e.g., facial expression, and appearances due to lighting condition, and head pose (even when dealing with "frontal view"), the use of a collection of images covering these variations and appearances will be of great help. We show that the expansion of the training set, by careful construction of synthetic images that capture all or most of the desired appearances can significantly improve the performance, especially when only one (or several, but very similar) real image(s) of a given individual are available. To have a better understanding of the issues involved, we address an inherently simpler problem, i.e., face/identity recognition/verification using only one eye and its associated eyebrow. Moreover, our experimental results indicate that the eye is rich in discriminative information, perhaps providing more information than what is normally utilized by humans. This wealth of information, however, can be exploited by machines for close-up images. Finally, we speculate that a similar improvement can be achieved when the training set is enriched with carefully generated synthetic images of the entire face. Issues concerning synthesis automation are also discussed.

14.0 Introduction

Over the past several years great advances have been made toward solving the complex problem of face recognition.[1,5,6,9–15] However, there are still challenges to be addressed. Most published studies in face recognition employ either the entire face (e.g.,[11]), or several facial features (e.g., eyes, nose, mouth, etc.)

simultaneously (e.g.,[8]). Of course, not only every feature possesses a certain amount of discriminating information, but also the relative positions of features with respect to others contain valuable discriminating significance. In general, by using a greater number of features, or a larger facial region, the likelihood of successful recognition is increased. Nevertheless, to gain a better understanding of the intricacies of the problem of face recognition, it is informative to study how accurately a computer can recognize individuals when only certain single feature, or certain facial regions, are used. For example, both from a theoretical and a practical point of view, it will be of interest to know how much information the eye alone can provide when a machine tries to recognize a person. More importantly, when dealing with a single feature, one deals with an inherently less complicated problem. Consequently, it may be simpler to understand the strengths and the weaknesses of the underlying approach.

In this Chapter, we investigate the capabilities of a face/identity recognition algorithm when using only one eye and the eyebrow, and particularly the improvement that can be achieved when synthetic eye images are used to augment the training set. Furthermore, we will discuss some of the practical issues that can arise in an automation mode. This Chapter presents further expansion of the work reported in.[4] We note that the problem we are addressing, i.e. eye recognition, is very different from iris recognition. Some aspects of eye recognition problem are discussed in.[2]

It is assumed that a single (or several, but very similar) close up image(s) of a person is available for use in the training set. Even in frontal view images, there will be variations such as head tilt, head rotation, gaze angle, the degree to which the eye is open, relative position of the eyebrow with respect to the eye, lighting condition, and certain expressions, etc., which can affect the recognition rate

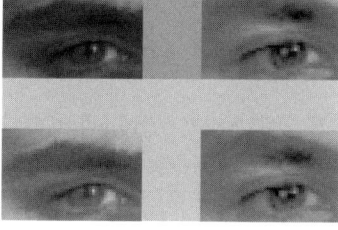

Figure 14.1. The left two photos show two people after our eyefinder program has located the center of their irises (bright spots in the eyes). The top right photos show their cropped-out right eyes after the original photos were rotated and scaled to a prespecified orientation and size. The bottom row shows the two eyes after intensity normalization, making their total brightness equal to each other.

when only one or very similar images of a given person are available. We propose generating synthetic images so that the training set already includes examples of the above-mentioned variations and appearances. Although anatomically imprecise, synthetic images may be adequate for an appearance-based system. We develop a system which crops one eye and the eyebrow out of a close-up image of a face. The system will then use the cropped-out window for a recognition task.

14.1 Approach

Synthetic images must be constructed from the available real image. To do so, we develop a number of operators such that when applied to an image, in sole, or in combination with others, the desired appearances are approximately obtained. The operators are applied to an eye and eyebrow, more precisely to a rectangular window covering an eye and the eyebrow. A relevant question is whether the process of generating synthetic images must be automated. Of course, if the system is intended for online learning, construction of the images must be automated. But, even in the case of off-line learning, a certain degree of automation can be very helpful.

The eye model   The iris is modeled with a circle; the center and the radius of this circle are determined during the process of locating the eye. Variations of the eyelids are governed by parabolic models. Normalized grey-level values in the window containing the eye and the eyebrow constitute the eye texture model.

14.1.1 Operators

Below is the list of the operators followed by their descriptions. We note that although the following operators may not produce precise anatomical changes, they may be adequate for an appearance-based system.

1. Lower eyeBrow (LB): pulls down the eyebrow (without any changes to the location or the shape of the eye).
2. Raise eyeBrow (RB): pushes up the eyebrow.
3. Tilt eyeBrow (TB): rotates the entire eyebrow around its middle so that the half closer to nose is lowered.
4. Arch eyeBrow (AB): bends the eyebrow downward while holding its middle portion fixed (AB is often used together with RB).
5. Lower Upper eyeLid (LUL): pulls down the upper eyelid so that the eye appears less open.
6. Raise Upper eyeLid (RUL): pushes up the upper eyelid so that the eye appears more open.
7. Raise Lower eyeLid (RLL): pushes up the lower eyelid so that the eye appears less open.

8. Lower Lower eyeLid (*LLL*): pulls down the lower eyelid so that the eye appears more open.
9. Turn Face Left (*TFL*): the face turns to the left without affecting the gaze angle.
10. Turn Face Right (*TFR*): the face turns to the right without affecting the gaze angle.
11. Eye Look Left (*ELL*): the eye looks to the left with no head rotation.
12. Eye Look Right (*ELR*): the eye looks to the right with no head rotation.
13. Cast Shadows (*CS*): creates shadows in the window in certain manners.

14.1.1.1 Notations

w_o is the original window covering the eye and the eyebrow, and contains *row* rows and *col* columns. $g_o[i][j]$ is the gray level value at the pixel on the ith row and the jth column in w_o. The pixel $i = 0$, $j = 0$ represents the top corner in w_o which is away from nose.

w_l is a larger window. It covers w_o in addition to r_t rows on top of it, and r_u rows just under it. It covers the same columns as w_o does. gl denotes gray level values in w_l. g_s denotes gray level values for the synthetic image.

14.1.1.2 Operators 1–4

LB operates on w_l to transform g_o's into g_s's as follows.

$$\text{for}(j = 0; j < col; j++)$$
$$\text{for}(i = i_j; i >= 0; i--)$$
$$g_s[i][j] = gl[i + r_t - m][j];$$

where i_j is the row index of a pixel on the jth column below the eyebrow or between the eye and the eyebrow, depending on j. m is a small positive integer which decides how far the eyebrow is lowered. Note that, in each column, m pixels move into w_o from w_l.

RB has a similar behavior, but m is a small negative integer. Here some pixels are pushed out of the window, and pixels with new gray level values (comparable to those between the eye and the eyebrow) are created to enlarge the area between the eye and the eyebrow. *TB* is a mixture of *LB* and *RB*, and *AB* is a variation of *LB*.

14.1.1.3 Operators 5–8

Opening the eye further usually requires the display of some or all of the covered portion of the iris. Hence, along with other subtasks, the (entire) iris must be constructed from the visible portion. We model the iris image with a circle. Fitting a small circle to quantized points with regression techniques is typically unreliable. We use a matched-filter like approach to find the iris, which we describe below. Let r denote its radius, and i_c and j_c the row and the column index of the pixel at its center. To calculate these parameters, we extract the visible portion of the iris. Let len_i denote the length of the iris on row i. While, j_c is calculated from len_i's directly, r and i_c will be calculated through fitting the model to len_i's. Because r is only several pixels long, the quantization error is significant and attempts should be made to reduce its impact. While fitting a circle to iris, we allow for 4 levels of quantization. Let f be the fraction (in the vertical direction) of the lowest iris pixel in column j_c which is actually covered by the iris. We then allow f to take on the values 1, .75, .5, or .25.

We allow the radius of the model circle r_M to vary for several pixels in the increments of 1/8 pixel. Furthermore, for each value of r_M, we allow 4 values for f (as mentioned above). For each set of r_M and f, we move the model circle along the vertical direction on the extracted iris and calculate the degree of mismatch between the corresponding segments of the model and the iris. The model configuration (specified by r_M and f) together with its placement (specified by i_c), which gives rise to the minimum discrepancy, will determine the values of r_M, f, and i_c. Once the entire iris is obtained, we can display it to the desired extent (see Fig. 14.2 for LLL). Operators LUL, RUL, and RLL can be similarly described.

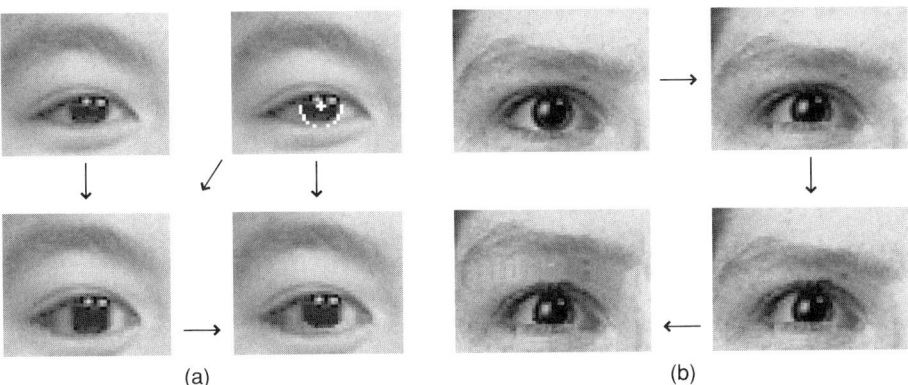

(a)   (b)

Figure 14.2. (a) Different stages showing how the operator Lower Lower eyeLid (LLL) acts. Top left: the real image. Top right: the model partial circle fit to the iris and its center, obtained by the eyefinder program. Bottom left: the iris and the lower part of the eye are stretched downward. Bottom right, i.e., the outcome of the operator LLL: the white of the eye has been pulled in so that the iris agrees with the model partial circle. (b) Cumulative effect of 3 operators to produce a synthetic image (stages showing how individual operators act have been omitted). Top left is the real image. Top right is the image after the application of the operator RLL (Raise Lower eyeLid). Bottom right is the outcome of the operator RUL (Raise Upper eyeLid) applied to the image on top of it. Bottom left is the outcome of the operator RB (Raise eyeBrow) applied to the image to its right. In short, the bottom left is generated to produce the appearance of an upward head tilt.

14.1.1.4 Operators 9–10

TFL and *TFR* operate as follows:

for($i = 0; i < row; i + +$)

　for($j = 0,\ count = 0.0; j < col/(1.0 + 0.05 * angle); count+ = 1.0 + 0.05 * angle$)

　　$g_s[i][j] = g_0[i][count]$

That is, these operators subsample the image in the horizontal direction. This image, since it will be smaller than the original, is then scaled to the desired size. Note then that an angle with a too great magnitude will cause the scaled image to be undesirable. These operators create synthetic images of the subject's face looking to one side or the other by giving the eye region a compacted appearance (see Fig. 14.3).

The appeal of these operators is in their simplicity which makes them dependable and robust. They do not require detail 3D computations and the possibility of its associated complexities, when accurate 3D information is not available.

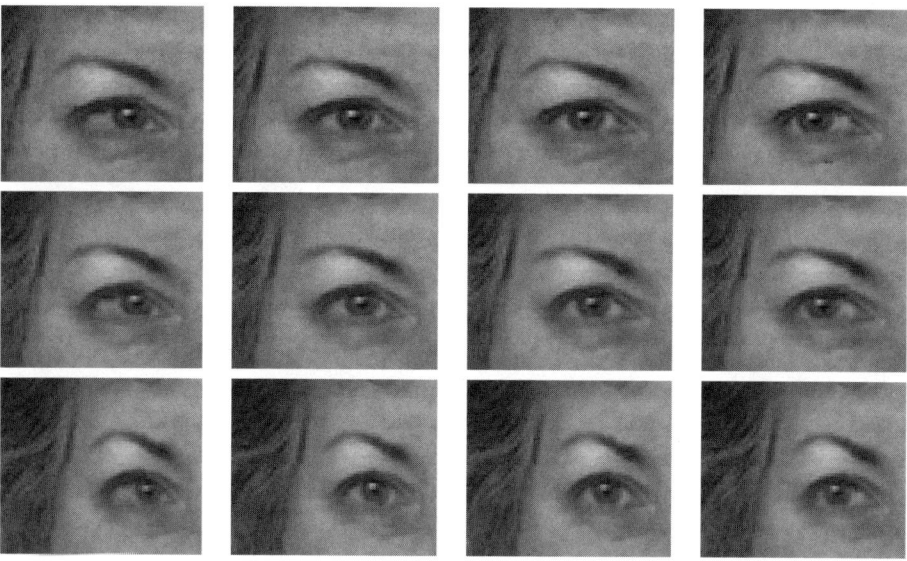

Figure 14.3. The third image from left on the first row is a real image, while the other eleven images are synthetic. The three synthetic images on the first row were generated through the application of *ELL* and *ELR*, i.e., Eye Look Left and Eye Look Right only. On the second row, each image has been derived from the image right above it by the application of a Turn Face operator. The third row is the same as the second row except that head rotation is more prominent. (Note that the windows shown in these images are larger than other windows so that they are visually more informative.)

14.1.1.5 Operator 11–12

Operators *ELL* and *ELR* act by translating the iris. The segmented iris is translated across the line from the end points of the upper eyelid in the image. The pixels that the iris used to occupy are filled in with the grey values of the closest "eye pixel" in the original image. The translated iris is then clipped against the upper eyelid to prevent any peculiar artifacts from arising in the final image.

This operator is quite straightforward and allows for accurate simulation of irises in different positions on the eye. Since lower eyelid segmentation cannot be performed with complete accuracy and consistency in an automated fashion, it is possible that the iris may overwrite a portion of the lower eyelid. Through many observations, however, the lack of lower eyelid segmentation has been shown not to be a significant problem, due to the fact that only a few pixels at most will be moderately affected. Also, in our experience, good iris segmentation is not nearly as important as accurate upper eyelid segmentation. (It is important, however, that the iris is overestimated and not underestimated, since then pieces of the iris will be left in the synthetic image.)

14.1.1.6 Operator 13

We confine ourselves only to the cases where the shadow is generated by the subject's own face. Furthermore, we exclude unusual cases such as the light source being below the face. Also disregarded is the case where the light source is to the side of the subject. This is because when a light casts shadows on only one side of the face, the other side of the face retains the property of uniform lighting. That is, the nose casts shadow over one of the subject's eyes, hence the unaffected eye can be used for recognition. The choice of what operator(s) would be needed was based on observations of existing databases, in particular the Yale Face Database B[7] and the FERET database. Introducing more operators and increasing the complexity of the shadow operators may slightly improve performance, however, the simplicity and generality of the operator described below offsets the gains.

The cast shadow operator (CS) creates a shadow region from the eyebrow to a set of computed points, which can be altered with an input parameter. These points are derived by reflecting all original upper eyelid points across the line connecting the first and the last upper eyelid points roughly approximating the eye socket. This operator simulates the situation where a light source is positioned above the subject. Our algorithm essentially wishes to estimate the boundary of the eye socket and place all pixels within that boundary in shadow (see Fig. 14.4 for an example). By using different parameter values for the lower shadow extant and the intensity of the shadow itself, it becomes easy to simulate a wide variety of shadows on the eye region caused by a light source moving back and forth or side to side above the head.

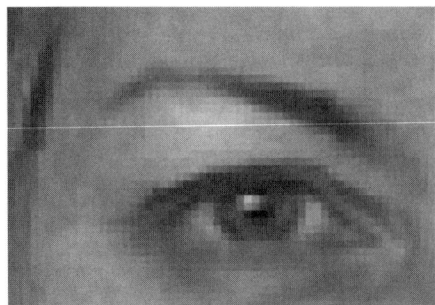 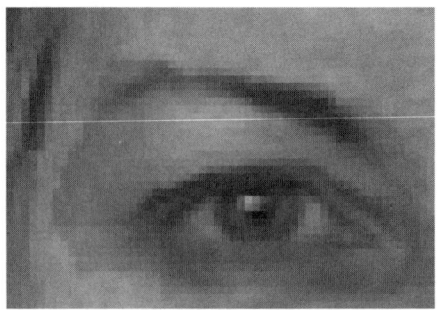

Figure 14.4. The left image is a real image. The right image is an example of Cast Shadow operator applied to the left image.

14.1.2 Intensity Normalization

To reduce the impact of variations in ambient light, we normalize the intensities in all the windows covering the eye so that they have the same total brightness. Intensity normalization should be performed so that the quality of the image is not (significantly) impacted. For details see.[3]

14.1.2.1 Window Size

We have experimented with different window sizes. We believe that the window should be picked so as to reduce shading that might be caused by certain lighting conditions. For example, Fig. 14.4 left, shows a window which is slightly too large, hence allowing unwanted lighting conditions. Whereas the window on the right is much better protected. The size of this window is 40×32 pixels, cropped out of a close-up image with the resolution of 256×384 pixels.

14.2 Experiments

There are different types of experiments with which one may try to assess the impact of the generated synthetic images. One way to do so would be to numerically quantify the improvement that a given synthetic image could provide. That is, suppose we have two real images of a given subject: in one image (call it I_1) the subject's face is straight toward the camera, while in the other one (call it I_2) the subject's head (face) is slightly rotated to the left. Using I_1, we make a synthetic image (call it I_3) with a left head rotation. Now, suppose according to a given metric, the distance between I_1 and I_2 is δ_{12} and the distance between I_3 and I_2 is $\delta_{23} > \delta_{12}$, then the creation of synthetic image I_3 has resulted in improvement. This is because if I_1 is assumed to be in the training set and I_2 in the test set, then, without I_3 added to the training set, the

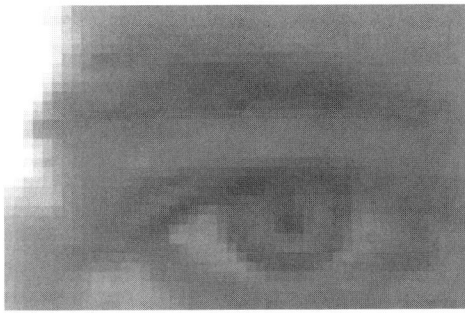
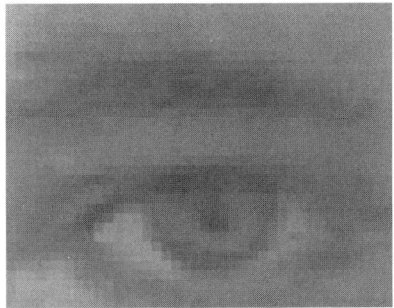

Figure 14.5. The window size at left is subject to many unwanted lighting conditions. The slightly smaller window at right is far simpler to handle.

likelihood that I_2 is recognized as the subject in question would be smaller. One could further quantify the improvement by calculating the ratio δ_{23}/δ_{12}. Of course, for a meaningful study the number of subjects for which the ratio is calculated must be large. A difficulty with this type of experiment is that ideally one would want the real images I_1 and I_2 to be different only because of a head rotation and not because of other factors such as expression, lighting condition, etc. The fact that, often, two images of a subject are different because of a number of factors, the computed improvement usually underestimates

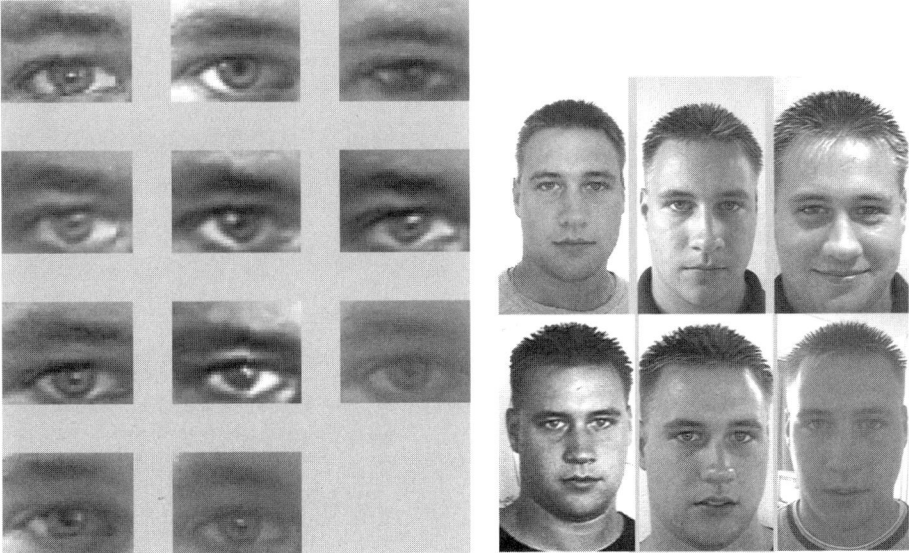

Figure 14.6. The same eye of the same person cropped out of his photos taken over a one-year period under different lighting conditions. Some of the full-face photos from which these eyes were cropped out are shown on the right.

the true effectiveness of the operator in question. We conducted this type of evaluation for operators TFL and TFR in conjunction with ELL and ELR, i.e., head rotation combined with iris translation (see Experiment I).

A different type of experiment was conducted to evaluate the impact of operators collectively on the recognition rate. For this study, we pick a single image (call it I_0) of a given subject. Ideally, in I_0, the subject looks straight into the camera with no head rotation or tilt, has a neutral expression, and has no shadows on the face (though quite often in our reported experiments this was not the case). We would then embed this image in a pool of some 100 images of other people, and use other available real images of the same subject as test images to see if their best match in the pool of 100 images (or so) is I_0 (correct recognition) or not (incorrect recognition). We then add several synthetic images, all derived from I_0, to the pool, and find out how many of the test images find I_0 or one of its derived synthetic images as the closest match (correct recognition) and how many do not (incorrect recognition). Higher (correct) recognition rate once

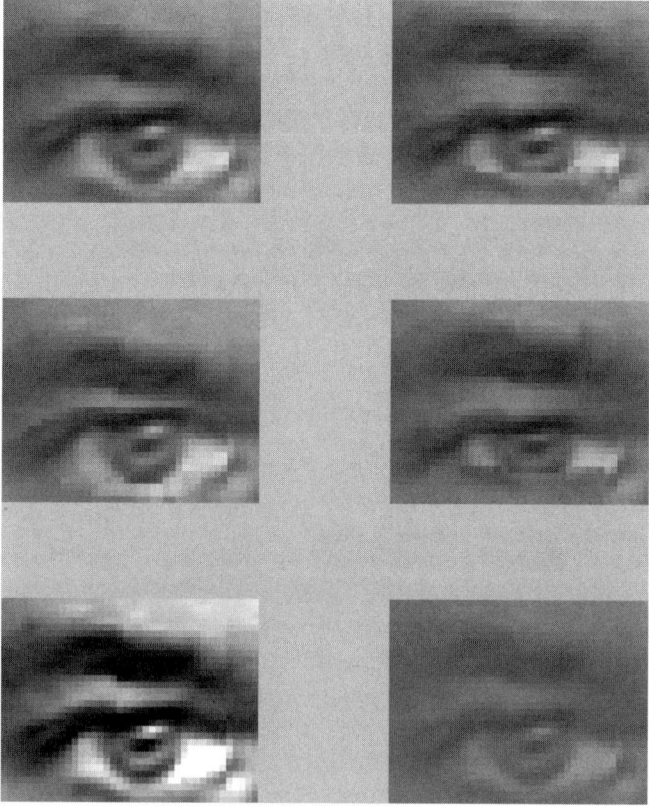

Figure 14.7. Top left is a real image. The other five are synthetic images which were automatically generated from the top left image.

synthetic images are used would reflect the usefulness of synthetic images (see Experiment II).

14.2.1 Experiment I

We tested the head rotation and iris translation operators on 110 images from the FERET database with the subject's head and eyes turned approximately 15° to their right. We chose this set of images because it would allow for better testing of both the head rotation and iris translation operators in tandem. We applied the aforementioned operators to the FERET image, where the head was not rotated and the subject was looking at the camera (although this was not precisely true in all cases). We then compared the synthetic images produced with the operators with the corresponding segmented eye region in the FERET image for that subject. We appropriately rotated and scaled the eye regions of each image so that the eyes were in approximately the same position.

Since all of the subjects did not necessarily turn their head exactly 15° and by no means had a gaze direction at that same angle, it was necessary to find the best synthetic image for each individual for comparison to the original segmented eye region. Thus we did a search (among synthetic images) from head rotation of 5–25°, as well as iris translation from $-25°$ to 25° (5° apart). We found the best match using the standard Euclidean metric.

Results   The original (without synthetic images) error on the average was 66.06 (with standard deviation of 60.18). With the synthetic images the average error reduced to 54.82 (with standard deviation of 51.96). This indicates an average of 17% improvement. We note that these results were obtained in the presense of other variations such as smile, raised eyebrow, etc. Such variations reduce the calculated improvement, hence the significance of the synthetic images are underrepresented.

14.2.1.1 Experiment II

Images used in the experiments were mostly from the FERET database, i.e., close-up frontal view images. However, because our FERET images provided no more than two images per person, we also included six "local" subjects from the Naval Research Laboratory and the Michigan State University, each providing 10 images taken in the same style as FERET images. Windows covering an eye and the eyebrow were cropped out automatically as follows. First, the two eyes were located using the eye locator program that we have developed. Next, the image was rotated, scaled, and translated so that the two eyes were located on prespecified pixels. Next, the window was cropped out and its intensity was normalized in the manner described earlier. The size of these images (or windows) was 40×32 pixels. Images were compared by calculating the sum of square differences of the gray-level values at corresponding pixels. We used this similarity measure because it is perhaps the most straightforward and commonly used similarity measure. Evaluation of the merits of different similarity measures is beyond the scope of this work.

14.2.2 Test 1: Real Images Only

We used 70 FERET subjects each providing 2 and 6 nonFERET subjects each providing 10 images. The test set had a total of 124 images, composed of 70 FERET images and 54 nonFERET images. The 70 FERET images belonged to 70 different subjects, whereas nonFERET images belonged to six subjects (each contributing 9 images). The training set was composed of 91 images, including 1 image, say S_0, of a test subject. For each image in the test set, its closest match (in the training set) was determined. If the closest match in the training set was S_0, i.e., if it belonged to the test subject, the answer would be regarded as correct; otherwise wrong. Of the 70 FERET test images, 59 (84%) of them found the correct match. Of the 54 nonFERET test images, 40 (74%) found the correct match.

14.2.3 Test 2: Real and Synthetic Images

We used S_0 and generated 8 synthetic images of S_0. These 8 synthetic images were generated as follows: Two CS operators were applied to the real image (causing two different synthetic lighting conditions). Operators LB and LLL in combination were applied creating the appearance of looking up into the camera (downward head tilt). Operators RLL, RUL, and RB in combination were applied creating the appearance of upward head tilt. See Fig. 14.2b. Operators RLL and LUL in combination were applied creating squint and possibly a slight smile. Operator CS was then applied to the above three synthetic images. These eight "derivatives" of S_0 were then added to the training set, while the same test set was used. If S_0 or one of its derivatives was the closest match, the answer was considered correct; otherwise wrong. We obtained correct answer for all of the FERET and nonFERET test images. The results are summarized in Table 14.1. We mention that we did not obtain error-free results when the generated synthetic images were fewer than eight. Results are further discussed in the Section 14.3.

14.3 Discussion

In general, humans do not need to closely examine people's eyes in order to recognize them. Furthermore, it is not an easy task for humans to recognize a given person in a photo if the presented photo displays only the eye and the eyebrow of the person. But, would this imply that the eye does not possess

Table 14.1. Summary of experimental results. For each subject eight synthetic images were produced.

Image source	Image collection info	Real images Error (percentage)	Real+Synthetic Error
FERET		11 (16%)	0
nonFERET set A	Images collected within a few days	2 (7%)	0
nonFERET set B	Images collected over many months	12 (44%)	0

unique and sufficient recognition information? Having so many recognition cues in their possession, humans have not had the need to specialize in recognition through the eye alone. Our experimental results are preliminary, nevertheless, they suggest that the eye is rich in discriminatory information—more than is normally utilized by humans. This wealth of information, however, can be exploited by machines.

Variations in the appearance of the eye and its associated eyebrow are caused by many factors, including head tilt, head rotation, gaze angle, the extent to which the eye is open, relative position of the eyebrow with respect to the eye, lighting condition, and certain expressions, etc. A versatile recognition system would have already seen examples of such variations. But, real images often do not provide an adequate number of these examples. Even if several photos of a given subject are available, they may not represent the entire space of possible eye/eyebrow variations. Furthermore, some of the real images may be too similar to each other and thus redundant. Synthetic images, on the other

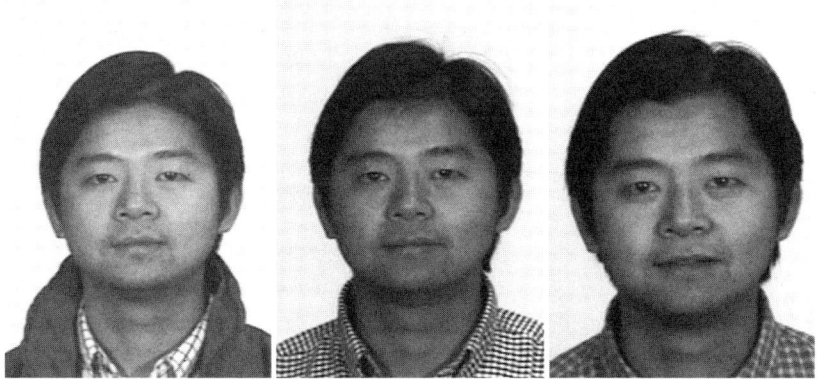

Figure 14.8. Some photos of a subject in nonFERET set A. Rotation, scaling, and intensity normalization will make (almost) identical eye images cropped out of these photos. That is, for images of this subject correct recognition did not require synthetic images (unlike the subject in Fig. 14.4).

hand-can enrich the training set so that it includes the desired variations. This has been indicated by our experimental results.

In Experiment I, we numerically quantified the improvement that a given synthetic image could provide. That is, suppose we had two real images of a given subject: in one image, I_1, the subject's face was straight toward the camera, while in the other one, I_2, the subject's face was slightly turned to the left. Using I_1 we made a synthetic image I_3 with a left head rotation. Now, suppose according to a given metric, the distance between I_1 and I_2 was δ_{12} and the distance between I_3 and I_2 δ_{23}. The ratio δ_{23}/δ_{12} would represent the improvement due to the created synthetic image, I_3.

In Experiment II, we evaluated the collective impact of synthetic images created to deal with a variety of facial variations/appearances on the recognition rate. As indicated in Table 14.1, for the nonFERET set A, i.e., for subjects whose images were taken within a few days, we obtained relatively good results even without synthetic images, while for set B subjects, synthetic images were greatly needed. The reason for such wide differences among these subjects appears to be as follows. The problem is much easier when images of the subject of interest are taken on the same day (or a few days apart), and under similar lighting conditions. Under such conditions, different images of the same subject can be so similar to each other that (small) variations in their eyes or eyebrows may not cause sufficient recognition difficulty and there would not be much need for synthetic images. However, when images are taken on different days, perhaps months apart, under considerably different conditions, then there will be large differences among them. Thus, variations such as head tilt, expression, etc., would add to the lack of similarity, resulting in incorrect recognition when only available real images are used. In these difficult cases, synthetic images, alleviating the differences in pose, expression, or lighting condition, are very helpful. Many, if not most, real life scenarios fall into the latter category where synthetic images can be of major help. Finally, we speculate that a similar improvement may be achieved when solving face recognition problem, i.e., when the training set is enriched with carefully generated synthetic images of the entire face.

Acknowledgments

This work has been supported by grants from Naval Research Laboratory and Office of Naval Research.

References

[1] Belhumeur PN, Hespanha J, and Kriegman D. (1997). "Eigenfaces versus fisherfaces: Recognition using class specific linear projection." *IEEE Trans. Pattern Anal. Mach. Intell.*, 19(7):711–720.

[2]   Hjelmas E and Wroldsen J. (1997). "Recognizing faces from the eyes only." IA: *Proc. Scandinavian Image Analysis Conference, 1997.*

[3]   Kamgar-Parsi Behrooz and Kamgar-Parsi, Behzad. (2002). "Recognizing eyes." in: *Naval Research Laboratory, NCARAI Technical Report.*

[4]   Kamgar-Parsi, Behrooz, Kamgar-Parsi, Behzad, and Jain AK. (2003). "Synthetic eyes." in: *Proc. 4th International Conference on Audio- and Video-Based Biometric Person Authentication.* pp. 412–420.

[5]   Kotropoulos C, Tefas A, and Pitas I. (2000). "Frontal face authentication using morphological elastic graph matching." *IEEE Trans. Image Process.* 9:555–560.

[6]   Moghaddam B, Wahid W, and Pentland A. (1998). "Beyond eigenfaces: Probabilistic matching for face recognition." in: *Proc. Int'l Conf. Automatic Face and Gesture Recognition.* Nara, Japan, April 1998. pp. 30–35.

[7]   Georghiades AS, Belhumeur PN, and Kriegman DJ. (2001). "From few to many: Illumination cone models for face recognition under variable lighting and pose." *IEEE Trans. Pattern Anal. Mach. Intell.* 23:643–660.

[8]   Penev PS. and Atick JJ. (1996). "Local feature analysis: A general statistical theory for object representation." *Network: Comput. Neural Syst.,* 7:477–500.

[9]   Sim T and Kanade T. (2001). "Combining models and exemplars for face recognition: An illuminating example." in: *Proc. CVPR 2001 Workshop on Models versus Exemplars in Computer Vision,* Dec. 2001.

[10] Kuang-Chil, L. Ho J, and Kriegman DJ. (2001). "Nine points of light: Acquiring subspaces for face recognition under variable lighting." in: *Proc. IEEE Computer Society Conference on Computer Vision and Pattern Recognition,* Dec. 2001. pp. 519–526.

[11] Turk M and Pentland A. (1991). "Eigenfaces for recognition." *J. Cognitive Neurosci.* 3(1): 71–86.

[12] Wang X and Tang X. (2004). "A unified framework for subspace face recognition." *IEEE Trans. Pattern Anal. Mach. Intell.* 26: 1222–1228.

[13] Wiskott L, Fellous JM, Kruger N, and von der Malsburg C. (1997). "Face recognition by elastic bunch graph matching." *IEEE Trans. Pattern Anal. Mach. Intell.* 19:775–779.

[14] Jian Y Zhang D, Frangi AF, and Jing-yu Y. (2004). "Two-dimensional PCA: A new approach to appearance-based face representation and recognition." *IEEE Trans. Pattern Anal. Mach. Intell.* 26:131–137.

[15] Zhang J, Yan Y, and Lades M. (1997). "Face recognition: Eigenfaces, elastic matching, and neural nets." in: *Proc. IEEE,* 85:1422–1435.

Hyperspectral Target Detection Based on Kernels

Heesung Kwon and Nasser M. Nasrabadi

U.S. Army Research Laboratoryz 2800 Powder Mill Road Adelphi, MD 20783
{hkwon, nnasraba}@arl.army.mil

Summary: In this chapter, linear signal or target detection algorithms are extended to nonlinear versions by using kernel-based methods. In kernel-based methods, learning is implicitly performed in a high-dimensional feature space where high order correlation or nonlinearity within the data are exploited. Nonlinear realization is mainly pursued to reduce data complexity in a high-dimensional feature space and consequently provide simpler decision rules for data discrimination.

A well-known anomaly detector, RX-algorithm, is extended to its nonlinear (kernel) version. Similarly, the conventional matched filter detector and the subspace matched filter detector are extended to their corresponding nonlinear versions using the ideas of kernel learning theory. Performance of all these linear methods and their kernel versions are compared on several hyperspectral images. Experimental results show that the kernel-based detection algorithms outperform the linear detection algorithms.

15.0 Introduction

Hyperspectral imagery can be used in reconnaissance and surveillance applications where objects of interest are detected and identified. This chapter describes techniques that are used to detect and identify military targets, camouflaged objects, and surface mines. Hyperspectral imagery provides a significant information about the spectral characteristics of the materials in the scene that can be used for anomaly and target detection.[1–9]

Typically, a hyperspectral spectrometer provides hundreds of narrow contiguous bands which can be exploited to detect and identify certain types of materials in the image. Hyperspectral sensors that exploit the reflective (or emissive) properties of objects can collect data in the visible and short-wave infrared (IR) regions (or the midwave and long-wave IR regions) of the spectrum. Collection of this data allows the algorithm to detect and identify targets of interest in a hyperspectral scene by exploiting the spectral signature of the materials.

The process of detecting and identifying a target in hyperspectral imagery consists of two stages. The first stage is an anomaly detector, which identifies spectral anomalies or a localized spectral difference. The second stage is to identify whether or not the anomaly is a target or a natural clutter. This stage can be achieved if the spectral signature of the target is known, which can be obtained from a spectral library or using an spectral subspace match filter designed from a set of training data.[1,5]

Almost all the anomaly and target detectors are based on a linear process that exploit the first- and second-order correlation of the data to identify anomalies or targets. For example, a well-known spectral anomaly detection algorithm was developed in[3] called RX-algorithm, which is now considered as a benchmark anomaly detector. RX-algorithm is based on exploiting the difference between the spectral signature of an input pixel with its surrounding neighbors.[8,10–13] This distance comparison is very similar to the Mahalanobis distance measure which is done by comparing the corresponding wavelengths (spectral bands) of two measurements normalized by the covariance matrix of background statistics. The convention RX's distance measure does not take into account the higher order relationships (higher order correlation) between the spectral bands at different wavelengths. The nonlinear relationships between different spectral bands within the target or clutter spectral signature needs to be exploited in order to better distinguish between target and background. Similarly, most of the target detection algorithms are based on linear matched (subspace) filters where the spectral characteristics of a target or a target subspace representing target information are assumed to be known. In spectral matched filtering[14–16] a linear (subspace) mixing model is assumed where the target spectral signature is used in conjunction with the covariance matrix of the background data to identify a specific target. In matched subspace filtering,[6,17] a subspace linear mixture model is used for target detection where the target and background signatures are represented by their corresponding linear subspaces. Both matched filtering and matched subspace filtering are based on linear mixture models that ignore the higher order correlation between the spectral bands.

In this chapter, RX-algorithm, matched filter, and matched subspace filtering techniques are extended to their corresponding nonlinear versions using kernels and their properties.[18–20] For example, a nonlinear version of RX-algorithm is formulated by transforming each spectral pixel into a high-dimensional feature space (could be potentially infinite dimension) by a nonlinear mapping function. The spectral pixel in the feature space now consists of possibly the original spectral bands and a nonlinear combination of the spectral bands of the original spectral signature. This way, the higher order correlation between spectral bands could be exploited by RX-algorithm. However, the nonlinear RX-algorithm cannot be implemented directly due to the high dimensionality of the feature space. But an efficient kernel-based nonlinear version of the RX-algorithm can be implemented by using kernel functions. Similarly, a nonlinear matched filter is implemented by defining a mixture model in the feature space which is equivalent to a nonlinear mixture model in the input space. It is then shown that the matched filter in the feature space can be efficiently implemented in terms of kernel functions. In fact, using kernel functions no explicit knowledge of the actual nonlinear mapping is necessary which means the actual algorithm is not computed explicitly in the feature space. This property is the major advantage of the kernel-based methods that reduce a nonlinear algorithm to a linear one in some high-dimensional feature space. In this chapter, it is also shown that a matched subspace detector can be extended to its corresponding nonlinear version using kernel-based methods.

The kernel methods have emerged as new nonlinear-based learning techniques that implicitly exploit the dot product of feature vectors generated by the nonlinear mapping of the input vectors using kernel representations. The implicit exploitation of nonlinear features through kernels provides crucial information about a given data which, in general, the learning methods based on linear models cannot achieve. In kernel methods, the learning is performed in a high-dimensional feature space where the complexity of the given data can be possibly reduced, subsequently generating simpler decision rules and improving generalization performance. Kernel-based versions of a number of feature extraction or pattern recognition algorithms have recently been proposed.[21–26]

This chapter is organized as follows. In Section 15.1 kernel feature space is defined. RX-algorithm and its nonlinear version (kernel RX-algorithm) is described in Section 15.2. In Section 15.3 linear matched filter and kernel matched filters are described. Similarly, in Section 15.4 linear subspace matched filter and its kernel version are reviewed. In Section 15.5, experimental results are provided comparing linear algorithm with their corresponding kernel version. Finally, conclusion is given in Section 15.6.

15.1 Kernel Feature Space and Kernel Methods

In this section, an introduction to kernel feature map and kernel learning is provided, which is used in the following sections to convert several linear target detection algorithms into their corresponding nonlinear versions which can then easily be implemented in terms of kernel functions. Suppose the input hyperspectral data is represented by the data space $(\mathcal{X} \subseteq \mathcal{R}^J)$ and $\mathcal{F}$ be a nonlinear feature space associated with $\mathcal{X}$ by a nonlinear mapping function Φ

$$\begin{aligned} \Phi &: \mathcal{X} \to \mathcal{F}, \\ \mathbf{x} &\mapsto \Phi(\mathbf{x}), \end{aligned} \tag{1}$$

where $\mathbf{x}$ is an input vector in $\mathcal{X}$ which is mapped into a potentially much higher dimensional feature space. Now any linear algorithm can be remodeled in this high-dimensional feature space by replacing the original input data $\mathbf{x}$ with the mapped data $\Phi(\mathbf{x})$. Kernel-based learning algorithms use an effective kernel trick (Eq. 2) to implement dot products in feature space in terms of kernel functions.[19] The kernel representation for the dot products in $\mathcal{F}$, known as kernel trick, is expressed as

$$\begin{aligned} \mathbf{k}(\mathbf{x}_i, \mathbf{x}_j) &= \; <\Phi(\mathbf{x}_i), \Phi(\mathbf{x}_j)> \\ &= \Phi(\mathbf{x}_i) \cdot \Phi(\mathbf{x}_j) \end{aligned} \tag{2}$$

where $\mathbf{k}$ is a positive definite kernel, such as a Mercer kernel.[19]

Implementing any linear algorithm (i.e., matched filter) in the feature space is equivalent to performing a nonlinear version of that algorithm (i.e., nonlinear matched filter) in the original data space. However, due to the high dimensionality of feature space $\mathcal{F}$, it is computationally not feasible to implement the algorithm in the feature space. Using the kernel trick it allows us to implicitly compute the dot products in $\mathcal{F}$ without mapping the input vectors into $\mathcal{F}$; therefore, in the kernel learning methods, the mapping Φ does not need to be identified. However, an appropriate kernel has to be defined which has a nonlinear mapping associated with it. Two commonly used kernels are the Gaussian Radial Bases Function kernel (RBF kernel): $\mathbf{k}(\mathbf{x}, \mathbf{y}) = \exp\left(\frac{-\|\mathbf{x}-\mathbf{y}\|^2}{c}\right)$ and Polynomial kernel: $((\mathbf{x} \cdot \mathbf{y}) + \theta)^d$. See[19] for detailed information about the properties of kernels and kernel-based learning.

15.2 Introduction to RX-Algorithm and Kernel RX-algorithm

15.2.1 RX Algorithm

Reed and Yu in[3] developed a generalized likelihood ratio test (GLRT), so-called RX anomaly detection, for multidimensional image data assuming that the spectrum of the received signal (spectral pixel) and the covariance of the background clutter are unknown. In the conventional RX algorithm, a nonstationary local mean is subtracted from each spectral pixel. The local mean $\boldsymbol{\mu}$ is obtained by sliding a double concentric window (a small inner window centered within a larger outer window) over every spectral pixel in the image, and the mean of the spectral pixels falling within the outer window. The size of the inner window is assumed to be the size of the typical target of interest in the image. The residual signal after mean subtraction is assumed to approximate a zero-mean pixel-to-pixel independent Gaussian random process.

Let each input spectral signal consisting of J spectral bands be denoted by $\mathbf{x}(n) = (x_1(n), x_2(n), \dots, x_J(n))^T$. Define $\mathbf{X}_b$ to be a $J \times N$ matrix of the

N reference background clutter pixels (or pixels in the outer window). Each observation spectral pixel is represented as a column in the sample matrix $\mathbf{X}_b$

$$\mathbf{X}_b = [\mathbf{x}(1)\mathbf{x}(2)\dots\mathbf{x}(N)]. \tag{3}$$

The two competing hypotheses that the RX algorithm must distinguish are given by

$$
\begin{aligned}
\mathbf{H}_0 &: \mathbf{x} = \mathbf{n}, && \text{Target absent} \\
\mathbf{H}_1 &: \mathbf{x} = a\mathbf{s} + \mathbf{n}, && \text{Target present}
\end{aligned}
\tag{4}
$$

where $a = 0$ under $\mathbf{H}_0$ and $a > 0$ under $\mathbf{H}_1$, respectively. $\mathbf{n}$ is a vector that represents the background clutter noise process, and $\mathbf{s}$ is the spectral signature of the signal (target) given by $\mathbf{s} = (s_1, s_2, \dots, s_J)^T$. The target signature $\mathbf{s}$ and background covariance $\mathbf{C}_b$ are assumed to be unknown. The model assumes that the data arises from two normal PDFs with the same covariance matrix but different means. Under $\mathbf{H}_0$ the data (background clutter) is modeled as $\mathcal{N}(0, \mathbf{C}_b)$ and under $\mathbf{H}_1$ it is modeled as $\mathcal{N}(s, \mathbf{C}_b)$. The background covariance $\mathbf{C}_b$ is estimated from the reference background clutter data. The estimated background covariance $\hat{\mathbf{C}}_b$ is given by

$$\hat{\mathbf{C}}_b = \frac{1}{N}\sum_{i=1}^{N}(\mathbf{x}(i) - \hat{\boldsymbol{\mu}}_b)(\mathbf{x}(i) - \hat{\boldsymbol{\mu}}_b)^T, \tag{5}$$

where $\hat{\boldsymbol{\mu}}_b$ is the estimated background clutter sample mean given by

$$\hat{\boldsymbol{\mu}}_b = \frac{1}{N}\sum_{i=1}^{N}\mathbf{x}(i). \tag{6}$$

Assuming a single pixel target $\mathbf{r}$ as the observation test vector, the results of RX-algorithm is given by

$$
\begin{aligned}
RX(\mathbf{r}) = (\mathbf{r} - \hat{\boldsymbol{\mu}}_b)^T \Bigg(&\frac{N}{N+1}\hat{\mathbf{C}}_b + \frac{1}{N+1} \\
&(\mathbf{r} - \hat{\boldsymbol{\mu}}_b)(\mathbf{r} - \hat{\boldsymbol{\mu}}_b)^T \Bigg)^{-1} (\mathbf{r} - \hat{\boldsymbol{\mu}}_b) \underset{H_0}{\overset{H_1}{\gtrless}} \eta,
\end{aligned}
\tag{7}
$$

where η is a threshold of the test. As $N \to \infty$, RX algorithm converges to

$$RX(\mathbf{r}) = (\mathbf{r} - \hat{\boldsymbol{\mu}}_b)^T \hat{\mathbf{C}}_b^{-1}(\mathbf{r} - \hat{\boldsymbol{\mu}}_b). \tag{8}$$

Equation (8) is the RX expression that is implemented in this chapter.

15.2.2 Kernel RX-algorithm

In this subsection, we first remodel the RX-algorithm in the feature space by assuming that the input data has already been mapped into a high-dimensional feature space via a nonlinear mapping Φ. The two hypotheses in the feature space are now

$$\mathbf{H}_{0_\Phi}: \Phi(\mathbf{x}) = \Phi(\mathbf{n}), \quad \text{Target absent}$$
$$\mathbf{H}_{1_\Phi}: \Phi(\mathbf{x}) = a\Phi(\mathbf{s}) + \Phi(\mathbf{n}). \quad \text{Target present} \tag{9}$$

The corresponding RX-algorithm in the feature space is now represented as

$$RX(\Phi(\mathbf{r})) = (\Phi(\mathbf{r}) - \Phi(\hat{\mu}_b))^T \hat{\mathbf{C}}_{b\Phi}^{-1}(\Phi(\mathbf{r}) - \Phi(\hat{\mu}_b)), \tag{10}$$

where $\hat{\mathbf{C}}_{b\Phi}$ and $\hat{\mu}_b$ are the estimated covariance and mean of the background clutter pixels in the feature space, respectively. The estimated covariance matrix for the mapped data $\Phi(\mathbf{X}_b) = \mathbf{X}_{b\Phi} := [\Phi(\mathbf{x}(1))\Phi(\mathbf{x}(2))\ldots\Phi(\mathbf{x}(N))]$ is given by

$$\hat{\mathbf{C}}_{b\Phi} = \frac{1}{N} \sum_{i=1}^{N} (\Phi(\mathbf{x}(i)) - \Phi(\hat{\mu}_b))(\Phi(\mathbf{x}(i)) - \Phi(\hat{\mu}_b))^T, \tag{11}$$

where $\Phi(\hat{\mu}_b)$ is the estimated background clutter sample mean given by

$$\Phi(\hat{\mu}_b) = \frac{1}{N} \sum_{m=1}^{N} \Phi(\mathbf{x}(n)). \tag{12}$$

The RX-algorithm given by Eq. (10) is now in the feature space which cannot be implemented explicitly due to the nonlinear mapping Φ that produces a data space of high dimensionality. In order to avoid implementing the Eq. 10 directly we need to kernelize it by using the kernel trick introduced in Section 15.1.

The estimated background covariance matrix can be represented by its eigenvector decomposition or so-called spectral decomposition[27] given by

$$\hat{\mathbf{C}}_{b\Phi} = \mathbf{V}_\Phi \Lambda_b \mathbf{V}_\Phi^T, \tag{13}$$

where Λ_b is a diagonal matrix consisting of the eigenvalues and $\mathbf{V}_\Phi$ is a matrix whose columns are the eigenvectors of $\hat{\mathbf{C}}_{b\Phi}$ in the feature space:

$$\mathbf{V}_\Phi = [\mathbf{v}_\Phi^1, \mathbf{v}_\Phi^2, \ldots, \mathbf{v}_\Phi^N], \tag{14}$$

where N is the maximum number of the eigenvectors with nonzero eigenvalues.

As shown in Appendix I the inverse of the estimated background covariance matrix can also be written in terms of its eigenvectors and eigenvalues as

$$\hat{\mathbf{C}}_{b\Phi}^{-1} = \mathbf{V}_\Phi \Lambda_b^{-1} \mathbf{V}_\Phi^T. \tag{15}$$

Each eigenvector $\mathbf{v}_\phi^j$ in the feature space, as shown in Appendix I, can be expressed as a linear combination of the input vectors $\Phi(\mathbf{x}_i)$ in the feature space

$$\mathbf{v}_\Phi^j = \sum_{i=1}^{N} \beta_i^j \Phi(\mathbf{x}_i) = \mathbf{X}_{b\Phi} \beta^j \tag{16}$$

where $\beta^j = (\beta_1^j, \beta_2^j, \ldots, \beta_N^j)^T$ and for all the eigenvectors

$$\mathbf{V}_\Phi = \mathbf{X}_{b_\Phi}\mathcal{B}, \tag{17}$$

where $\mathcal{B} = (\beta^1, \beta^2, \dots, \beta^N)^T$ are the eigenvectors of the kernel matrix (Gram matrix) $\mathbf{K}(X_{b_\Phi}, X_{b_\Phi})$ normalized by their corresponding eigenvalues as shown in the Appendix I.

Substituting Eq. (17) into (15) yields

$$\hat{\mathbf{C}}_{b_\Phi}^{-1} = \mathbf{X}_{b_\Phi}\mathcal{B}\Lambda_b^{-1}\mathcal{B}^T\mathbf{X}_{b_\Phi}^T. \tag{18}$$

Inserting Eq. (18) into (10) it can be rewritten as

$$RX(\Phi(\mathbf{r})) = (\Phi(\mathbf{r}) - \Phi(\hat{\mu}_b))^T\mathbf{X}_{b_\Phi}\mathcal{B}\Lambda_b^{-1}\mathcal{B}^T\mathbf{X}_{b_\Phi}^T(\Phi(\mathbf{r}) - \Phi(\hat{\mu}_b)). \tag{19}$$

The dot product term $\Phi(\mathbf{r})^T\mathbf{X}_{b_\Phi}$ in the feature space can be represented in terms of the kernel function

$$\begin{aligned}
\Phi(\mathbf{r})^T\mathbf{X}_{b_\Phi} &= (\mathbf{k}(\mathbf{x}(1), \mathbf{r})\ \mathbf{k}(\mathbf{x}(2),\mathbf{r})\ \dots,\ \mathbf{k}(\mathbf{x}(N),\mathbf{r})) \\
&= \mathbf{K}(\mathbf{r}, \mathbf{X}_b)^T \\
&= \mathbf{K}_r^T.
\end{aligned} \tag{20}$$

Similarly $\Phi(\hat{\mu}_b)^T\mathbf{X}_{b_\Phi}$,

$$\begin{aligned}
\Phi(\hat{\mu}_b)^T\mathbf{X}_{b_\Phi} &= (\mathbf{k}(\mathbf{x}(1), \hat{\mu}_b)\ \mathbf{k}(\mathbf{x}(2), \hat{\mu}_b)\ \dots,\ \mathbf{k}(\mathbf{x}(N), \hat{\mu}_b)) \\
&= \mathbf{K}(\hat{\mu}_b, \mathbf{X}_b)^T \\
&= \mathbf{K}_{\hat{\mu}_b}^T.
\end{aligned} \tag{21}$$

The above transformations in Eqs. (20) and (21) are referred to as the empirical kernel map.

Also using the properties of the Kernel PCA, as shown in Appendix I, we have the relationship

$$\hat{\mathbf{K}}_b^{-2} = \mathcal{B}\Lambda^{-1}\mathcal{B}^T, \tag{22}$$

where we denote the centered kernel (Gram) matrix $\hat{\mathbf{K}}_b = (\mathbf{K} - \mathbf{1}_M\mathbf{K} - \mathbf{K}\mathbf{1}_N + \mathbf{1}_N\mathbf{K}\mathbf{1}_N)$ where $(\mathbf{1}_N)_{ij} = 1/N$ is an $N \times N$ matrix and $\mathbf{K} = \mathbf{K}(\mathbf{X}_b, \mathbf{X}_b) = (\mathbf{K})_{ij}$ an $N \times N$ kernel matrix whose entries $\mathbf{K}(\mathbf{x}_i, \mathbf{x}_j)$ are the dot products $< \Phi(\mathbf{x}_i), \Phi(\mathbf{x}_j) >$. Substituting Eqs. (20), (21), and (22), into Eq. (19) and replacing $\mathbf{X}_{b_\Phi}$ with $\mathbf{X}_{b_\Phi} - \Phi(\hat{\mu}_b)$ in Eqs. (20) and (21) (due to centering) the kernelized version of the RX-algorithm is given by

$$\begin{aligned}
RX(\mathbf{K}(\mathbf{r})) &= (\hat{\mathbf{K}}(\mathbf{X}_b, \mathbf{r}) - \hat{\mathbf{K}}(\mathbf{X}_b, \hat{\mu}_b))^T\hat{\mathbf{K}}_b^{-2}(\hat{\mathbf{K}}(\mathbf{X}_b,\mathbf{r}) - \hat{\mathbf{K}}(\mathbf{X}_b, \hat{\mu}_b)) \\
&= (\hat{\mathbf{K}}_r - \hat{\mathbf{K}}_{\hat{\mu}_b})^T\hat{\mathbf{K}}_b^{-2}(\hat{\mathbf{K}}_r - \hat{\mathbf{K}}_{\hat{\mu}_b})
\end{aligned} \tag{23}$$

where $\hat{\mathbf{K}}_\mathbf{r}^T = \mathbf{K}_\mathbf{r}^T - \sum_{i=1}^N \mathbf{K}(\mathbf{x}_i,\mathbf{r})$ and $\hat{\mathbf{K}}_{\hat{\mu}_b}^T = \mathbf{K}_{\hat{\mu}_b}^T - \sum_{i=1}^N \mathbf{K}(\mathbf{x}_i, \hat{\mu}_b)$, respectively. The expression in Eq. (23) can now be implemented with no knowledge of the mapping function Φ. The only requirement is a good choice for the kernel function $\mathbf{k}$.

15.3 Linear-matched Filter and Kernel-matched Filter

15.3.1 Linear-matched Filter

In this section, we introduce the concept of linear spectral matched filter. The constrained least squares approach is used to derive the linear-matched filter. Let the input spectral signal $\mathbf{x}$ be $\mathbf{x} = [x(1), x(2), \ldots, x(J)]^T$ consisting of J spectral bands. We can model each spectral observation as a linear combination of the target spectral signature and noise

$$\mathbf{x} = a\mathbf{s} + \mathbf{n}, \tag{24}$$

where a is an attenuation constant (target abundance measure); when $a = 0$ no target is present and when $a > 0$ target is present. Vector $\mathbf{s} = [s(1), s(2), \ldots, s(J)]^T$ contains the spectral signature of the target and vector $\mathbf{n}$ contains the added background clutter noise.

We can design a linear-matched filter $\mathbf{w} = [w(1), w(2), \ldots, w(J)]^T$ such that the desired target signal $\mathbf{s}$ is passed through while the average filter output energy is minimized. Let us define $\mathbf{X}$ to be a $J \times N$ matrix of the N mean-removed reference pixels (centered) obtained from the input image. Let each centered observation spectral pixel to be represented as a column in the sample matrix $\mathbf{X}$

$$\mathbf{X} = [\mathbf{x}_1 \ \mathbf{x}_2 \ \ldots \ \mathbf{x}_N]. \tag{25}$$

The output of the filter for the input $\mathbf{x}_i$ is given by

$$y_i = \mathbf{w}^T\mathbf{x}_i = \mathbf{x}_i^T\mathbf{w} \tag{26}$$

The average output power of the filter for the reference data $\mathbf{X}$ is given by

$$\frac{1}{N}\sum_{i=1}^{N} y_i^2 = \mathbf{w}^T\left(\frac{1}{N}\sum_{i=1}^{N} \mathbf{x}_i\mathbf{x}_i^T\right)\mathbf{w} = \mathbf{w}^T\mathbf{C}\mathbf{w} \tag{27}$$

where $\mathbf{C}$ is the covariance matrix of the reference data. This constrained filter design is equivalent to a constrained least squares minimization problem, as was shown in,[28-30] which is given by

$$\min_{\mathbf{w}} \{\mathbf{w}^T\mathbf{C}\mathbf{w}\} \text{subject to } \mathbf{s}^T\mathbf{w} = 1. \tag{28}$$

The solution to this quadratic minimization problem was shown in[31] and was called Constrained Energy Minimization (CEM) filter given by

$$\mathbf{w} = \frac{\mathbf{C}^{-1}\mathbf{s}}{\mathbf{s}^T\mathbf{C}^{-1}\mathbf{s}}. \tag{29}$$

The covariance matrix $\mathbf{C}$ is usually estimated from the input image. Now the estimated covariance matrix $\hat{\mathbf{C}}$ for the mean-removed reference data is given by

$$\hat{\mathbf{C}} = \mathbf{X}\mathbf{X}^T. \tag{30}$$

The output of the linear-matched filter for a test input $\mathbf{r}$, given the estimated covariance matrix is given by

$$y(\mathbf{r}) = \mathbf{w}^T\mathbf{r} = \frac{\mathbf{s}^T\hat{\mathbf{C}}^{-1}\mathbf{r}}{\mathbf{s}^T\hat{\mathbf{C}}^{-1}\mathbf{s}}. \tag{31}$$

In[15,32] it was shown that using the generalized likelihood ratio test (GLRT), the same expression for the linear-matched filter in Eq. (31) can be obtained

$$\hat{a} = \frac{\mathbf{s}^T\mathbf{C}^{-1}\mathbf{r}}{\mathbf{s}^T\mathbf{C}^{-1}\mathbf{s}}. \tag{32}$$

where $\hat{a}$ represents the maximum likelihood estimate of the abundance measure a and when the estimated covariance matrix $\mathbf{C} = \hat{\mathbf{C}}$ is used in Eq. (32) this filter is referred to as the adaptive matched filter.[15] The CFAR behavior of this filter is given by

$$\alpha = \frac{|\mathbf{s}^T\hat{\mathbf{C}}^{-1}\mathbf{r}|^2}{\mathbf{s}^T\hat{\mathbf{C}}^{-1}\mathbf{s}}. \tag{33}$$

which is proportional to the estimated squared magnitude of the output matched filter referred in[16] as the signal-to-noise (SNR).

15.3.2 Kernel-matched Filter

Consider the linear model of the input data in a kernel feature space which is equivalent to a nonlinear model in the input space

$$\Phi(\mathbf{x}) = a_\Phi\Phi(\mathbf{s}) + \mathbf{n}_\Phi, \tag{34}$$

where Φ is the nonlinear mapping that maps the input data into a kernel feature space, a_Φ is an attenuation constant (abundance measure), the high-dimensional vector $\Phi(\mathbf{s})$ contains the spectral signature of the target in the feature space, and vector $\mathbf{n}_\Phi$ contains the added noise in the feature space.

Using the constrained least squares approach that was explained in the previous section, it can easily be shown that the equivalent matched filter $\mathbf{w}_\Phi$ in the feature space is given by

$$\mathbf{w}_\Phi = \frac{\hat{\mathbf{C}}_\Phi^{-1}\Phi(\mathbf{s})}{\Phi(\mathbf{s})^T\hat{\mathbf{C}}_\Phi^{-1}\Phi(\mathbf{s})}, \tag{35}$$

where $\hat{\mathbf{C}}_\Phi$ is the estimated covariance of pixels in the feature space. The estimated covariance is given by

$$\hat{\mathbf{C}}_\Phi = \mathbf{X}_\Phi\mathbf{X}_\Phi^T \tag{36}$$

assuming the sample mean has already been removed from each sample (centered), where $\mathbf{X}_\Phi = [\Phi(\mathbf{x}_1)\Phi(\mathbf{x}_2)\dots\Phi(\mathbf{x}_N)]$ is a matrix whose columns are the mapped input reference data in the feature space. The matched filter in the feature space in Eq. (35) is equivalent to a nonlinear matched filter in the input space and its output for the input $\Phi(\mathbf{r})$ is given by

$$y = \mathbf{w}_\phi^T \Phi(\mathbf{r}) = \frac{\Phi(\mathbf{s})^T \hat{\mathbf{C}}_\phi^{-1} \Phi(\mathbf{r})}{\Phi(\mathbf{s})^T \hat{\mathbf{C}}_\phi^{-1} \Phi(\mathbf{s})}. \tag{37}$$

The corresponding CFAR-matched filter in the feature space is given by

$$y = \mathbf{w}_\phi^T \Phi(\mathbf{r}) = \frac{|\Phi(\mathbf{s})^T \hat{\mathbf{C}}_\phi^{-1} \Phi(\mathbf{r})|^2}{\Phi(\mathbf{s})^T \hat{\mathbf{C}}_\phi^{-1} \Phi(\mathbf{s})}. \tag{38}$$

Due to the high dimensionality of the feature space the expressions in Eqs. (37) and (38) are not tractable. Therefore, we cannot directly implement them in the feature space. We need to convert these expressions in terms of the dot products of the input vectors in the feature space and then use the kernel trick to convert the dot products in the feature space in terms of the kernel function. We refer to this process as kernelizing the matched filter expression, and the resulting nonlinear matched filter is called the kernel-matched filter.

It was shown in subsection 15.2.2 that the inverse of the estimated background covariance matrix can be written as

$$\hat{\mathbf{C}}_\phi^{-1} = \mathbf{X}_\phi \mathcal{B} \Lambda^{-1} \mathcal{B}^T \mathbf{X}_\phi^T \tag{39}$$

Inserting Eq. (39) into Eq. (37) it can be rewritten as

$$y_\phi = \frac{\Phi(\mathbf{s})^T \mathbf{X}_\phi \mathcal{B} \Lambda^{-1} \mathcal{B}^T \mathbf{X}_\phi^T \Phi(\mathbf{r})}{\Phi(\mathbf{s})^T \mathbf{X}_\phi \mathcal{B} \Lambda^{-1} \mathcal{B}^T \mathbf{X}_\phi^T \Phi(\mathbf{s})}. \tag{40}$$

The dot product term $\Phi(\mathbf{s})^T \mathbf{X}_\phi$ in the feature space can be represented in terms of the kernel function

$$\Phi(\mathbf{s})^T \mathbf{X}_\phi = (\mathbf{k}(\mathbf{x}_1, \mathbf{s}), \mathbf{k}(\mathbf{x}_2, \mathbf{s}), \ldots, \mathbf{k}(\mathbf{x}_N, \mathbf{s}))$$
$$= \mathbf{K}(\mathbf{X}, \mathbf{s})^T = \mathbf{K}_\mathbf{s}^T. \tag{41}$$

Similarly,

$$\Phi(\mathbf{r})^T \mathbf{X}_\phi = (\mathbf{k}(\mathbf{x}_1, \mathbf{r}), \mathbf{k}(\mathbf{x}_2, \mathbf{r}), \ldots, \mathbf{k}(\mathbf{x}_N, \mathbf{r}))$$
$$= \mathbf{K}(\mathbf{X}, \mathbf{r})^T = \mathbf{K}_\mathbf{r}^T. \tag{42}$$

Also using the properties of the Kernel PCA as shown in Appendix I, we have the relationship

$$\mathbf{K}^{-2} = \mathcal{B} \Lambda^{-1} \mathcal{B}^T. \tag{43}$$

We denote $\mathbf{K} = \mathbf{K}(\mathbf{X}, \mathbf{X}) = (\mathbf{K})_{ij}$ an $N \times N$ Gram kernel matrix whose entries are the dot products $< \Phi(\mathbf{x}_i), \Phi(\mathbf{x}_j) >$. Substituting Eqs. (41), (42), and (43) into Eq. (40), the kernelized version of the matched filter is given by

$$y(\mathbf{K_r}) = \frac{\mathbf{K(X, s)}^T \mathbf{K}^{-2} \mathbf{K(X, r)}}{\mathbf{K(X, s)}^T \mathbf{K}^{-2} \mathbf{K(X, s)}} = \frac{\mathbf{K_s}^T \mathbf{K}^{-2} \mathbf{K_r}}{\mathbf{K_s}^T \mathbf{K}^{-2} \mathbf{K_s}}. \tag{44}$$

In the derivation of the kernel-matched filter, we assumed that the data has been already centered in the feature space by removing the sample mean. However, the sample mean cannot be directly removed in the feature space due to the high dimensionality of $\mathcal{F}$. That is, the expression for the kernel-matched filter needs to be derived in terms of the original uncentered input data. Therefore, the kernel matrix $\hat{\mathbf{K}}$ needs to be properly centered.[19] The effect of centering on the kernel-matched filter can be seen by replacing the uncentered $\mathbf{X}_\Phi$ with the centered $\mathbf{X}_\Phi - \mu_\Phi$ (where μ_Φ is the mean of the reference input data) in the estimation of the covariance matrix expression in Eqs. (36) as well as 41 and 42 for the empirical kernel mapping of the target and input data, respectively. The resulting centered $\hat{\mathbf{K}}$ is shown in[19] to be given by

$$\hat{\mathbf{K}} = (\mathbf{K} - \mathbf{1}_N \mathbf{K} - \mathbf{K} \mathbf{1}_N + \mathbf{1}_N \mathbf{K} \mathbf{1}_N), \tag{45}$$

where $(\mathbf{1}_N)_{ij} = 1/N$. The properly centered kernel-matched filter output is now given by

$$y(\mathbf{K_r}) = \frac{\hat{\mathbf{K}}_\mathbf{s}^T \hat{\mathbf{K}}^{-2} \hat{\mathbf{K}}_\mathbf{r}}{\hat{\mathbf{K}}_\mathbf{s}^T \hat{\mathbf{K}}^{-2} \hat{\mathbf{K}}_\mathbf{s}}, \tag{46}$$

where $\hat{\mathbf{K}}_\mathbf{s}^T = \mathbf{K}_\mathbf{s}^T - \sum_{i=1}^N \mathbf{K}(\mathbf{x}_i, \mathbf{s})$ and $\hat{\mathbf{K}}_\mathbf{r}^T = \mathbf{K}_\mathbf{s}^T - \sum_{i=1}^N \mathbf{K}(\mathbf{x}_i, \mathbf{r})$ which is obtained by replacing $\mathbf{X}_\Phi$ with $\mathbf{X}_\Phi - \mu_\Phi$ in 41, and 42, respectively.

15.4 Linear Subspace-matched Filter and Kernel Subspace-matched Filter

15.4.1 Linear Subspace-matched Filter

In order to jointly address the two main issues of subpixel target detection – the spectral variability and spectral mixing – especially, in airborne hyperspectral sensor applications, a linear subspace spectral mixing model has been used in which both target and background spectra are assumed to lie in two respective subspaces. The target pixel vectors are expressed as a linear combination of the target spectral signature and background spectral signature, which are represented by subspace target spectra and subspace background spectra, respectively. The detection is based on a hypothesis test in which two competing hypotheses ($\mathbf{H_0}$ and $\mathbf{H_1}$) are tested for an input spectrum to decide which hypothesis is best related to the input spectrum. The hyperspectral target detection problem in a J-dimensional input space is expressed as

$$\mathbf{H}_0\colon \mathbf{y} = \mathbf{B}\zeta + \mathbf{n}, \qquad\qquad\qquad \text{Target absent}$$

$$\mathbf{H}_1\colon \mathbf{y} = \mathbf{T}\theta + \mathbf{B}\zeta + \mathbf{n} = [\mathbf{TB}]\begin{bmatrix}\theta\\\zeta\end{bmatrix} + \mathbf{n}, \quad \text{Target present} \qquad (47)$$

where $\mathbf{T}$ and $\mathbf{B}$ represent orthogonal matrices whose J-dimensional column vectors span the target and background subspaces, respectively; θ and ζ are unknown vectors whose entries are coefficients that account for the abundances of the corresponding column vectors of $\mathbf{T}$ and $\mathbf{B}$, respectively; $\mathbf{n}$ represents Gaussian random noise ($\mathbf{n} \in \mathcal{R}^J$) distributed as $\mathcal{N}(0, \sigma^2\mathbf{I})$; and $[\mathbf{T}\ \mathbf{B}]$ is a concatenated matrix of $\mathbf{T}$ and $\mathbf{B}$. The numbers of the column vectors of $\mathbf{T}$ and $\mathbf{B}$, N_t and N_b, respectively, are usually smaller than $J(N_t, N_b < J)$. In this detection scenario, if $\mathbf{y}$ is from a target region it is represented by the hypothesis $\mathbf{H}_1$ and is labeled as target. If no target spectra are involved in the input spectrum $\mathbf{y}$, it is classified as background.

The matrices $\mathbf{T}$ and $\mathbf{B}$ spanning the target and background subspaces $<\mathbf{T}>$ and $<\mathbf{B}>$, respectively, are formed by first calculating the covariance matrices of target and background sample spectra and then taking only the significant eigenvectors of the covariance matrices in order to form the column vectors of the corresponding matrices. The numbers of the eigenvectors to be included in $\mathbf{T}$ and $\mathbf{B}$, N_t and N_b, respectively, are set such that they represent most of the energy of the original target and background subspaces.

15.4.2 GLRT for Target Detection

Given the linear subspace detection model and the two hypotheses about how the input vector is generated, as shown in Eq. (47), the likelihood ratio test (LRT) is used to predict whether the input vector $\mathbf{y}$ includes the target and is defined by

$$l(\mathbf{y}) = \frac{p_1(\mathbf{y}|\mathbf{H}_1)}{p_0(\mathbf{y}|\mathbf{H}_0)} \underset{H_0}{\overset{H_1}{\gtrless}} \eta, \qquad (48)$$

where $p_0(\mathbf{y}|\mathbf{H}_0)$ and $p_1(\mathbf{y}|\mathbf{H}_1)$ represent the class conditional probability densities of $\mathbf{y}$ given the hypotheses $\mathbf{H}_0$ and $\mathbf{H}_1$, respectively, and η is a threshold of the test. $p_0(\mathbf{y}|\mathbf{H}_0)$ and $p_1(\mathbf{y}|\mathbf{H}_1)$ can be expressed as Gaussian probability densities $\mathcal{N}(\mathbf{B}\zeta, \sigma^2\mathbf{I})$ and $\mathcal{N}(\mathbf{T}\theta + \mathbf{B}\zeta, \sigma^2\mathbf{I})$, respectively, since $\mathbf{n}_i$ is assumed to be Gaussian random noise. The LRT is derived from the Neyman–Pearson criterion where the probability detection P_D is maximized while the probability of false alarm P_F is kept a constant.[33] $l(\mathbf{y})$ is compared to η to make a final decision about which hypothesis best relates to $\mathbf{y}$.

The LRT includes the unknown parameters ζ and θ that need to be estimated using the maximum likelihood principle. The generalized likelihood ratio test is directly obtained from $l(\mathbf{y})$ by replacing the unknown parameters with their maximum likelihood estimates (MLEs) $\hat{\zeta}$ and $\hat{\theta}$.[6]

$$\hat{l}(\mathbf{y}) = \frac{\max\limits_{\zeta} p_1(\mathbf{y}|\mathbf{H}_1)}{\max\limits_{(\zeta,\,\theta)} p_0(\mathbf{y}|\mathbf{H}_0)}$$

$$= \left(\frac{\hat{\sigma}_1^2}{\hat{\sigma}_0^2}\right)^{-P/2} \exp\{-\frac{1}{2\hat{\sigma}_1^2}\|\hat{\mathbf{n}}_1\|^2 + \frac{1}{2\hat{\sigma}_0^2}\|\hat{\mathbf{n}}_0\|^2\} \underset{H_1}{\overset{H_0}{\gtrless}} \eta. \tag{49}$$

$\hat{\mathbf{n}}_i$, the MLEs of $\mathbf{n}_i$, for both the hypotheses $\mathbf{H}_0$ and $\mathbf{H}_1$ can be estimated using the least squares approximation where projection of the input vector onto a subspace is performed to provide the least squares solution to the linear subspace model Eq. (47) (Note that least squares of the additive noise model is equivalent to maximum likelihood estimation of the conditional likelihood ratio such as $l(\mathbf{y})$[34]:

$$\begin{aligned}
\hat{\mathbf{n}}_0 &= \mathbf{y} - \mathbf{B}\hat{\zeta}_0 = (\mathbf{I} - \mathbf{P}_{\mathbf{B}})\mathbf{y}, \\
\hat{\mathbf{n}}_1 &= \mathbf{y} - \mathbf{T}\hat{\theta}_1 - \mathbf{B}\zeta_1 = (\mathbf{I} - \mathbf{P}_{\mathbf{TB}})\mathbf{y},
\end{aligned} \tag{50}$$

where $\mathbf{P}_{\mathbf{B}} = \mathbf{B}(\mathbf{B}^T\mathbf{B})^{-1}\mathbf{B}^T = \mathbf{B}\mathbf{B}^T$ is a projection matrix associated with the N_b-dimensional background subspace $< \mathbf{B} >$;$\mathbf{P}_{\mathbf{TB}}$ is a projection matrix associated with the $(N_b + N_t)$-dimensional target and background subspace $< \mathbf{TB} >$

$$\mathbf{P}_{\mathbf{TB}} = [\mathbf{T}\ \mathbf{B}][[\mathbf{T}\ \mathbf{B}]^{\mathbf{T}}[\mathbf{T}\ \mathbf{B}]]^{-1}[\mathbf{TB}]^{\mathbf{T}}. \tag{51}$$

The columns of the matrix $[\mathbf{T}\ \mathbf{B}]$ are not orthogonal, but they are linearly independent meaning that $[\mathbf{T}\ \mathbf{B}]$ is a full-rank matrix. It can be easily found that the projection matrices $\mathbf{P}_{\mathbf{B}}$ and $\mathbf{P}_{\mathbf{TB}}$ hold the following properties[35]:

$$\mathbf{P}_{\mathbf{TB}}^T = \mathbf{P}_{\mathbf{TB}}, \mathbf{P}_{\mathbf{B}}^T = \mathbf{P}_{\mathbf{B}}, \tag{52}$$

$$\mathbf{P}_{\mathbf{TB}}^2 = \mathbf{P}_{\mathbf{TB}}, \mathbf{P}_{\mathbf{B}}^2 = \mathbf{P}_{\mathbf{B}}. \tag{53}$$

Therefore, using the properties of the projection matrices given in Eqs. (52) and (53), the energy of the noise signals can be written as

$$\begin{aligned}
\|\hat{\mathbf{n}}_1\|^2 &= (\mathbf{y} - \mathbf{P}_{\mathbf{TB}}\mathbf{y})^T(\mathbf{y} - \mathbf{P}_{\mathbf{TB}}\mathbf{y}) \\
&= \mathbf{y}^T\mathbf{y} - \mathbf{y}^T\mathbf{P}_{\mathbf{TB}}\mathbf{y} - \mathbf{y}^T\mathbf{P}_{\mathbf{TB}}{}^T\mathbf{y} + \mathbf{y}^T\mathbf{P}_{\mathbf{TB}}{}^T\mathbf{P}_{\mathbf{TB}}\mathbf{y} \\
&= \mathbf{y}^T\mathbf{y} - 2\mathbf{y}^T\mathbf{P}_{\mathbf{TB}}\mathbf{y} + \mathbf{y}^T\mathbf{P}_{\mathbf{TB}}\mathbf{y} = \mathbf{y}^T(\mathbf{I} - \mathbf{P}_{\mathbf{TB}})\mathbf{y}.
\end{aligned} \tag{54}$$

Similarly,

$$\|\hat{\mathbf{n}}_0\|^2 = (\mathbf{y} - \mathbf{P}_{\mathbf{B}}\mathbf{y})^T(\mathbf{y} - \mathbf{P}_{\mathbf{B}}\mathbf{y}) = \mathbf{y}^T(\mathbf{I} - \mathbf{P}_{\mathbf{B}})\mathbf{y}. \tag{55}$$

Since $\mathbf{n}_i$ is distributed as $\mathcal{N}(0, \sigma_i^2\mathbf{I})$, then $\hat{\sigma}_i^2$ is equal to $\frac{1}{J}\|\hat{\mathbf{n}}_i\|^2$[26]. Finally, the GLRT is obtained by taking $(J/2)$-root[6] after substituting Eqs. (54) and (55) and $\hat{\sigma}_i^2 = \frac{1}{J}\|\hat{\mathbf{n}}_i\|^2$ into Eq. (49):

$$\mathbf{L}_2(\mathbf{y}) = (\hat{l}(\mathbf{y}))^{2/J} = \frac{\|\hat{\mathbf{n}}_0\|^2}{\|\hat{\mathbf{n}}_1\|^2} = \frac{\mathbf{y}^T(\mathbf{I} - \mathbf{P}_{\mathbf{B}})\mathbf{y}}{\mathbf{y}^T(\mathbf{I} - \mathbf{P}_{\mathbf{TB}})\mathbf{y}} \underset{H_0}{\overset{H_1}{\gtrless}} \eta. \tag{56}$$

15.4.3 Nonlinear Models in Feature Space Based on Subspaces

In this subsection, the nonlinear hyperspectral detection problem based on the target and background subspaces is described in the feature space $\mathcal{F}$ as

$$\mathbf{H}_{0_\Phi} : \Phi(\mathbf{y}) = \mathbf{B}_\Phi \zeta_\Phi + \mathbf{n}_\Phi, \qquad\qquad\qquad \text{Target absent}$$

$$\mathbf{H}_{1_\Phi} : \Phi(\mathbf{y}) = \mathbf{T}_\Phi \theta_\Phi + \mathbf{B}_\Phi \zeta_\Phi + \mathbf{n}_\Phi = [\mathbf{T}_\Phi \mathbf{B}_\Phi]\begin{bmatrix} \theta_\Phi \\ \zeta_\Phi \end{bmatrix} + \mathbf{n}_\Phi, \quad \text{Target present}$$

$$(57)$$

where $\mathbf{T}_\Phi$ and $\mathbf{B}_\Phi$ represent full-rank matrices whose column vectors span target and background subspaces $< \mathbf{B}_\Phi >$ and $< \mathbf{T}_\Phi >$ in $\mathcal{F}$, respectively. In general, any sets of basis vectors that span the corresponding subspace can be used as the column vectors of $\mathbf{T}_\Phi$ and $\mathbf{B}_\Phi$. In the proposed method we use the significant eigenvectors of the target and background covariance matrices ($\mathbf{C}_{\mathbf{T}_\Phi}$ and $\mathbf{C}_{\mathbf{B}_\Phi}$) in $\mathcal{F}$ as the the column vectors of $\mathbf{T}_\Phi$ and $\mathbf{B}_\Phi$, respectively. $\mathbf{C}_{\mathbf{T}_\Phi}$ and $\mathbf{C}_{\mathbf{B}_\Phi}$ are based on the centered target and background sample sets ($\mathbf{Z}_\mathbf{T}$ and $\mathbf{Z}_\mathbf{B}$), respectively:

$$\mathbf{C}_{\mathbf{B}_\Phi} = \frac{1}{M} \sum_{i=1}^{M} \Phi(\mathbf{y}_i)\Phi(\mathbf{y}_i)^T, \text{ for } \mathbf{y}_i \in \mathbf{Z}_\mathbf{B},$$

$$\mathbf{C}_{\mathbf{T}_\Phi} = \frac{1}{N} \sum_{i=1}^{N} \Phi(\mathbf{y}_i)\Phi(\mathbf{y}_i)^T, \text{ for } \mathbf{y}_i \in \mathbf{Z}_\mathbf{T},$$

where N and M represent the number of training samples in $\mathbf{Z}_\mathbf{T}$ and $\mathbf{Z}_\mathbf{B}$, respectively. We now derive the GLRT of the nonlinear hyperspectral detection problem described by the model in (57):

$$\mathbf{L}_2(\Phi(\mathbf{y})) = \frac{\|\hat{\mathbf{n}}_{0_\Phi}\|^2}{\|\hat{\mathbf{n}}_{1_\Phi}\|^2} = \frac{\Phi(\mathbf{y})^T(\mathbf{P}_{\mathbf{I}_\Phi} - \mathbf{P}_{\mathbf{B}_\Phi})\Phi(\mathbf{y})}{\Phi(\mathbf{y})^T(\mathbf{P}_{\mathbf{I}_\Phi} - \mathbf{P}_{\mathbf{T}_\Phi \mathbf{B}_\Phi})\Phi(\mathbf{y})}, \qquad (58)$$

where $\mathbf{P}_{\mathbf{I}_\Phi}$ represents an identity projection operator in $\mathcal{F}$; $\mathbf{P}_{\mathbf{B}_\Phi} = \mathbf{B}_\Phi(\mathbf{B}_\Phi^T \mathbf{B}_\Phi)^{-1}\mathbf{B}_\Phi^T = \mathbf{B}_\Phi \mathbf{B}_\Phi^T$ is a background projection matrix; and $\mathbf{P}_{\mathbf{T}_\Phi \mathbf{B}_\Phi}$ is a joint target-and-background projection matrix in $\mathcal{F}$

$$\mathbf{P}_{\mathbf{T}_\Phi \mathbf{B}_\Phi} = [\mathbf{T}_\Phi \mathbf{B}_\Phi][[\mathbf{T}_\Phi \mathbf{B}_\Phi]^T[\mathbf{T}_\Phi \mathbf{B}_\Phi]]^{-1}[\mathbf{T}_\Phi \mathbf{B}_\Phi]^T$$

$$= [\mathbf{T}_\Phi \mathbf{B}_\Phi]\begin{bmatrix} \mathbf{T}_\Phi^T \mathbf{T}_\Phi & \mathbf{T}_\Phi^T \mathbf{B}_\Phi \\ \mathbf{B}_\Phi^T \mathbf{T}_\Phi & \mathbf{B}_\Phi^T \mathbf{B}_\Phi \end{bmatrix}^{-1} \begin{bmatrix} \mathbf{T}_\Phi^T \\ \mathbf{B}_\Phi^T \end{bmatrix} . \qquad (59)$$

Note 58 can not be implemented explicitly due to potential infinite dimensionality of Φ, therefore, 58 has to be kernelized in order to obtain an expression in terms of the kernel function $\mathbf{k}$.

15.4.4 Kernelizing MSD in Feature Space

To kernelize Eq. 58, first consider its numerator,

$$\|\hat{\mathbf{n}}_{0_\phi}\| = \Phi(\mathbf{y})^T(\mathbf{P}_{\mathbf{I}_\phi} - \mathbf{P}_{\mathbf{B}\phi})\Phi(\mathbf{y})$$
$$= \Phi(\mathbf{y})^T\mathbf{P}_{\mathbf{I}_\phi}\Phi(\mathbf{y}) - \Phi(\mathbf{y})^T\mathbf{B}_\phi\mathbf{B}_\phi^T\Phi(\mathbf{y}). \tag{60}$$

Using Eq. 74, $\mathbf{B}_\phi$ and $\mathbf{T}_\phi$ can be written as

$$\mathbf{B}_\phi = [\mathbf{e_b}^1 \mathbf{e_b}^2 \dots \mathbf{e_b}^{N_b}] = \Phi_{\mathbf{Z_B}}B, \tag{61}$$

$$\mathbf{T}_\phi = [\mathbf{e_t}^1 \mathbf{e_t}^2 \dots \mathbf{e_t}^{N_t}] = \Phi_{\mathbf{Z_T}}T, \tag{62}$$

where $\mathbf{e_b}^i$ and $\mathbf{e_t}^j$ are the significant eigenvectors of $\mathbf{C_{B_\phi}}$ and $\mathbf{C_{T_\phi}}$, respectively; $\Phi_{\mathbf{Z_B}} = [\Phi(\mathbf{y}_1)\Phi(\mathbf{y}_2)\dots\Phi(\mathbf{y}_M)], \mathbf{y}_i \in \mathbf{Z_B}$ and $\Phi_{\mathbf{Z_T}} = [\Phi(\mathbf{y}_1)\Phi(\mathbf{y}_2)\dots \Phi(\mathbf{y}_N)]$, $\mathbf{y}_i \in \mathbf{Z_T}$; the column vectors of B and T represent only the significant eigenvectors $(\beta^1, \beta^2, \dots, \beta^{N_b})$ and $(\alpha^1, \alpha^2, \dots, \alpha^{N_t})$ of the background kernel matrix $\mathbf{K}(\mathbf{Z_B}, \mathbf{Z_B}) = (\mathbf{K})_{ij} = \mathbf{k}(\mathbf{y}_i, \mathbf{y}_j), \mathbf{y}_i, \mathbf{y}_j \in \mathbf{Z_B}$ and the target kernel matrix $\mathbf{K}(\mathbf{Z_T}, \mathbf{Z_T}) = (\mathbf{K})_{ij} = \mathbf{k}(\mathbf{y}_i, \mathbf{y}_j), \mathbf{y}_i, \mathbf{y}_j \in \mathbf{Z_T}$, respectively.

Using Eq. (61), the projection of $\Phi(\mathbf{y})$ onto $\mathbf{B}_\phi$ becomes

$$\mathbf{B}_\phi^T\Phi(\mathbf{y}) = [\mathbf{e_b^1}\ \mathbf{e_b^2}\dots\mathbf{e_b}^{N_b}]^T\Phi(\mathbf{y}) = \begin{bmatrix} \beta^{1^T}\Phi_{\mathbf{Z_B}}^T\Phi(\mathbf{y}) \\ \beta^{2^T}\Phi_{\mathbf{Z_B}}^T\Phi(\mathbf{y}) \\ . \\ \beta^{b^T}\Phi_{\mathbf{Z_B}}^T\Phi(\mathbf{y}) \end{bmatrix} \tag{63}$$
$$= B^T\mathbf{K}(\mathbf{Z_B}, \mathbf{y}),$$

and, similarly, using Eq. (62), the projection onto $\mathbf{T}_\phi$ is

$$\mathbf{T}_\phi^T\Phi(\mathbf{y}) = [\mathbf{e_t}^1\mathbf{e_t}^2\dots\mathbf{e_t}^{N_t}]^T\Phi(\mathbf{y}) = \begin{bmatrix} \alpha^{1^T}\Phi_{\mathbf{Z_T}}^T\Phi(\mathbf{y}) \\ \alpha^{2^T}\Phi_{\mathbf{Z_T}}^T\Phi(\mathbf{y}) \\ . \\ \alpha^{N_t^T}\Phi_{\mathbf{Z_T}}^T\Phi(\mathbf{y}) \end{bmatrix} \tag{64}$$
$$= T^T\mathbf{K}(\mathbf{Z_T}, \mathbf{y}),$$

where $\mathbf{K}(\mathbf{Z_B}, \mathbf{y})$ and $\mathbf{K}(\mathbf{Z_T}, \mathbf{y})$ are column vectors whose entries are $\mathbf{k}(\mathbf{x}_i, \mathbf{y})$ for $\mathbf{x}_i \in \mathbf{Z_B}$ and $\mathbf{x}_i \in \mathbf{Z_T}$, respectively. Now using Eq. (64), $\Phi(\mathbf{y})^T\mathbf{B}_\phi\mathbf{B}_\phi^T\Phi(\mathbf{y})$ can be written as

$$\Phi(\mathbf{y})^T\mathbf{B}_\phi\mathbf{B}_\phi^T\Phi(\mathbf{y}) = \mathbf{K}(\mathbf{Z_B}, \mathbf{y})^T BB^T\mathbf{K}(\mathbf{Z_B}, \mathbf{y}). \tag{65}$$

The projection onto the identity operator $\Phi(\mathbf{y})^T\mathbf{P}_{\mathbf{I}_\phi}\Phi(\mathbf{y})$ also needs to be kernelized. $\mathbf{P}_{\mathbf{I}_\phi}$ is defined as $\mathbf{P}_{\mathbf{I}_\phi} := \mathbf{\Omega}_\Phi\mathbf{\Omega}_\phi^T$, where $\mathbf{\Omega}_\Phi = [\mathbf{e_q}^1\ \mathbf{e_q}^2\dots]$ is a matrix whose columns are all the eigenvectors with $\lambda \neq 0$ that are in the span of $\Phi(\mathbf{y}_i)$, $\mathbf{y}_i \in \mathbf{Z_T} \cup \mathbf{Z_B}$. From (Eq. 74), $\mathbf{\Omega}_\Phi$ can be expressed as

$$\mathbf{\Omega_\Phi} = [\mathbf{e_q}^1 \ \mathbf{e_q}^2 \ldots \ \mathbf{e_q}^{N_{bt}}] = \mathbf{\Phi_{Z_{TB}}}\mathit{\Delta}, \tag{66}$$

where $\mathbf{\Phi_{Z_{TB}}} = \mathbf{\Phi_{Z_T}} \cup \mathbf{\Phi_{Z_B}}$ and $\mathit{\Delta}$ is a matrix whose columns are the eigenvectors $(\mathbf{k^1}, \mathbf{k^2}, \ldots, \mathbf{k}^{N_{bt}})$ of the kernel matrix $\mathbf{K}(\mathbf{Z_{TB}}, \mathbf{Z_{TB}}) = (\mathbf{K})_{ij} = \mathbf{k}(\mathbf{y}_i, \mathbf{y}_j), \mathbf{y}_i, \mathbf{y}_j \in \mathbf{Z_T} \cup \mathbf{Z_B}$ with nonzero eigenvalues, normalized by the square root of their associated eigenvalues. Using $\mathbf{P_{I_\Phi}} = \mathbf{\Omega_\Phi}\mathbf{\Omega}_\Phi^T$ and (Eq. 66)

$$
\begin{aligned}
\Phi(\mathbf{y})^T \mathbf{P_{I_\Phi}} \Phi(\mathbf{y}) &= \Phi(\mathbf{y})^T \mathbf{\Phi_{Z_{TB}}} \mathit{\Delta}\mathit{\Delta}^T \mathbf{\Phi}_{Z_{TB}}^T \Phi(\mathbf{y}) \\
&= \mathbf{K}(\mathbf{Z_{TB}}, \mathbf{y})^T \mathit{\Delta}\mathit{\Delta}^T \mathbf{K}(\mathbf{Z_{TB}}, \mathbf{y}),
\end{aligned}
\tag{67}
$$

where $\mathbf{K}(\mathbf{Z_{TB}}, \mathbf{y})$ is a concatenated vector $[\mathbf{K}(\mathbf{Z_T}, \mathbf{y})^T \mathbf{K}(\mathbf{Z_B}, \mathbf{y})^T]^T$. The kernelized numerator of Eq. (58) is now given by

$$\|\hat{\mathbf{n}}_{0_K}\| = \mathbf{K}(\mathbf{Z_{TB}}, \mathbf{y})^T \mathit{\Delta}\mathit{\Delta}^T \mathbf{K}(\mathbf{Z_{TB}}, \mathbf{y}) - \mathbf{K}(\mathbf{Z_B}, \mathbf{y})^T \mathcal{B}\mathcal{B}^T \mathbf{K}(\mathbf{Z_B}, \mathbf{y}), \tag{68}$$

We now kernalize $\Phi(\mathbf{y})^T \mathbf{P_{T_\Phi B_\Phi}} \Phi(\mathbf{y})$ in the denominator of Eq. (58) to complete the kernelization process. Using Eqs. (59), (61), and (62)

$$
\begin{aligned}
\Phi(\mathbf{y})^T \mathbf{P_{T_\Phi B_\Phi}} \Phi(\mathbf{y}) &= \Phi(\mathbf{y})^T [\mathbf{T_\Phi} \ \mathbf{B_\Phi}] \begin{bmatrix} \mathbf{T}_\Phi^T \mathbf{T_\Phi} & \mathbf{T}_\Phi^T \mathbf{B_\Phi} \\ \mathbf{B}_\Phi^T \mathbf{T_\Phi} & \mathbf{B}_\Phi^T \mathbf{B_\Phi} \end{bmatrix}^{-1} \begin{bmatrix} \mathbf{T}_\Phi^T \\ \mathbf{B}_\Phi^T \end{bmatrix} \Phi(\mathbf{y}) \\[2mm]
&= [\mathbf{K}(\mathbf{Z}_T, \mathbf{y})^T \mathcal{T} \ \mathbf{K}(\mathbf{Z_B}, \mathbf{y})^T \mathcal{B}] \begin{bmatrix} \mathcal{T}^T \mathbf{K}(\mathbf{Z_T}, \mathbf{Z_T})\mathcal{T} & \mathcal{T}^T \mathbf{K}(\mathbf{Z_T}, \mathbf{Z_B})\mathcal{B} \\ \mathcal{B}^T \mathbf{K}(\mathbf{Z_B}, \mathbf{Z_T})\mathcal{T} & \mathcal{B}^T \mathbf{K}(\mathbf{Z_B}, \mathbf{Z_B})\mathcal{B} \end{bmatrix}^{-1} \\[2mm]
&\times \begin{bmatrix} \mathcal{T}^T \mathbf{K}(\mathbf{Z_T}, \mathbf{y}) \\ \mathcal{B}^T \mathbf{K}(\mathbf{Z_B}, \mathbf{y}) \end{bmatrix}.
\end{aligned}
\tag{69}
$$

Finally, substituting Eqs. (65), (68), and (69) into (58), the kernelized GLRT is given by

$$\mathbf{L_{2K}} = \frac{\mathbf{K}(\mathbf{Z_{TB}}, \mathbf{y})^T \mathit{\Delta}\mathit{\Delta}^T \mathbf{K}(\mathbf{Z_{TB}}, \mathbf{y}) - \mathbf{K}(\mathbf{Z_B}, \mathbf{y})^T \mathcal{B}\mathcal{B}^T \mathbf{K}(\mathbf{Z_B}, \mathbf{y})}{\mathbf{K}(\mathbf{Z_{TB}}, \mathbf{y})^T \mathit{\Delta}\mathit{\Delta}^T \mathbf{K}(\mathbf{Z_{TB}}, \mathbf{y}) - [\mathbf{K}(\mathbf{Z_T}, \mathbf{y})^T \mathcal{T} \mathbf{K}(\mathbf{Z_B}, \mathbf{y})^T \mathcal{B}] \mathit{\Lambda}_1^{-1} \begin{bmatrix} \mathcal{T}^T \mathbf{K}(\mathbf{Z_T}, \mathbf{y}) \\ \mathcal{B}^T \mathbf{K}(\mathbf{Z_B}, \mathbf{y}) \end{bmatrix}}, \tag{70}$$

where $\mathit{\Lambda}_1 = \begin{bmatrix} \mathcal{T}^T \mathbf{K}(\mathbf{Z_T}, \mathbf{Z_T})\mathcal{T} & \mathcal{T}^T \mathbf{K}(\mathbf{Z_T}, \mathbf{Z_B})\mathcal{B} \\ \mathcal{B}^T \mathbf{K}(\mathbf{Z_B}, \mathbf{Z_T})\mathcal{T} & \mathcal{B}^T \mathbf{K}(\mathbf{Z_B}, \mathbf{Z_B})\mathcal{B} \end{bmatrix}$.

15.5 Experimental Results

15.5.1 Simulation Results for Kernel RX-algorithm

The centered kernel matrix $\hat{\mathbf{K}}_b$ can be estimated either globally or locally. The global estimation must be performed prior to detection and normally needs a large amount of data samples to successfully represent all the background types present in a given data set. In this chapter, to globally estimate $\hat{\mathbf{K}}_b$, we need to

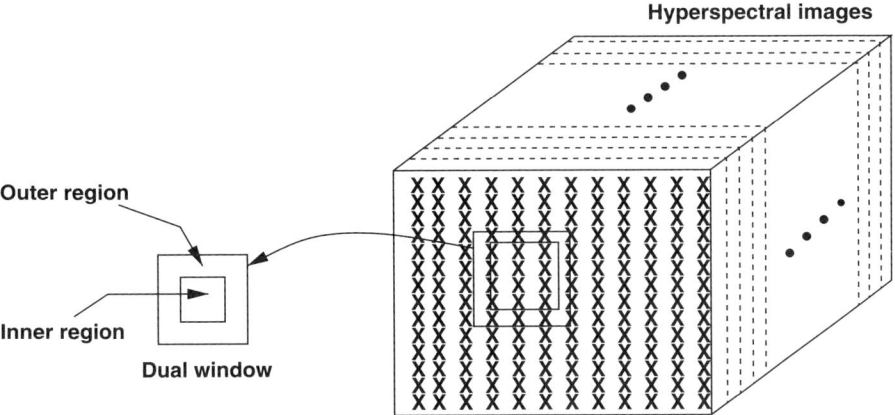

Figure 15.1. Example of the dual concentric windows in the hyperspectral images.

use all the spectral vectors in a given test image. A well-known data clustering algorithm, k-means[36], is used on all the spectral vectors in order to generate a significantly less number of spectral vectors (centroids) from which $\hat{\mathbf{K}}_b$ is estimated. By using a small number of distinct background spectral vectors a manageable kernel matrix is generated where a more efficient kernel RX-algorithm is now implemented.

For local estimation of $\hat{\mathbf{K}}_b$ we use local background samples, which are from the neighboring area of the pixel being tested. For each test pixel location, a dual concentric rectangular window is used to separate a local area into two regions – the inner-window region (IWR) and the outer-window region (OWR), as shown in Fig. 15.1; the local kernel matrix and the background covariance matrix are calculated from the pixel vectors in the OWR. The dual concentric windows naturally divide the local area into the potential target region – the IWR – and the background region – the OWR – whose local statistics in the original and nonlinear feature domain are compared using the conventional RX- and kernel RX-algorithms, respectively. The size of the IWR is set to enclose targets to be detected whose approximate size is based on prior knowledge of the range, field of view (FOV), and the dimension of the biggest target in the given data set. Similarly, the size of the OWR is set to include sufficient statistics from the neighboring background.

We apply both the kernel RX- and conventional RX-algorithms to two HYDICE images, the Forest Radiance I (FR-I) image and the Desert Radiance II (DR-II) image, as shown in Fig. 15.2. FR-I includes total 14 targets and DR-II contains six targets along the road; all the targets are military vehicles. A HYDICE imaging sensor generates 210 bands across the whole spectral range $(0.4$–$2.5\,\mu\text{m})$, but we only use 150 bands by discarding water absorption and low signal to noise ratio (SNR) bands; the bands used are

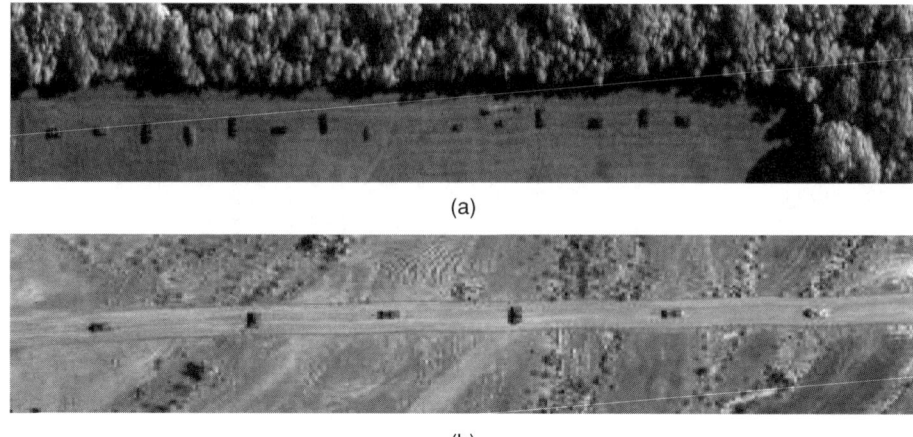

(a)

(b)

Figure 15.2. Sample band images (48th) from HYDICE images: (a) the Forest Radiance I image and (b) the Desert Radiance II image.

the 23rd–101st, 109th–136th, and 152nd–194th. Gaussian RBF kernel, $\mathbf{k}(\mathbf{x}, \mathbf{y}) = \exp\left(\frac{-\|\mathbf{x}-\mathbf{y}\|^2}{c}\right)$, was used to implement the kernel-RX algorithm; the value of c was set to 40. The sizes of the IWR and OWR used for the local kernel and covariance matrix estimations were 5×5 and 15×15 pixel areas, respectively. For the global kernel matrix estimation, the number of the representative spectral vectors obtained from the k-means procedure was set to 600.

We first used the local dual-window for the local covariance and kernel matrix estimations for the conventional RX- and kernel RX-algorithms, respectively, and performance between the two algorithms was compared. We also estimated the global kernel matrix and the performance between the kernel RX-algorithms implemented with the local and global kernel matrices was compared.

Figures 15.3 and 15.4 show the anomaly detection results of both the kernel RX and the conventional RX using the local dual window applied to the FR-I and DR-II images, respectively. The kernel RX detected most of the targets with a few false alarms while the conventional RX generated much more false alarms and missed some targets; especially, in the case of FR-I the conventional RX missed seven successive targets from the left.

The receiver operating characteristics (ROC) curves representing detection probability P_d versus false alarm rates N_f, were also generated to provide quantitative performance comparison for the FR-I and DR-II images between the two algorithms based on the local dual window, as shown in Figs. 15.5 and 15.6, respectively. For ROC curves generation, based on the ground truth information for the HYDICE images, we obtain the coordinates of all the rectangular target regions. Every target pixel inside the target regions is then considered as a target candidate to be detected. P_d and N_f are defined as

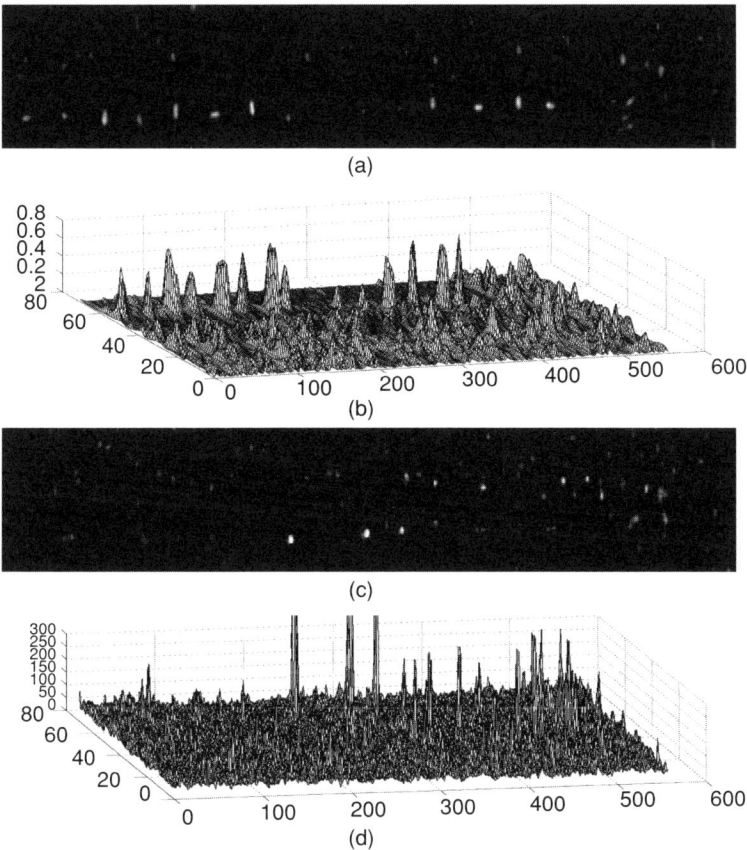

Figure 15.3. Detection results for the Forest Radiance I image using kernel RX and conventional RX; (a) Kernel RX, (b) 3D plot of (a), (c) RX, and (d) 3D plot of (c).

$$P_d := \frac{N_{hit}}{N_t} \text{ and } N_f := \frac{N_{miss}}{N_{tot}}, \tag{71}$$

where N_{hit} represents the number of target pixels detected given a certain threshold; N_t represents the total number of target pixels in the images; N_{miss} represents the number of background pixels detected; and N_{tot} represents the total number of pixels in the images. P_d becomes one only when all the individual target pixels within a target are detected; perfect detection is, therefore, difficult to achieve and the values of P_d for both the kernel RX and conventional RX are usually less than one. For both the FR-I and DR-II images the kernel RX showed significantly improved performance over the conventional RX.

We also generated the global kernel matrix and the performance of the kernel RX using the local and global kernel matrices, and compared in terms of ROC curves, as shown in Figs. 15.3 and 15.8. The global method provided

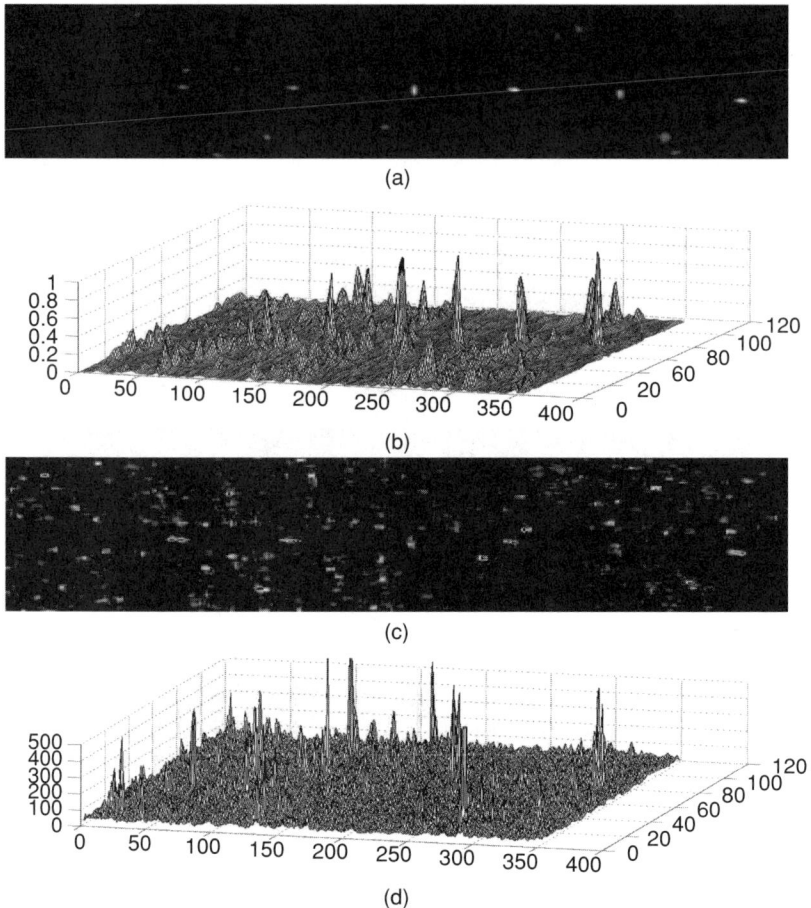

Figure 15.4. Detection results for the Desert Radiance II image using kernel RX and conventional RX; (a) Kernel RX, (b) 3D plot of (a), (c) RX, and (d) 3D plot of (c).

slightly improved performance over the local method for the HYDICE images that were tested.

15.5.2 Simulation Results for the Kernel-matched Filter

We implemented both the proposed kernel-matched filter detector (KMFD) described in Eq. (46) and the conventional-matched filter detector (MFD) described in (Eq. 33) to detect targets of interest (military vehicles) in the HYDICE images. We implemented KMFD with four different kernel functions, each kernel function being associated with its corresponding feature space. The four different kernels used were 1) the Gaussian RBF kernel, $\exp\left(\frac{-\|\mathbf{x}-\mathbf{y}\|^2}{30}\right)$,

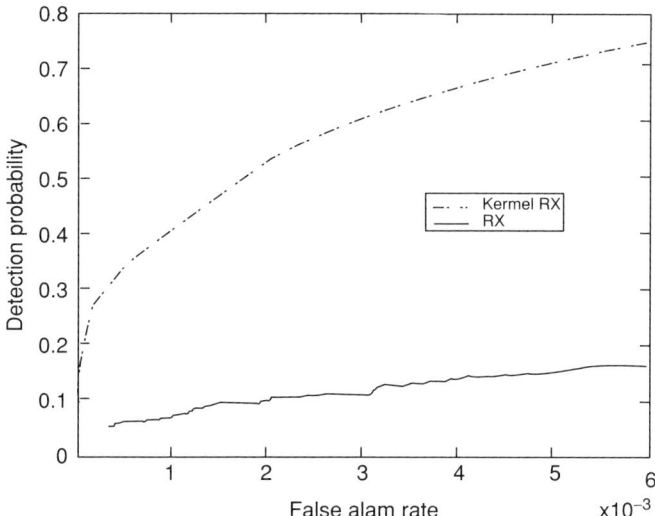

Figure 15.5. ROC curves obtained by the kernel RX and RX based on the local dual window for the Forest Radiance I image.

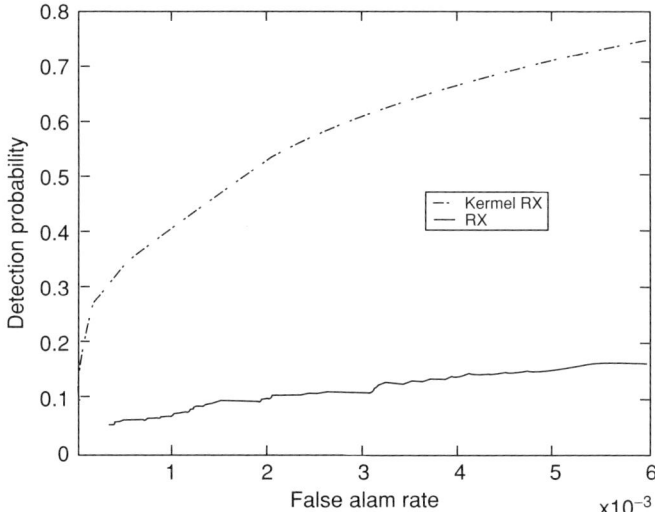

Figure 15.6. ROC curves obtained by the kernel RX and RX based on the local dual window for the Desert Radiance II image.

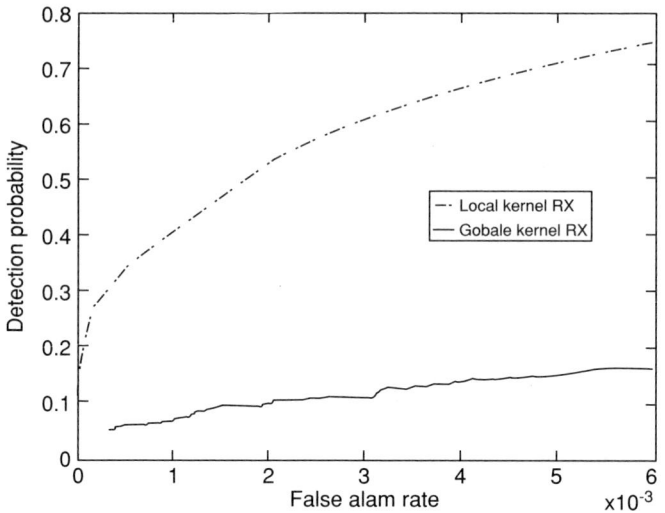

Figure 15.7. ROC curves obtained by the kernel RX using the local and global kernel matrices for the Forest Radiance I image.

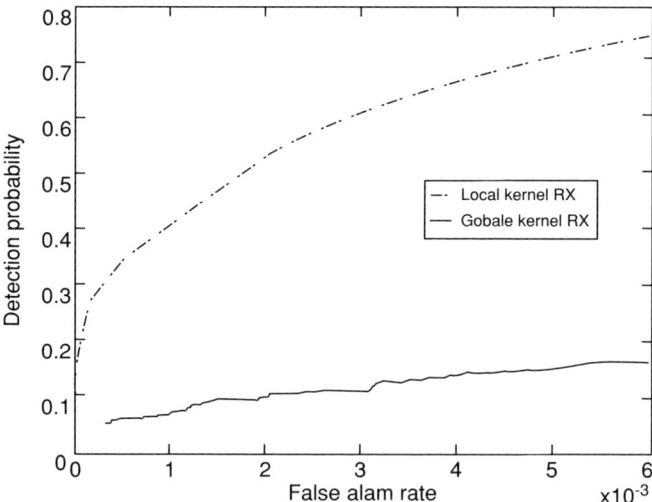

Figure 15.8. ROC curves obtained by the kernel RX using the local and global kernel matrices for the Desert Radiance II image.

2) inverse multiquadric kernel: $\frac{1}{\sqrt{\|\mathbf{x}-\mathbf{y}\|^2+1}}$, 3) spectral angle-based kernel, $\frac{\mathbf{x}\cdot\mathbf{y}}{\|\mathbf{x}\|\|\mathbf{y}\|}$, and 4) 5th order polynomial kernel, $((\mathbf{x}\cdot\mathbf{y})+1)^5$.

All the pixel vectors in a test image are first normalized by a constant, which is a maximum value obtained from all the spectral components of the spectral vectors in the corresponding test image, so that the entries of the normalized pixel vectors fit into the interval of spectral values between zero and one. The rescaling of pixel vectors was mainly performed to effectively utilize the dynamic range of the Gaussian RBF kernel. The rescaling does not affect the performance of KMFDs or MFD when the other three kernel functions are used.

Figures 15.9 and 15.10 show the detection results for the DR- II and FR-I images using KMFD with the four different globally estimated kernels and MFD with the globally estimated covariance matrix, respectively. The corresponding ROC curves for the detection results in Figs. 15.9 and 15.11 are shown in Figs. 15.11 and 15.12, respectively. For DR-II, as shown in Fig. 15.11, KMFD with any type of kernels could detect all the targets at a very low false alarm rate ($N_f \approx 3 \times 10^{-4}$), while conventional MFD detected all the targets at a much higher false alarm rate ($N_f \approx 3 \times 10^{-3}$).

For the FR-I image the background is much more complex than that of DR-I. It includes the tree area where most irregular illumination effects occur; the long shadowy transition and the region filled mostly with grass. KMFD using any kernels except the polynomial kernel still showed improved detection results over the conventional MFD, as shown in Fig. 15.10 and the ROC plots in Fig. 15.12.

Locally estimated kernel and covariance matrices using the dual concentric window were also used to implement KMFD (with the Gaussian RBF kernel) and MFD, respectively. The detection results for DR-II and FR-I are shown in Fig. 15.13 and 15.14, respectively. The Gaussian RBF kernel was chosen because it was the kernel (along with the inverse multiquadric kernel) that showed consistent performance for both DR-II and FR-I when the global kernel estimation was used. The ROC curves for the detection results in Fig. 15.13 and 15.14 are shown in Fig. 15.15 and 15.16, respectively.

MFD with the locally estimated covariance matrix produced, in general, flattened detection values in the background regions compared to MFD with the globally estimated covariance matrix. However, it suppressed some of the targets detected by the global MFD, lowering detection performance. KMFD with the locally estimated Gaussian RBF kernel matrix effectively produced lower filter output values in the background regions, particularly, in the tree and transition areas, significantly reducing potential false alarms, while still generating higher target values than those of the background.

15.5.3 Simulation Results for the Kernel-matched Subspace Filter

We compare the detection performance between the MSD in Eq.(56) defined in the original input domain and the KMSD Eq.(70) using the two test HYDICE

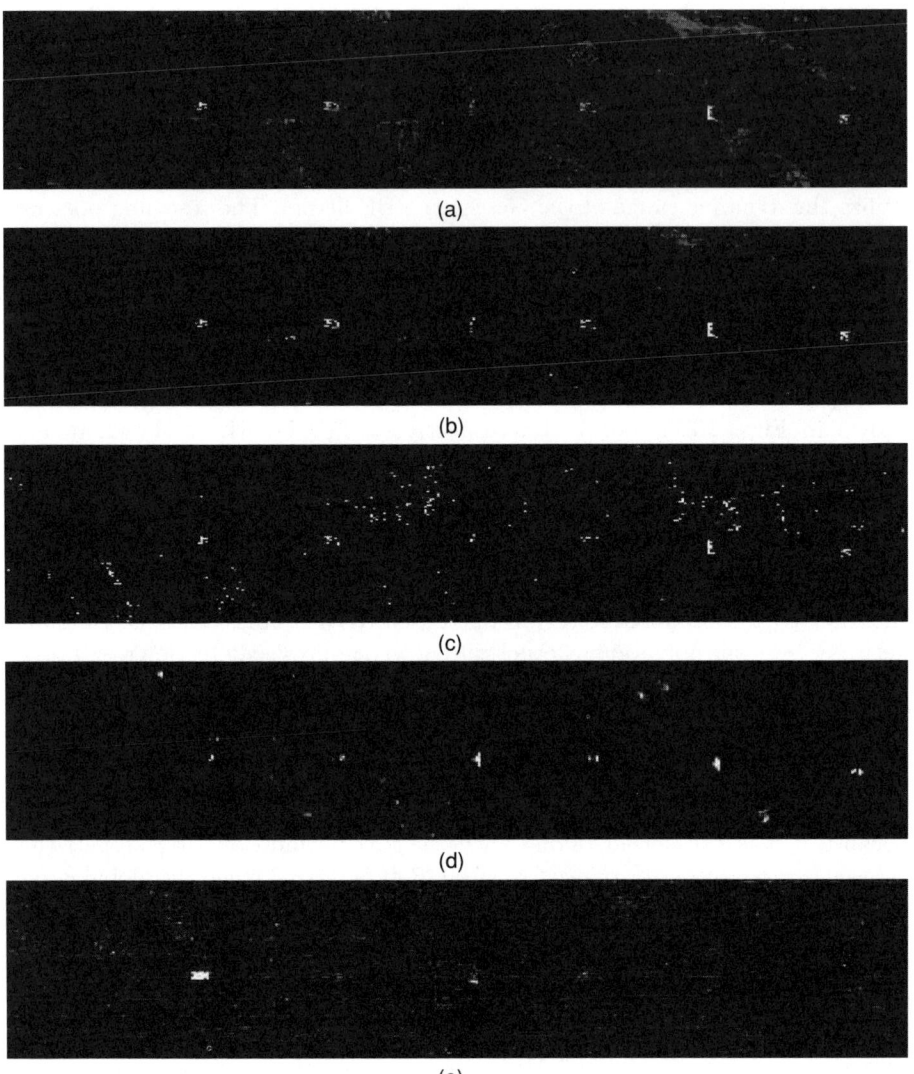

Figure 15.9. Detection results for the Desert Radiance II image using the kernel-matched filter detectors (KMFD) and the matched filter detector (MFD), with globally estimated kernels and covariance matrices, respectively. (a) KMFD with the Gaussian RBF kernel, (b) KMFD with the inverse multiquadric kernel, (c) KMFD with the spectral angle-based kernel, (d) KMFD with the polynomial kernel, and (e) MFD in the original input domain.

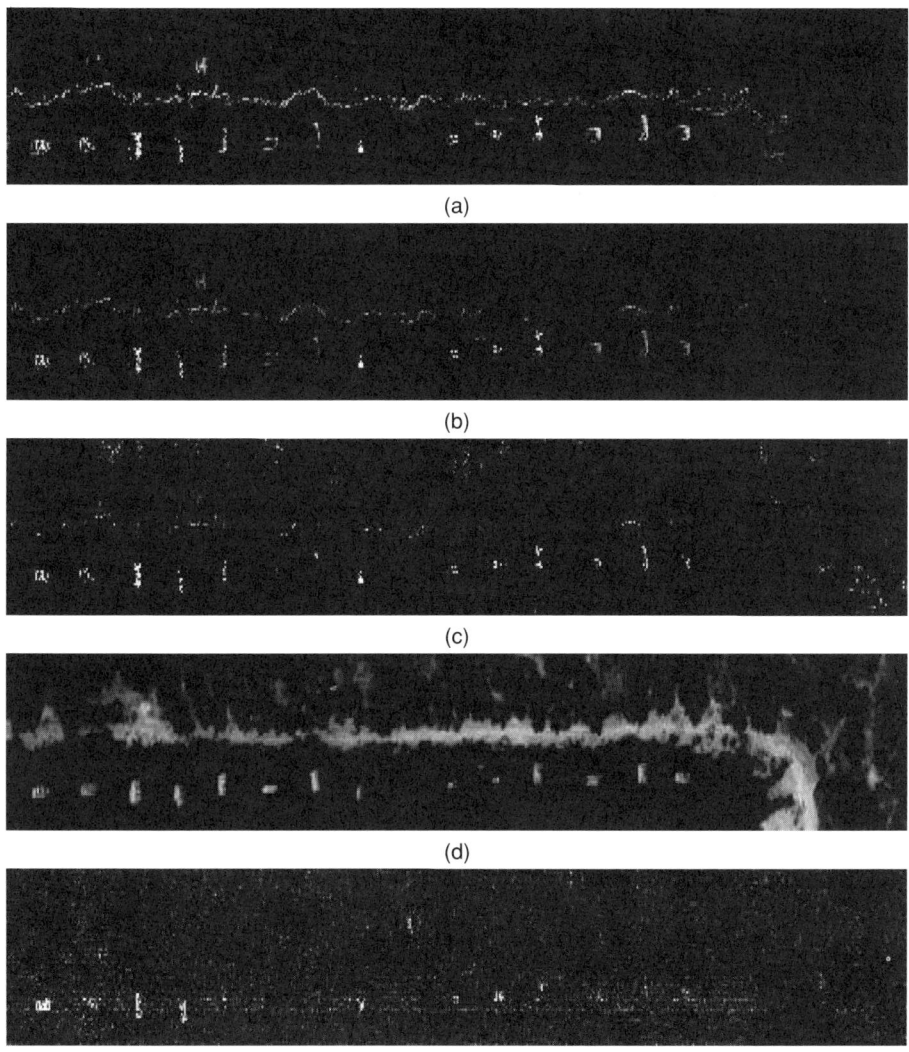

(a)

(b)

(c)

(d)

(e)

Figure 15.10. Detection results for the Forest Radiance I image using the kernel-matched filter detector (KMFDs) and the matched filter detector (MFD), with globally estimated kernels and covariance matrices, respectively. (a) KMFD with the Gaussian RBF kernel, (b) KMFD with the inverse multiquadric kernel, (c) KMFD with the spectral angle-based kernel, (d) KMFD with the polynomial kernel, and (e) MFD in the original input domain.

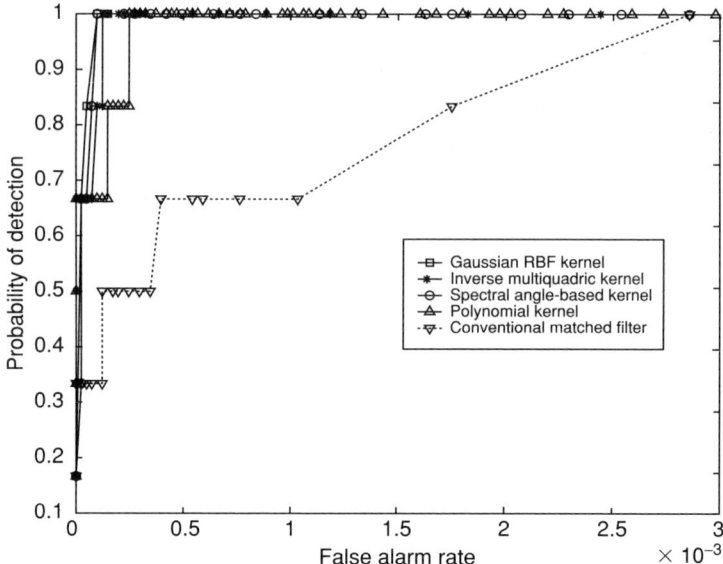

Figure 15.11. ROC curves obtained from the detection results for the Desert Radiance II image shown in Fig. 15.9.

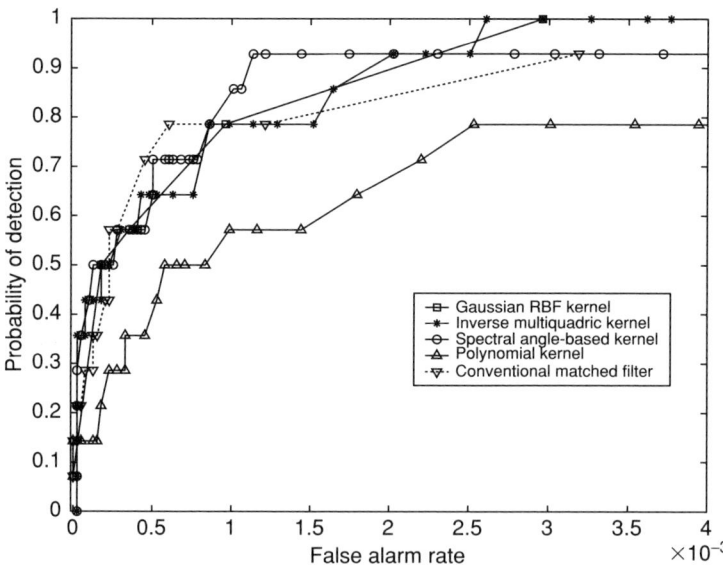

Figure 15.12. ROC curves obtained from the detection results for the Forest Radiance I image shown in Fig. 15.10.

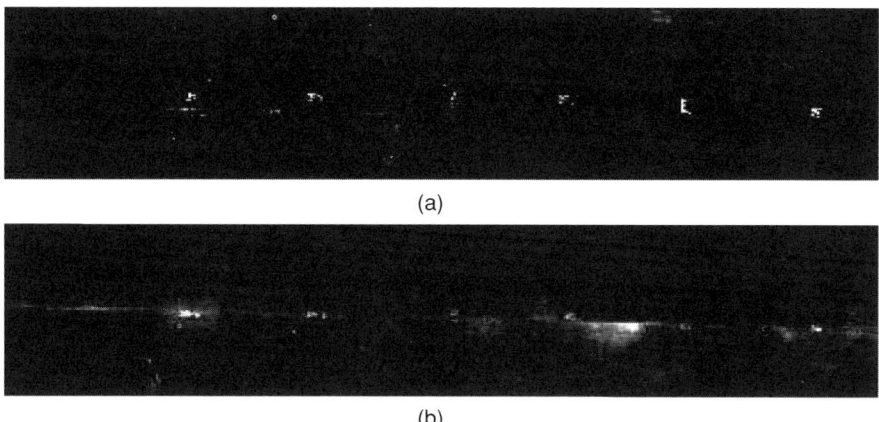

(a)

(b)

Figure 15.13. Detection results for the Desert Radiance II image using KMFD and MFD with locally estimated Gaussian RBF kernel and covariance matrices, respectively: (a) KMFD with the Gaussian RBF kernel and (b) MFD.

images. The Gaussian RBF kernel, $\mathbf{k}(\mathbf{x},\mathbf{y}) = \exp\left(\frac{-\|\mathbf{x}-\mathbf{y}\|^2}{c}\right)$, was used to implement the kernel-based detectors; the value of c was set to 40.

Figures 15.17–15.20 show the detection results including the ROC curves generated by applying KMSD and MSD to both the DR-II and FR-I test images. In implementing KMSD and MSD, the background samples were obtained from outside the test images to estimate the background subspace. Due to a lack of available target samples (the test images include all the targets, military vehicles, present in the given data sets), the target samples were

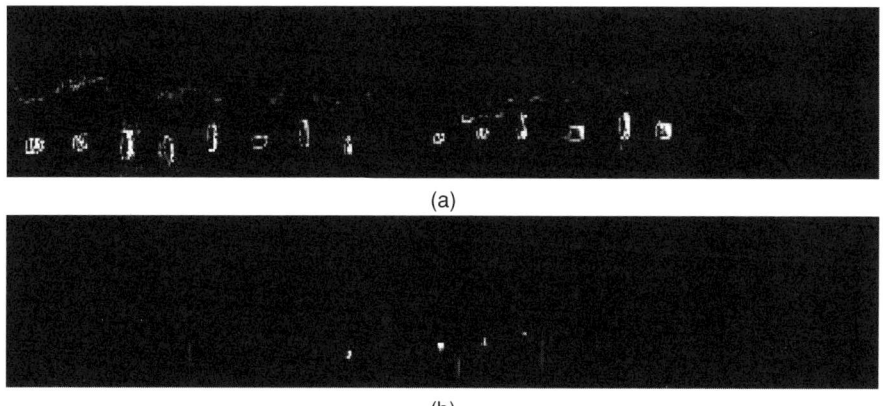

(a)

(b)

Figure 15.14. Detection results for the Forest Radiance I image using KMFD and MFD with locally estimated Gaussian RBF kernel and covariance matrices, respectively: (a) KMFD with the Gaussian RBF kernel and (b) MFD.

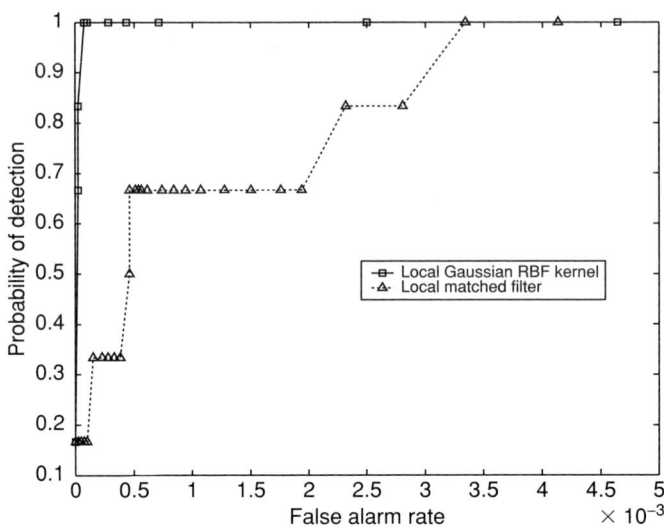

Figure 15.15. ROC curves obtained from the detection results for the Desert Radiance II image shown in Fig. 15.13.

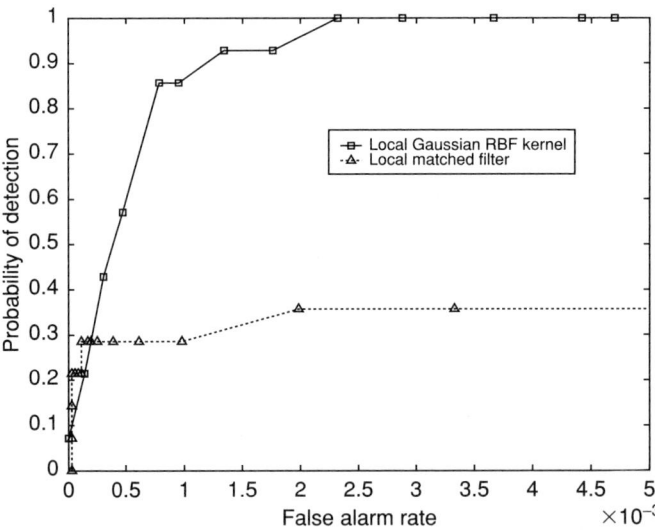

Figure 15.16. ROC curves obtained from the detection results for the Forest Radiance I image shown in Fig. 15.14.

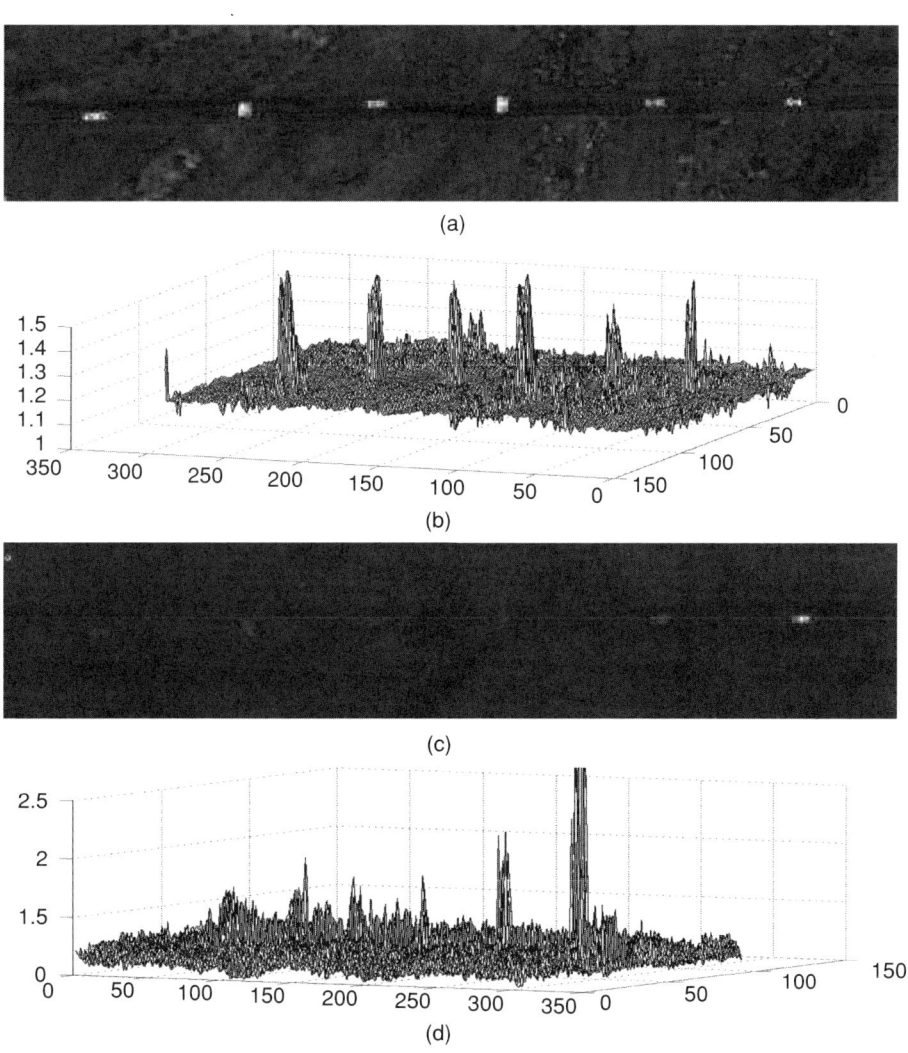

(a)

(b)

(c)

(d)

Figure 15.17. Detection results for the Desert Radiance II image using the kernel-matched subspace detector (KMSD) and the matched subspace detector (MSD): (a) KMSD, (b) 3D plot of (a), (c) MSD, and (d) 3D plot of (c).

collected from one of the targets in each HYDICE test set: the right most target in the DR-II image and the left most target in the FR-I image, as shown in Fig. 15.2.

For the DR-II image, KMSD detected the majority of the target pixels with high contrast in GLRT values with respect to those of the background while MSD found a relatively small number of the target pixels with low contrast,

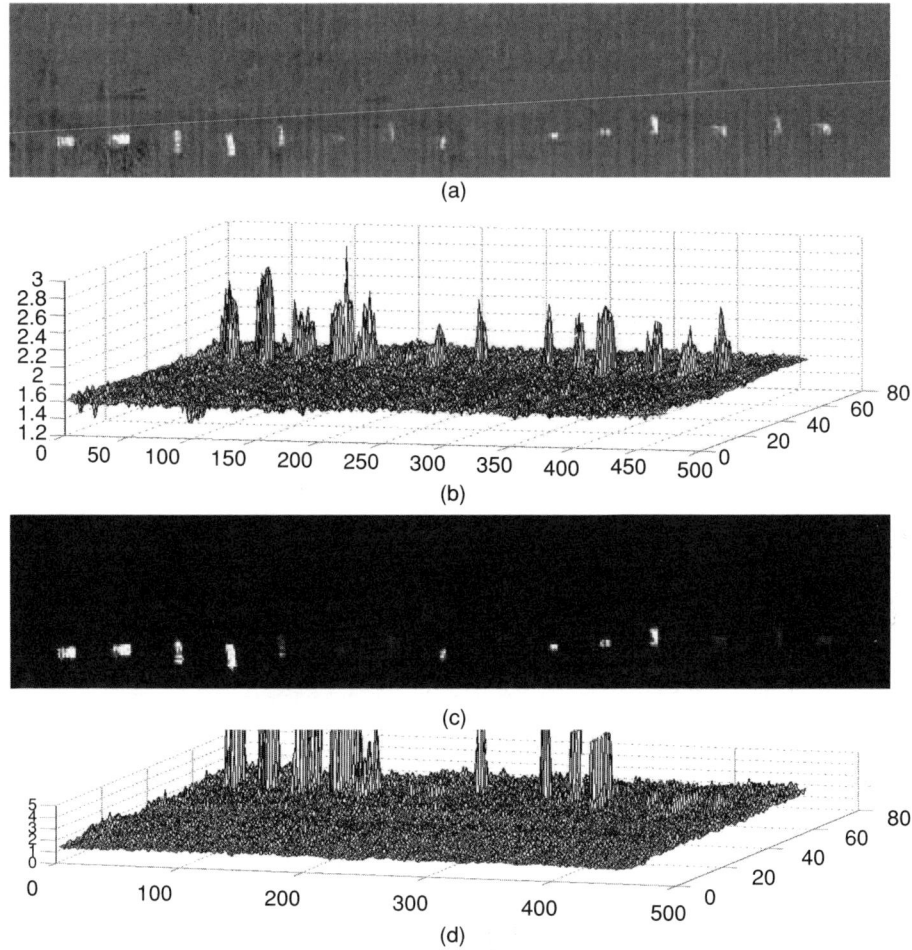

Figure 15.18. Detection results for the Forest Radiance I image using KMSD and MSD: (a) KMSD, (b) 3D plot of (a), (c) MSD, and (d) 3D plot of (c).

especially, for the first four targets from the left, as shown in Fig. 15.17. The ROC curves in Fig. 15.19 also showed the superiority of KMSD over MSD for the DR-II image.

Both KMSD and MSD were also applied to the FR-I image and the detection results were shown in Fig. 15.18 and the corresponding ROC plots in Fig. 15.20. The targets in the FR-I image are relatively difficult to detect mainly because the target spectral variability is in a wider range than that of the DR-II image; some targets show considerably different spectral characteristics than those of the target samples; and some targets are heavily shadowed. Nevertheless, KMSD detected, at least, a small portion of the target pixels for every target region in the images, while MSD missed some of the targets (6th,

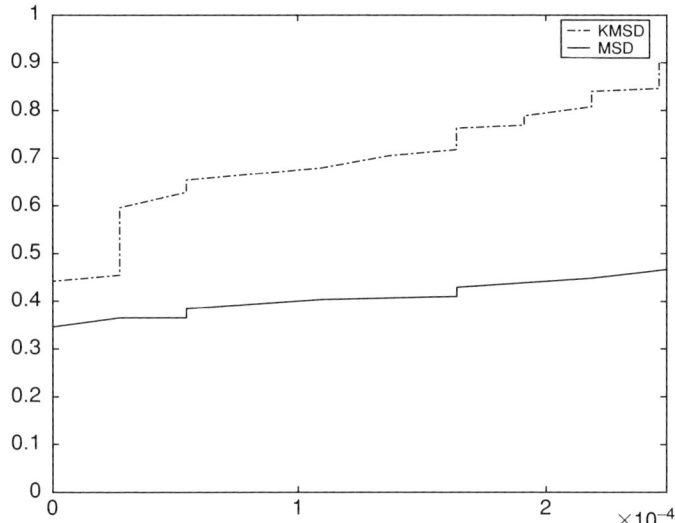

Figure 15.19. ROC curves obtained by KMSD and MSD for the Desert Radiance II image.

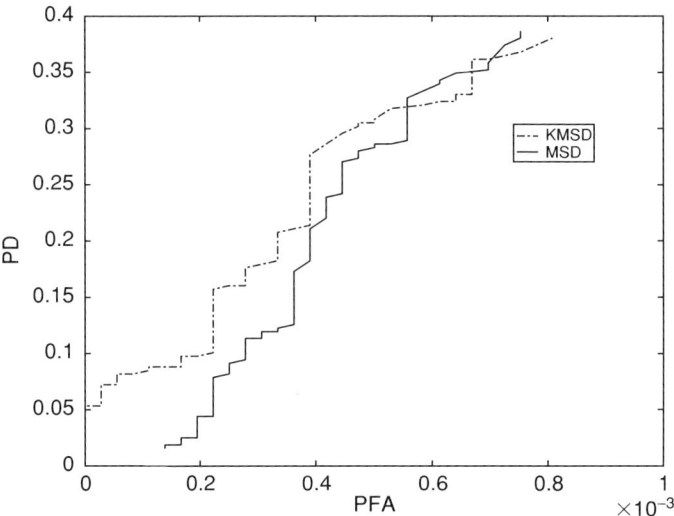

Figure 15.20. ROC curves obtained by KMSD and MSD for the Forest Radiance I image.

7th, and 12th through 14th from the left), as shown in Fig. 15.18. While most of the target GLRT values generated by KMSD are distributed within a relatively narrow range, MSD produced significantly lower GLRT values for the missed targets than those of the other targets, as shown in Fig. 15.5; this is mainly because the missed targets show low reflectivity over the whole spectral range making them similar to shadow areas. It can be said that even though KMSD was trained with a relatively small number of target samples, it provides better generalization than MSD in representing various unseen targets in the given database. In conclusion, for the FR-I image KMSD provided improved overall detection performance over MSD; note that the performance of KMSD at the low false alarm rates was noticeably better than other rates. Given the experimental results based on our limited hyperspectral data set, it can be said that if the target subspace is properly estimated to represent the target spectral variability, KMSD is expected, in general, to perform better than MSD.

15.6 Conclusions

We have extended the conventional RX-algorithm, matched filter detector, and matched subspace detector in the original input space to a nonlinear feature space by kernelizing the corresponding nonlinear expressions for the three algorithms. The detection results show that the kernel-based nonlinear detection methods are quite suitable for identifying underlying structures of complex data such as hyperspectral data, thus they are more powerful in discriminating targets of interest.

For example, kernel RX showed superior detection performance over the conventional RX algorithm given the two HYDICE images tested; kernel RX produced a lot fewer false alarms and detected most of the targets. This is mainly because the high order correlations between the spectral bands are exploited by the kernel RX algorithm. KMFD, the kernel counterpart of MFD, was implemented with several different kernels, each with different characteristics. In general, KMFD with almost all the kernels showed a superior detection performance when compared to the conventional MFD for the HYDICE images tested in this chapter. Kernel-matched subspace detectors were applied to hyperspectral subpixel target detection, and the detection performance was compared to the original matched subspace detectors. The detection results based on the given HYDICE images confirmed that the kernel-based learning, which was implicitly performed in the nonlinear feature space, was a powerful approach to understand the underlying structures of the given data set and representing unseen targets based on the limited training samples.

In this chapter, only a small number of HYDICE images were tested for the performance comparison between kernel RX and conventional RX, and also for the performance of kernel RX associated with the local and global kernel matrix estimations. However, a large hyperspectral database needs to be used to generate more general as well as accurate comparison between the two algorithms and the local and global estimations in the future.

15.7 Acknowledgments

This research was sponsored by the U.S. Army Research Laboratory (ARL) and was accomplished under the ARL/ASEE postdoctoral fellowship program, contract DAAL01-96-C-0038. The views and conclusions contained in this chapter are those of the authors and should not be interpreted as representing the official policies, either expressed or implied, of ARL or the U.S. government. The U.S. government is authorized to reproduce and distribute reprints for government purposes, notwithstanding any copyright notation herein.

15.8 Appendix I. (Kernel PCA)

In this Appendix we will represent derivation of Kernel PCA and its properties providing the relationship between the covariance matrix and the corresponding Gram matrix. Our goal is to prove Eqs. 18 and 43. To drive the Kernel PCA, consider the estimated background clutter covariance matrix in the feature space and assume that the input data has been normalized (centered) to have zero mean. The estimated covariance matrix in the feature space is given by

$$\hat{\mathbf{C}}_\Phi = \frac{1}{N}\mathbf{X}_\Phi\mathbf{X}_\Phi^T. \tag{72}$$

The PCA eigenvectors are computed by solving the eigenvalue problem

$$\begin{aligned}
\lambda\mathbf{v}_\Phi &= \hat{\mathbf{C}}_\Phi\mathbf{v}_\Phi \\
&= \frac{1}{N}\sum_{i=1}^{N}\Phi(\mathbf{x}_i)\Phi(\mathbf{x}_i)^T\mathbf{v}_\Phi \\
&= \frac{1}{N}\sum_{i=1}^{N}<\Phi(\mathbf{x}_i),\mathbf{v}_\Phi>\Phi(\mathbf{x}_i).
\end{aligned} \tag{73}$$

where $\mathbf{v}_\Phi$ is an eigenvector in $\mathcal{F}$ with a corresponding nonzero eigenvalue λ. Equation (73) indicates that each eigenvector $\mathbf{v}_\Phi$ with corresponding $\lambda \neq 0$ are spanned by $\Phi(\mathbf{x}_1),\dots,\Phi(\mathbf{x}_N)$ – i.e.,

$$\mathbf{v}_\Phi = \sum_{i=1}^{N}\beta_i\Phi(\mathbf{x}_i) = \mathbf{X}_\Phi\beta, \tag{74}$$

where $\mathbf{X}_\Phi = [\Phi(\mathbf{x}_1)\Phi(\mathbf{x}_2) \ \dots \ \Phi(\mathbf{x}_N)]$ and $\beta = (\beta_1, \beta_2,\dots,\beta_N)^T$. Substituting Eqs. (74) into (73) and multiplying with $\Phi(\mathbf{x}_n)^T$, $n = 1,\dots,N$, yields

$$\lambda \sum_{i=1}^{N} \beta_i < \Phi(\mathbf{x}_n), \Phi(\mathbf{x}_i) >$$

$$= \frac{1}{N} \sum_{i=1}^{N} \beta_i \Phi(\mathbf{x}_n) \Phi(\mathbf{x}_i) \Phi(\mathbf{x_i})^T \sum_{i=1}^{N} \Phi(\mathbf{x}_i) \tag{75}$$

$$= \frac{1}{N} \sum_{i=1}^{N} \beta_i < \Phi(\mathbf{x}_n), \sum_{j=1}^{N} \Phi(\mathbf{x}_j) < \Phi(\mathbf{x}_j), \Phi(\mathbf{x}_i) >> ,$$

for all $n = 1, \dots, N$

We denote by $\mathbf{K} = \mathbf{K}(\mathbf{X}, \mathbf{X}) = (\mathbf{K})_{ij}$ the $N \times N$ kernel matrix whose entries are the dot products $< \Phi(\mathbf{x}_i), \Phi(\mathbf{x}_j) >$. Equation (73) can be rewritten as

$$N\lambda\beta = \mathbf{K}\beta, \tag{76}$$

where β turns out to be the eigenvectors with nonzero eigenvalues of the kernel matrix $\mathbf{K}$. Note that each β needs to be normalized by the square root of its corresponding eigenvalue.

From the definition of PCA in the feature space Eq. 73 and the Kernel PCA in Eq. 76 we can now write the eigenvector decomposition of the estimated covariance matrix as

$$\hat{\mathbf{C}}_\Phi = \mathbf{V}_\Phi \Lambda \mathbf{V}_\Phi^T, \tag{77}$$

where $\mathbf{V}_\Phi = [\mathbf{v}_\Phi^1 \ \mathbf{v}_\Phi^2 \ \dots \ \mathbf{v}_\Phi^N]$ and Λ is a diagonal matrix with its diagonal elements being the eigenvalues of $\hat{\mathbf{C}}_\Phi$. Similarly, the kernel matrix eigen decomposition is given by

$$\mathbf{K} = \mathcal{B} \Omega \mathcal{B}^T, \tag{78}$$

where $\mathcal{B} = [\beta^1 \ \beta^2 \ \dots \ \beta^N]$ are the eigenvectors of the kernel matrix and Ω is a diagonal matrix with diagonal values equal to the eigenvalues of the kernel matrix $\mathbf{K}$. Using inverse matrix properties of invertible matrices the inverse background covariance matrix $\hat{\mathbf{C}}_\Phi^{-1}$ and inverse Gram matrix $\mathbf{K}^{-1}$ can be also written as

$$\hat{\mathbf{C}}_\Phi^{-1} = \mathbf{V}_\Phi \Lambda^{-1} \mathbf{V}_\Phi^T \tag{79}$$

and

$$\mathbf{K}^{-1} = \mathcal{B} \Omega^{-1} \mathcal{B}^T, \tag{80}$$

respectively. The eigenvalues of the covariance matrix in the feature space and the eigenvalues of the kernel matrix are related by

$$\Lambda = \frac{1}{N} \Omega. \tag{81}$$

Substituting Eqs. 81 into 80, we obtain the relationship

$$\mathbf{K}^{-1} = \frac{1}{N} \mathcal{B} \Lambda^{-1} \mathcal{B}^T, \tag{82}$$

where N is a constant representing the total number of background clutter samples which can be ignored.

In the derivation of the kernel PCA we assumed that the data has already been centered in the feature space by removing the sample mean. However, the sample mean cannot be directly removed in the feature space due to the high dimensionality of $\mathcal{F}$. That is the kernel PCA needs to be derived in terms of the original uncentered input data. Therefore, the kernel matrix $\hat{\mathbf{K}}$ needs to be properly centered.[19] The effect of centering on the kernel PCA can be seen by replacing the uncentered $\mathbf{X}_\Phi$ with the centered $\mathbf{X}_\Phi - \boldsymbol{\mu}_\Phi$ (where $\boldsymbol{\mu}_\Phi$ is the mean of the reference input data) in the estimation of the covariance matrix expression in Eq. 72. The resulting centered $\hat{\mathbf{K}}$ is shown in[19] to be given by

$$\hat{\mathbf{K}} = (\mathbf{K} - \mathbf{1}_N\mathbf{K} - \mathbf{K}\mathbf{1}_N + \mathbf{1}_N\mathbf{K}\mathbf{1}_N), \tag{83}$$

where $(\mathbf{1}_N)_{ij} = 1/N$. In the above Eqs. 78, 80, and 82 the kernel matrix $\mathbf{K}$ needs to be replaced by the centered kernel matrix $\hat{\mathbf{K}}$.

References

[1]   Manolakis D and Shaw G. (2000). "Detection algorithms for hyperspectral imaging applications." *IEEE Signal Process. Mag.*, 19(1):29–43.
[2]   Harsanyi JC and Chang C-I. (1994). "Hyperspectral image classification and dimensionality reduction: An orthogonal subspace projection approach." *IEEE Trans. Geosci. Remote Sensing*, 32(4):779–785.
[3]   Reed IS and Yu X. (1990). "Adaptive multiple-band CFAR detection of an optical pattern with unknown spectral distribution." *IEEE Trans. Acoustics, Speech Signal Process.*, 38(10):1760–1770.
[4]   Chang C-I, Zhao X-L, Althouse MLG, and Pan JJ. (1998). "Least squares subspace projection approach to mixed pixel classification for hyperspectral images." *IEEE Trans. Geosci. Remote Sensing*, 36(3):898–912.
[5]   Healey G and Slater D. (1999). "Models and methods for automated material identification in hyperspectral imagery acquired under unknown illumination and atmospheric conditions." *IEEE Trans. Geosci. Remote Sensing*, 37(6):2706–2717.
[6]   Scharf LL and Friedlander B. (1994). "Matched subspace detectors." *IEEE Trans. Signal Process.*, 42(8):2146–2157.
[7]   Schweizer SM and Moura JMF. (2001). "Efficient detection in hyperspectral imagery." *IEEE Trans. Image Process.*, 10:584–597.
[8]   Stein DWJ, Beaven SG, Hoff LE, Winter EM, Schaum AP, and Stocker AD. (2002). "Anomaly detection from hyperspectral imagery." *IEEE Signal Process. Mag.*, 19:58–69.
[9]   Kwon H, Der SZ, and Nasrabadi NM. (2001). "An adaptive unsupervised segmentation algorithm based on iterative spectral dissimilarity measure for hyperspectral imagery." in: *Proc. SPIE*, 4310:144–152.
[10] Stein DWJ. (2001). "Stochastic compositional models applied to subpixel analysis of hyperspectral imagery." in: *Proc. SPIE*, 4480:49–56.
[11] Kwon H. Der SZ, and Nasrabadi NM. (2003). "Adaptive anomaly detection using subspace separation for hyperspectral images." *Opt. Eng.* 42(11):3342–3351.

[12] Chang C-I and Chiang S-S. (2002). "Anomaly detection and classification for hyperspectral imagery." *IEEE Trans. Geosci. Remote Sensing*, 40(6):1314–1325.

[13] Yu X and Reed IS. (1993). "Comparative performance analysis of adaptive multispectral detectors." *IEEE Trans. Signal Process.*, 41(8):2639–2656.

[14] Manolakis D, Shaw G, and Keshava N. (2000). "Comparative analysis of hyperspectral adaptive matched filter detector." in: *Proc. SPIE*, 4049:2–17.

[15] Robey FC, Fuhrmann DR, and Kelly EJ. (1992). "A CFAR adaptive matched filter detector." *IEEE Trans. Aerospace and Elect. Syst.*, 28(1):208–216.

[16] Kraut S, Scharf LL, and McWhorter T. (2001). "Adaptive subspace detectors." *IEEE Trans. Signal Process.*, 49(1):208–216.

[17] Thai B and Healey G. (2002). "Invariant subpixel material detection in hyperspectral imagery." *IEEE Trans. Geosci. Remote Sensing*, 40(3):599–608.

[18] Vapnik VN. (1999). *The nature of statistical learning theory*. Springer.

[19] Schölkopf B and Smola AJ. (2002). *Learning with Kernels*. MIT.

[20] Müller KR, Mika S, Rätsch G, Tsuda K, and Schölkopf B. (2001). "An introduction to kernel-based learning algorithms." *IEEE Trans. Neural Networks*, (2):181–202.

[21] Lu J, Plataniotis KN, and Venetsanopoulos AN. (2003). "Face recognition using kernel direct discriminant analysis algorithm." *IEEE Trans. Neural Networks*, 14(1):117–126.

[22] Paclik P, Pekalska E, and Duin RPW. (2001). "A generalized kernel approach to dissimilarity-based classification." *J. Mach. Learning*, 2:175–211.

[23] Ruiz A and Lopez-de Teruel E. (2001). "Nonlinear kernel-based statistical patten analysis." *IEEE Trans. Neural Networks*, 12:16–32.

[24] Schölkopf B, Smola AJ, and Müller K-R. (1999). "Kernel principal component analysis." *Neural Comput.* (10):1299–1319.

[25] Baudat G and Anouar F. (2000). "Generalized discriminant analysis using a kernel approach." *Neural Comput.* (12):2385–2404.

[26] Girolami M. (2002). "Mercer kernel-based clustering in feature space." *IEEE Trans. Neural Networks*, 13(3):780–784.

[27] Joliffe IT. (1986). *Principal Component Analysis*. Springer-Verlag.

[28] Scharf LL. (1991). *Statistical Signal Processing*. Addison-Wesley.

[29] Chang C-I and Heinz DC. (2000). "Constrained subpixel target detection for remotely sensed imagery." *IEEE Trans. Geosci. Remote Sensing*, 38(3):1144–1159.

[30] Settle JJ. (1996). "On the relationship between spectral unmixing and subspace projection." *IEEE Trans. Geosci. Remote Sensing*, 34(4):1045–1046.

[31] Harsanyi JC. (1993). *"Detection and Classification of Subpixel Spectral Signatures in Hyperspectral Image Sequences."* Ph.D. dissertation, Dept. Elect. Eng., Univ. of Maryland, Baltimore County.

[32] Kraut S and Scharf LL. (1999). "The CFAR adaptive subspace detector is a scale invariant-invariant GLRT." *IEEE Trans. Signal Process.*, 47(9):2538–2541.

[33] Van Trees HL. (1968). *Detection, Estimation, and Modulation Theory*. Wiley.

[34] Hastie T, Tibshirani R, and Friedman J. (2001). *The Elements of Statistical Learning*. Springer.

[35] Strang G. (1986). *Linear algebra and its applications*. Harcourt Brace.

[36] Jain AK, Murty MN, and Flynn PJ. (1999). "Data clustering: A review." *ACM Comput. Surveys*, 31(3):264–323.

Detecting 3D Location and Shape of Distorted 3D Objects Using LADAR Trained Optimum Nonlinear Filters

Seung-Hyun Hong and Bahram Javidi

Department of Electrical and Computer Engineering, University of Connecticut, 06269-2157, USA. *shhong@engr.uconn.edu* bahram@engr.uconn.edu

16.0 Introduction

Correlation approach[1–14] has been widely used in many pattern recognition problems. To be distortion tolerant[1,8–12], the filter must recognize the target viewed from various angles, perspectives, scales, and illuminations. Therefore, a training data set of reference targets is needed. Many composite filters have been proposed to perform distortion-tolerant pattern recognition.

Optimum nonlinear distortion-tolerant filter is obtained by optimizing[1,10,12] the filter's discrimination capability and noise robustness. In this chapter, we introduce a novel optimum nonlinear distortion-tolerant filter to detect targets placed in nonoverlapping (disjoint) background noise. The filter maintains fixed output peaks for the members of the true class training target set. The nonlinear filter is derived to minimize the mean of the output energy in the presence of disjoint background noise and additive overlapping noise. The output energy in response to the input scene that may include the false class objects is minimized, which improves discrimination. The filter is applied to the recognition of 3D objects using LADAR[13–16] range data.

A range camera is a device that can acquire a 2D image of distance measurements, as measured from a plane or a single point on the camera. In a LADAR range image, the distance to the 3D object is recorded over a quantized range. For display purposes, the distances are often coded in gray scale, such that the darker pixels represent closer object pixels to the sensor. A range image may also be displayed using a random color for each quantized distance. Therefore, LADAR camera measures range information and encodes in the form of a 2D image. It is possible to apply 2D correlation filters for recognition of objects based on range data to detect the target. Also a 2D encoded range image can be preprocessed using rendering technique to have a better detection performance. The preprocessed image is still a 2D image. However, to detect the 3D location of a target from a sensor, 3D data processing is needed. We present a mapping technique to show how 2D encoded LADAR range image can be converted into 3D space to produce a binary 3D profile. Three-dimensional filtering is applied to this data to detect the object and its 3D coordinates.

This chapter is organized as follows. In Section 16.1, we introduce a novel optimum distortion tolerant nonlinear filter with disjoint background noise by setting up and solving the minimization problem. In Section 16.2, converting a 2D LADAR image into a 3D binary profile image is described. In Section 16.3, we carry out the performance tests of the 3D optimum filter using LADAR data. The conclusions are presented in Section 16.4. In Appendix A, the derivation of the mean squared absolute value of the noise in Fourier domain is shown.

16.1 Analysis

For simplicity one-dimensional discrete notations are used throughout our analysis. Let $r_i(t)$ denote one of the distorted reference targets where $i = 1, 2, \ldots, T$; T is the size of reference target set. The input image $s(t)$ is

$$s(t) = \sum_{i=1}^{T} v_i r_i(t - \tau_i) + n_b(t) \left[w(t) - \sum_{i=1}^{T} v_i w_{ri}(t - \tau_i) \right] + n_a(t) w(t), \qquad (1)$$

where v_i is a binary random variable that takes a value of 0 or 1. v_i indicates whether the target $r_i(t)$ is present in the scene or not. For simplicity in the analysis, the probabilities that v_i takes a value 0 or 1 are the same for all i, that is, $p(v_i = 1) = 1/T$, $p(v_i = 0) = 1 - 1/T$. If $r_i(t)$ is one of the reference targets, $n_b(t)$ is the nonoverlapping background noise with mean m_b, $n_a(t)$ the overlapping additive noise with mean m_a, $w(t)$ the window function for entire input scene, $w_{ri}(t)$ the window function for the reference target $r_i(t)$, and τ_i a uniformly distributed random location of the target in the input scene, whose probability density function (pdf) is $f(\tau_i) = w(\tau_i)/d$ (d is the area of the region of support of the input scene). $n_b(t)$ and $n_a(t)$ are wide-sense stationary random processes and statistically independent of each other. The output of the shift invariant filter is

$$o(t) = \sum_{\tau=0}^{M-1} h(\tau - t)^* s(\tau), \qquad (2)$$

where $h(t)$ is the impulse response of the distortion-tolerant filter, * denotes complex conjugate, and M is the number of sample points. The filter is designed so that when the input to the filter is one of the reference targets ($r_i(t)$), then the output of the filter is

$$o_i(0) = \sum_{t=0}^{M-1} h(t)^* r_i(t) = C_i \tag{3}$$

where C_i is a positive real constant. Adopting discrete notations for simplicity, Eq. (3) can be stated in a Fourier domain expression as:

$$\sum_{k=0}^{M-1} H(k)^* R_i(k) = M C_i, \tag{4}$$

where $H(k)$ and $R_i(k)$ are the discrete Fourier transforms of $h(t)$ and $r_i(t)$, respectively. Equation (4) is the constraint imposed on the filter. To obtain noise robustness, we minimize the output energy due to the disjoint background noise and additive noise. We can define the noise including disjoint background noise and additive noise as $n(t) = n_b(t)\left\{ w(t) - \sum_{i=1}^{T} v_i w_{ri}(t - \tau_i)\right\}$ $+ n_a(t) w(t)$. The mean of the output energy due to the input noise as a Fourier domain expression is:

$$E\left[\frac{1}{M} \sum_{k=0}^{M-1} |H(k)|^2 |N(k)|^2\right] = \frac{1}{M} \sum_{k=0}^{M-1} |H(k)|^2 E|N(k)|^2, \tag{5}$$

where E is the expectation operator, and $N(k)$ is the Fourier transform of $n(t)$. The expected value of the absolute squared of the noise Fourier transform is expressed as:

$$E|N(k)|^2 = \frac{1}{MT} \sum_{i=1}^{T} \left(\Phi_b^0(k) \otimes \left\{ |W(k)|^2 + |W_{ri}(k)|^2 - 2\frac{|W(k)|^2}{d} \mathrm{Re}(W_{ri}(k)) \right\} \right)$$

$$+ \frac{1}{M} \Phi_a^0(k) \otimes |W(k)|^2 + \frac{1}{T} \sum_{i=1}^{T} \left(\begin{array}{c} m_b^2 \left\{ \begin{array}{c} |W(k)|^2 + |W_{ri}(k)|^2 \\ -2\frac{|W(k)|^2}{d} \mathrm{Re}(W_{ri}(k)) \end{array} \right\} \\ +2 m_a m_b |W(k)|^2 \mathrm{Re}\left\{ 1 - \frac{W_{ri}(k)}{d} \right\} \end{array} \right)$$

$$+ m_a^2 |W(k)|^2 \tag{6}$$

where $\Phi_b^0(k)$ is the power spectrum of the zero-mean stationary random process $n_b^0(t)$, and $\Phi_a^0(k)$ is the power spectrum of the zero-mean stationary random process $n_a^0(t)$. $W(k)$ and $W_{ri}(k)$ are the discrete Fourier transforms of $w(t)$ and $w_{ri}(t)$, respectively. $\otimes$ denotes a convolution operator and $\mathrm{Re}(\bullet)$ denotes real value. In Appendix A, Eq. (6) is derived. To obtain the discrimination capability, the output energy due to the input scene is minimized. The output energy due to the input scene as a Fourier domain expression is

$$\frac{1}{M} \sum_{k=0}^{M-1} |H(k)|^2 |S(k)|^2, \tag{7}$$

where $S(k)$ is the Fourier transform of $s(t)$. Thus, we minimize a linear combination of the output energy due to the input noise and the output energy due to the input scene under the filter constraint:

$$\frac{w_n}{M} \sum_{k=0}^{M-1} |H(k)|^2 E|N(k)|^2 + \frac{w_d}{M} \sum_{k=0}^{M-1} |H(k)|^2 |S(k)|^2, \tag{8}$$

where w_n and w_d are the positive weights of the noise robustness capability and discrimination capability, respectively. Let $a_k + jb_k$ be the kth element of $H(k)$, $c_{ik} + jd_{ik}$ be the kth element of $R_i(k)$, and $D(k) = \left(w_n E|N(k)|^2 + w_d |S(k)|^2 \right)/M$. With these notations, our constraint can be written as follows,

$$\begin{aligned}
\sum_{k=0}^{M-1} H(k)^* R_i(k) &= \sum_{k=0}^{M-1} (a_k - jb_k)(c_{ik} + jd_{ik}) \\
&= \sum_{k=0}^{M-1} [(a_k c_{ik} + b_k d_{ik}) + j(a_k d_{ik} - b_k c_{ik})] = MC_i
\end{aligned} \tag{9}$$

Since MC_i is a real constant, we can separate the real and imaginary parts, and obtain the following set of real constraints

$$\sum_{k=0}^{M-1} (a_k c_{ik} + b_k d_{ik}) = MC_i \quad for \; i = 1,2,\ldots, T \tag{10a}$$

$$\sum_{k=0}^{M-1} (a_k d_{ik} - b_k c_{ik}) = 0 \quad for \; i = 1,2,\ldots, T. \tag{10b}$$

Thus, the problem is to minimize

$$\frac{w_n}{M} \sum_{k=0}^{M-1} |H(k)|^2 E|N(k)|^2 + \frac{w_d}{M} \sum_{k=0}^{M-1} |H(k)|^2 |S(k)|^2 = \sum_{k=0}^{M-1} (a_k^2 + b_k^2) D(k) \tag{11}$$

under the $2T$ constraints given by Eqs. (10a) and (10b). We use the Lagrange multiplier to solve this minimization problem. Let the function to be minimized with the Lagrange multipliers be

$$\begin{aligned}
J \equiv &\sum_{k=0}^{M-1} (a_k^2 + b_k^2) D(k) + \sum_{i=1}^{T} \lambda_{1i} \left(MC_i - \sum_{k=0}^{M-1} a_k c_{ik} - \sum_{k=0}^{M-1} b_k d_{ik} \right) \\
&+ \sum_{i=1}^{T} \lambda_{2i} \left(0 - \sum_{k=0}^{M-1} a_k d_{ik} + \sum_{k=0}^{M-1} b_k c_{ik} \right)
\end{aligned} \tag{12}$$

We must find a_k, b_k and λ_{1i}, λ_{2i} that satisfy Eqs. (10a) and (10b), and the following two equations, which are the derivatives of J with respect to a_k, b_k and set to zeros,

$$\frac{\partial J}{\partial a_k} = 2a_k D(k) - \sum_{i=1}^{T} \lambda_{1i} c_{ik} - \sum_{i=1}^{T} \lambda_{2i} d_{ik} = 0 \tag{13a}$$

$$\frac{\partial J}{\partial b_k} = 2b_k D(k) - \sum_{i=1}^{T} \lambda_{1i} d_{ik} + \sum_{i=1}^{T} \lambda_{2i} c_{ik} = 0. \tag{13b}$$

Note that the second derivatives of J are $\frac{\partial^2 J}{\partial a_k \partial a_l} = \frac{\partial^2 J}{\partial b_k \partial b_l} = 2D(k)\delta_{kl}$, where δ_{kl} is Kronecker delta, i.e., $\delta_{kl} = \begin{cases} 1, & k = l \\ 0, & k \neq l \end{cases}$. Therefore, J has a minimum value with respect to a_k and b_k. Solving Eqs. (13a) and (13b), we obtain values for a_k and b_k that minimize J and satisfy the required constraints,

$$a_k = \frac{\sum_{i=1}^{T} (\lambda_{1i} c_{ik} + \lambda_{2i} d_{ik})}{2D(k)} \tag{14a}$$

$$b_k = \frac{\sum_{i=1}^{T} (\lambda_{1i} d_{ik} - \lambda_{2i} c_{ik})}{2D(k)}, \tag{14b}$$

where $\lambda_{1i}, \lambda_{2i}$ must satisfy the constraints. In order to obtain $\lambda_{1i}, \lambda_{2i}$, we substitute a_k and b_k given by Eqs. (14a) and (14b) into Eqs. (10a) and (10b), and obtain

$$\sum_{k=0}^{M-1} \frac{1}{2D(k)} \sum_{i=1}^{T} \left[\lambda_{1i}(c_{ik} c_{pk} + d_{ik} d_{pk}) + \lambda_{2i}(d_{ik} c_{pk} - c_{ik} d_{pk}) \right]$$
$$= MC_p \; for \; p = 1, 2, \ldots, T \tag{15a}$$

$$\sum_{k=0}^{M-1} \frac{1}{2D(k)} \sum_{i=1}^{T} \left[\lambda_{1i}(c_{ik} d_{pk} - d_{ik} c_{pk}) + \lambda_{2i}(d_{ik} d_{pk} + c_{ik} c_{pk}) \right] = 0 \tag{15b}$$
$$for \; p = 1, 2, \ldots, T$$

We introduce the following additional notations to complete the derivation,

$$\boldsymbol{\lambda}_1 \equiv [\lambda_{11} \quad \lambda_{12} \quad \cdots \quad \lambda_{1T}]^t$$
$$\boldsymbol{\lambda}_2 \equiv [\lambda_{21} \quad \lambda_{22} \quad \cdots \quad \lambda_{2T}]^t$$
$$\mathbf{C} \equiv [C_1 \quad C_2 \quad \cdots \quad C_T]^t$$

$$A_{x,y} \equiv \sum_{k=0}^{M-1} \frac{\mathrm{Re}[R_x(k)]\mathrm{Re}[R_y(k)] + \mathrm{Im}[R_x(k)]\mathrm{Im}[R_y(k)]}{2D(k)} = \sum_{k=0}^{M-1} \frac{c_{xk} c_{yk} + d_{xk} d_{yk}}{2D(k)}$$

$$B_{x,y} \equiv \sum_{k=0}^{M-1} \frac{\mathrm{Im}[R_x(k)]\mathrm{Re}[R_y(k)] - \mathrm{Re}[R_x(k)]\mathrm{Im}[R_y(k)]}{2D(k)} = \sum_{k=0}^{M-1} \frac{d_{xk} c_{yk} - c_{xk} d_{yk}}{2D(k)}$$

where superscript t is the matrix transpose, and $\mathrm{Re}(\bullet)$, $\mathrm{Im}(\bullet)$ denote the real and imaginary parts, respectively. Let $\mathbf{A}$ and $\mathbf{B}$ be $T \times T$ matrices whose elements at (x, y) are $A_{x,y}$ and $B_{x,y}$, respectively. Equations (15a) and (15b) can be written as

$$\boldsymbol{\lambda}_1^t \mathbf{A} + \boldsymbol{\lambda}_2^t \mathbf{B} = M\mathbf{C}^t \tag{16a}$$

$$-\boldsymbol{\lambda}_1^t \mathbf{B} + \boldsymbol{\lambda}_2^t \mathbf{A} = \mathbf{0}^t. \tag{16b}$$

Solving Eqs. (16a) and (16b), we obtain,

$$\boldsymbol{\lambda}_1^t = M\mathbf{C}^t \left(\mathbf{A} + \mathbf{B}\mathbf{A}^{-1}\mathbf{B} \right)^{-1} \tag{17a}$$

$$\boldsymbol{\lambda}_2^t = M\mathbf{C}^t \left(\mathbf{A} + \mathbf{B}\mathbf{A}^{-1}\mathbf{B} \right)^{-1} \mathbf{B}\mathbf{A}^{-1}. \tag{17b}$$

Using Eqs. (13a) and (13b), we obtain the kth element of the distortion-tolerant filter $H(k)$,

$$
\begin{aligned}
a_k + jb_k &= \frac{1}{2D(k)} \sum_{i=1}^{T} \left[\lambda_{1i}(c_{ik} + jd_{ik}) + \lambda_{2i}(d_{ik} - jc_{ik}) \right] \\
&= \frac{1}{2D(k)} \sum_{i=1}^{T} (\lambda_{1i} - j\lambda_{2i})(c_{ik} + jd_{ik})
\end{aligned}
\tag{18}
$$

For simplicity we have chosen both w_n and w_d in $D(k)$ as $M/2$. Therefore, the optimum nonlinear distortion-tolerant filter $H(k)$ is

$$
H(k) = \sum_{i=1}^{T} (\lambda_{1i} - j\lambda_{2i}) R_i(k) \Big/
\left(
\begin{array}{l}
\frac{1}{MT} \sum_{i=1}^{T} \left(\Phi_b^0(k) \otimes \left\{ |W(k)|^2 + |W_{ri}(k)|^2 - 2\frac{|W(k)|^2}{d}\mathrm{Re}[W_{ri}(k)] \right\} \right) \\
+\frac{1}{M}\Phi_a^0(k) \otimes |W(k)|^2 + \frac{1}{T}\sum_{i=1}^{T} \left(m_b^2 \left\{ \begin{array}{l} |W(k)|^2 + |W_{ri}(k)|^2 \\ -2\frac{|W(k)|^2}{d}\mathrm{Re}[W_{ri}(k)] \end{array} \right\} \\ +2m_a m_b |W(k)|^2 \mathrm{Re}\left[1 - \frac{W_{ri}(k)}{d} \right] \right) \\
+m_a^2 |W(k)|^2 + |S(k)|^2
\end{array}
\right)
\tag{19}
$$

λ_{1i} and λ_{2i} are obtained in Eqs. (17a) and (17b).

16.2 Converting the 2D Encoded LADAR Image to 3D Binary Space

With a LADAR range camera, it is possible to rapidly capture true 3D data. Basically, the LADAR acquires survey shots that cover a field of view from the viewpoint of the LADAR sensor location. Range is determined based on either time of flight or the phase shift of the signal reflected from the objects in the field of view. Substantial LADAR development work has been undertaken for applications involving battlefield assessment and for autonomous vehicle navigation. Real-time applications based on LADAR sensor data have recently become possible owing to advances both in imaging sensor acquisition rate and computer processing speed.

Acquired LADAR range image has distance information from a sensor and it is encoded and recorded in 2D data form. In other words, each pixel in LADAR range image represents the distance from a LADAR sensor to a point on the object. Therefore, pixels that have the equal distance from a LADAR sensor have the same gray scale values. We can form L contours from an $M \times N$ 2D LADAR range image, where M and N are the pixel sizes of the LADAR range image, respectively, and L is the size of digitized distance information of range image. We can convert this $M \times N$ 2D LADAR range image into a 3D $M \times N \times L$ binary space. Each 2D contour image of size $M \times N$ forms one level of equal distance from the sensor, and the pixel value of that contour on that particular depth level is marked as '1' and the rest of the pixels on that particular level are marked as '0's. This is done for entire L contour range images. This set of 2D binary contour images form a 3D binary space representing a profile of the object. Therefore, each depth level represents a different distance from a range sensor. Illustration of the conversion process is shown in Fig. 16.1.

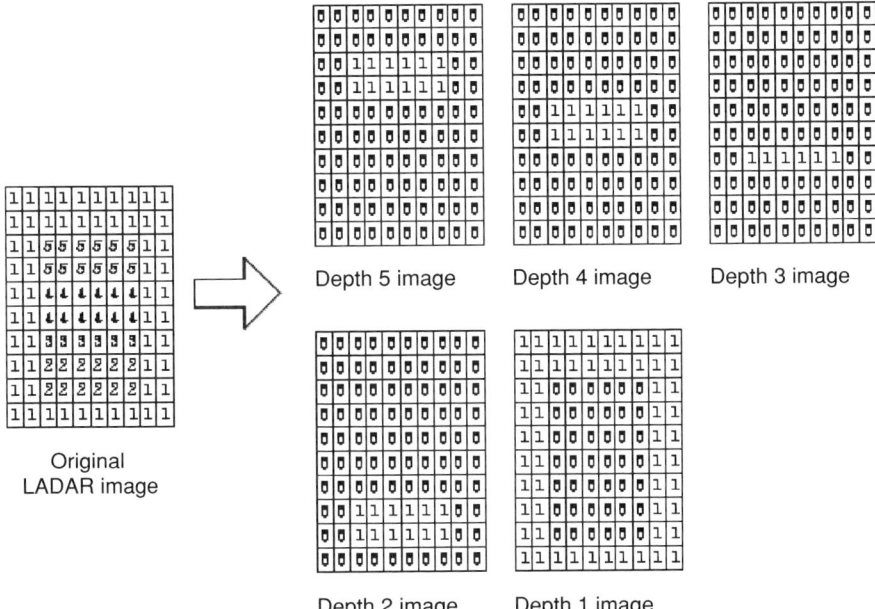

Depth 5 image   Depth 4 image   Depth 3 image

Original
LADAR image

Depth 2 image   Depth 1 image

Figure 16.1. Illustration of converting 2D encoded LADAR image to 3D binary space.

16.3 Computer Simulations

The 3D correlation can be done by inverse 3D Fourier transform of multiplication of two 3D Fourier transformed functions. We set up two simulations. First, the filter is single-target trained to test the performance of the 3D optimum nonlinear filter with a disjoint background noise model. Then, the filter is trained with multiple training targets with different out-of-plane perspectives (azimuth and elevation) to detect the true class training and true class nontraining target sets in the input LADAR range image. For both of the simulations, we have no additive noise. Therefore, the variance of the noise is zero throughout the simulations. Instead, a real LADAR range image was used as an input image. The input LADAR range image has false objects (tanks) to test the detection performance of the 3D optimum nonlinear distortion-tolerant filter.

16.3.1 Test of the Optimum Nonlinear Filter
Using LADAR Data

We use a synthetically generated LADAR M60 tank image as the reference target. The distance is 900 m, elevation is 70°, and azimuth is 30°. Target size is 55×25 pixels. We show the correlation outputs at the target depth level and two adjacent levels. To make sure that the peak at the target depth level is dominant over all the output peaks, we find depth level that has the next highest peak.

Figure 16.2 shows the 2D encoded reference target range image. This 2D range LADAR image is converted to a 3D binary profile, and the 3D optimum nonlinear filter is trained with it. The size of the converted 3D binary profile of target range image is $55 \times 25 \times 2$ pixels. Figure 16.3 is the 2D-encoded input LADAR range image. Size of the converted 3D input binary profile is $128 \times 128 \times 104$ pixels. A target is located in the right bottom corner at the depth level of 25 in the input scene. The converted 3D LADAR binary profile of Fig. 16.2 is applied to the 3D nonlinear optimum filter without the background noise model (see Ref.[12], which can be extended to 3D). The filter outputs at the

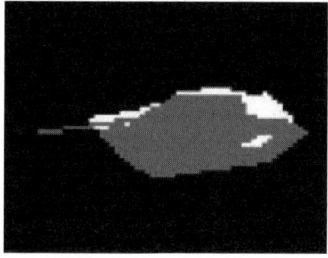

Figure 16.2. 2D encoded LADAR range image used as a training target.

Figure 16.3. 2D encoded input LADAR range image with a training target at the right bottom corner.

depth level of 25, 24, and 26 are shown in Figs. 16.4(a)–16.4(c). We can observe that the output peak is dominant only at the location of the target and at the depth level of the target. However, the peak is not sharp, because the used target image is relatively small and it has only two depth levels.

Figures 16.5(a), (b), and (c) are the outputs of the 3D optimum nonlinear distortion-tolerant filter in Eq. (19) at the depth level of 25, 24, and 26, respectively. Again, only Fig. 16.5(a) shows a dominant and very sharp peak at the location of the target depth level. Difference between Figs. 16.5(a), (b), and (c) and Figs. 16.4(a), (b), and (c) can be observed clearly. We have better discrimination performance to detect the target LADAR range image of Fig. 16.2 with the optimum nonlinear filter with the background noise model. In both cases, the second highest peak appears at the adjacent depth levels of the target levels.

(a)   (b)   (c)

Figure 16.4. Optimum nonlinear filter outputs without the background noise model at the location of the training target. The input LADAR range image in Fig. 16.3 is applied to the filter in Ref.[12] (a) Correlation output without the background noise model at the target depth level of 25. (b) Correlation output without the background noise model at the target depth level of 24. (c) Correlation output without the background noise model at the target depth level of 26.

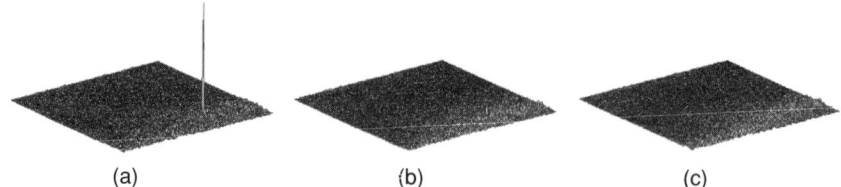

(a)   (b)   (c)

Figure 16.5. Optimum nonlinear filter outputs with the background noise model at the location of the training target. The input LADAR range image in Fig. 16.3 is applied to the filter in Eq. (19). (a) Correlation output of the optimum nonlinear filter with the background noise model at the target depth level of 25. (b) Correlation output of the optimum nonlinear filter with the background noise model at the target depth level of 24. (c) Correlation output of the optimum nonlinear filter with the background noise model at the target depth level of 26.

16.3.2 Test of Optimum Nonlinear Distortion Tolerant Composite Filter Using LADAR Data

We use a synthetically generated LADAR M60 tank image in Section 16.3.1 as the reference target. Figure 16.6 shows nine of the 2D encoded true class target LADAR range images. These 2D LADAR range images are converted to 3D binary profiles, and they are trained to the 3D optimum nonlinear filter. The sizes of converted 3D binary profiles are $52 \times 24 \times 2, 55 \times 23 \times 2,$ $55 \times 22 \times 2,$ $52 \times 26 \times 2, 55 \times 25 \times 2, 55 \times 24 \times 2, 51 \times 29 \times 2, 55 \times 27 \times 2,$ and $55 \times 25 \times 2.$ They have different perspectives. Azimuth varies from $60°$ to $80°$ for every $10°$, and elevation varies from $20°$ to $40°$ for every $10°$. Figure 16.7 is the 2D encoded input LADAR range image. Size of the converted 3D binary profile of the input image is $128 \times 128 \times 104$. The input image has a true class training target and a true class nontraining target. A training target is located at the right bottom

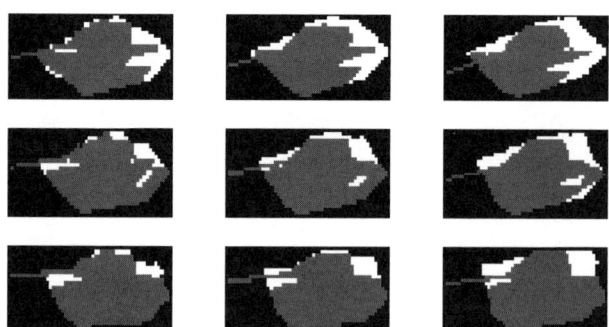

Figure 16.6. Nine 2D encoded LADAR images of true class training target set. Azimuth and elevation of true class training targets are $60°$, $70°$, $80°$ and $20°$, $30°$, and $40°$, respectively, and the distance of the targets from the LADAR range sensor is 900 m.

Figure 16.7. 2D encoded input LADAR image with training target at bottom right-hand corner and a true class nontraining target at upper right corner. The distance, elevation, and the azimuth of the training target are 900 m, 70°,and 30°, respectively. The distance, elevation, and the azimuth of the true class nontraining target with respect to the LADAR range sensor are 1000 m, 75°, and 35°, respectively.

corner at the depth level of 25 in the input scene. The distance, elevation, and the azimuth of the training target with respect to the LADAR range sensor are 900 m, 70°, 30°, respectively. A true class nontraining target is located at the right upper corner at the depth level of 74. The distance of the true class nontraining target is 1000 m, elevation is 75°, and azimuth is 35°. Because the true class nontraining target is farther from the LADAR range sensor than the training targets are, its size (41 × 19 pixels) is smaller than those of the training targets.

The converted 3D binary space of the true class nontraining target LADAR range image has two depth levels. Therefore, the true class nontraining target used in the test is distorted in terms of out-of-plane rotation (in azimuth and elevation) and scale, which makes it very challenging to detect the true class nontraining target. Figures 16.8(a), (b), and (c) are the outputs of the 3D optimum nonlinear distortion-tolerant filter in Eq. (19) at the depth level of 25, 24, and 26, respectively. Only at the target depth level [Fig. 16.8(a)] a dominant and very sharp peak appears. Figures 16.9(a), (b), and (c) are the outputs of the 3D optimum nonlinear filter at the depth level of 74, 73, and 75, respectively. Figure 16.9(a) shows a dominant peak at the location of a true class nontraining target. Though the true class nontraining target is distorted in perspective (azimuth and elevation) and the scale, the output peak is sharp. The maximum 3D correlation output peak appears at the depth level of 25, on which a training target exists, and the second highest peak appears at the depth level of 74, on which a true class nontraining target exists. Third height peak appears at the adjacent depth level of the training target, which is at depth level of 26. Therefore, we can easily threshold the output level to detect the 3D location of the training and distorted true class nontraining targets.

(a)   (b)   (c)

Figure 16.8. Optimum nonlinear filter outputs with the background noise model at the location of the training target. The input LADAR range image in Fig. 16.7 is applied to the filter in Eq. (19). (a) Correlation output of the optimum nonlinear filter with the background noise model at the target depth level of 25. (b) Correlation output of the optimum nonlinear filter with the background noise model at the target depth level of 24. (c) Correlation output of the optimum nonlinear filter with the background noise model at the target depth level of 26.

16.4 Conclusion

We have presented a 3D optimum nonlinear distortion-tolerant filter with disjoint background noise model to detect the 3D target location. The nonlinear filter is derived to minimize the mean of the output energy in the presence of disjoint background noise and additive noise, and the output energy due to the input scene that may include false objects, while maintaining a fixed output peak for the members of the true class target training set. This filter is applied to 3D LADAR range image, which was reconstructed from the 2D encoded LADAR range image. Computer simulations are presented to illustrate the filter's enhanced performance due to the consideration of disjoint background noise model compared to the optimum nonlinear filter without the background noise model. A distortion-tolerant filter is designed using a set of out-of-plane rotated targets. The proposed optimum nonlinear filter is able to detect the 3D coordinates of both true class training target set and true class nontraining target set.

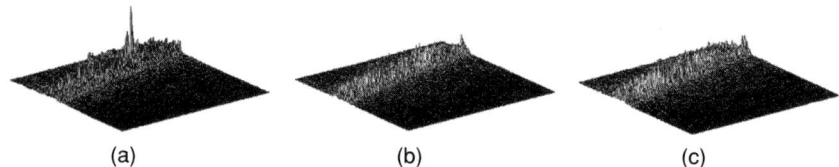

(a)   (b)   (c)

Figure 16.9. Optimum nonlinear filter outputs with the background noise model at the location of the true class nontraining target. The input LADAR range image in Fig. 16.7 is applied to the filter in Eq. (19). (a) Correlation output of the optimum nonlinear filter with the background noise model at the target depth level of 74. (b) Correlation output of the optimum nonlinear filter with the background noise model at the target depth level of 73. (c) Correlation output of the optimum nonlinear filter with the background noise model at the target depth level of 75.

The filter outputs are obtained to detect the target location as well as the target distance from the input LADAR range image. For the LADAR images used in our simulations, 3D optimum nonlinear distortion-tolerant filter with the background noise model detects the 3D location of LADAR range targets very well, even when the target is relatively small compared with background.

References

[1] Javidi B. (2002). *Image Recognition and Classification, Algorithms, Systems, and Applications*. Marcel Dekker. New York.

[2] Goodman JW. (1996). *Introduction to Fourier Optics*, 2nd edn. McGraw-Hill, New York.

[3] VanderLugt A. (1992). *Optical Signal Processing*. Wiley. New York.

[4] Javidi B and Horner JL. (1994). *Real Time Optical Information Processing*. Academic.

[5] Turin JL. (1960). "An introduction to matched filters." *IRE Trans. Inf. Theory*, IT-6:311–329.

[6] Javidi B. (1989). "Nonlinear joint power spectrum based optical correlation." *Appl. Opt.*, 28:2358–2367.

[7] Refreigher P, Laude V, and Javidi B. (1994). "Nonlinear joint-transform correlation: An optimal solution for adaptive image discrimination and input noise robustness." *Opt. Lett.*, 19:405–407.

[8] Mahalanobis A. (1996). "Review of correlation filters and their application for scene matching." *Optoelectronic Devices and Systems for Processing, Critical Reviews of Optical Science Technology*, CR 65. SPIE Press, pp. 240–260.

[9] Casasent D and Psaltis D. (1976). "Position, rotation and scale invariant optical correlation." *Appl. Opt.*, 15:1795–1799.

[10] Javidi B and Wang J. (1995). "Optimum distortion-invariant filter for detecting a noisy distorted target in nonoverlapping background noise." *J. Opt. Soc. Am. A.*, 12:2604–2614.

[11] Javidi B and Painchaud D. (1996). "Distortion-invariant pattern recognition with Fourier-plane nonlinear filters." *Appl. Opt.*, 35:318–331.

[12] Hong S and Javidi B. (2002). "Optimum nonlinear composite filter for distortion-tolerant pattern recognition." *Appl. Opt.*, 41:2172–2178.

[13] Prona MT, Mahalanobis A, and Zachery KN. (1999). "LADAR automatic target recognition using correlation filters." in: *Proc. SPIE, Automatic Target Recognition IX*, 3718, pp. 388–396.

[14] Prona MT, Mahalanobis A, and Zachery KN. (2000). "System level evaluations of LADAR ATR using correlation filters." in: *Proc. SPIE*, 4050, Aerosense Orlando.

[15] Mahalanobis A and Novel A. (2002). "Performance of multidimensional algorithms for target detection in LADAR imagery." in: *Proc. SPIE*, 4789:134–147.

[16] Chang S, Rioux M, and Domey J. (1997). "Face recognition with range images and intensity images." *Opt. Eng.*, 35(4):1106–1112.

Appendix A

In this appendix, we derive Eq. (6). The input image is

$$s(t) = \sum_{i=1}^{T} v_i r_i(t - \tau_i) + n_b(t) \left[w(t) - \sum_{i=1}^{T} v_i w_{ri}(t - \tau_i) \right] + n_a(t) w(t), \quad \text{(A1)}$$

where v_i is a binary random variable that takes a value of 0 or 1. v_i indicates whether the target $r_i(t)$ is present in the scene or not. For simplicity in the analysis, the probabilities that v_i takes a value 0 or 1 are equal for all i, that is, $p(v_i = 1) = 1/T, p(v_i = 0) = 1 - 1/T$, and we assume that only one of the training targets appears in the input scene at one time. We make following assumptions.

The disjoint background noise is represented by $n_b(t) [w(t) - \sum_{i=1}^{T} v_i w_{ri} (t - \tau_i)]$, where $n_b(t)$ is the wide-sense stationary random process with mean m_b ($n_b(t) = n_b^0(t) + m_b$, $n_b^0(t)$ is zero-mean wide-sense stationary random process), and $w(t)$ and $w_{ri}(t)$ are window functions of the entire input scene and the training target $r_i(t)$, respectively. τ_i is the uniformly distributed random location in the input scene, whose pdf is $f(\tau_i) = w(\tau_i)/d$, where d is the area of the input scene ($d = W(0) = \sum_t w(t)$) and $R_{n_b^0}(t)$ and $\Phi_b^0(k)$ are the autocorrelation and the power spectrum of $n_b^0(t)$, respectively.

The detector or sensor noise is represented by $n_a(t) w(t)$, where $n_a(t)$ is wide-sense stationary random process with mean m_a ($n_a(t) = n_a^0(t) + m_a$, $n_a^0(t)$ is zero-mean wide-sense stationary random process). $R_{n_a^0}(t)$ and $\Phi_a^0(k)$ are the autocorrelation and the power spectrum of $n_a^0(t)$, respectively.

The random variables $n_b(t)$, $n_a(t)$, v_i, and τ_i are statistically independent of each other.

Equation (A1) can be rewritten as

$$s(t) = \sum_{i=1}^{T} v_i \{ r_i(t - \tau_i) + n_b(t)[w(t) - w_{ri}(t - \tau_i)] + n_a(t) w(t) \}$$
$$- \left(\sum_{i=1}^{T} v_i - 1 \right) \{ n_a(t) w(t) + n_b(t) w(t) \} \quad \text{(A2)}$$

Since the second term in Eq. (A2) is zero with probability of 1, it can be omitted. Therefore,

$$s(t) = \sum_{i=1}^{T} v_i \{ r_i(t - \tau_i) + n_b(t)[w(t) - w_{ri}(t - \tau_i)] + n_a(t) w(t) \} \quad \text{(A3)}$$

We define $n_i(t)$ as $n_i(t) = n_b(t) w_{ci}(t) + n_a(t) w(t)$, where $w_{ci}(t)$ is the complementary window function of the training target ($w_{ci}(t) = w(t) - w_{ri}(t - \tau_i)$).

The other notations used in this appendix are as follows. $\otimes$ denotes a convolution operator, E[•] an expectation operator, F[•] a discrete Fourier transform operator, and the capital letters stand for the discrete Fourier transforms of each corresponding lower cases. The absolute value square of the discrete Fourier transform of the complementary window of each target is

$$
\begin{aligned}
|W_{ci}(k)|^2 &= \left| W(k) - W_{ri}(k) \exp\left(\frac{-j2\pi k\tau_i}{M}\right) \right|^2 \\
&= |W(k)|^2 + |W_{ri}(k)|^2 - 2\mathrm{Re}\left(W(k) \exp\left(\frac{j2\pi k\tau_i}{M}\right) W_{ri}^*(k) \right)
\end{aligned}
\tag{A4}
$$

The expected values of exponential of the random location is

$$
E_{\tau_i}\left[\exp\left(\frac{-j2\pi k\tau_i}{M}\right)\right] = \frac{W(k)}{d}
\tag{A5a}
$$

and

$$
E_{\tau_i}\left[\exp\left(\frac{j2\pi k\tau_i}{M}\right)\right] = \frac{W(k)^*}{d}
\tag{A5b}
$$

There are four statistically independent random variables, i.e., random location of the target, additive noise, disjoint background noise, and the probability of the appearance of the targets in the input scene. The expected value of the absolute squared of the noise Fourier transform is,

$$
E|N(k)|^2 = E_{v_i} E_{\tau_i} E_{n_a} E_{n_b} \left| \sum_{i=1}^{T} v_i N_i(k) \right|^2 = \frac{1}{T} \sum_{i=1}^{T} E_{\tau_i} E_n |N_i(k)|^2,
\tag{A6}
$$

where we define $E_n = E_{n_a} E_{n_b}$. The expected value of absolute squared of the discrete Fourier transformed noise is,

$$
E_n |N_i(k)|^2 = E_n\left[\mathbf{F}\left(\begin{array}{c} (n_b(t) w_{ci}(t) + n_a(t) w(t)) \\ \otimes (n_b(-t) w_{ci}(-t) + n_a(-t) w(-t)) \end{array} \right) \right]
\tag{A7}
$$

With the zero-mean noise notations, Eq. (A7) becomes

$$
\begin{aligned}
E_n |N_i(k)|^2 = E_n[\mathbf{F} - &\{ (n_b^0(t) + m_b) w_{ci}(t) \} \otimes \{ (n_b^0(-t) + m_b) w_{ci}(-t) \} \\
&+ \{ (n_b^0(t) + m_b) w_{ci}(t) \} \otimes \{ (n_a^0(-t) + m_a) w(-t) \} \\
&+ \{ (n_a^0(t) + m_a) w(t) \} \otimes \{ (n_b^0(-t) + m_b) w_{ci}(-t) \} \\
&+ \{ (n_a^0(t) + m_a) w(t) \} \otimes \{ (n_a^0(-t) + m_a) w(-t) \}]
\end{aligned}
\tag{A8}
$$

Since $n_b(t)$, $n_a(t)$ are statistically independent of each other, Eq. (A8) becomes

$$
\begin{aligned}
E_n |N_i(k)|^2 = {}& \mathbf{F}\Big\{ R_{n_b^0}(t) \times [w_{ci}(t) \otimes w_{ci}(-t)] \Big\} \\
& + \mathbf{F}\Big\{ R_{n_a^0}(t) \times [w(t) \otimes w(-t)] \Big\} \\
& + m_b^2 |W_{ci}(k)|^2 + m_a^2 |W(k)|^2 \\
& + 2 m_a m_b \mathrm{Re}\{ W_{ci}(k)\, W^*(k) \}
\end{aligned}
\tag{A9}
$$

After performing discrete Fourier transform, Eq. (A9) is

$$
\begin{aligned}
E_n |N_i(k)|^2 = {}& \frac{1}{M} \Phi_b^0(k) \otimes
\left\{
\begin{array}{l}
|W(k)|^2 + |W_{ri}(k)|^2 \\
-2\mathrm{Re}\big(W(k) \exp\big(\tfrac{j 2\pi k \tau_i}{M}\big) W_{ri}^*(k) \big)
\end{array}
\right\} \\
& + \frac{1}{M} \Phi_a^0(k) \otimes |W(k)|^2 + m_a^2 |W(k)|^2 \\
& + m_b^2
\left\{
\begin{array}{l}
|W(k)|^2 + |W_{ri}(k)|^2 \\
-2\mathrm{Re}\big(W(k) \exp\big(\tfrac{j 2\pi k \tau_i}{M}\big) W_{ri}^*(k) \big)
\end{array}
\right\} \\
& + 2 m_a m_b \mathrm{Re}\Big\{ |W(k)|^2 - W_{ri}(k)\, W^*(k) \exp\big(\tfrac{-j 2\pi k \tau_i}{M}\big) \Big\}
\end{aligned}
\tag{A10}
$$

Therefore, taking an expectation operation of $E_n |N_i(k)|^2$ with respect to τ_i is

$$
\begin{aligned}
E_{\tau_i} E_n |N_i(k)|^2 = {}& \frac{1}{M} \Phi_b^0(k) \otimes \left\{ |W(k)|^2 + |W_{ri}(k)|^2 - 2\frac{|W(k)|^2}{d} \mathrm{Re}(W_{ri}(k)) \right\} \\
& + \frac{1}{M} \Phi_a^0(k) \otimes |W(k)|^2 + m_a^2 |W(k)|^2 \\
& + m_b^2 \left\{ |W(k)|^2 + |W_{ri}(k)|^2 - 2\frac{|W(k)|^2}{d} \mathrm{Re}(W_{ri}(k)) \right\} \\
& + 2 m_a m_b |W(k)|^2 \mathrm{Re}\left\{ 1 - \frac{W_{ri}(k)}{d} \right\}
\end{aligned}
\tag{A11}
$$

Therefore, the expected value of the absolute squared of the noise Fourier transform becomes

$$
\begin{aligned}
E |N(k)|^2 = {}& \frac{1}{MT} \sum_{i=1}^{T} \left(\Phi_b^0(k) \otimes
\left\{
\begin{array}{l}
|W(k)|^2 + |W_{ri}(k)|^2 \\
-2\frac{|W(k)|^2}{d} \mathrm{Re}(W_{ri}(k))
\end{array}
\right\} \right) \\
& + \frac{1}{M} \Phi_a^0(k) \otimes |W(k)|^2 \\
& + \frac{1}{T} \sum_{i=1}^{T} \left(
\begin{array}{l}
m_b^2 \left\{ |W(k)|^2 + |W_{ri}(k)|^2 - 2\frac{|W(k)|^2}{d} \mathrm{Re}(W_{ri}(k)) \right\} \\
+2 m_a m_b |W(k)|^2 \mathrm{Re}\left\{ 1 - \frac{W_{ri}(k)}{2d} \right\}
\end{array}
\right) \\
& + m_a^2 |W(k)|^2
\end{aligned}
$$

Planar Microoptical Systems for Correlation and Security Applications

Stefan Sinzinger,[1] Jürgen Jahns,[2] Vincent R. Daria,[3] and Jesper Glückstad[3]

[1]Technische Universitat Ilmenau, Technische Optik, P.O. Box 100565, 98684 Ilmenau, Germany, stefan.sinzinger@tu-ilmenau.de
[2]Fernuniversitat Hagen, Optische Nachrichtentechnik, Universitatstr. 27, 58084 Hagen, Germany, jahns@fernuni-hagen.de
[3]Ris National Laboratory, Optics and Plasma Research Department, PO Box 49, DK4000 Roskilde, Denmark, daria@risoe.dk, gluckstad@risoe.dk

17.0 Introduction

Throughout the chapters of this book a variety of concepts and demonstrations of optical systems for security applications are presented. Several of these concepts are based on optical correlator or spatial filtering systems. In this chapter we supplement these contributions by focusing on the realization of the necessary optical or optoelectronical systems in compact integrated microoptical systems. To this end we specifically address the concept of planar integrated free-space optics (PIFSO). We start in Section 17.1 by introducing the basics of PIFSO mainly focussing on design considerations necessary for planar integrated imaging systems. In Section 17.2, we address the planar integration of optical correlators working on discrete input arrays, e.g., for optical interconnections. In Sections 17.3 and 17.4, planar integrated optical correlators and phase contrast systems are investigated for security applications. A summary and conclusions of the chapter are presented in Section 17.5.

17.1 Planar Integrated Free-space Optics (PIFSO)

17.1.1 Basic Concept

Miniaturization, a dominant trend in the fabrication of integrated circuits, is also a key factor in optics. There, a strong tendency can be observed from macro-optical components and systems based on discrete mounting techniques toward microoptical systems technologies. The interest in miniaturization and integration is motivated by several reasons: firstly, small size and light weight are a must for many applications. This is the case, for example, for any mobile application for obvious reasons, and/or if the optics is part of a highly integrated overall system. The latter case is valid, for example, for optical interconnection, i.e., data communications inside a computer. Besides the reasons just stated, cost is also usually an important driving force for systems integration. In conventional optomechanics, a significant part of the cost goes into the mounts, the alignment, and the packaging. These expenses can be significantly reduced by integration.

The key to integration is the technological development in the area of microfabrication techniques since the 1970s. Footed on the basic lithographic process, a variety of techniques for the fabrication of microoptical elements have been developed.[1] Diffractive and refractive elements can be implemented, to be used in transmission or reflection. Microfabrication techniques allow one the realization of the optics in practically any kind of material: glass, polymers, and semiconductors. Finally, precise replication techniques have become available as a means for low cost and mass production.

One can distinguish between three main approaches to the integration of free-space optical systems: the stacked approach due to Iga et al.[2], the planar approach due to Jahns and Huang[3] and the micromechanics approach based on the use of Silicon micromaching, demonstrated by Wu.[4] Here, we will discuss the basic concept of PIFSO and its applications to imaging and spatial filtering. The PIFSO is based on the motivation to make as much use as possible of existing lithographic fabrication and packaging technologies, as they are well known from the fabrication of electronic integrated circuits. Hence, a ("planar") layout results which is suitable for batch processing, replication, 2D and surface mounting. This is schematically represented by Fig. 17.1.

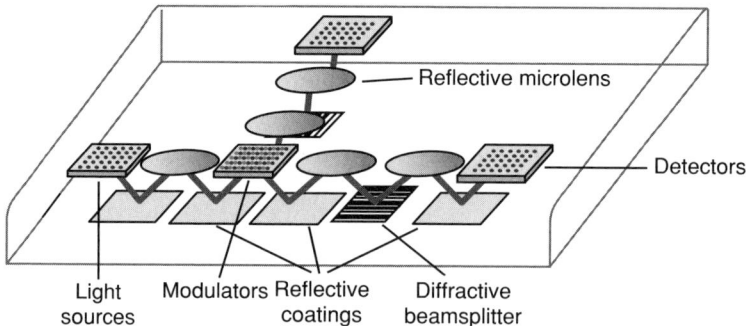

Figure 17.1. Schematic representation of planar-integrated free-space optics (PIFSO). Charactistic for this concept are the 2D layout of the microoptical elements on a transparent substrate and the light propagation along a folded optical axis inside the substrate.

The optical elements are placed on one or both surfaces of a thick and transparent substrate (for example, SiO_2, plastic, or Si). They are batch fabricated by lithographic techniques such as etching, milling, etc. (see Ref.[1]) with submicron precision. The elements may be diffractive and/or refractive and/or reflective. Except for input and output to the substrate, the light signal travels inside the substrate along a zigzag line which represents a tilted and folded optical axis. The tilt angle may range between a few degrees up to more than $45°$. The substrate thickness is on the order of a few millimeters. Hence, it is important to note that despite the light propagation inside a substrate similar to waveguide optics, it is not confined to waveguide dimensions. Rather, the propagation is characterized by a laterally homogeneous medium, as it is typical for free-space optics. In order to keep the light inside the substrate, reflective coatings of the elements are required, either metallic or dielectric or, if the tilt angle of the optical axis is large enough, total internal reflection can be used.

Due to the elimination of mechanical mounts and supporting structures, the volume (and thus weight) of a PIFSO system is considerably reduced in comparison with a conventional optomechanical system. The substrate serves a threefold purpose: as a monolithic block for integration of the passive optics, as a propagation medium for the light signal(s), and as an "open platform" for the integration of devices. Due to the monolithic integration, the passive optics is "perfectly" aligned with lithographic precision. Since the light travels mostly inside the substrate, it is protected against disturbing influences like dust and humidity. These features strongly enhance the aspect of applicability under various environmental influences. The 2D layout of PIFSO systems supports the integration of a variety of devices using surface-mounting techniques, like, for example, flip-chip bonding. Chips with light sources (such as arrays of vertical cavity surface-emitting laser diodes), detectors, modulators (like liquid crystal devices), etc., can be integrated to build up complex and highly functional systems.

Practical aspects of hybrid integration, in general, include issues like thermal and mechanical stress to the bonds, capsuling in order to cover sensitive surfaces for processing and for operation. Experimental systems built of fused silica substrates and optoelectronic chips as well as fiber mounts have been demonstrated.[5] Further work in this area dealing with specific aspects of materials, packaging and thermal management are described in.[6–8] An overview of the technological aspects of PIFSO systems is given in.[9] Here, we want to focus on the optical design since it is relevant for the implementation of the spatial filtering systems that will be described in the latter sections of this chapter.

17.1.2 Design Considerations

One inherent property of PIFSO systems is the propagation of light signals along an oblique optical axis. Planes that represent optical signals and elements are not perpendicular to the optical axis (Fig. 17.2). It has been shown from the

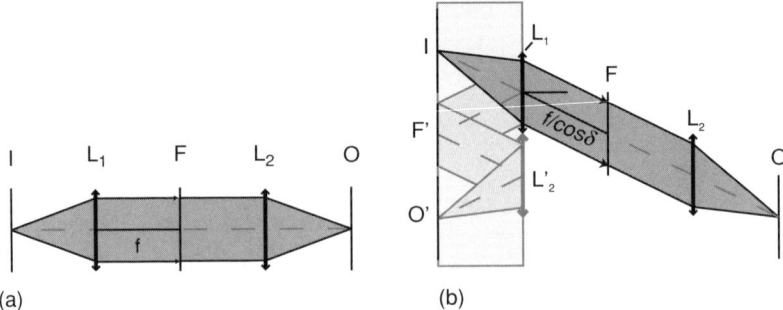

Figure 17.2. 4*f*-imaging systems with (a) untilted and (b) tilted optical axis. By folding the optical axis, one arrives at the PIFSO concept. *I*: input plane; L_1 and L_2: lenses; *F*: filter plane; *O*: output or image plane. The characters with a prime denote the corresponding planes in the folded setup.

investigation of self-imaging in planar optics that, for free-space propagation of signals along an oblique optical axis, aberrations occur that are not present in systems with Cartesian symmetry.[10] Even if these aberrations can be neglected for a specific application, the paraxial behavior of the system depends on the orientation of the axis. Therefore, previous investigations of planar optical systems often contain a derivation of the parabolic approximation, which usually is the starting point for an analysis of conventional optical systems. In other words, compared with conventional optical systems, in which a well-established theory of components and systems can be used, the design of planar optics requires more theoretical efforts. Therefore, a paraxial theory of planar optical systems has been established.[11] The results can be used as tools for the analysis of planar systems, similar to the paraxial theory of conventional optical systems. We will present here briefly the results of these considerations (Section 17.1.2). These results are applicable in general, no matter if the input signal is continuous or discrete. In Section 17.1.2, we will discuss a specific design approach for discrete input signals. For such special situations, one can find suitable ways of engineering the systems design to minimize aberration. A specific approach that has been shown to be very useful in this regard is the concept of "hybrid imaging."[12,13] Hybrid imaging represents a combination of conventional 4*f*-imaging system with "microchannel imaging." It has been shown that this approach allows one to efficiently implement imaging systems for a large number of discrete channels,[14] and that it can also be conveniently applied to the implementation of discrete correlators, as we shall discuss in Section 17.2.

17.1.2.1 Paraxial Theory of PIFSO Imaging Systems

We consider a simple optical imaging system in PIFSO configuration and its unfolded version (Figs. 17.3a and b). In this case paraxial propagation means that we consider a potentially large tilt angle ϑ of the optical axis, but a small

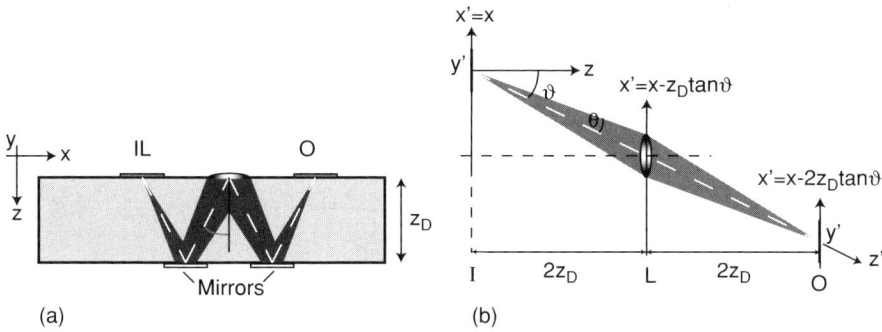

Figure 17.3. (a) Planar Intergrate Free-space Optics (PIFSO) imaging system, here a simple 2f-2f-imaging setup is considered. (b) Unfolded version with a tilt angle ϑ of the optical axis.

numerical aperture represented by the angle θ. The separation between the different planes is denoted as Δz and the focal length of the lens as f where $\Delta z = 2f$.

As in the case of a conventional optical imaging system, the complex amplitude $u(x, y, z + \Delta z)$ in a specific z-plane can be derived by a convolution operation:

$$u(x, y, z + \Delta z) = \int u_0(\xi, \eta, z) h(x - \xi, y - \eta, \Delta z)\, d\xi\, d\eta \qquad (1)$$

Here, $h(x - \xi, y - \eta, \Delta z)$ is the point spread function of the tilted optical system. It can be derived as

$$h(x - \xi, y - \eta, \Delta z) \approx \frac{1}{i\lambda} \frac{\exp\left[i\frac{2\pi}{\lambda}\sqrt{x^2 + y^2 + z^2}\right]}{\sqrt{x^2 + y^2 + z^2}} \cos\vartheta \qquad (2)$$

By using a coordinate transformation ($x' = x - \Delta z \tan\vartheta$ and $y' = y$) and a Taylor's series expansion, one can derive the impulse response for the coordinates centered around the tilted optical axis:

$$h(x', y', \Delta z) \approx \frac{\cos^2\vartheta}{i\lambda\Delta z} \exp\left[i\frac{2\pi}{\lambda}\frac{\Delta z}{\cos\vartheta}\right] \exp\left[\frac{i\pi\cos\vartheta}{\lambda\Delta z}\left(x'^2\cos^2\vartheta - y'^2\right)\right] \qquad (3)$$

With the help of the parabolic approximation of light propagation it is straightforward to calculate the transmission function $L(x', y')$ of an ideal parabolic lens for PIFSO systems. The wave emitted from a point source located in the front focal point of a positive lens is converted into a plane wave propagating parallel to the optical axis z'. Neglecting constant factors, one obtains:

$$h(x', y') L(x', y') = 1 \qquad (4)$$

from which follows:

$$L(x',\,y') \approx \exp\left[-\frac{i\pi\cos\vartheta}{\lambda}\left(x'^2\cos^2\vartheta - y'^2\right)\right].$$ (5)

This expression describes an astigmatic lens which is elongated along the x-axis to compensate for the inherent astigmatism of the propagation along the tilted optical axis. The lines of equal phase represent ellipses with an eccentricity that is given by the factor $\cos^2\vartheta$.

It is of interest to determine the space-bandwidth product (SBP) of a PIFSO system designed with lenses according to Eq. 5. The SBP denotes the number of channels that can be transmitted with a certain image quality, satisfying, for example, the Rayleigh criterion. Two factors have to be considered in order to determine the SBP: the first is the optical performance or, in other words, the aberrations that occur. Second, the 2D layout with elements side by side imposes certain geometrical constraints (Fig. 17.4). After a double pass through the substrate, the lateral offset of the light beam (given as $2z_D\tan\vartheta$, with z_D as the substrate thickness) must be sufficiently large. As can be understood from the Figure, the aperture of the imaging lens, A_x, must not become arbitrarily small, otherwise diffraction blur would become dominant. On the other hand, for increasing tilt angle ϑ, aberrations increase. Hence, there exists an optimal angle for the tilt of the optical axis in order to optimize the SBP, which was shown to be quite large ($> 50°$) in.[11] There, an expression for the SBP of a simple integrated imaging system is also derived:

$$\text{SBP} = \left(\frac{z_D}{\lambda}\sin^2\vartheta\right)^2$$ (6)

This expression represents an upper bound for the number of optical channels that can be transmitted. It is interesting to note, that Eq. 6 is quite similar to the well-known expression that gives the number of modes in an optical multimode waveguide. If we insert numerical values into Eq. 6 (for example with $z_D = 1000\lambda$ and $\vartheta = 30°$), we obtain: SBP $= 500^2$. Such values can actually be

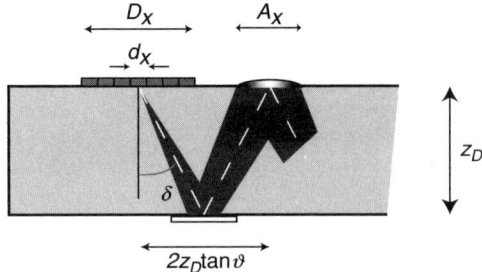

Figure 17.4. Geometrical constraints of PIFSO imaging system.

achieved, as we will show in the next section, which means that PIFSO systems can actually be used for the implementation of practical systems.

17.1.2.2 Imaging Systems for Discrete Objects

The considerations presented in Section 17.1.2 apply to continuous input objects. For certain applications, however, the input objects are discrete. A typical and important case where discrete input signals occur is the area of optical interconnection for datacom and telecommunications. In this case the input objects are, for example, arrays of light sources (laser diodes or light emitting diodes), modulators or optical fibers. Output devices may be detector arrays, modulators, or arrays of waveguides. Since the coupling efficiency of the optical setup is an important parameter, it is necessary that the geometry of the image matches the array geometry of the detecting device array. This has to be warranted for large array dimensions. In other words, optical interconnection tasks often require the distortion-free imaging of a large field of discrete sources.

In a conventional imaging setup (such as discussed above), the field size is related with the magnitude of the aberrations. In particular, for microoptics with its usually small apertures, it is virtually impossible to achieve high resolution without distortion over a large optical field. Therefore, alternative imaging concepts have to be considered. Two of them will be discussed here.

The first straightforward approach to the imaging of discrete objects is shown in Fig. 17.5. The simple idea is to use arrays of microlenses (for collimation at the input and focusing at the output side) so that for every source a separate optical channel is provided. Obviously, the requirements of a large field is fulfilled by extending the array to the desired size. The image is at the

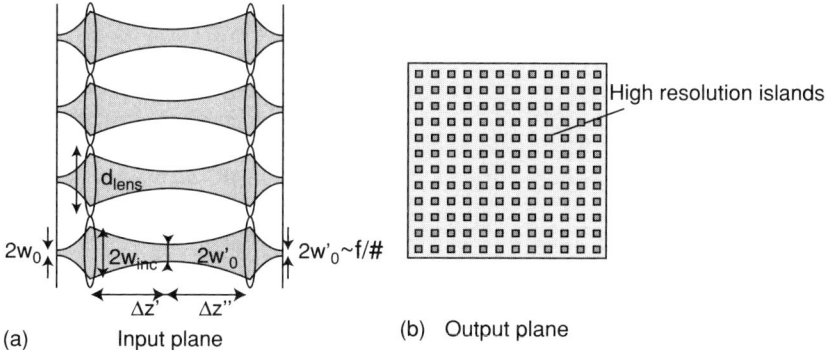

Figure 17.5. Microchannel imaging: (a) optical setup with two microlens arrays. Each data channel consists of an individual imaging system. The beamlets are represented by isophotes, i.e. the $1/e$-boundaries of the assumed Gaussian intensity profile. (b) Input field consisting of individual spots.

same time free of distortion. This concept of "microchannel imaging" has been demonstrated for various situations and using different microoptical technologies.[15–17] One additional interesting feature is that it can also be applied to the implementation of space-variant interconnection patterns.

A specific problem of this approach, however, is that there is a direct trade-off between the transmission distance, Δz, and the pitch, d. The transmission distance is limited by crosstalk, which, in turn, is caused by diffraction at the individual apertures. It is a straightforward calculation to show that the maximum transmission distance is given approximately as

$$\Delta z_{\max} = \frac{d_{\mathrm{lens}}^2}{\lambda} \tag{7}$$

For practical values of $d_{\mathrm{lens}} = 100\,\mu\mathrm{m}$ and $\lambda = 1\,\mu\mathrm{m}$, we obtain: $\Delta z_{\max} = 10\,\mathrm{mm}$.

In order to achieve larger transmission distances and maintain (to some extent) the advantages of microchannel imaging, the concept of "hybrid imaging" has been suggested. The term "hybrid" refers to the combination of conventional imaging (using, for example, a $4f$-system) with microchannel imaging. The basic setup is shown in Figure 17.6.

Here, the imaging task is split up between the lenslets in the two arrays and the central imaging lenses in the $4f$ setup. The lenslets in array A1 precollimate the optical beams, the lenslets of A2 are used to focus to a small spot size. The collimation by the lenslets in A1 will, of course, not be achieved perfectly due to diffraction at the lenslets, however, it reduces the numerical aperture (NA) of the beamlets. The $4f$ setup forms an image of array L1 onto array A2. Note, that this task does not require very high resolution, since the lenslet diameter will, in general, be one order of magnitude larger than the spot diameter at the output. This is one of the main features of the hybrid imaging setup: the reduced NA of the input beams also reduces the requirements to the NA of

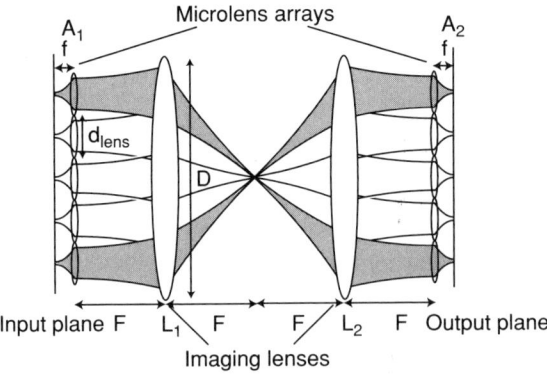

Figure 17.6. Hybrid imaging system: Combination of conventional imaging with microchannel imaging.

the 4f setup. This, in turn, means that aberrations are reduced, too. Hence, it becomes possible to image large fields, larger than with conventional designs. A simple analysis of the properties of the hybrid imaging systems shows that hybrid imaging is particularly advantageous for very dilute arrays (i.e., where the size of the sources is much smaller than the diameter of the lenslets).

The hybrid imaging concept was successfully used in combination with the lens design considerations of Section 17.1.2 for PIFSO imaging systems[14,18] and discrete correlation.[19] The latter work will be described in the next section.

17.2 Optical Correlation with PIFSO Systems

Spatial filtering or correlation is a well-established technique for optical signal processing. Correlation can be performed on a 2D complex optical amplitude distribution $s(x',y')$ in an optical 4f-system (so-called VanderLugt correlator; Fig. 17.7). To this end, the input distribution is Fourier transformed optically to generate the spatial frequency spectrum in the filter or Fourier plane (ν_x, ν_y).[20] In this plane an arbitrary complex filter function $F(\nu_x, \nu_y)$ can be applied through spatial filtering with a complex transmission function. In the output plane $o(x,y)$ of the system the correlation integral between the input distribution $s(x',y')$ and the Fourier transform of the filter function $F(\nu_x, \nu_y)$ is observed:

$$o(x, y) = \int \int s(x', y')f(x - x', y - y')dx'dy' \qquad (8)$$

Thus, if the Filter function is, e.g., "matched" to the spectrum $S(\nu_x, \nu_y)$ of the input distribution, i.e.,

$$F(\nu_x, \nu_y) = S^*(-\nu_x, -\nu_y) \qquad (9)$$

the so-called autocorrelation peak results:

$$\int \int s(x', y')s^*(x' - x, y' - y)dx'dy' = \delta(x, y). \qquad (10)$$

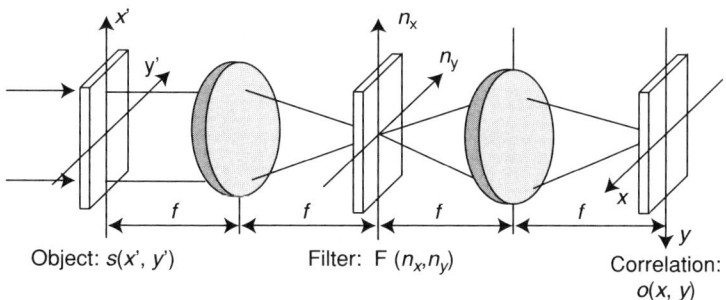

Object: $s(x', y')$ Filter: $F(n_x, n_y)$ Correlation: $o(x, y)$

Figure 17.7. Optical correlation with a 4f setup (VanderLugt correlator).

This simple example illustrates the potential of matched filtering or feature extraction though analog optical spatial filtering. More complex filters can be applied for a variety of applications such as real-time processing of synthetic aperture radar images, associative memories, and neural networks. A large variety of optical systems for similar applications have been demonstrated in laboratory experiments. Due to difficult environmental conditions, such as poor and varying lighting conditions, analog optical image processing techniques are, however, still only used in niche applications. In other application areas such as optical interconnections or homeland security, the environmental conditions can be controlled with relatively high precision. Thus optical correlation approaches become interesting for these areas.

17.2.1 Discrete Correlation for Interconnection

Optical correlation on discrete input/output arrays is an important technique which can be applied, e.g. for address decoding in optical communication systems in context with optical data buses. In principle this is the extension of the concept of code division multiple access to spatially multiplexed data channels.[21] To this end each pulse in the data stream is encoded with a spatially encoded address code. In spatially multiplexed optical communication systems this data code contains the information about the location of the receiver of the particular piece of data. Fast and parallel switching between the various receiver channels can be achieved through optical correlators.

For spatially multiplexed optical communication systems generally arrays of active light sources (such as vertical cavity surface emitting laser diodes (VCSELs)) or VCSEL-based smart pixel arrays are used. Due to their potentially high modulation speed and good signal-to-noise ratio VCSEL arrays have advantages over passive modulator arrays. Since the individual sources of a VCSEL array are mutually incoherent, for applications in optical interconnections we have to deal with mutually incoherent light signals which are to be correlated. The architecture of the VanderLugt correlator can also be used in combination with such mutually incoherent light sources. In this case the output signal is generated as an incoherent superposition of the shifted copies of the filter point spread function:

$$o(x',y') = \sum_{i,j} |s_{i,j}|^2 |f^*(-x-x_i, -y-y_j)|^2 \qquad (11)$$

$|s_{i,j}|^2$ represent the intensities of the individual sources and $f_{i,j} = |f^*(-x-x_i, -y-y_j)|^2$ are the diffraction intensities of the filter function.

17.2.2 Planar Integration of Discrete Optical Correlators

Several implementations of planar integrated joint transform correlators have been suggested in the literature.[22-24] Due to the need for coherent input and

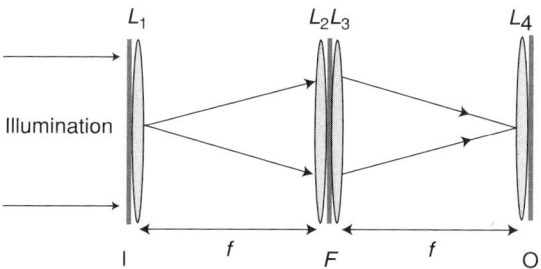

Figure 17.8. Optical system for correlation based on the "light pipe" configuration: L_1, L_4: field lenses; L_2, L_3: transforming lenses; I, O: input/output planes.

output signals these architectures are, however, not useful for application with smart pixel arrays. Compact realizations of discrete optical correlators based on the concept of PIFSO have been demonstrated by Eckert et al.[19,25] For a compact realization it is possible to reduce the length of the correlator system by adopting the so-called "light pipe" configuration (Fig. 17.8). To this end, an additional set of field lenses (L_1 and L_2) is used immediately behind/in front of the input/output planes. For spatially incoherent source arrays, these field lenses, which correct for phase errors, can principally be omitted since the intensities rather than amplitudes of the diffraction orders are correlated (Eq. 11). However, they are also used to improve the uniformity of the illumination of the Fourier transforming lenses.

Due to the symmetry such a configuration can be folded, according to the concept of PIFSO, to result in a very compact systems. Figure 17.9 shows two possible configurations. In Fig. 17.9a) we additionally made use of the fact that the input channels in the smart pixel arrays are not densely packed so that input and output signals can be interlaced. For the correlator system this means that only a small shift between the input and the output plane is necessary for a separation between the input and output. The necessary lateral

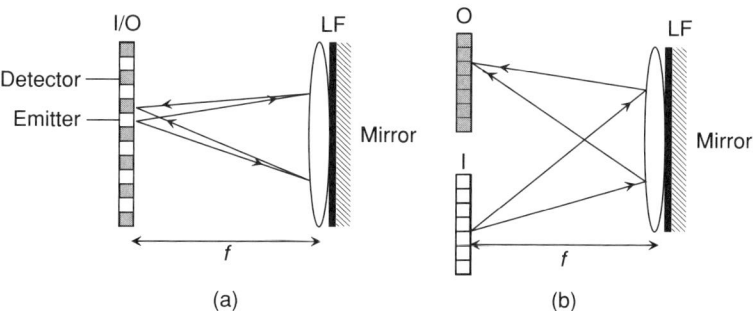

Figure 17.9. Folded versions of the "light pipe" correlator for incoherent input signals: a) on-axis configuration for interlaced I/O signals (and b) off-axis configuration.

shift can be integrated in the filter design. No further optimization of the optical system is necessary. It is, however, also possible to adjust the design in order to achieve full separation between input and output array. In this case the optical system needs to be optimized for good performance along the oblique optical axis as described in Section 17.1.

17.2.3 Experimental Demonstration of Integrated Optical Correlators

17.2.3.1 On-axis Integrated Correlators

Both versions of integrated microoptical correlator systems shown in Fig. 17.9 have been realized and demonstrated experimentally. To this end we exploited the potential of diffractive microoptics, i.e., specifically the possibility, to integrate complex optical functionality in single diffractive elements. Figure 17.10 shows the schematic of the on-axis system which has been fabricated lithographically. In this case a single complex diffractive optical element (DOE) is used to perform the correlation on the discrete smart pixel array aligned at the opposite surface of the transparent substrate. We used a fused silica substrate with a thickness of 6 mm. A comparison with Fig. 17.9 a) shows that the complex DOE combines the functionality of the beam splitter as well as the reflective lenses.

For such an implementation most of the design effort is focussed on the complex DOE.[25] Since it combines the functionality of the diffractive correlation filter as well as the Fourier lenses specific care is necessary in order to avoid errors due to undersampling of the phase profile. The numerical aperture of the lens elements depends on the pixel size and extension of the smart pixel array used for signal input/output. The spatial frequency ν_L required to implement the diffractive lens element with a specific numerical aperture NA is:

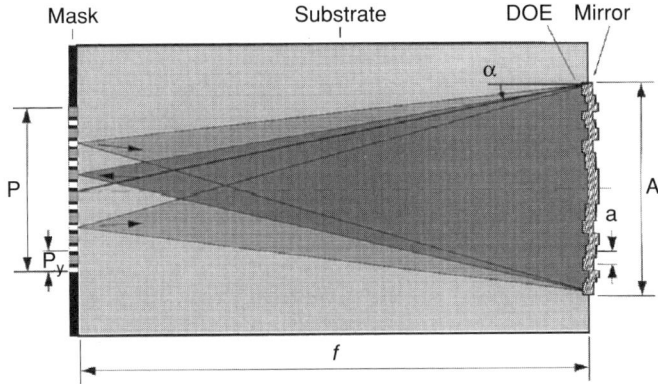

Figure 17.10. Compact realization of the on-axis correlator using a complex diffractive optical.

$$\nu_L = 2\frac{NA}{\lambda n} \tag{12}$$

where n is the refractive index of the substrate at the design wavelength λ.

Since the DOE at the same time performs as correlation filter the complexity is significantly increased compared to the pure lens element. The maximum spatial frequency ν_F of the filter element is determined by the coordinates of the pixels at the edge of the I/O array:

$$\nu_F = \frac{\text{Max}(x_o, y_o)}{\lambda f} \tag{13}$$

An upper bound of the spatial frequency $\nu_{\max}$ which occurs in the complex DOE can thus be found from:

$$\nu_{\max} = \nu_L + \nu_F \tag{14}$$

From the sampling theorem we find an upper bound of the pixel size in the DOE:

$$a \leq \frac{1}{2\nu_{\max}} \tag{15}$$

Due to the Gaussian illumination of the DOE the periods at the edge of the lens element are illuminated with significantly less intensity and contribute less to the optical power of the lens. The total efficiency as well as noise and cross talk in the system is therefore reduced. Thus, the introduction of an undersampling factor s is justified[25] so that the minimum feature size in the complex DOE can be calculated from:

$$a \leq \frac{1}{2s\,\nu_{\max}} \tag{16}$$

For the experimental demonstrations an I/O plane with 8×8 pixels and a pixel pitch of $p_x = p_y = 65\,\mu\text{m}$ has been assumed. The correlation filter was designed to generate a 5×5 pattern using a modified iterative algorithm according to Gerchberg-Saxton[26,27]. For the experimental realization one period of the filter function, consisting of 32×32 pixels was replicated to form a DOE with 640×640 pixels and a pixel size of $1.25\,\mu\text{m}$. The phase profile of this quantized DOE was added to the phase profile of a quantized and pixelated lens element of equal extension in order to form the phase profile of the complex DOE for the integrated correlator system. Figure 17.11 a) shows a grey level image of the resulting element. A value of 66% was theoretically calculated for the diffraction efficiency with a noise level of less than 0.5%. In the experiment this value could not be reached due to significant fabrication problems. The measured value for the diffraction efficiency was 47%. The complex DOE was fabricated on a fused silica glass substrate (thickness 6 mm) to form a compact integrated correlator system. The correlation experiments show decent performance of the integrated correlator system[25].

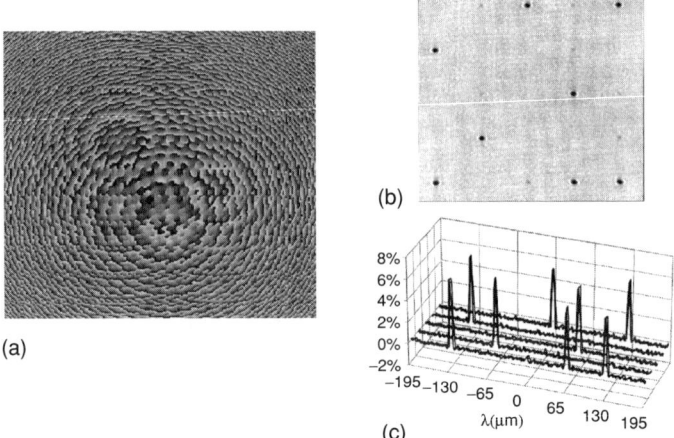

Figure 17.11. a) Grey level representation of the complex diffractive optical element consisting of 640 × 640 pixels. The filter in the functionality of the Fourier lenses as well as the correlation filter; b) point spread function of the integrated optical correlator: CCD image (inverted and overexposed to show the noise orders); and c) line scans through the intensities showing the good signal-to-noise ratio.

17.2.3.2 Off-axis Integrated Correlators

Two different architectures of VanderLugt correlators integrated in PIFSO have been demonstrated. Figure 17.12 shows the schematic of an integrated correlator system based on the light pipe configuration shown in Fig. 17.8. Such a system has been fabricated lithographically in a quartz substrate (thickness: $t = 8\,\mathrm{mm}$). In order to avoid alignment errors all alignment sensitive elements are fabricated on the same surface of the substrate. Due to the oblique optical axis the system needs to be specifically corrected for aberrations. To this end,

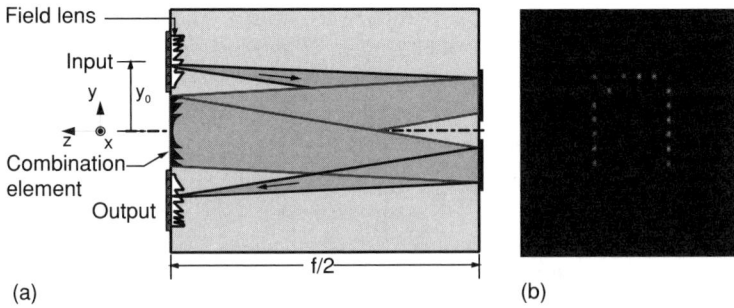

Figure 17.12. (a) Schematic of an integrated correlator in the off-axis configuration and (b) psf of the system.

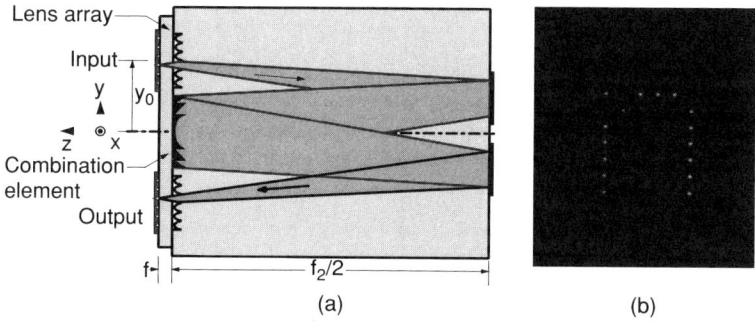

(a)   (b)

Figure 17.13. (a) Schematic of an hybrid integrated correlator using microlens array to achieve tight focussing of the diffraction orders and (b) psf of the system showing the good contrast and quality of the focus spots.

the focal length of the lens elements was corrected by a factor of $\cos\alpha$ and $\cos^3\alpha$ in x- and y-direction, respectively[11,23], where α is the tilt angle with respect to the surface normal. In this case a relatively small angle of $\alpha = 2.93°$ was sufficient to avoid an overlap of the I/O planes after a zig-zag propagation. The diffractive combination element (CE) again combines the functionality of the Fourier lenses and the correlation filter in one single element. In our experiment it consists of 819×817 pixels with approximately $(1.42\,\mu m)^2$. Due to the astigmatic correction, however, neither the individual pixels nor the overall element are exactly quadratic. Six periods of a periodic diffractive optical element (96 pixels/period), which was designed to generate the correlation filter, were superposed with the diffractive lens structure (NA ≈ 0.038) to form the CE. Figure 17.12 shows the point-spread function of the integrated correlator. Good contrast and low background noise indicate the good quality of the correlator system. In order to improve the quality of the focussing of the individual diffraction spots the correlator system can be integrated with microlens arrays. The configuration as well as the psf of such a correlator following the idea of hybrid imaging is shown in Fig. 17.13.

17.3 Planar Optically Integrated Correlators for Security Applications

Security applications are an important application area for which optical correlators can be extremely useful. Similar to the application in optical interconnects well-defined illumination conditions are helpful for the potential application of optical correlator systems in this field. Since the introduction on the MasterCard[TM] in the early 1970s optical security features have had enormous success. Today protection holograms, kinegrams[TM], or diffractive identification devices can be found on a huge variety of products, from banknotes to entrance tickets etc.[28] Due to the proliferation of the know-how and technology necessary for the fabrication of those devices the level of protection is constantly decreasing. In order to keep a high level of security it will therefore be necessary to encode and verify increasingly complex information in such devices. For example serial numbers or also biometric data could be encoded in the optical security features. Optical correlators can then become very useful for readout and verification of this information. In this case a potential integration of the optical systems will be crucial for the successful application in real-world systems.

The schematic configuration of an integrated optical correlator for verification, e.g., of kinegrams[TM] is shown in Figure 17.14 a).[29] The optical system represents a folded $4f$ configuration. A plane illumination wave is coupled into the system and after the first zig-zag propagation is modulated by a phase grating representing the decryption key. In the configuration shown in the figure this phase grating is imaged via a folded $4f$ imaging system onto the kinegram[TM] to be verified. After an additional Fourier transformation the correlation between the kinegram[TM] pattern and the decryption key appears in the output plane. Thus, in this configuration the correlation is performed in the spatial domain and is consequently sensitive to positioning errors. The same configuration could conceptually be implemented for correlation in the Fourier domain, which would result in shift invariant correlation. Due to the high level of integration a variety of correlator systems could in principle be implemented simultaneously without additional fabrication cost. Such

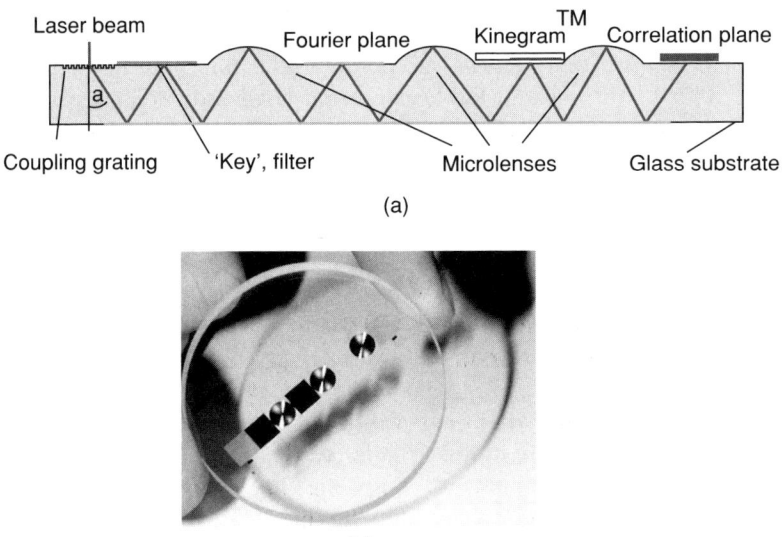

(a)

(b)

Figure 17.14. (a) Schematic of the planar integrated optical correlator for kinegram[TM] verification; and (b) Photograph of the system fabricated lithographically.

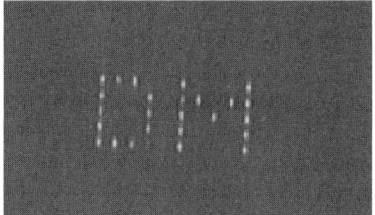
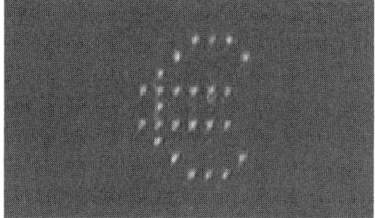

Figure 17.15. Diffraction patterns of the optical phase elements used for the demonstration experiment: (a) diffraction pattern of the integrated decryption key and the "real" kinegram^{TM} and (b) diffraction pattern of the forged kinegram^{TM}.

multiplexed correlation systems could, e.g., be useful in order to reduce problems with the system's sensitivity to relative rotations between the decryption filter and the kinegram.

Figure 17.14 b) shows the planar optical realization of the system described above. All optical components have been fabricated lithographically as multiple phase level diffractive optical elements. The coherent plane wave is coupled into the planar substract (thickness: 12 mm) through a diffraction grating (binary; period: 2.13 μm) to propagate inside the substrate at an angle of 11.77°. Along the zig-zag path between the reflection-coated substrat surfaces the wavefront is modulated by the decryption key and imaged by the diffractive lenses onto the kinegram^{TM} aligned above the area between lenses 2 and 3 where no reflection coating is applied. The Fourier transforming lenses are implemented as 4 phase level diffractive lenses and optimized for imaging along the oblique optical axis as described previously. The decryption key for our demonstration experiments was formed by a 4 phase level DOE with a period of 330 μm and a pixel size of app. 10 μm. The grating was optimized to generate the "DM" sign shown in Fig. 17.15a. A second DOE with the EURO sign (Fig. 17.16b) as diffraction pattern was used as forged kinegram for our experiment. Figure 17.16 shows the result of our correlation experiments. When the "real" kinegram, i.e., the "DM" DOE, was aligned above the integrated system, a strong autocorrelation peak was measured in the output plane (Fig. 17.16b). This peak shows intensity values seven times as high as the maximum peaks in cross correlation which appears in the output if the wrong DOE is aligned above the system (Fig. 17.16a). Thus, a verification of the kinegram can be performed by a simple thresholding operation. This clearly demonstrates the potential of those planar integrated systems for kinegram verification.

17.4 Generalized Phase Contrast Method for Phase-only Optical Cryptography

Cryptography entails recording or transmission of concealed information where only the application of a correct key enables the comprehension of the original

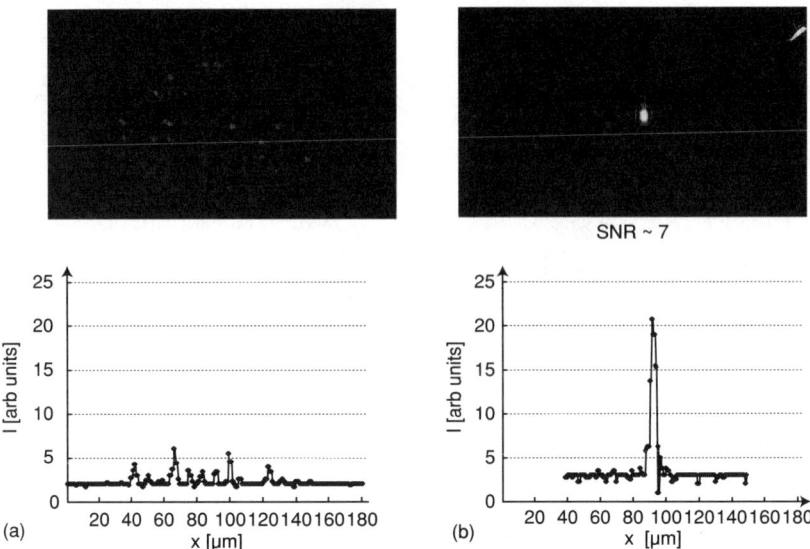

Figure 17.16. Experimental results of the correlation experiment: a) intensity distribution and line scan of the crosscorrelation; b) results of autocorrelation.

information. The art of cryptography dates back to ancient times where secret information was transmitted in terms of symbols and sketches. Through the years, cryptography has evolved and the medium by which it has been implemented changed depending on the state of science of a particular era. In the Renaissance period, secured information was believed to be transmitted by incorporating hidden symbolisms in the paintings of famous painters. Now, the current state of technology rests on specialized electronic data processing machines and computers.

Rapid technological advancements have exploited light as a medium for data transport and storage. Using light to encode digital information has proven as a highly efficient technology that has radically revolutionized modern-day data communications. Hence, it is a natural course to incorporate optical cryptographic techniques into contemporary optical data communication and storage systems. Moreover, as pointed out in the previous section, optical cryptographic methods can be future solutions to problems related to intellectual property protection, product authentication, falsified bankcards, identification cards, and other similar predicaments.

Cryptographic techniques based on the use of light exploit the coherent nature of a laser beam. These techniques have been shown to yield efficiently ciphered information in addition to fast decryption via parallel optical processing.[30] Javidi et al. have proposed various optical cryptographic schemes involving the use of phase masks for: (1) encrypting amplitude information based on the double-phase encoding scheme[31,32]; (2) encryption of phase-encoded information[33]; and (3) holographic storage of encrypted information.[34] These schemes require the recording of encrypted masks containing both amplitude and phase information.

Optical cryptography can also be achieved by operating on a single lossless parameter that allows for full optical reversibility: the phase[35] or polarization[36–38] of a coherent light carrier. Phase-only cryptography is based on the direct superposition of a phase mask containing the original data and an encrypting phase key and vice versa.[39–41] This encryption process also implies that all operating light fields in general have at least a full 2π-phase cycle of modulation. Since the optical phase is undetectable by the eye or by standard light-capturing devices, an encrypted phase array is invisible in addition to its incomprehensible format. Upon decryption, visualizing an invisibly decrypted field can be achieved by an efficient conversion of the field into a high-contrast intensity image. The phase contrast technique proposed by Nobel Laureate Frits Zernike[42] can only view phase images correctly which have less than $\pi/3$ phase modulation. Since a decrypted phase can have a much larger phase stroke, a phase-only cryptographic approach relies heavily on an optimized visualization of the decrypted phase information. This can be accomplished by the Generalized Phase Contrast (GPC) method,[43–45] which allows for the full visualization of phase patterns with large phase strokes in addition to making full utilization of all of the light power within a designated area.

This section describes the fundamentals of phase-only optical cryptography and the visualization of decrypted information using the GPC method. This section also explores the feasibility of implementing the GPC method in a miniaturized device by using planar-integrated free-space optics (PIFSO).[46,47] Such miniaturized and robust GPC systems are significantly better suited for real-world applications in optical cryptography and even provide a direct interface to micro-opto–electro-mechanical devices.

17.4.1 Phase-only Optical Cryptography

Phase-only optical cryptography entails the encryption of a 2D phase distribution $O(x, y)$ of an original amplitude image, $o(x, y)$, with a random phase pattern, $R(x, y)$ to yield an encrypted field, $E(x, y)$, given by:

$$E(x, y) = O(x, y)R(x, y) = \exp\left[i2\pi(o(x, y) + r(x, y))\right], \qquad (17)$$

where $o(x, y)$ and $r(x, y)$ are 2D matrices with element values normalized within the interval $[0; 1]$. To decrypt the field $O(x, y)$, $E(x, y) = \exp\left[i2\pi e(x, y)\right]$ is applied with a decrypting key generated using the complex conjugate of the encrypting phase, $R^*(x, y)$, given by:

$$O(x, y) = E(x, y)R^*(x, y) = \exp\left[i2\pi(e(x, y) - r(x, y))\right]. \qquad (18)$$

Figure 17.17 illustrates the encryption and decryption procedure showing the patterns of $O(x, y)$, $R(x, y)$, $E(x, y)$, and $R^*(x, y)$. The matrices in this case

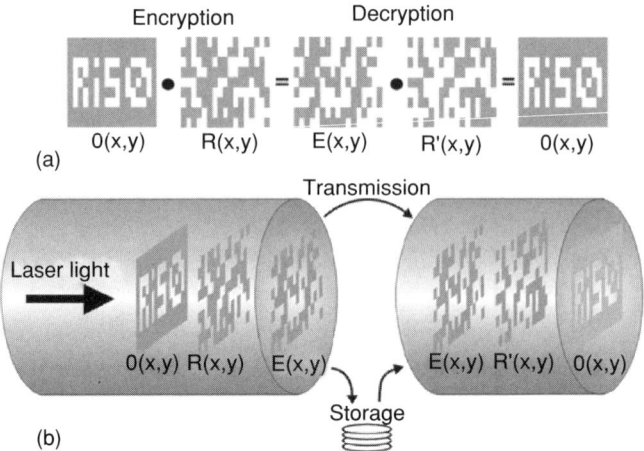

Figure 17.17. Phase-only optical cryptography.

represent phase-encoded information which is visually incomprehensible. However, for clarity of the illustration, in Fig. 17.17, the fields are shown as grey level patterns. The grey and white pixels indicate the relative phase shifts rather than the amplitudes of the light fields.

The geometry of the system for implementing this phase-only optical encryption is shown in Fig. 17.17b. A plane-polarized wavefront is incident on the original phase image to set the field $O(x, y)$. Aligning the field $O(x, y)$ with $R(x, y)$ results in the encrypted field $E(x, y)$. The encrypted information can, in principle, be transmitted or stored optically into appropriate recording devices. Phase-only optical decryption is achieved by aligning the decrypting key, $R^*(x, y)$, to reproduce the field $O(x, y)$, which is maintained to have phase-only modulation and is therefore indiscernible to the naked eye.

17.4.2 The Generalized Phase Contrast (GPC) Method

The GPC method plays a vital role in phase-only optical cryptography as it is used to visualize the phase encoded fields. The fundamental concept of the GPC method can be traced back to Zernike's phase contrast method, which describes the conversion and direct mapping of a weak incident phase perturbation into an intensity distribution with good contrast. This is achieved by the interference of a phase shifted on-axis low-frequency component with directly transmitted spatial varying terms essentially containing the spatial phase information. In Zernike's approach, the incident field with a weak phase distribution can be represented by a truncated Taylor expansion. In this case the low-frequency term is assumed to be constant and real valued, while the spatially varying terms are approximated by the second term of the Taylor expansion. Due to the unbalanced interference of low- and high-spatial frequency terms, this approach does not provide optimized output intensities of arbitrary phase distributions.

The GPC method resolves this limitation by using a more elaborate analytic model of the process. This enhanced approach allows an analytic determination of the exact working parameters where any variation in the incident phase, weak or strong, yields visible information by optimized light throughput.

The GPC method is based on the spatial filtering architecture illustrated in Fig. 17.18. It consists of a 4f-imaging setup with a phase contrast filter (PCF) in the Fourier plane between the two lenses. The output intensity distribution is produced from the input phase pattern by the on-axis phase-filtering operation in the spatial frequency domain. The first lens performs a spatial Fourier transform, so that directly propagating light is focused onto this onaxis filtering region whereas spatially varying phase information generates light scattered to locations outside this central region. By application of the phase filter, the input phase distribution is converted into an intensity pattern in the back focal plane of the second lens. The GPC method works best in cases where there is a large separation between the on-axis, low spatial frequency light, and the higher spatial frequencies in the Fourier plane. Moreover, if the spatial average value of the incident phase-modulated wavefront is carefully matched to the relative phase shift of the PCF, an essentially lossless phase-to-intensity mapping of the input light can be achieved. To understand the phase-to-intensity conversion via the GPC method, we analyze the relationship between the decrypted phase values and the output intensity. The expression for the intensity at the output $I(x', y')$ of the optical setup shown in Fig. 17.18 can be expressed as[45].

$$I(x', y') = |e^{i\tilde{\phi}(x', y')}\mathrm{circ}(\frac{r'}{\Delta r}) + g(r')|\bar{\alpha}|(e^{i\theta} - 1)|^2 \tag{19}$$

with

$$\bar{\alpha} = A^{-1} \int\!\!\int_A e^{i\phi(x,y)}\mathrm{d}x\mathrm{d}y = |\alpha|e^{i\phi_{\bar{\alpha}}}$$

$$\tilde{\phi}(x', y') = \phi(x, y) - \phi_{\bar{\alpha}}$$
$$r' = \sqrt{x'^2 + y'^2}$$

$$\tag{20}$$

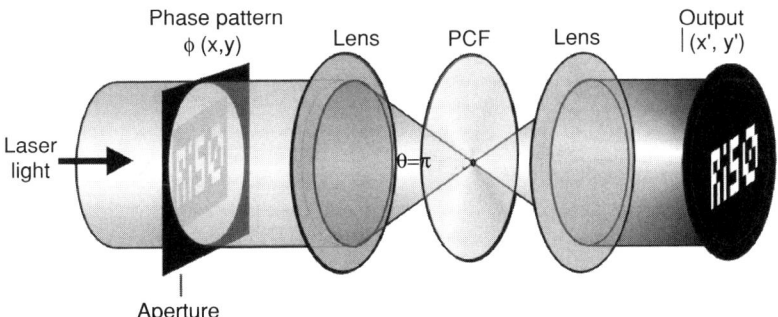

Figure 17.18. Phase-only optical cryptography.

The object dependent term, $\bar{\alpha}$, is the spatial average of the input wavefront with absolute value $|\bar{\alpha}|$ and phase $\phi_{\bar{\alpha}}$. While in Zernike's approach this term is equal to one, it is described as a complex term $|\bar{\alpha}|e^{i\phi_{\bar{\alpha}}}$ in the framework of the GPC analysis. The input phase pattern, $\phi(x,\ y)$, corresponding to the decrypted phase is here truncated with a circular aperture area $A = \pi(\Delta r)^2$. The intensity at the output also depends on the zero-order phase shift θ of the filter and the spatial profile $g(r')$ of the synthetic reference wave (SRW). This SRW is generated by diffraction at the phase-step in the on-axis region of the filter. The interference between the on-axis term carrying the filter parameters and the unfiltered phase information generates the intensity distribution at the output described by Eq. 19.

The output of the GPC system is strongly influenced by the parameter $\eta = \frac{R_1}{R_2}$ which relates the radius of the PCF ($R1$) to the radius of the mainlobe of the Airy function ($R2 = 0.61\frac{\lambda f}{\Delta r}$). $R2$ results from the spatial Fourier transform, by the first lens with focal length f, of an incident monochromatic light source with wavelength λ truncated with a circular input aperture of radius Δr. For η-values smaller than 0.63 (and operating within the central region of the image plane), the higher order spatial terms are insignificant and the SRW can be expressed[45] as:

$$K = g(r' = 0) = 1 - J_0(1.22\pi\eta). \tag{21}$$

Hence, Eq. 19 can further be simplified to give:

$$I(x',\ y') = |e^{i\bar{\phi}(x',\ y')} + K|\alpha|(e^{i\theta} - 1)|^2. \tag{22}$$

For phase visualization of decrypted phase patterns which are not assured to be regular, fine tuning of η in the region $[0.4 - 0.63]$ and correspondingly K within $[0.5–1]$ provides an efficient operating regime with minimum loss. Moreover, setting $\theta = \phi = \pi$ yields optimized and a nearly loss-less conversion of an input binary phase to its corresponding intensity pattern at the output.

17.4.3 Miniaturization of the GPC Method Via Planar Integrated Microoptics

We applied the concept of PIFSO for the miniaturization and integration of the optical spatial filtering system necessary to implement GPC imaging according to Fig. 17.18. The 4f-imaging setup is implemented with two diffractive microlenses ($L1$ and $L2$) aligned on the surfaces of a glass substrate (thickness: 12 mm) as shown in Fig. 17.19 (top view). Figure 17.19 (side view) shows the folded beam path of the integrated system. A perpendicularly incident wavefront is coupled into the substrate through a binary phase grating. The coupling gratings are fabricated with a period of 2.13 μm resulting in a deflection angle of 11.77° inside the substrate for the design wavelength $\lambda = 633$ nm. The distance from the coupling grating to $L1$ is equivalent to the focal length indicating that the object plane is located at the surface of the input grating.

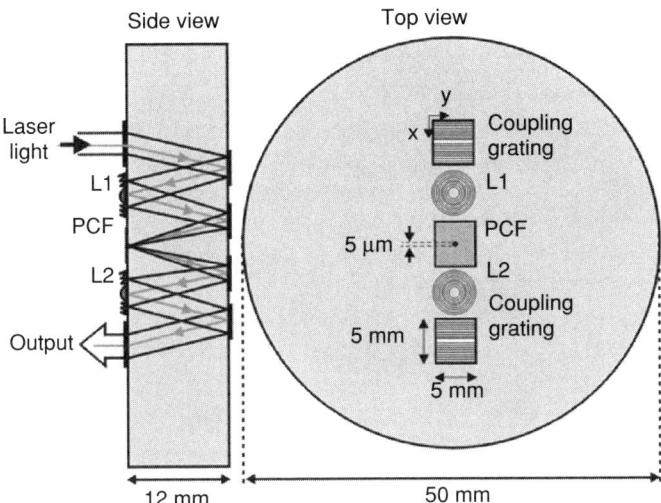

Figure 17.19. Miniaturized GPC system.

$L1$ focuses the beam to the spatial Fourier plane where a reflection coated PCF is fabricated on the substrate to perform a π-phase shift of the on-axis region of the focused light. The filter is designed for operation at $\lambda = 633$ nm and is etched as a hole with radius $R_1 = 2.5\,\mu$m into the substrate. Figure 17.20 shows an atomic force microscope image of the PCF. An anisotropic etching process is used to form a steep-edged cylindrical hole. After the PCF, the reverse Fourier transform is performed in the second half of the symmetric system. For optimized imaging behaviour along the tilted optical axis, the microlenses are slightly elliptic in shape and have different focal lengths ($f_x = 25.58$ mm and $f_y = 24.51$ mm) in the two perpendicular lateral directions with f-numbers $f/ \approx 5$.

The diffractive implementation of the integrated GPC imaging systems causes some practical problems. The quality of finite aperture diffractive optics affects the resolution as well as the size of the image field. The use of DOEs, although favored for practical reasons, such as compactness and compatibility with standard photolithography techniques, influences the efficiency and spatial resolution of the optical system because of discrete phase quantization. Furthermore, undesired higher diffraction orders can cause disturbing interference. However, for phase-only cryptography, the relevance of these undesired factors, such as low-light throughput, is not so high. Phase-only cryptography basically requires efficient visualization of the decrypted spatial phase modulation. Hence, a virtually lossless light propagation provided by the GPC method compensates for the inherently low throughput of planar-integrated microoptical devices.

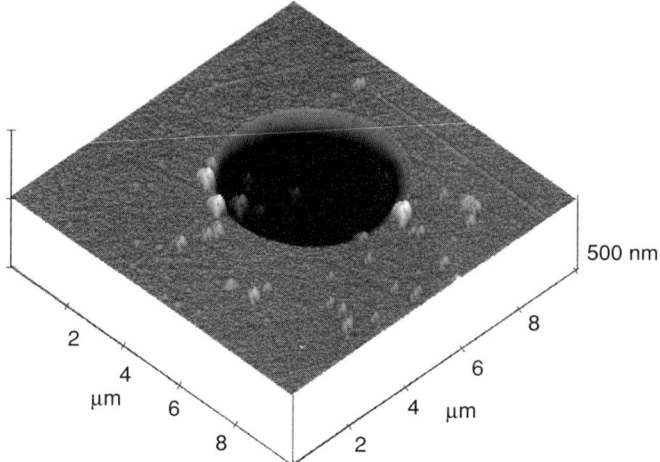

Figure 17.20. Phase contrast filter fabricated on an integrated planar-optical device.

17.4.4 Miniaturized GPC System for Phase-only Optical Decryption

The performance of the planar-optically integrated GPC (PO–GPC) system for phase-only optical decryption is tested with the experimental system shown in Fig. 17.21. The external macrooptical system is composed of three imaging steps. In the first step the decrypting key is imaged onto the phase mask with the information to be encrypted. The second imaging step is for coupling the decrypted field to the input grating of the integrated system. The third imaging step after the GPC, is for magnifying the intensity distribution to the output to the CCD camera. The decrypting key information is encoded using a phase-

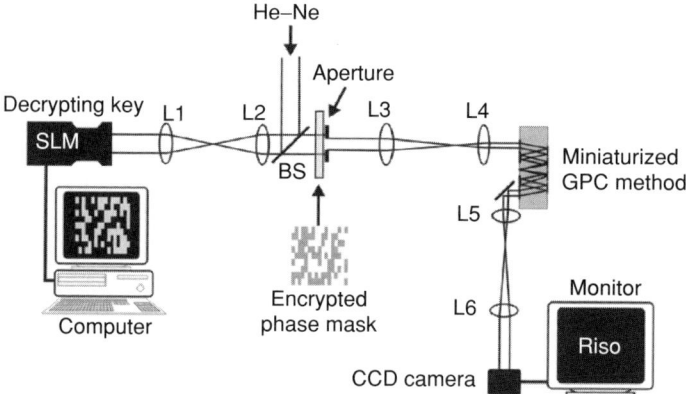

Figure 17.21. System for testing phase-only optical decryption using the miniaturized GPC method.

only spatial light modulator (SLM), which is illuminated by an expanded He–Ne laser beam ($\lambda = 633$ nm). The SLM is a parallel-aligned nematic liquid crystal type (Hamamatsu Photonics), which can modulate a phase range of at least 2π at $\lambda = 633$ nm. The SLM is optically addressed by an XGA-resolution (768×768 pixels) liquid crystal projector element and controlled from the video output of a computer. Lenses $L1$ and $L2$ project the phase image of the decrypting key to the encrypted phase pattern facilitating phase-only optical decryption. The decrypted phase pattern is scaled and directed to the PO–GPC system via lenses $L3$ and $L4$. The truncating circular aperture placed just after the decrypting key governs the central spot size of the beam at the filtering region at the Fourier plane.[45] Lenses $L5$ and $L6$ project the image of the resulting intensity distribution at the output of the PO–GPC system to the CCD camera. The encrypted phase mask is fabricated as a photoresist pattern on an optical flat where the phase-shifting pixels of 0 and π constitute the 17×9-pixel 2D phase key. The size of each pixel is approximately $176 \times 333\,\mu$m.

Figure 17.22 shows the successful decryption of a 17×9 pixel phase pattern using the setup shown in Fig. 17.21. The embedded information consists of 4 5×3 pixel letters depicting the word "RIS." The decrypting key encoded at the SLM is imaged through the macro-optical setup (lenses $L1$ to $L4$) and visualized using the PO–GPC system as a high contrast intensity pattern as shown in Fig. 17.22a. Inserting the phase mask with the encrypted information after the first 4f-lens setup ($L1$ and $L2$) while the SLM is turned off (to set no phase modulation of the decrypting key) visualizes the encrypted information as a contrasted intensity pattern at the output as shown in Fig. 17.22b. As the SLM is operated to encode the decrypting key, Fig. 17.22c shows the high-contrast image resulting from the phase-only decryption and subsequent visualization of the originally phase-encrypted information. It is important to note that the high diffraction orders caused by the binary coupling gratings[46] do not affect the intensity pattern in the field of view of the output. The use of a larger PCF on the planar-integrated microoptics, results in better contrast images, however, with a slightly smaller input aperture diameter. The details of this optimization have been discussed in a previous work.[47] The low quality of

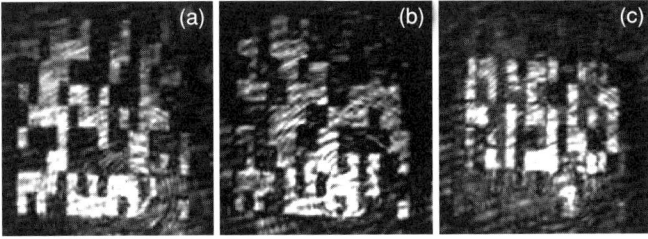

Figure 17.22. Performance of the phase-only optical decryption using the miniaturized GPC method.

visualization is due to tilt and alignment errors for both the encrypted and the key patterns. A slight tilt of the decrypting phase mask will result in uneven phase visualization as shown in Fig. 17.22b. Such error propagates through the decryption process results in poor visualization of some of the pixels in Fig. 17.22c.

It should be noted that the pixels of the phase mask are relatively large and cannot truly be used as basis for demonstrating the imaging limitations of the integrated system. The imaging performance of the planar-optical device can resolve feature sizes smaller than $10\,\mu$m.[29] Theoretically, the resolution of the PO–GPC system can be estimated depending on the operating numerical aperture (NA ≈ 0.28) to resolve features as small as $2.3\,\mu$m. Considering aberrations, it is safe to assume that the miniaturized GPC system can handle decryption and visualization of a phase-encrypted information with 300×300 pixels having pixel sizes of $5\,\mu$m. This assumption, however, only covers the imaging performance of the PO–GPC system. A further limitation on the number of pixels can be attributed to the current state of technology of SLMs.

The current "proof-of-principle" experimental setup consists of an external macro-optical setup necessary to scale both the encrypted and the key patterns to the appropriate sizes for imaging using the PO–GPC system. To include the entire optical setup into a fully integrated microoptical system, both the encrypted and the key patterns have to be fabricated at the appropriate scale. Implementing the system into a fully integrated planar-optical device will ease up alignment and tilt problems, which are common difficulties encountered when using discrete optical components.

Figure 17.23 shows the intended implementation of the whole opto-electronic setup in PIFSO using a two-stage 4f-lens setup. An image of a phase-encrypted pattern is projected on the decrypting phase key pattern using the first imaging step. An encrypted phase-only pattern recorded on a bankcard, a passport, a currency note, etc., can be instantly verified for authenticity by subjecting it to this planar integrated system. The phase-only key can be

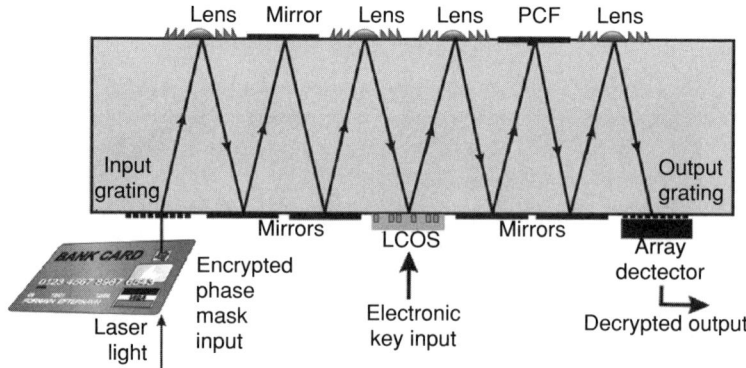

Figure 17.23. Fully integrated phase-only optical decrypting system.

dynamically encoded on a compact, electronically controlled Liquid Crystal on Silicon (LCOS) SLM. Moreover, the functionalities of the encrypted pattern and the key pattern can be interchanged such that the phase key can be written on a static phase mask to decrypt a dynamic stream of opto-electronic data. The successful decryption of the encrypted data is only achieved when both the encrypted pattern and the key pattern are correctly aligned. An initial calibration cycle by a nonmechanical alignment of the two phase-only patterns can be achieved by an automated electronic scrolling of the pattern encoded on the LCOS–SLM. The decrypted phase data is then converted into an intensity pattern using the GPC method via the second 4f-setup with a PCF at the Fourier plane. The intensity pattern at the output can subsequently be recorded using a detector array and then transformed to a second medium of transmission.

17.5 Conclusions

In this chapter we focussed on the potential of microoptically integrated systems for security applications. Adopting the concept of planar-integrated free-space optics we demonstrated a variety of systems architectures of optical correlators which are usefull for this area of applications. The experiments with microoptical systems clearly show that optical systems for applications in homeland security can be miniaturized and integrated to extremely compact and rugged devices with state-of-the-art microoptics and lithographic technology. Especially the planar interfaces of PIFSO offer the chance for hybrid integration of a variety of devices such as opto-electronics or micromechanics. Thus, in combination with further technological improvement, the door seems to be open for real-world applications.

References

[1] Sinzinger S, Jahns J. (2003). *Microoptics*, 2nd edn. Wiley-VCH, Weinheim (2003).
[2] Iga K, Oikawa M, Misawa S, Banno J, and Kokubun Y. (1982). *Appl. Opt.*, **21**:3456.
[3] Jahns J and Huang A (1989). *Appl. Opt.*, **28**:1602.
[4] Wu MC (1997). *Proc. IEEE* **85**:1833.
[5] Gruber M, ElJoudi E, Sinzinger S, and Jahns J (2001). *Appl. Opt.* **40**:2902.
[6] Jahns J, Lee YH, Burrus CA, Jewell JL (1992). *Appl. Opt.*, **31**:592.
[7] Acklin B and Jahns J (1994). *Appl. Opt.*, **33**:1391.
[8] Gimkiewicz Ch. and Jahns J (1997). *Proc. SPIE*, **3226**:56.
[9] Gruber M and Jahns J (2004). Planar-integrated free-space optics: From components to systems. in: *Microoptics: From Technology to Applications*, Jahns J, Brenner K.-H. Springer, New York.
[10] Testorf M, Jahns J, Khilo NA and Goncharenko AM (1996). *Opt. Comm.* eds. **129**:167.
[11] Testorf M and Jahns J (1997). *J. Opt. Soc. Am. A.*, **14**:1569.

[12] Lohmann AW (1991). *Opt. Comm.*, **86**:365.
[13] Jahns J, Sauer F, Tell B, Brown-Goebeler KF, Feldblum AY, Nijander CR and Townsend WP (1994). *Opt. Comm.*, **109**:328.
[14] Sinzinger S and Jahns J (1997). *Appl. Opt.*, **36**:4729.
[15] Jahns J and Daschner W (1990). *Opt. Comm.*, **79**:407.
[16] Hutley MC, Savander P and Schrader M (1992). *Pure Appl. Opt.*, **1**:337.
[17] McCormick FB (1993). Free-space optical interconnection systems. in: *Photonics in Switching, vol. II*, ed. Midwinter JE Academic, Boston.
[18] Jahns J and Acklin B (1993). *Opt. Lett.*, **18**:1594.
[19] Eckert W, Arrizón V, Sinzinger S and Jahns J (2000). *Opt. Comm.*, **186**:83.
[20] Vander Lugt AB (1992). *Optical Signal Processing*. Wiley, New York.
[21] Weiner AW and Salehi JA (1993). Optical Code-Division Multiple Access. in: *Photonics in Switching II*, ed. Midwinter JE Academ, Boston.
[22] Gosh AK, Lapis MB and Aossey D (1991). *Electron. Lett.*, **27**:871.
[23] Reinhorn S, Amitai Y, and Friesem AA (1997). *Opt. Lett.*, **22**:925.
[24] Song SH, Jeong J.-S, Park S and Lee E.-H (1997). *Opt. Comm.*, **143**:287.
[25] Eckert W, Arrizón V, Sinzinger S and Jahns J (2000). *Appl. Opt.*, **39**:759.
[26] Gerchberg RW and Saxton WO (1972). *Optik*, **35**:237.
[27] Arrizón V and Testorf M (2000). *J. Opt. Soc. Am. A.*, **17**:2157.
[28] van Renesse RL (1994). *Optical document security*. Artech House, Norwood.
[29] Sinzinger S (2002). *Opt. Comm.*, **209**:69.
[30] Javidi B (1997). *Phys. Today*, **50**(3):27.
[31] Javidi B and Horner J (1994). *Opt. Eng.*, **33**:1752.
[32] Refregier P and Javidi B (1995). *Opt. Lett.*, **20**:767.
[33] Towghi N, Javidi B and Lou Z (1999). *J. Opt. Soc. Am. A.*, **16**:1915.
[34] Javidi B and Nomura T (2000). *Opt. Lett.*, **25**:28.
[35] Glückstad J (1998). Phase contrast scrambling. Int. PCT pat. WO 002339A1.
[36] Mogensen PC and Glückstad J (2000). *Opt. Commun.*, **173**:177.
[37] Mogensen PC, Eriksen RL and Glückstad J (2001). *J. Optics A.*, **3**:10.
[38] Eriksen RL, Mogensen PC and Glückstad J (2001). *Opt. Commun.*, **187**:325.
[39] Glückstad J (1999). Image decrypting common path interferometer. in: *Optical Pattern recognition X*, eds. Casasent DP, Chao T. eds. (Proc. SPIE, **3715**:1999).
[40] Mogensen PC and Glückstad J (2000). *Opt. Lett.*, **25**:566.
[41] Mogensen PC and Glückstad J (2001). *Appl. Opt.*, **40**:1226.
[42] Zernike F (1955). *Science*, **121**:345.
[43] Glückstad J (1996). *Opt. Commun.*, **130**:225.
[44] Glückstad J (2000). Phase contrast imaging. U.S. patent 6,011,874.
[45] Glückstad J and Mogensen PC (2001). *Appl. Opt.*, **40**:268.
[46] Daria V, Glückstad J, Mogensen PC, Eriksen RL, and Sinzinger S (2002). *Opt. Lett.*, **27**:945.
[47] Daria V, Eriksen RL, Sinzinger S and Glückstad J (2003). *J. Opt. A: Pure Appl. Opt.*, **5**:211.
[48] Glückstad J, Daria VR and Rodrigo PJ (2003). *Opt. Lett.*, **28**:1075.

Optical Waveguide-mode Resonant Biosensors

D. Wawro,[1] S. Tibuleac,[2] and R. Magnusson[3]

[1]Resonant Sensors Incorporated 202 E, Border st, #201 Arlington, Texas 76010
[2]Movaz Networks, One Technology Parkway South, Norcross, Georgia 30092
[3]Department of Electrical and Computer Engineering, University of Connecticut, 371 Fairfield Road, Unit 2157, Storrs, Connecticut 06269–2157

18.0 Introduction

It has been suggested that by changing the refractive index and/or thickness of a resonant waveguide grating, its resonance frequency can be changed, or tuned[1]. This idea has clear applications for biosensors as the buildup of the attaching biolayer can be monitored in real time, without use of chemical fluorescent tags, by following the corresponding resonance wavelength shift with a spectrometer[2,3]. Thus, the association rate between the analyte and its designated receptor can be quantified; in fact, the characteristics of the entire binding cycle, involving association, disassociation, and regeneration can be registered[4]. Similarly, small variations in the refractive indices of the surrounding media, or in any of the waveguide grating layers, can be measured; for example, in Ref.[5] results for resonance switching using a variation of the grating dielectric constant were presented. A new class of highly sensitive bio- and chemical sensors is thus enabled. This sensor technology is broadly applicable to medical diagnostics, drug development, industrial process control, genomics, environmental monitoring, and homeland security.

The sensors described in this chapter are based on resonating periodic waveguides that produce sharp resonance spectra on broadside illumination with an optical source[1–19]. The guided-mode resonance occurs when the illuminating wave is phase matched to a leaky waveguide mode by an appropriate periodic layer. Subwavelength structures (period Λ smaller than the incident wavelength λ) admit only the zeroth propagating diffraction orders thus promoting efficient energy exchange between the transmitted and reflected zero-order waves as illustrated in Fig. 18.1.

For such a device, a peak in the reflection spectrum occurs at a particular wavelength, angle, and polarization, when the incident electromagnetic wave

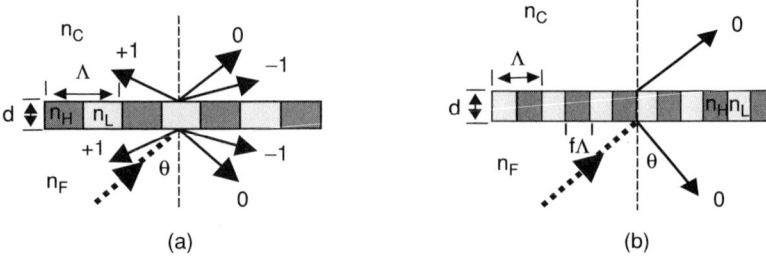

Figure 18.1. A single-layer waveguide grating: (a) low spatial frequency admits multiple diffracted orders and (b) subwavelength grating period admits only zero-order waves (cover refractive index n_C, fiber refractive index n_F, d grating thickness, grating high refractive index n_H, grating low refractive index n_L, angle of incidence θ, grating period Λ, fill factor f).

couples to the leaky mode. As the coupling range is typically small, these resonating elements exhibit high parametric sensitivity rendering them extremely responsive to small amounts of trace chemicals or biological molecules. The theoretically predicted high reflection efficiencies and narrow spectral lines have been verified experimentally[15]. These resonant elements are simple in design requiring only a few layers one of which being a high-quality, scatter-free subwavelength periodic layer. Suggested applications include optical filters[5], modulators[19], mirrors for vertical-cavity lasers[20], and sensors.[1-3,8,21]

The general utility of the guided-mode resonance effect for optical sensing applications has been known for some time.[1,5] Experimental implementation and testing has been conducted with resonance elements deposited on bulk substrates.[2,3,8,21] The spectral characteristics of the sensor including center wavelength, line width, and sideband response depend on the physical parameters of the device such as grating period, fill factor, refractive indices, and thicknesses of the grating and homogeneous layers as well as on the refractive indices of the substrate and cover media. Variation in any of these parameters provides a spectral shift of the reflection peak and the attendant transmission notch.

In general, for spectral filtering applications, the most stable guided-mode resonance filter is sought to prevent an unwanted resonance shift due to small parameter fluctuations. In contrast, for spectroscopic sensing applications, it is desired to promote this resonance sensitivity by creating a device that responds well to small parametric changes. To further broaden the applicability of this technology, a resonant biosensor can be integrated on the tip of an optical fiber. The incident wave and the resonant information-bearing reflected wave thus propagate transversely confined in a fiber waveguide. In principle, the fiber-tip resonant sensor functions like the bulk sensor with free-space input light; the main differences lie in fabrication methods and applicability.

18.1 Resonant Sensor Modeling

The sensor spectral response can be computed using rigorous coupled-wave analysis (RCWA) assuming plane waves incident on structures with infinite dimensions in the plane of the grating.[22] Therefore, employing the RCWA method, we have written efficient computer codes to solve the general multilayer diffraction problem underlying resonant sensors. Furthermore, increased efficiency and flexibility in design is achieved with a genetic algorithm search and optimization method integrated with the RCWA code.[23] This inverse approach allows specification of the desired resonant sensor characteristics to conversely obtain the corresponding layer-system parameters. With these computation tools, sensors can be designed to exhibit specific sensitivity, resolution, and operational dynamic range. Diffractive elements recorded on optical fiber tips possess a finite number of grating periods. However, since our experiments involve multimode fibers with large core diameters (diameter $D_F > 1000\lambda$) the plane-wave model may be used to a good approximation for this work.

Numerous experimental configurations can be envisioned to make use of the waveguide grating resonance in a sensing application. A possible sensor test setup is depicted in Fig. 18.2. By capturing the reflection spectrum from the guided-mode resonance sensor, the response can be monitored remotely and in real time. Since dielectric materials are used in the fabrication of resonant sensors, there are many potential configurations available. A single-layer resonant waveguide grating sensor is illustrated in Fig. 18.3. This sensor can be optimized to enhance sensitivity to specific parameters, such as refractive index of a solution or gas, and/or solid phase deposits on the sensor surface. In this research, refractive index and thickness values are chosen to model typical bioselective agents. Biochemical recognition reactions, such as antigen-antibody, enzyme-substrate, or ligand-receptor reactions allow selective sensing of complex biological molecules such as proteins, viruses, or drugs.[24] For antigen/

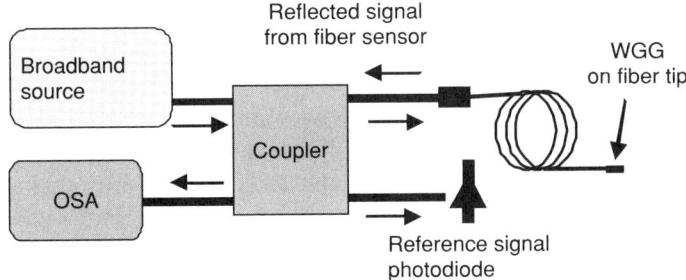

Figure 18.2. Schematic of a test setup to measure the spectral reflectance from a fiber-tip guided-mode resonance device (OSA: optical spectrum analyzer; WGG: waveguide grating).

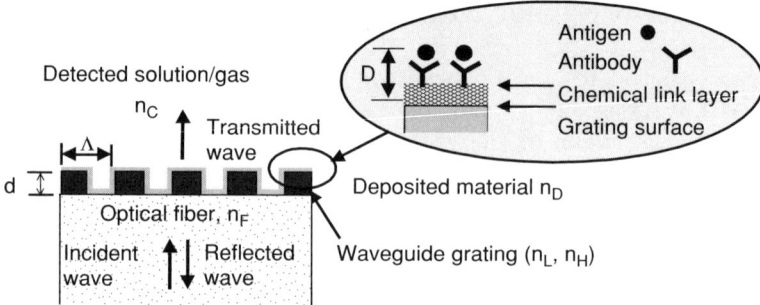

Figure 18.3. Generic single-layer fiber-tip reflection filter diagram used in sensor design. Refractive index of cover region is the same as that of the grating, $n_L = n_C$. An analyte-receptor example is shown in the inset for thickness sensing applications.

antibody reactions, the sensor surface is activated by chemically attaching a layer of antibodies via a linkage layer bound to the substrate (Fig. 18.3). The antibodies will bind only to its matched antigen. In this embodiment, the antigen is the analyte to be detected. As a monolayer of antigen accumulates on the grating surface, the refractive index and thickness change can be monitored. The rate of mass accumulation can be used, for example, to quantify concentrations of a specific drug in fluids or tissue. Typical attached biolayer thickness ranges from ~1 nm for small molecules to greater than 50 nm for viruses.

The spectral response for a single-layer sensor designed for use in a liquid environment is indicated in Fig. 18.4. This sensor can be fabricated with Si_3N_4 and patterned by plasma etching to create the diffractive layer. One-dimensional resonant waveguide grating structures have separate reflectance peaks for TE (electric vector normal to the page in Fig. 18.1) and TM polarized incident waves.[5,6] In this case, two peaks are available for each sensing measurement. This feature can be exploited to resolve both thickness and refractive index changes simultaneously if both are varying; alternatively, it is useful for improving the accuracy of the measurement.

The calculation shows that this design can resolve an average refractive index change of 3×10^{-5} refractive index units (RIU) assuming a spectrometer resolution of 0.01 nm. A nearly linear wavelength shift is maintained (Fig. 18.5) for a wide refractive index change of the medium in contact with the grating structure ($n_C = n_L = 1.3 - 1.8$), making this a versatile sensor with a large dynamic range. The sensitivity of a biosensor is defined as the measured response (such as peak wavelength shift) for a particular amount of material that is detected [24]. This indicates the maximum achievable sensitivity to the analyte under detection. Sensor ressolution includes realistic component limitations such spectroscopic equipment resolution, power meter accracy, bioselective agent response and peak shape or linewidth. Linewidth is the full width at half maximum (FWHM) of the reflected peak response. This property affects the accuracy of spectroscopic sensors as a narrow line typically permits improved resolution of wavelength shifts; resonant waveguide

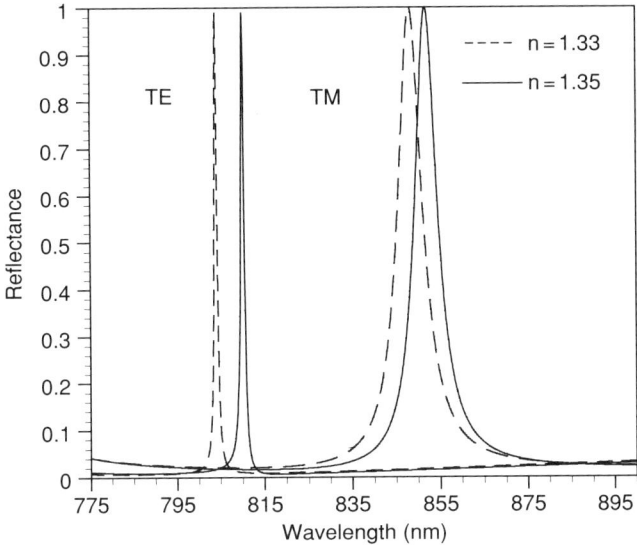

Figure 18.4. Refractive index sensing in water. Calculated TE and TM polarization spectral response of a fiber-tip reflection sensor using a Si_3N_4 diffractive element. The peak wavelengths shift as the refractive index of the detected liquid varies from 1.33 to 1.35. The sensor layout is depicted in Fig. 18.3. Physical parameters of the waveguide grating are as follows: grating period $\Lambda = 530\,nm$, fill factor $f = 0.5$, thickness $d = 470\,nm$, refractive index of the grating layer $n_H = 2.0$ (Si_3N_4),$n_L = n_C = 1.33$ and 1.35. The refractive index of the optical fiber $n_F = 1.45$.

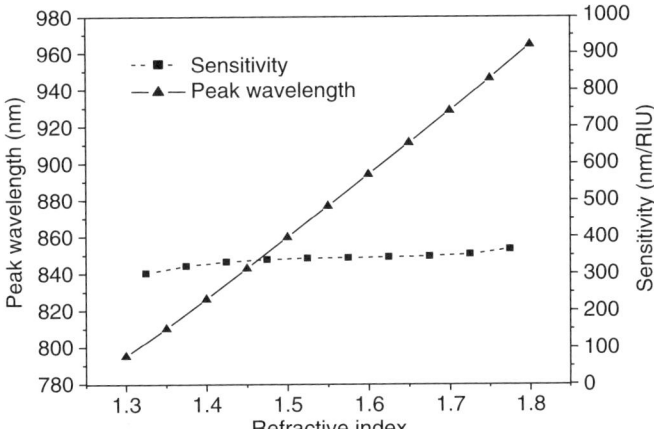

Figure 18.5. Calculated TE polarization resonance wavelength shift for large dynamic range sensing. The fiber-tip structure described in Fig. 18.4 has an average resonance peak shift of 338.7 nm per refractive index unit (RIU), from $n_C = 1.3$ to 1.8.

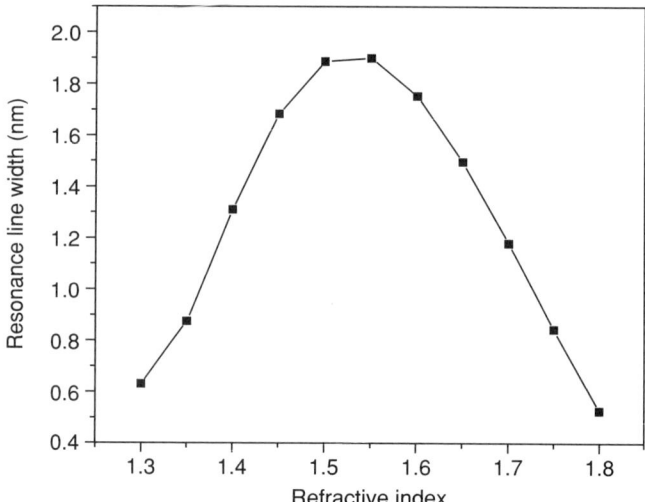

Figure 18.6. Calculated TE polarization peak line width changes over the refractive index detection range. Physical parameters are described in Fig. 18.4.

grating sensors typically have narrow line widths. Figure 18.6 illustrates the resonance line width changes for the Si_3N_4 design associated with output liquid index variation. While the average refractive index of the waveguide grating layer increases over the detection range, the grating modulation is simultaneously decreasing. A maximum resonance line width is reached around $n_L = n_C = 1.53$.

By monitoring the reflected power at a fixed wavelength, an intensity mode of detection can, in principle, be used within a reduced dynamic range (Fig. 18.7). Small changes in the reflected intensity are calibrated to a specific refractive index or thickness change. For the sensor in Fig. 18.4, calculated refractive index changes of 1.9×10^{-5} RIU ($n_L = 1.3287 - 1.3297$) can be resolved by intensity mode detection. This assumes an optoelectronic system that can resolve changes in optical power to 1%. To further optimize this mode of detection, the sensor could be designed with an increased line width to improve the dynamic range of the sensor. Additionally, the line shape can be made asymmetrical by selecting appropriate thin film and waveguide thickness values. An asymmetrical line shape offers practical benefits since it can be used to eliminate the ambiguity caused by the same intensity value being measured for two different values of the parameter to be measured. Besides the spectral-shift mode, and the intensity mode of operation, a resonant waveguide grating based sensor can operate in a polarization mode. In general, the TE and TM peaks of a resonant structure occur at different wavelengths. By design, the two peaks can be made to coincide resulting in a polarization insensitive waveguide grating. However, changes in refractive index or thickness of the layers of this polarization insensitive structure would shift the TE and TM peaks apart. A polarization-mode waveguide grating sensor could operate with a light source with an appropriate wavelength and a polarizer on the return path of the

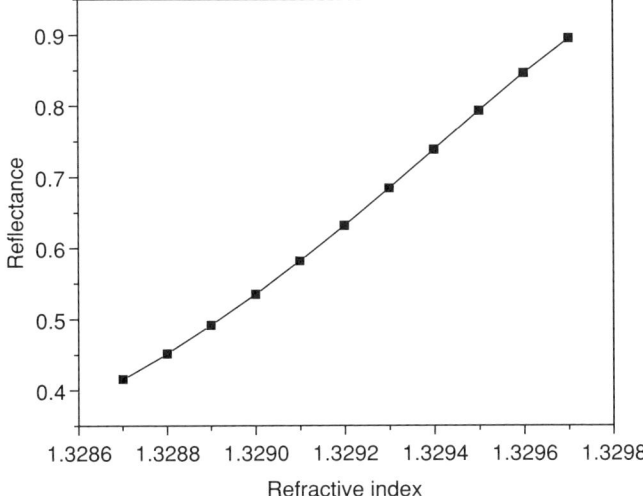

Figure 18.7. Calculated reflectance response for a fixed incident wavelength ($\lambda_{\text{inc.}} = 804\,\text{nm}$, TE polarization) due to a refractive index variation in n_C for the fiber-tip device described in Fig. 18.4.

signal. Changes in refractive index or thickness that would occur in a particular sensing application would produce varying levels of reflected intensity in each state of polarization.

While resonant sensors can monitor tiny refractive index changes, they can also be used to detect thickness changes at the sensor surface, as shown in Fig. 18.8. For thickness detection, the degree of resonant central wavelength shift is attributed to three parameter changes: the increase in grating thickness, the refractive index of the attached layer, and a change in grating fill factor. The grating fill factor can contribute significantly to the resonance shift. In Fig. 18.8, peak shifts are shown both without a fill factor change ($\Delta f = 0$) and with a fill factor change ($\Delta f \neq 0$). When material is attached only to the top and bottom surface of the grating, the fill factor stays constant. In this case the resonance peak shifts an average of 0.004 nm per angstrom of added material (Fig. 18.9). Alternatively, when the added material coats the entire grating surface including the sidewalls ($\Delta f \neq 0$), the sensitivity is increased by 400% to an average of 0.016 nm shift per angstrom. Average thickness changes of 0.7 angstroms can be resolved, with spectrometric equipment resolution of 0.01 nm. Figure 18.10 illustrates the resonance line width changes over the sensed thickness range. This line width increase occurs similarly for both cases without a fill factor change and with a fill factor change due to the relatively small index difference between the cover index $n_C = 1.33$ and the added material index $n_D = 1.4$ in this example.

Resonant sensors can be designed to operate within any particular wavelength range of interest. The materials should be chosen to exhibit low optical losses at

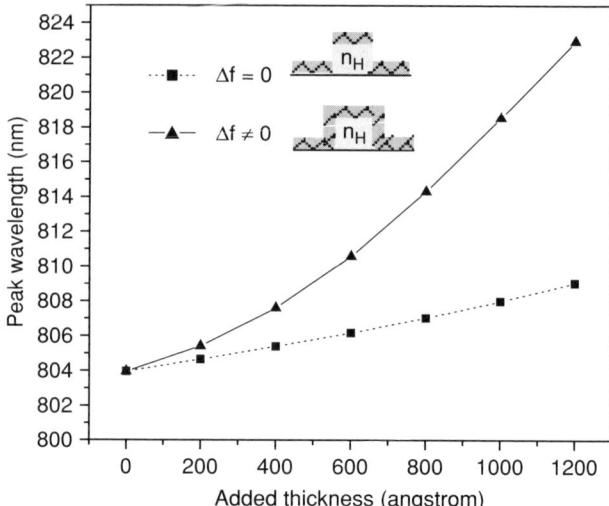

Figure 18.8. Thickness sensing in water. Calculation of TE polarization peak wavelength shift for detection of deposited material at the sensor surface in water ($n_D = 1.4$, $n_C = n_L = 1.33$). Peak shifts are shown both without a fill factor change ($\Delta f = 0$) and with a fill factor change ($\Delta f \neq 0$). Other physical parameters are described in Fig. 18.4.

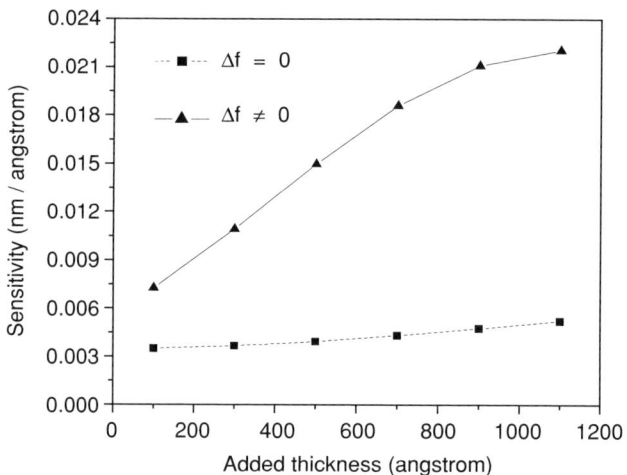

Figure 18.9. Calculation of the spectral sensitivity to deposited material for the sensor described in Fig. 18.8. The resonance peak shifts an average of 0.016 nm per angstrom of material deposited at the sensor surface for ($\Delta f \neq 0$). Sensitivity is reduced to 0.004 nm/ angstrom when the fill factor stays constant ($\Delta f = 0$).

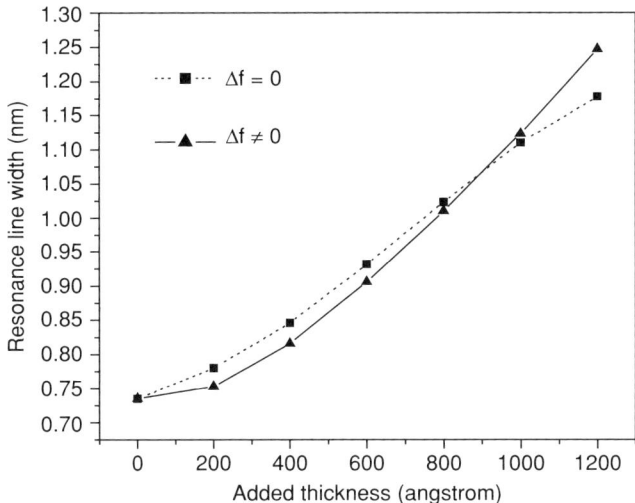

Figure 18.10. Calculation of the TE polarization resonance line width over the deposited thickness detection range. The line width increases as the added layer thickness increases. Physical parameters are described in Fig. 18.8.

the central wavelength of the sensor. Thus, a thickness detection sensor for use in an air environment is described in Fig. 18.11 for operation around 1550 nm. Peak shifts are illustrated both without a fill factor change ($\Delta f = 0$) and with a fill factor change ($\Delta f \neq 0$). As shown in Figs. 18.12 and 18.13, when the added layers cause a fill factor increase ($\Delta f \neq 0$), the sensitivity increases in a relatively linear manner over the detection range, while the line width increase is exponential. The resonance peak shifts an average of 0.07 nm per angstrom of added material deposited at the surface for this case. When the fill factor does not change ($\Delta f = 0$) with the addition of material, the sensitivity and line width increases are minimal with average sensitivity reduced to 0.004 nm per angstrom. Due to the high index modulation in this design, average thickness changes of 0.15 angstroms can be resolved, with spectrometric equipment resolution of 0.01 nm. This sensor might be used for the detection of airborne contaminants, such as those that may be encountered in homeland security applications.

18.2 Sensor Fabrication

Practical implementation of fiber-tip waveguide grating sensors entails fiber preparation, thin-film deposition, and diffraction grating fabrication. A cleaving tool is used to obtain a flat, optical quality endface. A thin layer of

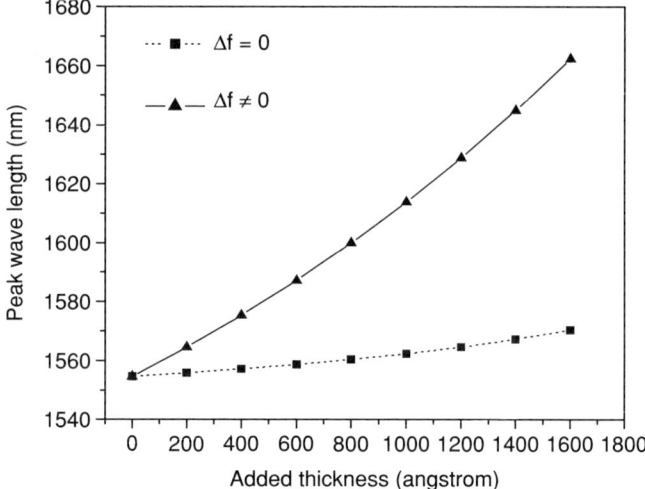

Figure 18.11. Thickness sensing in air. Calculated TE polarization spectral response of a reflection sensor designed to operate in the 1550 nm wavelength range. Peak shifts are shown both without a fill factor change ($\Delta f = 0$) and with a fill factor change ($\Delta f \neq 0$). Physical parameters are as follows: grating period $\Lambda = 907$ nm, fill factor $f = 0.5$ (without added layer), thickness $d = 1100$ nm, refractive indices of the grating layer $n_H = 3.2$ and $n_C = n_L = 1.0$ (air). Refractive index of the optical fiber $n_F = 1.45$. Refractive index of the material to be detected is $n_D = 1.4$. The sensor layout is depicted in Fig. 18.3.

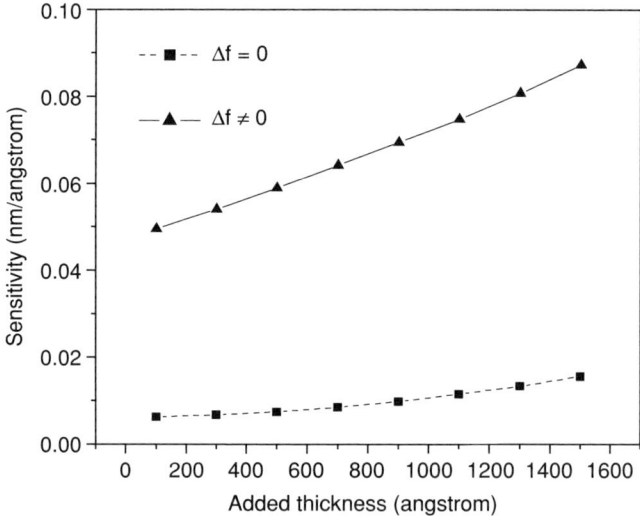

Figure 18.12. Calculation of the spectral sensitivity to deposited material for the sensor described in Fig. 18.11. The resonance peak shifts an average of 0.07 nm per angstrom of material deposited at the sensor surface for ($\Delta f \neq 0$). Sensitivity is reduced to 0.01 nm/angstrom when the fill factor stays constant ($\Delta f = 0$).

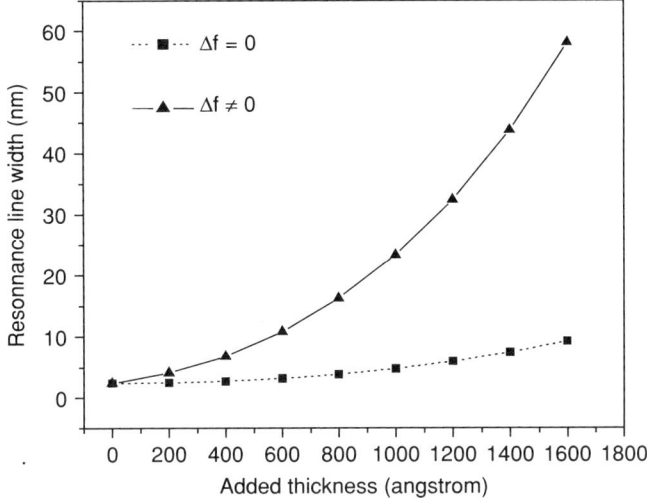

Figure 18.13. Calculation of the TE polarization resonance line width over the deposited thickness detection range. The line width increases significantly as the added layer thickness increases for ($\Delta f \neq 0$). When the fill factor is constant ($\Delta f = 0$), changes in the resonance line width are minimal over the detection range. Physical parameters are described in Fig. 18.11.

photoresist is deposited on the fiber endface so that diffraction gratings can be recorded using a UV Ar+ laser interferometer (Fig. 18.14). Exposure and development times are optimized by iterations. Figure 18.15 (a) shows the diffraction pattern emerging from a large-period fiber-tip grating under white-light illumination. A scanning electron micrograph of a diffraction grating with 800 nm period on a fiber-tip with a 100 μm core diameter is depicted in Fig. 18.15(b). On coupling a He–Ne laser beam ($\lambda = 633$ nm) into the fiber, this device produces ± 1 diffracted orders containing $\sim$50% of the total output power. With similar methods, gratings with periods of $\sim$530 nm were recorded on fiber endfaces with 6.7 μm core diameters. The ± 1 transmitted diffraction orders were measured to contain $\sim$10% of the total power coupled out of the fiber at a wavelength of 442 nm (He–Cd laser).

To fabricate the sensors, deposition of dielectric thin films is required to create a waveguide grating device. In this work, films of Si_3N_4 are deposited by sputtering on clean, uncoated optical fiber endfaces. The fibers are mounted in the chamber along with a reference substrate to monitor thickness deposition. The test substrate thickness, refractive index, absorption, index grading, and surface roughness of the deposited film are measured using a spectroscopic ellipsometer. A layer of photoresist is deposited on the coated fiber endfaces and on a reference substrate and a submicron diffraction grating is recorded on each.

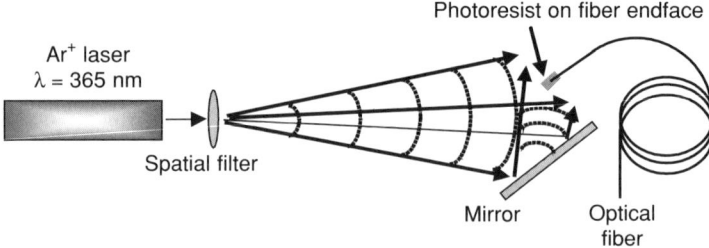

Figure 18.14. Single beam UV laser interference system to record the grating pattern.

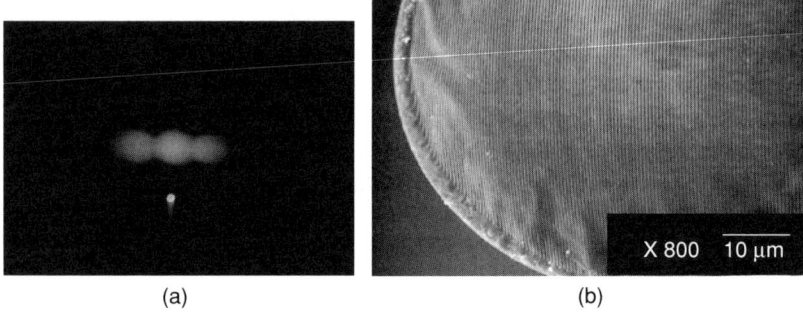

(a)   (b)

Figure 18.15. (a) Output intensity pattern of a multimode optical fiber with 1.2 μm period grating recorded on the endface using a white-light testing source. The color spectrum associated with the ±1 diffracted orders was clearly visible during the experiment. (b) Scanning electron micrograph of an 800 nm period photoresist grating recorded on a multimode fiber endface.

The reference element is tested first to determine performance of the sensor design in a bulk format before continuing to the fiber-tip device. The resulting two-layer waveguide grating device is shown in Fig. 18.16 and is used to determine the sensitivity of the guided-mode resonance peak to a refractive index change in the cover medium. The calculated spectral response is depicted in Fig. 18.16; it is comparable with the experimentally obtained results in Fig. 18.17 with particularly good agreement found in the resonance wavelength. The device is tested using a tunable Ti–sapphire laser ($\lambda = 730 - 900$ nm) at normal incidence in air; and after immersion in a reservoir of water. The transmission notch is shown to shift approximately 10 nm for a cover refractive index change from $n_C = 1.0$ to $n_C = 1.33$. Decreased efficiency in the water environment is partially due to scattering from bubbles formed in the water reservoir.

To investigate the resonance peak response after layering biomaterials on the surface of a resonant waveguide grating sensor, experiments are performed with Bovine Serum Albumin (BSA). A two-layer resonant device is fabricated using a SiO_2 grating layer on a HfO_2 waveguide by methods described in reference.[25]

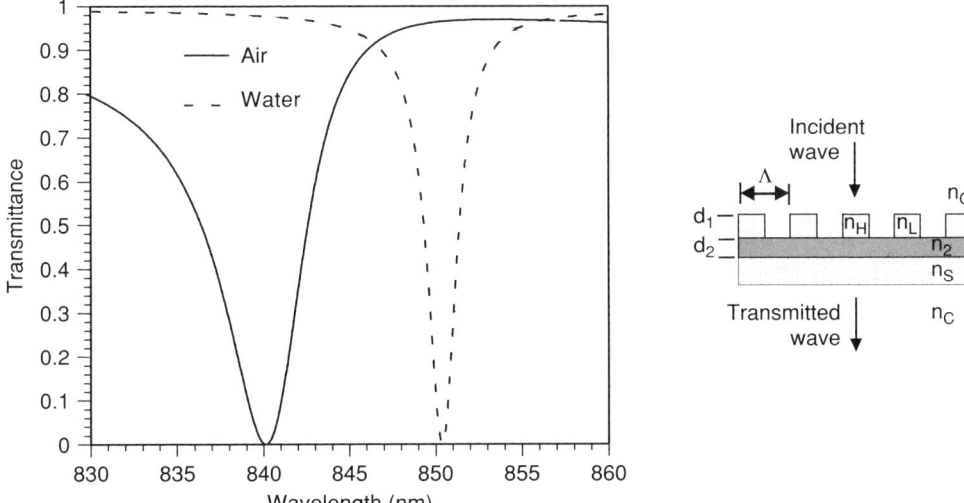

Figure 18.16. Calculated TE polarization spectral shift for the two-layer GMR sensor shown in the inset. The physical parameters of the waveguide grating are as follows: grating period $\Lambda = 510\,\mathrm{nm}$, fill factor $f = 0.5$, $d_1 = 300\,\mathrm{nm}$, $d_2 = 200\,\mathrm{nm}$ $n_H = 1.62$ (photoresist), $n_L = n_C = 1.0$ (air) and $n_L = n_C = 1.33$ (water), $n_S = 1.453$ (substrate).

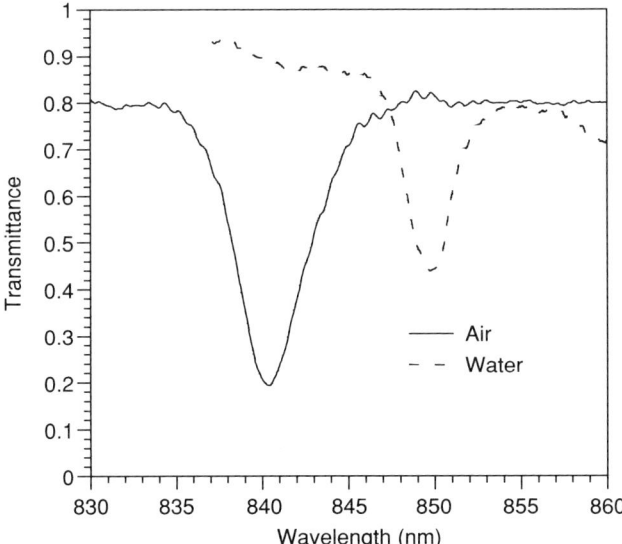

Figure 18.17. Experimentally determined spectral shift for the device described in Fig. 18.16. Testing is performed with a tunable Ti:sapphire laser with TE polarization at normal incidence in air, and immersed in water. A transmission notch shift of ~10 nm results as the cover index changes from 1.0 to 1.33. Line width in air = 5.8 nm, line width in water = 4.5 nm.

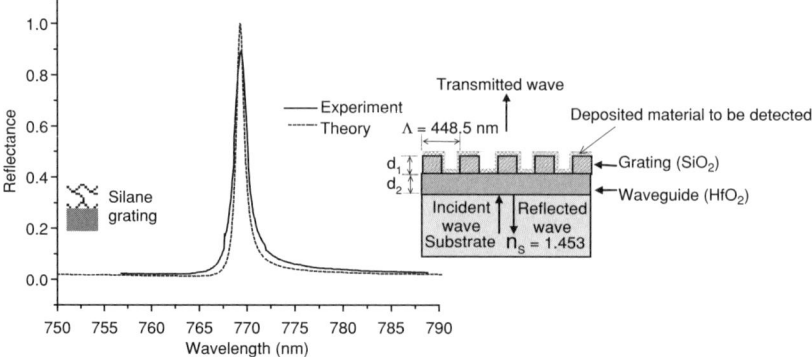

Figure 18.18. Calculated and measured spectral response for a two-layer resonant sensor device coated with a reactive silane layer. Grating parameters are: cover index $n_c = 1.0$, grating index $n_{1H} = 1.453(SiO_2)$, $n_2 = 1.977(HfO_2)$, substrate index $n_s = 1.453(SiO_2)$, $d_1 = 110\,nm$, $d_2 = 202\,nm$, fill factor $f = 0.58$, $\Lambda = 448.5\,nm$.

The clean grating surface is first chemically modified with amine groups by treating with a 3% solution of aminopropyltrimethoxysilane (Sigma) in methanol. Figure 18.18 depicts the TE polarization spectral response of the silanated resonant sensor in air. A solution of BSA, (100 mg/ml, Sigma) is then rinsed over the surface of the sensor for 60 s, and subsequently rinsed in PBS for 90 s. After allowing to drip dry for 1 min, a spectral peak shift of 6.4 nm is measured due to the BSA layer deposited on the surface ($\sim$38 nm thick) as shown in Fig. 18.19. Typical biomaterials have minimal absorption in the near-IR wavelength range.

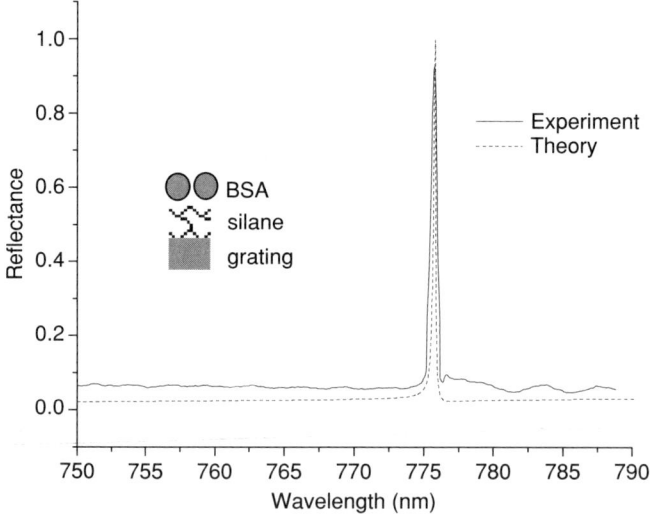

Figure 18.19. Calculated and measured spectral response from the sensor described in Fig. 18.18 for an attached bovine serum albumin layer $\sim$38 nm thick.

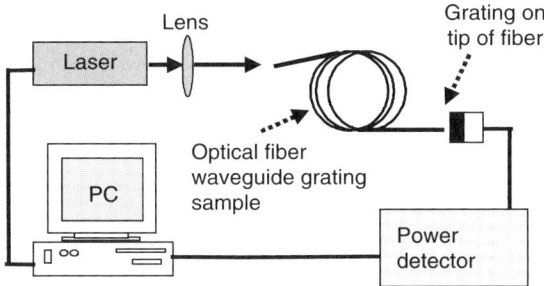

Figure 18.20. Test setup used to measure the spectral response of a fiber-tip waveguide grating.

Final experiments apply the setup depicted in Fig. 18.20. Spectral measurements made with the tunable laser coupled into the fiber as shown indicate a split guided-mode resonance notch of ~18% in the transmitted power, measured at the output of the optical fiber. The resonance split occurs as the fiber was cut at a small angle. Figure 18.21 illustrates preliminary measured results; the slow variation in transmitted power is due to variation in input coupling efficiency across the testing wavelength range. The low resonance efficiency is

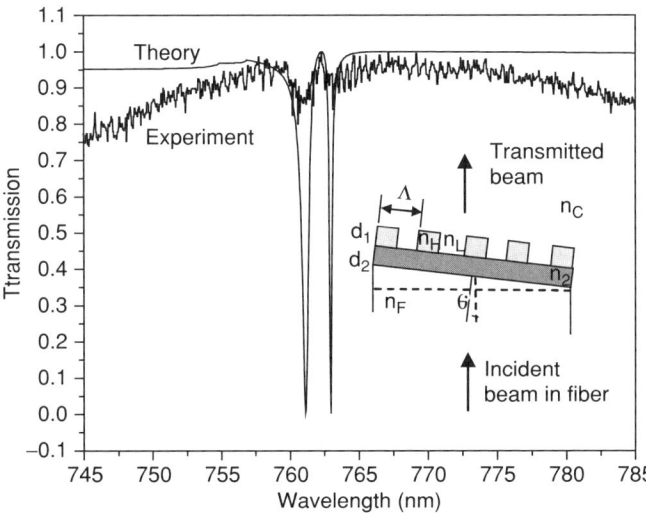

Figure 18.21. Calculated and measured transmission data for a fiber-tip guided-mode resonance device with the following parameters: grating period $\Lambda = 520.2$ nm, $d_1 = 200$ nm, $d_2 = 380$ nm, $n_L = n_C = 1.0$ (air), $n_H = 1.62$ (photoresist), $n_2 = 2.05$ (Si$_3$N$_4$), $n_F = 1.453$ (fiber), $\theta = 0.11°$, fill factor $f = 0.3$. A split TM-polarization resonance notch of ~18% is detected at $\lambda_{\text{res}} = 761.1$nm.

attributed, in large measure, to scattering due to imperfect fiber cleaves and to rough silicon nitride films. Improved fabrication processes and designs will ameliorate these problems.

18.3 Conclusions

Resonant waveguide grating bio- and chemical sensors in bulk or integrated fiber format are potentially useful for high-resolution sensing as discussed in this chapter. This sensor is realized by utilizing a fundamental resonance effect that occurs in waveguide gratings where slight changes in refractive index or thickness provide a detectable shift of the central wavelength associated with the reflection peak or transmission notch. Integrated fiber-tip biosensors have numerous potential applications in clinical diagnostics, surgical tools, food industry, environmental monitoring, and industrial sensing; real-time, remote operation can be implemented.

Numerous results illustrating the application of the guided-mode resonance effect for refractive index and thickness sensing are presented in this chapter. Using numerical computational methods, it is possible to design resonant sensors with high sensitivity for a particular thickness and/or refractive index range. The sensors can be designed to have a large operational range, while maintaining high sensitivity over the full sensing range. The sensitivity is enhanced by stationing the resonant leaky mode in a layer near the region of molecular attachment or index change. The sensor design examples presented possess a sensor refractive index spectral resolution of 3×10^{-5} RIU for a refractive index range of n $= 1.3$ to n $= 1.8$, and average thickness resolutions as low as 0.7 angstroms in water, and 0.15 angstroms in air (assuming spectroscopic resolution of 0.01 nm). Experimental results presented include a bulk sensor in air and water and preliminary data with a fiber-tip probe.

The fiber-tip resonant sensors under study are fabricated with low-loss dielectric materials such that bulk absorption losses are not a physical limitation; however, care must be taken to minimize creation of scattering centers in the process of fabrication. The sensors typically have narrow, well-defined resonance shapes that can provide accurate, high-resolution measurements. Separate resonance locations for TE and TM polarizations are available for detection; thus, accuracy and reliability are enhanced. Using simultaneous TE and TM sensing, both thickness and refractive index are found. This can be used to detect the background density of a solution during an analyte-receptor biolayer attachment. Furthermore, the binding rate between an analyte and its matching receptor is directly monitored, with no fluorescent/absorbance tags or special microwell plates required for operation. Since the sensing element is located on the fiber tip, highly accurate proximity sensing is possible with minimal sample volume. Other sensor designs might include the use of biopolymers in the fabrication. For example, stimuli-responsive hydrogels can be integrated in the sensor to detect small changes in pH, temperature, pressure, and antigen/antibody reactions. These biopolymers undergo abrupt changes in refractive index and thickness in direct response to external stimuli such as pH[26], temperature[26], pressure[27], and the presence of a specific antigen.[28] Optical fiber sensor arrays can be readily implemented to simultaneously detect multiple analytes, such as DNA sequences, or repeat sampling to increase detection precision.

Acknowledgments

This work was supported in part by the Texas Advanced Technology Program under grant number 003656–042.

References

[1]   Magnusson R and Wang SS. (1993). *"Optical guided-mode resonance filter."* US patent number 5,216,680.
[2]   Wawro D, Tibuleac S, Magnusson R, and Liu H. (2000). "Optical fiber endface biosensor based on resonances in dielectric waveguide gratings." in: *Biomedical Diagnostic, Guidance, and Surgical-Assist Systems II, Proc. SPIE*, 3911:86–94.
[3]   Tibuleac S, Wawro D, and Magnusson R. (1999). "Resonant diffractive structures integrating waveguide gratings on optical fiber endfaces," in: *Proc. 1999 IEEE LEOS 12th Annual Meeting*, San Francisco, CA. 2:874–875.
[4]   Cooper M. (2002). *Nature Reviews. Drug Discovery*, 1:515–528.
[5]   Wang SS and Magnusson R. (1993). "Theory and applications of guided-mode resonance filters." *Appl. Opt.*, 32(14):2606–2613.
[6]   Magnusson R and Wang SS. (1992). "New principle for optical filters." *Appl. Phys. Lett.*, 61(9):1022–1024.
[7]   Norton S, Morris GM, and Erdogan T. (1998). "Experimental investigation of resonant-grating filter line shapes in comparison with theoretical models." *J. Opt. Soc. Am. A.*, 15(2):464–472.
[8]   Kikuta H, Maegawa N, Mizutani A, Iwata K, and Toyota H. (2001). "Refractive index sensor with a guided-mode resonant grating filter." *Optical Engineering for Sensing and Nanotechnology, Proc. SPIE*, 4416:219–22.
[9]   Shin D, Tibuleac S, Maldonado TA, and Magnusson R. (1998). "Thin-film optical filters with diffractive elements and waveguides." *Opt. Eng.*, 37(9):2634–2646.
[10] Magnusson R, Shin D, and Liu ZS. (1998). "Guided-mode resonance Brewster filter." *Opt. Lett.*, 23(8):612–614.
[11] Brundrett D, Glytsis E, and Gaylord TK. (1998). "Normal-incidence guided-mode resonant grating filters: Design and experimental demonstration." *Opt. Lett.*, 23(9):700–702.
[12] Rosenblatt D, Sharon A, and Friesem AA. (1997). "Resonant grating waveguide structures." IEEE *J. Quanton Electron.*, 33(11):2038–2059.
[13] Tamir T and Zhang S. (1997). "Resonant scattering by multilayered dielectric gratings." *J. Opt. Soc. Am. A.*, 14(7):1607–1616.
[14] Avrutsky IA, Svakhin AS, and Sychugov VA. (1989). "Interference phenomena in waveguides with two corrugated boundaries." *J. Mod. Opt.*, 36(10):1303–1320.

[15] Liu ZS, Tibuleac S, Shin D, Young PP, and Magnusson R. (1998). "High-efficiency guided-mode resonance filter." *Opt. Lett.*, 23(19):1556–1558.

[16] Peng S and Morris GM. (1996). "Experimental demonstration of resonant anomalies in diffraction from two-dimensional gratings." *Opt. Lett.*, 21(8):549–551.

[17] Gale MT, Knop K, and Morf RH. (1990). "Zero-order diffractive microstructures for security applications." *Optical Security and Anticounterfeiting Systems, Proc. SPIE*, 1210:83–89.

[18] Mashev L and Popov E. (1985). "Zero-order anomaly of dielectric coated gratings," *Opt. Commun.*, 55(6):377–380.

[19] Sharon A, Rosenblatt D, Friesem AA, Weber HG, Engel H, and Steingrueber R. (1996). "Light modulation with resonant grating-waveguide structures." *Opt. Lett.*, 21(19):1564–1566.

[20] Magnusson R, Young PP, and Shin D. (2000). "Vertical-cavity laser and laser array incorporating guided-mode resonance effects and method for making the same." US patent number 6,154,480.

[21] Cunningham B, Li P, Lin B, and Pepper J. (2002). "Colorimetric resonant reflection as a direct biochemical assay technique." *Sens. Actuators B.*, 81(2–3):316–328.

[22] Gaylord TK and Moharam MG. (1985). "Analysis and applications of optical diffraction by gratings." *Proc. IEEE*, 73(5):894–937.

[23] Tibuleac S, Magnusson R, Maldonado TA, Shin D, and Zuffada C. (1997). "Direct and inverse techniques of guided-mode resonance filter designs," in: *Proc. 1997 IEEE AP-S Int. Symp.*, Montreal, Canada. 4:2380–2383.

[24] Cunningham A. (1998). *Introduction to Bioanalytical Sensors.* Wiley, New York.

[25] Priambodo PS, Maldonado TA, and Magnusson R. (2003). "Fabrication and characterization of high-quality waveguide-mode resonant optical filters." *App. Phys. Lett.*, 83:3248–3250.

[26] Okano T. (1998). *Biorelated Polymers and Gels: Controlled Release and Applications in Biomedical Engineering.* Academic, San Diego.

[27] Harmon M, Jakob T, Knoll W, and Frank C. (2002). "A surface plasmon resonance study of volume phase transitions in N-isopropylacrylamide gel films." *Macromolecules,* 35(15):5999–6004.

[28] Miyata T, Asami N, and Uragami T. (1999). "A reversibly antigen-responsive hydrogel." *Nature,* 399:766–769.

Chapter 19

Improved Optical Document Security Techniques Based on Volume Holography and Lippmann Photography

Hans I. Bjelkhagen

OpTIC Technium Centre for Modern Optics Ffordd William Morgan St Asaph Business Park St Asaph LL17 0JD North Wales, UK Tel: +44-1745 535130 Fax: +44-1745 535101 E-mail: hansholo@aol.com

Abstract: Optical variable devices (OVDs), such as holograms, are now common in the field of document security. Up until now mass-produced embossed holograms or other types of mass-produced OVDs are used not only for banknotes but also for personalized documents, such as passports, ID cards, travel documents, driving licenses, credit cards, etc. This means that identical OVDs are used on documents issued to individuals. Today, there is need for a higher degree of security on such documents and this chapter covers new techniques to make improved mass-produced or personalized OVDs.

The introduction of volume holography offers a possibility to apply monochrome and full color reflection holograms in the field of document security. A presentation of the technique and recording materials used for volume and color holography is provided. Another technique, interferential photography or Lippmann photography, represents a new type of OVD, which belongs to the interference security image structures. In this type of photography, color is recorded in a photosensitive film as a black-and-white interference structure. The technique offers additional advantages over holographic labels for unique security applications. The application of the Lippmann OVD for document security and counterfeit-resistant purposes is presented here.

19.0 Introduction

Holography has been used for protecting security documents for more than 20 years. The first application was for protecting credit cards starting with the VISA card. The hologram type used in the beginning and until today has been the transmission "rainbow" or Benton hologram.[1] The advantage of this hologram type is that it can be mass-produced by an embossing technique. After the introduction of high-quality color photocopiers and scanners, many countries started using embossed holograms or kinegrams attached to banknotes to make banknotes more difficult to fake.

The hologram is an example of an OVD in which the image appearance (content and/or color) changes when illumination and observation direction varies. This feature makes it impossible to copy such a device by a photocopier or computer scanner. Although not possible to copy in such a way, there are many places around the world where embossed holograms can be illegally copied or regenerated and then attached to fake security documents. At a recent conference, Dr. Bilorus of the National Bank of Ukraine, stated that some high quality counterfeit € 200 banknotes had been intercepted.[9] The embossed hologram on these banknotes was exceptionally good and almost impossible to distinguish from the genuine one. He concluded that the embossed hologram could no longer provide banknotes with protection from counterfeiting and a new technology was required to replace it.

This demonstrates why there is a need to find improved techniques to make security documents more difficult to counterfeit. Improvements in the origination technique for embossed holograms have been achieved by the introduction of e-beam lithography. In this technology the beam is used to record the microrelief on a special medium sensitive to electrons. It is possible to obtain very high image resolution (a few tens of nanometers) with this technique. The equipment for e-beam technology is very expensive, which will add to the security of the production process. OVDs based on e-beam technology can contain microcripts that are not possible to record with the optical mastering techniques.

A book edited by Ruud van Renesse is an excellent source for information on optical document security including OVDs.[13] Recent patents on holograms and OVDs issued between 1999 and 2003 are listed in a database from Honnorat Researches & Services.[7]

There is interest in finding more secure OVDs particularly for personalized documents, such as passports. By introducing volume holograms, there are possibilities to make a new type of holographic OVD. A volume hologram has the recorded interference pattern located within the light-sensitive emulsion and cannot be mass-produced by conventional embossing methods. In addition, a volume reflection hologram can be recorded with red, green, and blue laser wavelengths, which can create full-color holograms. More sophisticated equipment and recording materials are needed for producing such holograms, which make them more secure. They are also more suitable for producing personalized OVDs compared to embossed holograms which require an expensive mastering process.

19.1 Hologram Recording

To record a hologram, a micropattern caused by interference between the object beam and the reference beam is recorded in a light-sensitive high-resolution material. The quality of a holographic image depends on a number of factors, such as, the geometry and stability of the recording setup, the coherence of the laser light, the reference and object beam ratio, the type of hologram produced, the size of the object and its distance from the recording material, the recording material and the emulsion substrate used, the processing technique applied, as well as the reconstruction conditions. A material must comply with certain requirements to be suitable for the recording of holograms.[2] The most important of these concerns is the resolving power of the material. The recording material must be able to resolve the highest spatial frequencies of the interference pattern created by the maximal angle θ between the reference and the object beams in the recording setup (Fig. 19.1). If λ is the wavelength of the laser light used for the recording of a hologram, n the refractive index of the emulsion, then the closest separation d_e between the fringes in the interference pattern created by the angle θ between the reference and the object beams in the recording setup is

$$d_e = \frac{\lambda}{2n \sin{(\theta/2)}}. \qquad (1)$$

For high-quality holograms the resolution limit of the material must be higher than the minimum value obtained according to the above formula. One example of the resolving power needed in a practical situation using an emulsion with a refractive index of $n = 1.62$ is the following: A ruby laser with the

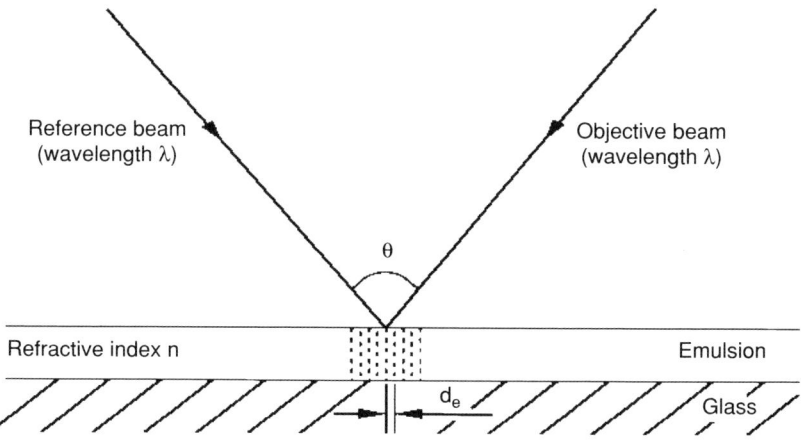

Figure 19.1. Demand on resolution for recording a hologram.

wavelength $\lambda = 694\,\text{nm}$ and a recording geometry with the maximum angle $\theta = 25°$ between the beams are used. This gives $d_e \sim 1\,\mu\text{m}$, which corresponds to $v = 1/d = 1000$ lines/mm; this is the minimum resolving power required. Close to its resolution limit the material will exhibit a low Modulation Transfer Function (MTF) and will thus make a low-quality hologram with poor fringe contrast and low signal-to-noise ratio. For a reflection hologram recorded in blue light ($\lambda = 400\,\text{nm}$) with an angle of $180°$ between the beams, a minimum resolving power of 7600 lines/mm is needed.

This shows that the demand on the holographic recording setup as well as the recording material is much higher when recording reflection holograms. For color holograms at least three laser wavelengths are needed for the recording making them more difficult to manufacture.

19.2 Plane and Volume Holograms

There are many ways of categorizing holograms. One valid criteria is the thickness of the recording layer (as compared to the interference fringe spacing within the layer), i.e., the layer coated on the material substrate. Holograms can thus be classified into "*thin*" or "*thick*" (sometimes also called "*plane*" or "*volume*" holograms, respectively). To distinguish between the two types, the *Q-parameter* is normally used; it is defined in the following way:

$$Q = 2\pi\lambda\, d/(n\Lambda^2) \qquad (2)$$

where λ is the wavelength of the illuminating light, d the thickness of the layer, n the refractive index of the emulsion, and Λ the spacing between the recorded fringes. A hologram is considered thick if $Q \sqsupseteq 10$, and thin when $Q \# 1$. Holograms with Q-values between 1 and 10 are sometimes treated as thin and at other times as thick.

The complex amplitude transmittance, $T_a(x)$, of a holographic recording material can be written in a general manner as

$$T_a = *t_a(x) * \exp\left[-\alpha(x)d\right] \cong \exp\left[-i\varphi_t(x)\right] = *t_a(x) * \exp\left[-\alpha(x)d\right]$$
$$\cong \exp\left[i2\pi nd/\lambda(x)\right], \qquad (3)$$

where α is the absorption constant of the material, d the thickness, n the refractive index of the material, and λ the wavelength of the laser light. The amplitude transmittance T_a is the square root of transmittance T.

$$T_a = \%T = 10^{-D/2}. \qquad (4)$$

In a phase hologram ($\alpha = 0$, $* t_a(x)* = 1$) either n or d changes with the exposure. For the phase hologram (simple grating) the phase factor is

$$\varphi_t = (2\pi/\lambda)\, nd, \quad \Delta\varphi_t = (2\pi/\lambda)[d\Delta n + (n-1)\Delta d]. \qquad (5)$$

If the hologram is thin, $d.0$: Phase variations are then caused by surface relief variations only

$$\Delta\varphi_t = (2\pi/\lambda)(n-1)\Delta d. \tag{6}$$

The masters for embossed holograms are recoded in transmission holograms of the thin type, where the recorded interference fringes generate a relief pattern. The recording material for the master is a photoresist plate. For mass-production, the recorded relief pattern (converted into nickel shims) is embossed into plastic foil. Since the hologram is of the transmission type it needs to be provided with a reflecting layer to illuminate and observe it from the same side (important for most security documents). In this way, the transmission hologram is acting as a reflection hologram. This fact explains why most embossed holograms on security documents have a mirror-like appearance.

If on the other hand a hologram is thick (volume hologram) and is supposed to have negligible surface relief ($\Delta d = 0$), phase variations are caused by index variations only

$$\Delta\varphi_t = (2\pi/\lambda)d\Delta n. \tag{7}$$

This type of hologram can be used to make OVDs, which can be recorded as a reflection hologram. Therefore, there is no need to use reflection foil behind this type of hologram on security documents.

The recording material for such security holograms is most often of the photopolymer type. Monochrome reflection volume holograms can be recorded in photopolymer film sensitized to a particular laser wavelength, most often a blue or green wavelength.

19.3 Holograms Suitable for Improved OVDs

Up until today, the transmission rainbow hologram has been the main type used for producing OVDs. The mass-produced mirror-backed embossed OVD has been used for banknotes and personalized security documents. Improvements over the years have occurred by more complex masters, the introduction of Dot-Matrix technique, and recently direct *e*-beam lithography.

In this section, OVDs based on volume phase holograms of the reflection type are discussed. The advantage of such OVDs is that they are more suitable for personalized documents. However, if cost-effective mass-production techniques can be developed it may also be possible to apply them to banknotes and other similar documents where identical security devices are required. Most likely the first application of volume holograms will be for personalized documents such as passports, ID cards etc. A higher cost of the security device may be acceptable here compared to the mass-produced embossed OVDs used for banknotes.

19.3.1 Monochrome Volume Holograms of the Reflection Type

A monochrome reflection hologram is recorded with a single laser wavelength with the object beam entering from one side of the holographic recording material and the reference beam from the opposite side. This requires a very stable setup when CW lasers are used for the recording. If a pulsed laser is employed almost any type of object can be recorded since no particular stability is required. For the recording of such holograms both silver halide emulsions as well as photopolymer materials have been used. There are a very few manufacturers of monochrome silver halide materials in the market today. In regard to photopolymer materials, E.I. du Pont de Nemours & Co. (DuPont) is the main commercial producer of such materials. Xetos AG in Germany has started to manufacture a photopolymer material for volume holography. The application of this material is for the document security market.

Monochrome volume holograms have been around for a long time but only recently been introduced in the document security field. In the past, Polaroid produced MIRAGE holograms for journal covers, product packages, tickets for sport events, baseball cards, etc. These holograms were produced on Polaroid's own photopolymer materials.

Dai Nippon Printing Co. Ltd. (DNP) in Japan has introduced the SECURE IMAGE™ LABEL, which is of the monochrome Lippmann-type reflection hologram shown in Fig. 19.2. It can contain a high-quality 3D image on a black background. For the document market DNP has introduced this original design and quality that make it difficult to duplicate and easier to authenticate. Only a few manufacturers in the world can handle the mass production of this type of hologram, as it requires advanced and difficult replicating techniques. These holograms are recorded in DuPont photopolymer materials, which are

Figure 19.2. SECURE IMAGE™ from DAI Nippon.

under complete distribution control. The SECURE IMAGE™ demonstrates its efficiency in anticounterfeiting by utilizing a combination of reuse–protection techniques such as "broken seal" or "self-destructing labels."

In addition, DNP proposes the use of this type of OVD as an effective measure of brand protection to prevent and eliminate the increasingly sophisticated copying, counterfeiting, and pirating of high-end products. In cooperation with Nippon Paint Company, DNP developed a new photopolymer transfer foil (SECURE IMAGE™ FOIL), which makes it possible to transfer photopolymer volume holograms by a heat-transfer method for mass production. Another DNP development is the Virtugram® technique where computer-generated security holograms are recorded in photopolymer materials.

Germany introduced a new passport, which has a transparent volume reflection hologram laminated over the page with biographical data. The hologram is a unique 2D monochrome hologram recorded of the normal photograph on the passport. This is an important improvement over earlier passport protection with identical mass-produced embossed holograms attached to all passports.

A new German personal identification card (ID card) was introduced in 2004. The new ID cards (Identigrams) have enhanced security features intended to hinder fraud and to enable computerized identification of cardholders. They are covered with a film-like layer containing multiple holograms. The holograms, among them a second photo of the cardholder, the federal eagle, and a capital letter "D" the size of a quarter, become visible when the card is tilted in the light. Moreover, the name of the cardholder and the ID number appear as iridescent, 3D figures on the lower edge of the document.

The new hologram technology was created in cooperation with the German federal ministries of the interior and of justice. The process of creating the holograms requires sophisticated equipment and expertise, making the IDs virtually impossible to copy or counterfeit, according to federal sources. The old IDs that carry only the photograph and the signature of the bearer, were too easily misused, according to the German authorities. In light of new global terrorism threats, identification procedures are more critical than ever, especially at borders and airports. The German government's ID project was underway long before the events of September 11.

In addition to improved OVD features on security documents many countries are considering other new security measures. Among them are the listing or encoding of biometric characteristics, i.e., distinguishing biological traits, such as an electronic "print" of a cardholder's finger, hand, iris, or voice that can be recognized using computer-aided technologies on the passport or ID. Several German companies specialize in the manufacture of such identification technologies, now used in private industries requiring rigorous scrutiny of personnel, such as nuclear power plants.

DuPont Authentication Systems (DAS), a joint venture between DuPont and Keystone Technologies/Label Systems, Inc. in the USA, manufactures advanced security techniques based on DuPont's photopolymers. The Izon™ technology sets a new standard for overt authentication security devices with

Figure 19.3. The IZON label for AMD's PIB products.

its full parallax deep 3D imaging. Authentication and identification security products using Izon^{TM} allow the viewer to literally look around the sides of the object in the image as if it were real, delivering instant visual verification of the authenticity of the document or product being inspected. Since Izon^{TM} material and technology is only provided by DAS and is maintained within a rigorously monitored secured supply chain, the verification is extremely reliable. Izon^{TM} products are available in a variety of constructions, including permanent or tamper-evident security labels and tamper-resistant seals, and as component elements. DAS can also provide Izon^{TM} tamper-evident labels that reveal a 3D image at one angle of view and becomes completely transparent at another. An example of a product protected with the Izon^{TM} technology is ADM's holographic label on its boxed processor (Processor-In-a-Box, or PIB) packaging. The label, shown in Fig. 19.3, provides customers an easy, effective way to authenticate AMD PIB products. The gold-green label consists of a 3D, full-parallax view surrounding the AMD logo, which shows dots arranged in a pattern: from one dot to two dots, three dots, and four dots as it is viewed from different angles.

19.3.2 Color Volume Holograms of the Reflection Type

Color reflection holograms can be recorded using red, green, and blue (RGB) laser wavelengths[3]. For a Denisyuk-type or single-beam reflection hologram the different laser beams pass through the same beam expander and spatial filter. The setup for recording such a hologram is shown in Fig. 19.4. The object is illuminated through the recording holographic plate. The light reflected from the object constitutes the object beam of the hologram. The reference beam is formed by the three expanded laser beams. This "white" laser beam illuminates both the holographic plate and the object itself through the plate. Each of the three primary laser wavelengths forms its individual interference pattern in the emulsion, all of which are recorded simultaneously during the exposure. In this

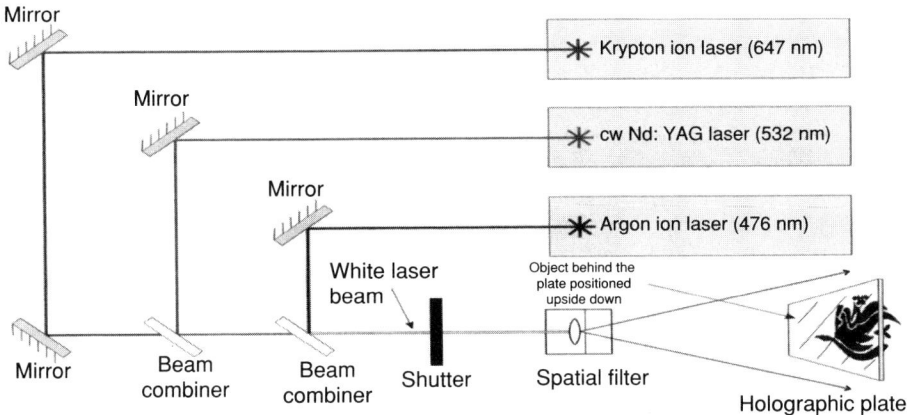

Figure 19.4. Setup for recording a volume color hologram of the reflection type.

way, three holographic images (a red, a green, and a blue image) are superimposed upon one another in the emulsion.

Three laser wavelengths are employed for the recording, for example: 476 nm, provided by an argon ion laser, 532 nm, provided by a CW frequency-doubled Nd–YAG laser, and 647 nm, provided by a krypton laser. Two dichroic filters are used for combining the three laser beams. The "white" laser beam goes through a spatial filter, illuminating the object through the holographic plate. By using the dichroic filter beam combination technique it is possible to perform simultaneous exposure recording, which makes it possible to control independently the RGB ratio and the overall exposure energy in the emulsion. The RGB ratio can be varied by individually changing the output power of the lasers, while the overall exposure energy is controlled solely by the exposure time. The recording materials needed to record such holograms have to be panchromatic and of very high resolution. There are two types of materials in the market that can be used.

19.3.2.1 Silver Halide Materials

To be able to record high-quality color reflection holograms it is necessary to use extremely low light-scattering recording materials,e.g., ultrafine-grain silver halide emulsions (grain size about 10 nm). Currently, the only producer of a commercial holographic panchromatic ultrafine-grain silver halide material is the Micron branch of the Slavich photographic company located outside Moscow.[15]

19.3.2.2 Photopolymer Materials

The color holography photopolymer material from DuPont is another alternative recording material for color holograms.[11,12] In particular, this type of material is suitable for mass production of color holograms. It has special advantages of easy handling and dry processing (only UV-curing and baking). The DuPont color photopolymer material has a coated film layer thickness of about 20 μm. However, the panchromatic polymer material is only supplied to specially selected and approved hologram producers because of its application in the field of document security. To obtain the right color balance, the RGB sensitivity depends on the particular material, but typically red sensitivity is lower than green and blue sensitivities. It is difficult to obtain high red-sensitivity of photopolymer materials. Simultaneous exposure is the best recording technique for photopolymer materials. Holograms can be recorded manually, but in order to produce large quantities of color holograms, a special machine is required. For hologram replication the scanning technique can provide the highest production rate. In this case, three scanning laser lines are needed. The output power in the scanning lines can be adjusted in such a way that all three simultaneously can scan the film.

As the first company in the world, Dai Nippon (DNP) introduced the TRUE IMAGE® LABEL, which is a full-color Lippmann-type reflection hologram.[8,14] It can contain a color 3D image on a black background. DNP has developed a mass-replication technique for these hologram labels. The holograms are recorded in DuPont panchromatic photopolymer materials. An example is shown in Fig. 19.5.

Figure 19.5. TRUE IMAGE® from DAI Nippon.

19.4 A New Type of OVD Based on Lippmann Photography

The new type of OVD presented here is based on an old photographic technique known as interferential photography or Lippmann photography.[10] Gabriel Lippmann was awarded the Nobel Physics Prize for his invention in 1908. This type of photography was the first technique that could record color photographs directly in the camera. Although the technique was unparalleled at the time, it also had several limitations: the camera needed a special type of isochromatic film, the image could not be copied, and the color image switched between negative and positive, depending on the viewing direction. Oddly enough, these limitations actually underline the strengths of the photographic security feature described below.

A modern Lippmann photograph may be categorized as a new type of OVD.[5] It can be applied to individually issued security documents, such as ID cards, passports, credit cards, driving licenses, and other documents that require a high degree of security. Although a Lippmann photograph is similar to a hologram, it allows a unique recording of each document to be made, thus offering a higher level of security than mass-produced holograms. The recording of Lippmann photographs requires a special type of photosensitive recording material, which is in contact with a reflecting layer. Use can be made of modern panchromatic photopolymer materials for holography or ultrahigh-resolution silver halide emulsions, which, once recorded and processed, may be laminated to the security document. Special equipment is needed to record the image. Lippmann photographs are nearly impossible to copy and certainly cannot be reproduced by means of conventional photography or with the use of color photocopiers. In addition, there is no interest in copying an existing Lippmann OVD on a document since it is unique to that document only.

19.4.1 Brief Description of the Recording Technique

To record a Lippmann photograph on photopolymer material, the photosensitive layer has to be rather thin (in the order of a few micrometers only). Moreover, the light-sensitive layer must be coated onto a flexible transparent base, and a special type of reflecting foil has to be laminated to the photosensitive polymer layer (it should be noted that contact between the two must be perfect). Experimental photopolymer materials have been manufactured by DuPont to prove the concept of recording Lippmann photographs in modern photopolymer materials.[4] The polymer film, which, as indicated, is laminated to the reflecting foil can only be exposed in a special camera. Once the document has been recorded, the reflecting foil is detached from the photopolymer film, which is developed with the assistance of strong white light, or UV light. The brightness of the image is subsequently increased by heat-treating the film. The development of the DuPont type of photopolymer film does not involve

Figure 19.6. Passport with a Lippmann OVD attached in the upper right corner.

any liquid agents, and the entire process is therefore "dry." As a result the technology can easily be incorporated in machines that record and process Lippmann security labels. Once processed, the transparent photopolymer label is laminated to the security document.

As the processed polymer film contains no dyes or fading chemicals, the film's archival stability is expected to be very high. The Lippmann OVD simply consists of a piece of plastic material onto which the information is recorded as an optical phase structure (refractive index variations within the photopolymer layer).

Figure 19.6 depicts a sample US passport having a Lippmann OVD attached in the upper right corner. Figure 19.7a shows a close-up of the Lippmann OVD in color, when observed perpendicular to the passport page. In Fig. 19.7b, the Lippmann OVD appears as a negative when viewed at an angle. Figure 19.7c illustrates the color shift when the OVD is illuminated and observed at oblique angles.

19.4.2 Advantages of the Lippmann OVD

Lippmann photography offers a new type of optical security device that is unique, and can be individually produced for each security document issued. A Lippmann OVD offers the following advantages:

- Equipment capable of automatically recording and processing Lippmann OVDs can be manufactured for use by issuers and manufacturers of security documents.
- The recording process is uncomplicated and does not require a specially-equipped laboratory.

(a)

Figure 19.7a. Lippmann OVD in color.

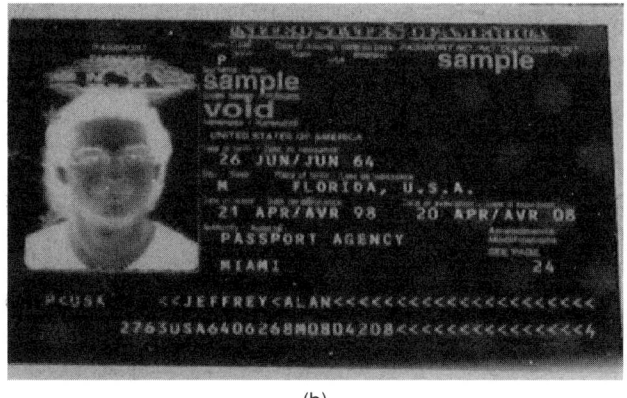

(b)

Figure 19.7b. Lippmann OVD as a negative.

(c)

Figure 19.7c. Lippmann OVD illuminated and observed under oblique angles.

- Access to the photosensitive film used for recording purposes (for example the special photopolymer film), can be strictly controlled by the film's manufacturer. Only approved manufacturers and issuing authorities can order the requisite material from the film manufacturer (e.g., like currency printers ordering banknote paper).
- The Lippmann OVD has a very high archival stability.
- The Lippmann OVD is Bragg sensitive. In other words, it changes color, depending on the angle of illumination and observation. It also switches between a positive and negative image. These features are extremely important because the effects are easily to discern when inspecting the Lippmann OVD.
- The Lippmann OVD cannot be copied by means of conventional color photography. Neither can it be reproduced using color photocopiers or scanners.
- Since the resolution of the Lippmann OVD is extremely high, a reduced image of, for example, the biographical data page, can be laminated to the security document (occupying only a small area of it). In such instances, the information recorded in the high-resolution Lippmann image may need to be magnified before detailed inspection can take place.

Lippmann OVDs can be used to make passports more secure and counterfeit resistant. In this case a Lippmann OVD can be recorded of the biographical data page, and could include specific bearer-related information, including the bearer's signature, and a reproduction of the bearer's (conventional color) photograph. The reduced Lippmann image is laminated to the page, in an appropriate position. The color shift displaced by the Lippmann OVD indicates that it is a genuine Lippmann photograph, as opposed to a conventional photograph. In addition, all information recorded in the OVD may be compared to the corresponding information in the document itself. Lippmann OVD-protected documents are extremely difficult to tamper with. The most important advantage is that there really is no point in trying to copy a Lippmann OVD; it is unique to a particular document and cannot be applied to other documents. Although the authenticity of a Lippmann OVD can be easily discerned with the naked eye, it is also possible to develop automatic inspection equipment capable of checking the iridescence of the image or comparing the information recorded in the document to the information "stored" in the Lippmann OVD.

The application of Lippmann photography to security documents is protected by a US patent[6] and European patents are pending. Although DuPont has expressed an interest in developing and manufacturing the special photopolymer film needed to create Lippmann OVDs, there is still a requirement for parties capable of designing and developing the requisite recording and processing equipment. Moreover, additional work is needed to improve the quality of the recording material used for Lippmann OVDs.

19.5 Conclusion

Today, there is worldwide interest in improving document security in general. In England, as in many other countries, where OVDs have been introduced to protect banknotes, use is often made of kinegrams (similar to an embossed hologram). Even though personalized documents, such as the new UK passport, also contain OVDs, identical holograms are used in all passports. The same applies to credit cards, which despite being personalized carry identical holograms. As described above, Germany recently introduced a new type of passport and ID card that contain individually-made volume holograms. There is no doubt that the need to make personalized documents more difficult to fake can only be catered for by personalizing the security features they contain.

The worldwide market for document security is large and growing rapidly. There are many personalized documents that would benefit from a higher degree of security, including drivers' licenses, travel documents (Schengen Visa), corporate ID cards, ID cards for pilots, people working in nuclear power plants, military personnel, police officers, custom inspectors, etc. All are potential candidates for more secure ID documents.

As far as Lippmann OVDs are concerned, the way forward is to introduce a complete recording/processing system that may be used to manufacture the requisite Lippmann OVDs in-house. Today, mass-produced, embossed holograms and other types of OVDs (e.g., kinegrams) are made by different holographic manufacturers, which deliver their products to the issuing parties for inclusion in the various security documents. Since the Lippmann OVD is unique to a particular document, there is no need to produce them in advance. There is, therefore, a distinct requirement for special recording/processing equipment that can be sold or leased to issuing authorities or companies. The special machine that is similar to a photocopier enables issuers to record the Lippmann OVD when the document is prepared. As a consequence, the issuer has full control over the issuing process and the security that surrounds it.

References

[1] Benton A. (1969). "Hologram reconstructions with extended incoherent sources." *J. Opt. Soc. Am.*, 59:1545–1546A.
[2] Bjelkhagen HI. (1993). *Silver Halide Recording Materials for Holography and Their Processing,* Springer Series in Optical Sciences, vol. 66. Springer-Verlag, Heidelberg, New York.
[3] Bjelkhagen HI, Jeong TH, Vukicevi D. (1996). "Color reflection holograms recorded in a panchromatic ultrahigh-resolution single-layer silver halide emulsion." *J. Imaging Sci. Technol.*, 40:134–146.
[4] Bjelkhagen HI. (1997). "Lippmann photographs recorded in DuPont color photopolymer material." in: *Practical Holography XI and Holographic Materials III.,* Benton SA; Trout TJ. eds. Proc. SPIE 3011:358–366.

[5] Bjelkhagen HI. (1999a). "A new optical security device based on 100-year-old photographic technique." *Opt Eng.,* 38:55–61.

[6] Bjelkhagen HI. (1999b). *"Secure photographic method and apparatus."* US patent No. 5, 972, 546.

[7] Honnorat A. (2004). Honnorat Recherches & Services, 23 rue President Favre, F-74000 Annecy, France.

[8] Kodama D, Watanabe M, and Ueda K. (2001). "Mastering process for color graphics arts holograms." in: *Practical Holography XV and Holographic Materials VII.*, Benton SA; Stevenson SH; Trout TJ. eds. Proc. SPIE, 4296:198–205.

[9] Lancaster I. (2004). "High quality Euro hologram counterfeits revealed at conference." *Holography News,* 18 (3):1–7.

[10] Lippmann G. (1891). "La photographie des couleurs." *C R Hebd Seances Acad Sci,* 112:274–275.

[11] Steijn KW. (1996). "Multicolor holographic recording in Dupont holographic recording film: Determination of exposure conditions for color balance." in: *Holographic Materials II.* Proc. SPIE 2688, Trout TJ. ed. pp. 123–134.

[12] Trout TJ, Gambogi WJ, Stevenson SH. (1995). "Photopolymer materials for color holography." in: *Applications of Optical Holography.* Proc. SPIE 2577, Honda T. ed. pp. 94–105.

[13] van Renesse RL. (Ed.). (2005). *Optical Document Security,* 3rd edn. Artech House, Boston, London.

[14] Watanabe M, Matsuyama T, Kodama D, Hotta T. (1999). "Mass-produced color graphic arts holograms." in: *Practical Holography XIII.* Proc. SPIE 3637, Benton SA. ed. pp. 204–212.

[15] Zacharovas SJ, Ratcliffe DB, Skokov GR, Vorobiov SP, Kumonko PI, Sazonov YuA. (2000). "Recent advances in holographic materials from Slavich." in: *HOLOGRAPHY 2000,* Proc. SPIE, 4149, Jeong TH; Sobotka WK. eds. pp 73–80.

Index

Page numbers with t and f represent table and figures respectively

2D correlation filters, 319
2D Gabor filter, 119
2D shape recognition of microorganisms,
    138
  based on image intensity, 138
3-bit code, 215
3D object recognition, distortion-
    tolerant
  composite correlation filters, 14–16
  neural network, 16, 18–20
3D object recognition, experiments, 128
3D
  amplitude distribution, 10, 12
  auto-stereoscopic display, 155
  computations, 272
  correlation techniques, 10
  filtering, 320
  Fourier transformed functions, 326
  image-based recognition system of
      microorganisms, benefits of, 140
  imaging with SEOL digital
      holography, 147
  light distribution, 11
  microorganism reconstruction and
      feature extraction, 147, 149
  modified Dynamic Link
      Association (DLA), 120,
      125–126
  object recognition, 1–2, 14
  optical code, 179

optical monitoring system, 140
recognition technique, 21
reference object, 14
scene reconstruction, 48
scene, artificial views constructed of,
    164f
security codes, 179
spatial bandwidth, 132
spatial frequency, 132
topology, 140
transformation, 11
4f-system, 342
5th order polynomial kernel, 305

A

Active imaging systems, 175
  wavelength-hopped laser
      system, 175
Active optical tags, 179
Adaptive matched filter, 291
ADM's holographic label, 388;
    see also Izon^{TM} technology
Advantage of multiple cameras
  nonoverlapping fields, 206
  overlapping fields, 206
Affine transformation, 256
AI-based fusion techniques,
    examples of, 245
Airy disk, 54
Alleles, 192

All-optical implementation, 156
Ambient light, 274
Amplitude
 distribution, 90
 reconstruction, 54
Analog optical spatial filtering, 344
Angular pixel resolution, 184
ANN, unsupervised, 25–26, 34, 38
ANNs, suitable for nonuniform
  quantization, 32–33
Appearance-based system for synthetic
  images, 269
Artificial neural network (ANN), 230
 clustering algorithms, 32
Astigmatic lens, 340
Astigmatism, 340
ATR-Tracker interaction
 Mini model, 213
 Tank model, 213
Attenuation constant, 290–291
Autocorrelation
 of the integral image, 164
 peak, 164
Automated fashion, 273
Automatic
 passenger counting system, 245
 scaling, 54
Automatic target detection (ATD),
  205–206
Automatic target recognition
  (ATR), 205
Automation mode, 268
Axial displacement, 68

B
Back propagation neural network, 230,
  231f, 240
Background model for a pixel, 247–248
Background noise, 49
 model, 330
Background subspace, 294–296, 309
Bandpass filter, 119
Bandwidth, 56
 management approach, 183, 192
 manager module, 194–195, 198
 reduction, 198, 198f–200f
Bank of linear correlation filter
 correlation plane, 206
 input test image, 206

Batch
 fabrication by lithographic techniques,
   337
 processing, 336
Beam
 expander, 27, 62, 88
 splitter, 27, 48, 62
Bidirectional reflectance distribution
  functions (BRDF), physical
  models, 248
Binary random variable, 320
Biochemical recognition reactions, 365
Biometric data, 350
Bovine Serum Albumin (BSA), 374
Bragg effect, 178
Burrows-Wheeler (BW), 30, 32, 42
BW technique, 30
BW, see Burrows-Wheeler (BW)

C
Camera geometry, 207–208
Cameraplane, 28
Camouflaged objects, 283
Carrier frequency, 119
Cartesian coordinates, 10
Cast shadows, 270
CCD
 camera, high-definition, 49, 76,
   141, 157
 detector, 8
 recording plane, 59
 sensors, 3, 75
 TV, 49
CFAR, 291–292
Chromosome, 191–192
Classification
 operation, 190, 198, 201
 output transmission, 192
Cleaving tool, use of, 371
Client-server model, 212
Close-up images of, 267, 269, 274
CNR INOA, 58
Coarse matching stage, 120
Coarsest angular resolution, 198
Code sequence, authentic, 176–177
Codebook vector, 34
Coherent imaging applications
 fingerprint recognition, 48
 forensics, 48

object recognition, 48
ranging, 48
Coherent light, 1, 7, 47, 88, 143
Coherent speckle noise, 76
Collimated illumination, 176
Collimating lens, 176
Color holograms, 383
Color reflection holograms recording,
    388–389
Color reflection holograms recording,
    materials used
  photopolymer materials, 389–390
  silver halide materials, 389
Comparison between linear and nonlinear
    correlation, 236–237
Complex conjugate, 50
Complex optical field, 62
Composite filter, 14–16, 18, 319
Computation time, 11
Computational load, 149
Computer processing speed, 325
Computer simulations, using LADAR
    data, 326, 330
  test of optimum nonlinear filter,
    326–328
  test of optimum nonlinear distortion
    tolerant composite filter, 328–330
Concatenated
  matrix, 294
  vector, 298
Conditional likelihood ratio, 295
Confusion disk, 63
Constrained least squares, 290–291
Constraint vector, 215
Construction set, 224
Controlling
  image resolution, 75–80
  image size, 70–75
Conventional
  optical microscope, 62
  matched filter detector (MFD), 302,
    305, 306f–307f, 309f, 314
  RX algorithms, 299
Convolution, 5
  operator, 321, 333
  transformation method, 55
Cooperative Coevolution (CC) algorithm
  credit assignment, 251
  interdependability, 251

population diversity, 251
problem decomposition, 251
Cooperative coevolutionary paradigm,
    252
Corner cube reflectors, see Retroreflective
    reflectors
Correct recognition, 276, 279f
Correlation filtering approach
  classification, 206
  target detection, 206
Correlation of holograms
  measurement of small rotations, 9–10
  object recognition, 8–9
Correlation
  approach, 319
  credibility function, 253–254
  filter, 206
  lengths, 7
  measurement, 224
  pattern recognition technique, 214
  plane, 227, 233, 235–236
Cross-correlation
  of integral image, 164
  peak, 164, 167, 170
Cryptographic techniques using light, 352
Cryptography, 352
Cryptosystems, 25

D
Data
  space, 285–286, 288
  vector, 34
Decryption, 25–26, 28, 42
  and reconstruction, 28–29
Defocus aberration, 61
Deformable mirror devices, 179
Detection probability, 300, 303f–304f
DFT algorithm, 56
DHM, see Digital holographic microscope
Dichroic filters, 389
Dichromatic model, 249
Dielectric materials, 249
Dielectric thin films, deposition of, 373
Diffraction, 342
  integral, 87
  limit, 54
Diffractive
  microoptics, 346
  optical element (DOE), 346

Digital hologram, 4, 8, 10–13, 91
  compression, 25–26
  encryption, 27–28
Digital holographic interferometry
  classical holography, 60
  deformation measurement, 60
  optical path, 60
  refractive index, 60
Digital holographic
  microscope, 61, 63
  microscopy, 60, 66
Digital Holography (DH), 138, 179, 181f
  compensation of aberrations, 60–61
  convolution approach, 55–56, 70
  performance and limitations, 56–57
  reconstruction methods, 52–56
  theory and principle of operation,
      49–52
Digital Holography, reconstruction
      methods
  convolution transformation method,
      55–56
  Fresnel Transformation Method
      (FTM), 52–55
Digital
  3D model, 155
  camera, high-definition, 155
  data, 90
  processing, 49
Digitally reconstructed holographic
      image segmentation, 143
Discrete Fourier transform operator, 333
Discrete mounting techniques, 335
Discrimination capability, 322
Display applications, 25–26, 34, 38
Distortion
  tolerance, 14, 206
  free imaging, see Optical
      interconnection tasks
Distortion-tolerant
  filter, 320
  object recognition, 128–129
  tolerant, 113
Double exposure holography, 61
Dual concentric windows, 299, 299f
DuPont Authentication Systems (DAS),
      387
DuPont photopolymer materials, 385
  color, 390

Dynamic Link Association (DLA), 114,
      145
  3D Modified, 120, 125–126
  coarse matching stage, 114
  fine matching stage, 114
Dynamic sensor fusion, 265

E
e-beam lithography, 382
Eigenvalues, 288–289, 298, 316
Elastic Graph Matching (EGM), 114
  with sequential and recursive
      realization, 126–127
Electrical storage, 90
Electron microscope images, 48
Electrooptical devices, 1
Elemental image, 155–161, 160f–162f,
      164–165
  central, 161
Embossed holograms, 384
Embossing techniques, 178
Emissivity of the human body, 262
Empirical kernel map, 289, 293
Encryption and decryption experiments
  decryption, 91–92
  encryption, 89–90
  experimental system, 88–89
  recording digital hologram, 89
Encryption, 25–31, 42, 85–92
  wavefront, 90
Environmental degradation effects, 178
Equiprobability, 32
Etching, see Batch fabrication by
      lithographic techniques
Euler angles, 120
Evaluation of nonuniform quantization
      techniques, 33–38
Expectation operator, 333
External noise factors
  environmental fluctuation, 138
  environmental vibration, 138
Eye locator program, 277
Eye model
  location of eye, 269
  normalized greylevel values, 269
  variations of eyelids, 269
Eye
  finder program, 268f, 271f
  lid segmentation, 273

pixel, 273
recognition, 268
region, 272–273, 277
socket, 273

F

Face/identity recognition
   algorithm, 268
   /verification of, 267–268, 280
False alarm probability, 128
Fast Fourier transformation (FFT), 151
Fatigue effect, 33
Feature vector, 139
   extraction, 119–120, 143–145
FERET
   Cast shadow operator (CS), 273, 274f
   database, 273, 377
   image, 277–278, 279t
   subjects, 278
FFT algorithm, 142
Field of view (FOV), 189, 206, 210, 299
Filamentous
   microorganisms, 139
   structures, 145
Filter, generic single-layer fiber-tip
      reflection, 366f
Filtering applications, spectral, 364
Fine matching stage, 120
First-order correlation, 284
Fitness function, 191–196, 198, 201t
   thermal channel, 252
   video channels, 252
Flat earth geometry, 184
Flip-chip bonding, 337
FLIR system thermal camera, 255
Focal resolution, 62
Focus tracking during dynamic
      recording of digital holograms in,
      67–70
Fourier
   domain expression, 321
   lenses, 346, 349
Fourier transform, 1, 5, 8, 12, 14, 343
   discrete, 321, 334
   inverse, 166
   noise, 321, 333
   reverse, 357
   spatial, 355
Fourier transformation, 350

lenses, 345, 351
Frame rate, 184, 187, 189, 191, 195, 198,
      194t–197t, 200f
Free-space optics, properties of, 1
   high space-bandwidth product, 1
   massive parallelism, 1
Fresnel diffraction
   field, 138
   integral, 90
Fresnel
   approximation, 10–11
   diffraction, 20, 26, 42, 55
   integral, 60
   propagation, 26, 28, 32, 142
   reconstruction algorithm, 54
   region, 76
   transform, 54
   transformation, 142
   Transformation Method (FTM), 52–55
   treatment, 60
Fresnel–Kirchhoff integral, 5
Fringe projection, 61
FTM, *see* Fresnel, Transformation
      Method (FTM)
Full-face photos, 275f
Full-rank matrix, 295
Future netted system
   effective strike operation, 204
   initial target detection, 204

G

Gabor filter, Gaussian-envelope in, 144
Gabor filtering
   2D, 113
   3D, 113
Gabor
   3D, 114
   based wavelets, 114, 143–144
   coefficient, 120, 144
   feature, 113
   filter, 144
   jets, 114, 119, 120
   kernels, 119, 137, 144–145
Gabor-based wavelets, 144
   impulse response (or kernel), 144
Gaussian
   envelope, 119, 144
   form kernels, 143
   illumination, 347

Gaussian (*continued*)
models, 250
parameters, 253
probability density, 294
Radial Bases Function kernel (RBF
kernel), 286
random noise, 294
random process, 286
RBF kernel, 300, 302, 305, 306f–310f,
309
white noise, 238
Gaze
angle, 268, 270, 279
direction, 277
Generalized
likelihood ratio test (GLRT), 286, 291,
294–296, 298, 311, 314
phase contrast method, *see* GPC
method
Generation number, 198, 201t
Genes, 191–193
Genetic
algorithm (GA), 183, 191–196, 198, 201,
201t
evolutionary process, 247
optimization algorithms, 192
Genotype, 191
Gerchberg-Saxton, modified iterative
algorithm of, 347
German personal identification card, 387
Global
coordinate system, 187
label, 212
recognition peak, 165
GMR sensor, TE polarization spectral
shift for the two-layer, calculation
of, 375f
GPC method, 353–356
using planar integrated microoptics,
miniaturization of, 356–357
Gram
kernel matrix, 299
matrix, 289, 315–316
Grating
fill factor of resonant sensor,
calculation of, 369, 370f
thickness of resonant sensor,
calculation of, 369, 370f
Gray scale, 319, 325

GRIN lens, 176
Groundtruth data, 253
Guided-mode resonance effect, 363
application of, in optical sensing, 364

H
Half-wave plate, 63
Head rotation, 242f, 268, 270, 274–277,
279–280
Head tilt, 268, 279–280
downward, 278
upward, 271f, 278
Height distribution, 62
He-Ne laser, 158
High
dimensional feature space, 283, 285–287
frequency spatial variations, 7
order correlation, 283, 314
resolution sensors, 183
Histogram analysis, 139, 143
Hologram recording, 383–384
digital, 2–4
experimental setup for, 2
materials for, 385
phase-shift technique, 4
Hologram, 2, 7–11, 14–16, 18–19
CCD camera, 2
for improved OVDs, 385
fringes, 68
interference pattern, 2
Mach-Zehnder interferometer, 2
plane, 56
recording on charge coupled device
(CCD) cameras, 49
window extraction in, 5–6
Holograms, types of, 381
volume holograms, 382
volume reflection holograms, 382
Holographic image, 151
quality of, 383
Holographic
interferometry, 48, 61, 70
sensor, computational, 140
speckle, 26
Holography as a metrological tool, 48
Holography, 1–2, 156, 381
limitations, 48
property, 5
Homeland security, 47, 344

Hough transform, 211
Huffman coding, 30
Human aided recognition of
    microorganisms, 140–152
  recognition of *Sphacelaria* Alga, 149
  recognition of *Tribonema aequale* Alga,
    149–152
Hybrid encryption system, 86
  decryption, 87, 90
  encryption, 86, 89
Hybrid imaging, 338, 342–343
Hybrid integration, practical aspects of
  capsuling for processing and operation,
    337
  thermal and mechanical stress to the
    bonds, 337
Hybrid optical encryption, 85
  of a 3D object, 92–93
HYDICE
  images, 299–300, 300f, 302, 314
  imaging sensor, 299
  test set, 311
Hydrogels, use of, 378–379
Hyperspectral imagery, 283
Hypothesis test, 293

I

Identigrams, 387
Illumination effects, 305
Image
  acquisition systems, 63
  compression, 25–26
  encryption, 137
Imaging
  sensor acquisition rate, 325
  systems for discrete objects, design
    considerations of, 341–343
Imaging, based on radiation
  coherent, 47
  incoherent, 47
Incorrect recognition, 276
Incremental angle, 238
In-line digital holograms, 26
Inner-window region (IWR), 299–300
Input irradiance distribution, 12
Input
  parameter, 273
  pixel, 284
  vector, 34

Inspecting microstructure by, 63–66
Integral imaging technique, 155, 171
  principle of, 156–158
Integral photography, 1
  scheme, *see* Integral imaging technique
Integrated
  circuits, fabrication of, *see*
    Miniaturization
  fiber-tip biosensors, potential
    applications of, 378
  optical correlators, planar, security
    application of, 349–351
Integrated optical correlators,
    experimental demonstration of
  off-axis integrated correlators,
    348–349
  on-axis integrated correlators, 346–348
Intel Web-cam, 255
Intensity
  distribution, 54
  normalization, 268f, 274, 279f
  profiles, 54
Interference
  angle, 56
  filter, 176, 177f
  fringes, 59
  microscopy, 66
  pattern, 49, 50, 141
Interferogram, 3–4, 28, 142
Interferometric
  fringe pattern, 48
  measurements, 66
Inverse multiquadric kernel, 305,
    306f–308f
Iris
  recognition, 268
  segmentation, 273
  translation, 276–277
Isolated planar objects, 161f
ISP framework, 205
Izon^{TM} technology, 387–388

J
Joint power spectrum, 164

K
Kernel feature map and kernel learning
  input hyperspectral data, 285
  kernel functions, 285

Kernel feature map and kernel learning
    (*continued*)
  linear target detection algorithm, 285
  Mercer kernel, 285
  nonlinear mapping function, 285
Kernel RX-algorithm, 299–300
  background clutter pixels, 288
  covariance matrix, 288
  eigenvector decomposition, 288–289
  nonlinear mapping, 287
Kernel
  based methods, 283, 285
  matched filter detector (KMFD), 302,
      305, 306f–307f, 309f, 314
Kernel-matched filter, 290–293, 302
  attenuation constant, 291
  CFAR-matched filter, 292
  equivalent matched filter, 291
  high dimensional vector, 291
  kernel feature space, 291
  nonlinear matched filter, 292
  kernel function, 292
Key benefits of netted and distributed
      systems for security, surveillance
  adaptive self-adjusting properties, 205
  efficient resource usage, 205
  improved ATD/R performance,
      205–206
  improved strike capability, 205
  increased tolerance, 204
Kinegrams^{TM}, 349–351
k-means clustering algorithm, 34
KMSD, 305, 309, 311f–313f, 311–312, 314
KNIGHT
  human detection, 205
  video processing system, 213
KNIGHT tracking system, 212
  activity detection, 209
  object detection, 209
  shadow removal, 209
  tracking object classification, 209
Kodak Megaplus CCD camera, 27
Kohonen competitive network, 32–34,
      37–43
Kronecker
  delta, 323
  product, 215
kth law, 14
  nonlinear correlation effect, 167

kth-law nonlinear correlation for pose
      estimation
  2D Fourier transform, 232
  Fourier-plane SDF filter, 233
  Fourier domain, 232
  maximum correlation peak, 233
  nonlinear factor k, 232
  shift invariance, 233
k-tuple SDF, 214

L
LADAR image, 2D encoded, convertion
      of, 324–325
LADAR
  autonomous vehicle navigation, 325
  battlefield assessment, 325
  camera measures, 319
  range data, 319
  range image, 319, 326, 328f
  range sensor, 328f
  sensor location, 324
Lagrange multiplier, 322
Laser tagging by wavelength division
      multiplexing, 177f
Last In First Out (LIFO), 253
Lateral resolution, 5
Least squares approximation, 295
Lempel
  Ziv (LZ77), 30, 32, 42–43
  Ziv-Welch (LZW), 30, 32, 42–43
Levenberg-Marquardt algorithm
  artificial neural network (ANN), 230
  single-layer ANN, 230
  standard deviation, 230
  two-layer back propagation neural
      network, 230, 231f
Light field, 11
Light pipe
  configuration, 345
  correlator, 345f
Light source, 273
Lighting condition, 267–268, 274–275,
      278–280, 275f
Likelihood ratio test (LRT)
  Gaussian probability density, 294
  Gaussian random noise, 294
  Neyman-Pearson criterion, 294
  maximum likelihood estimates (MLEs),
      294

Linear
  algorithm, 285–286
  correlation, 224
  dependency assumption, 239
  dependency, 226, 239
  discriminant analysis, 133
  estimation, 226, 230
  matched (subspace) filters, 284–285,
    290–291
  mixture model, 284
  phase factor, 6, 8
  (subspace) mixing model, 284
  weighted composite filter, 225, 240
Linear correlation for pose estimation
  column vectors, 224, 227
  composite correlation filter, 223–224
  correlation peak value, 226–227
  correlation value, 224, 226–227
  F15 airplane, 224–225, 225f
  in-plane rotation, 225
  matrix transposition operation, 224
  out-of-plane rotation, 225
  pose parameters, 224
  pose vector, 224, 226
  recognition flag, 225
  synthetic-discriminant function (SDF)
    filter, 224
  transformation matrix, 226
Linear-matched filter, 290–291
  attenuation constant, 290
  Constrained Energy Minimization
    (CEM), 290
  covariance matrix, 290
  noise, 290
  target spectral signature, 290
Linnick interferometer, 61
Lippmann OVD holograms, use in
    passports, 395
Lippmann OVD
  advantages of, 392, 394
  kinegrams, 395
  photosensitive film, 394
  recording technique, 391–392
Liquid crystal display, 179
Liquid Crystal on Silicon (LCOS) SLM,
    361
Lithium niobate structure, 64
Lithographic fabrication, 336
Local

kernel matrix, 299
  spectral analysis, 143
Localized spectral difference, 284
Longitudinal shift, 2
Lossy compression of encrypted digital
    holograms, 29–32
Lower shadow extant, 273
Low-level disposable sensor, 204
LZ77 algorithm, 30
LZ77, see Lempel-Ziv (LZ77)
LZW, see Lempel-Ziv-Welch (LZW)

M
MACH separation metric, 217
Mach-Zehnder
  configuration, 63
  interferometer, 27, 58–59, 63, 139
Mahalanobis distance measure, 284
Mapping technique, 320
MasterCard^{TM}, 349
Matched subspace filtering techniques,
    284
Maximum average correlation height
    (MACH), 206, 216
Maximum likelihood
  estimates (MLEs), 294–295
  principle, 294
Mean Absolute Error (MAE), 128,
    130
Median
  fitness value, 198, 201t
  plane, 5
Micro-electro-mechanical-system
    (MEMS), 65, 179
  structures, surface profiles of, 65
Metric, 274, 277, 280
Microchannel imaging, 338, 341f, 342
MEMS, see Micro-electro-mechanical-
    system (MEMS)
Microfabrication
  process, 69
  techniques, 336
Microlens array, 1, 155, 156f, 157–158,
    161–162, 165, 341
Micromachined beams, 65
Microoptical
  systems, 335
  technologies, 342
Micro-optics, 178

Microscopic applications, 60
Microstructures
  circular fringes, 63
  fabrication, 65
  functionality, 65
  inspecting by DHM, 63
  reliability, 65
  surface morphology, 65
  topographic characterization, 63
Military vehicles, 299, 302, 309
Milling, *see* Batch fabrication by
      lithographic techniques
Miniaturization, 335
  problem, 320
MIRAGE holograms, 385
Mixture of Gaussians, representation of,
      247–248
Mode of detection, 368
Modulation speed, 344
Monochrome reflection hologram
      of reflection type, recording of,
      386
Moore-Penrose pseudoinverse, 228
Morphological traits, 138
Mosaicing operation, 183–184, 193, 198,
      185f, 193f, 194t
Motion-based classification, 210
Moving object detection
  advantages, 243
  shortcomings, 243–244
Moving object detection, categories of
  feature-based methods, 243
  featureless methods, 243
Moving target density, 189, 195
MSD, 312, 313f, 314
MSTAR data set, 214, 217
Multidimensional image data, 286
Multiple
  exposures, 138
  fiber optics links, 176
  interferogram recordings, 138
Multisensor fusion and integration
      approaches, paradigms used
  Artificial Intelligence (AI), 245–246
  data structure, 245–246
  physics, 245–246
  statistical, 245–246
Multisensor fusion, advantages, 244
Multiwavelength DH (MWDH), 70

N
Natural clutter, 284
NBPP, 197t, 198
NBPROI, 197t, 198
Near-field approximation, 56
Negative response (code bit 0), 218
Network architecture
  Internet Protocol (IP), 211
  Transmission Control Protocol (TCP),
      211
Network integration of tracking and
      ATD/R
  hoc network, 211
  synchrotech adapter, 211
  wireless card from MeshLAN, 211
  wireless peer-to-peer, 211
Network-centric application, 214
Neural network, 14, 16, 19–21, 230
Neutral
  density filter, 63
  expression, 276
Neyman-Pearson criterion, 294
Noise robustness capability, 322
Noise/clutter suppression, 178
NonFERET, 278, 279t, 279f, 280
Nonlinear
  correlation, 14, 232
  mixture model, 285
Non-overlapping pixel data, 184
Nonstationary local mean, 286
Normalized rms (NRMS), 35, 38
NRMS, *see* Normalized rms (NRMS)
Numerical reconstruction, 49
Nyquist limit, 54–55

O
Object beam, 62
Object recognition system, 8, 25, 113
  2D, 113
  3D, 113
  an experiment, 8–9
Optical information
  systems, 137
  technologies, 85
Optical
  axis, 142
  codes, 175, 178, 180f
  coherent microscopy and imaging, 48
  communication systems, 344

correlation on discrete input/output
  arrays, 344
correlation techniques, 1
correlator systems, 335
cryptography, 353
field, 51
ID tags, 175, 178
identification phase code, 177–178
implementation, 25
interconnection tasks, 341
multimode waveguide, 340
processing, application of, 1
security technologies, 175
setup,coupling efficiency of, 341
signal processing, *see* Spatial filtering
system aperture, 54
waveforms, 176
wavefront, 48
Optimum nonlinear distortion-tolerant
    filter, 319, 324
  3D, 326–327, 329
Optimum
  parameter values, 195, 198
  reduction regime, 183
  rule quantity, 187
Optoelectronic devices, 49
Optoelectronical systems, 335
Optomechanical system, 337
Optomechanics, 336
Orthogonal matrices, 294
Outer-window region (OWR), 299–300
OVD, 382
  based on Lippmann photography, 391
  based on volume phase holograms of the
      reflection type, 385

P
Panchromatic polymer material, 390
Parabolic
  approximation, 338
  phase factor, 63
Parallax, 85, 91
Paraxial approximation (PA), 51
Pattern recognition
  problems, 319
  systems, 137
Pavlidis et al., automatic passenger
    counting system, 245
Peak response of a biosensor, 366

Peak to sidelobe ratio (PSR), 207
Percent-of-Saving, 185, 187, 189–191
Performance Analysis, receiver operating
    characteristic (ROC) curves, 262,
    265
Performance characteristic function and
    filter synthesis
  ATR algorithm, 216
  2D Fourier transforms, 216
  MACHfilter algorithm, 216
  mean and spectral variance, 216
Phase shifting, 57–59, 138
  algorithm, 57–58
  technique, 4–5, 89
Phase
  contrast filter (PCF), 355
  distribution, 61
  map, 64
  masks, fabrication of, 178
Phase-only optical decryption, 354
  miniaturized GPC system for, 358–361
Phase-only
  cryptography, 353, 357
  filter, 12
Phase-shift
  digital holography, 27
  holography, 152
  interferometry (PSI), 25–26, 28
Phase-unwrapping methods, 62
Phenotype, 191
Photo
  detector array, 176
  sensor, 138
Photographic holography, 51, 61
Photographic
  materials, 49
  media, 49
  plate, 48, 58
  transparency, 58
Photopolymer, 178
Physical Models of Reflectance, 248–249
  Dichromatic, 248
  Lambertian, 248
  Phong, 248
  Ward, 248
Piezoelectric transducer mirror, 58
PIFSO imaging systems
  optical correlation of, 343–349
  paraxial theory of, 338–341

PIFSO imaging systems (*continued*)
  space-bandwidth product (SBP),
    calculation of, 340
  transmission function of ideal parabolic
    lens, calculation of, 339
Pixel, 4–5, 7–10, 13–14
  vectors, 293, 299, 305
Planar integrated free-space optics
    (PIFSO), 335–336
  basic concept, 335–337
  design considerations, 337–343
Planar
  integrated imaging systems, 335
  integration of discrete optical
    correlators, 344–345
  integrated free-space optics (PIFSO),
    353, 356
Plane
  and volume holograms, 383
  wave, 141
  wave model, 365
Polarized light, 140
Polarizing beam splitter, 63
Polynomial kernel, 286, 305, 306f–308f
Polysilicon cantilevers, 65
Potential infinite dimensionality, 296
Potentially infinite dimension, 284
PPLN structure, 64
Pressure, 198
Principle of triangulation, 158, 160, 171
Probability
  density, 294
  function (pdf), 320
  of observing a background pixel, 248
Processor-In-a-Box (PIB), 388
Projection matrix, 295–296
Propagation distances, 12
PZT stage, 88

Q
Quantization of encrypted digital
    holograms, 38–42

R
Range
  camera, 319
  images, 156
  finder system, 2
Raster, 76

Rayleigh
  criterion, 340
  Sommerfield diffraction integral, 51
RCWA code, 365
Real
  optical system, 92
  time acquisition, 66
Receiver operating characteristic (ROC),
    187
Recognition of object by pose estimation
  construction set, 233
  nonlinear filter, 233
  optimum nonlinear filter, 232
  recognition flag RF, 232
  two-layer ANN, 232
Recognition task of
  facial image data, 267
  frontal view, 267
  head pose, 267
  images, 268, 277
  natural facial variations, 267
Recognition
  rate, 268, 276, 280
  system, 279
Reconstructed 3D scenes, correlation of
  comparison between 2D and 3D
    correlation, 170–171
  principle, 165–167
  recognition of a 3D object using
    nonlinear correlation, 167
  three-dimensional object localization,
    167–170
Reconstructed
  holographic images, 140
  images, enhancement of, 7, 70
Reconstruction
  algorithm, 49
  pixels (RP), 67, 70
Recording
  distance, 56
  environment, 142
  geometry, 60
  medium, 48
  plane, 58
  sensor array, pixel area of, 57
Recurrent Motion Images (RMIs), 210
Recursive method, 126
Reference wave, 48
Reflective coatings, 337

Refractive index
    of the attached layer, calculation of,
        369, 370f
    sensing in water, calculation of,
        366f–368f
    units (RIU), 366
Region of Interest Extraction, 184–185
Regions of Interest (ROI)
    clutter density, 185, 187
    dynamic, 183, 188–190, 192, 195, 200f
    dynamic ROI transmission, 192
    extraction operation, 183, 185
    operating characteristic (ROC) curve
        of, 187
    static ROI transmission, 192
    target area density, 185, 187, 189
Regression techniques, 270
Relaxation parameter, 116
Remote authentication/verification
        system, 179
Replication techniques as a means for
        low cost, 336
    mass production, 336
Resonance switching phenomenon, 363
Resonant waveguide grating sensors,
        366
    application of, 371
Resonant
    central wavelength shift, calculation of,
        369–371
    device, two-layer, fabrication of, 374
    sensor functions, fiber-tip, 364
    sensor modeling, 365–371
    sensor, silanated, TE polarization
        spectral response of, 376f
Retroreflective
    optical ID tags, 175, 178
    reflective reflectors, 177
RGB combination, 75
RGM, 139
    process, 150
Rigid Graph Matching (RGM), 120, 137,
        145–146
    with Rotation-tolerant Property, 120,
        125–126
Rigorous coupled-wave analysis (RCWA),
        365
ROI extraction algorithm
    algorithm internal parameter, 187

complementary cumulative distribution
        function, 187
    false alarm probability, 187
    Image/Scene Metric, 187
    inverse complementary cumulative
        distribution function, 187
Ronchi grating, 73–74
Rotation
    tolerance, 14–15
    invariant property, 144
    tolerant object recognition, 129–130
    tolerant property, 114
RX anomaly detection, 286
RX-algorithm
    background clutter data, 287
    background clutter noise process, 287
    background covariance, 287
    Gaussian random process, 286
    Generalized Likelihood Ratio Test
        (GLRT), 286
    spectral pixel, 286

S
Sacrificial layer, 65
SAR images
    BTR (class-2 object), 217–218
    of target, 214
    T72 (class-1 object), 217–218
SBP, see Space-bandwidth product (SBP)
Scalar diffraction theory, 51
Scene-related parameters, 194–195,
        197–198, 201
Seamless tracking across network,
        205–206
Search algorithms
    brute force: depth first, breadth first,
        250
    genetic algorithms (GA), 250
    gradient methods: neural networks, 250
    heuristic methods: best first, beam
        search, A*, 250
Second-order correlation, 284
SECURE IMAGE^{TM}
    FOIL, 387
    LABEL, 385–386
Segmentation techniques, 238
Self-configuring network, 214
Self-organizing map (SOM), 32–38
SEM, 78

Semi-transparent objects, 143
Sensing modalities, 244
Sensor fusion architecture, physical models
  reflectance models, 246, 246f
  thermal models, 246, 246f
Sensor fusion system, advantages
  context-based adaptation, 244
  evolutionary-based approach for fusion, 244
  physical models, 244
Sensor fusion system, features of, 244
  consistent data representation, 244
Sensor network and algorithm, parameters settings between
  bit rates, 205
  modality, 205
  resolution, 205
  sensor geometry, 205
Sensor
  angular resolution, 193, 198, 200f
  fabrication, 371–378
  fusion system, 243
  high-resolution detail-oriented sensor, 204
  low-resolution simple sensor, 204
  polarization-mode waveguide-grating, 368
  processing techniques, 192
  resolution, 189, 191, 194t–197t, 200f
  spectral characteristics of, 364
  technology, 363
SEOL digital holography, 138
Sequential method, 126
Shift invariant
  filter, 320
  linear, 11
Shift-invariant 3D object recognition
  experimental results, 10–12
  principle, 12–13
Short wave infrared (SWIR), 245
Signal to noise ratio (SNR), 291, 299, 344
Silicon substrate, 65
Similarity function, 146
Single-exposure online (SEOL) digital holography, 137, 140–143
  benefits, 139, 142
  stages, 139
Single-layer waveguide grating, 364f

Site map, 208
SNR, see Signal to noise ratio (SNR)
Soft competition, 33
Solid-state pyroelectric sensors, for infrared, 49
SOM, see Self-organizing map (SOM)
Space-bandwidth product (SBP), 138, 340
Spatial
  coherence, 140, 238
  filter, 88
  frequency spectrum, 343
  light modulator (SLM), 179, 359
  registration, 256
  resolution, 4, 63
Spatial filtering, 343
  architecture, see Generalized phase contrast (GPC) method
  systems, 335, 337
Speckle
  diameter, 54
  methods, 48
  metrology, 61
  noise, 14, 21, 35
Spectral bands
  angle-based kernel, 305, 306f–308f
  anomaly detection algorithm, 284
  bands, 284, 286, 290, 314
  decomposition, 288
  library, 284
  matched filtering, 284
  pixel, 284, 286–287, 290
  signature, 284, 287, 290–291, 293
  subspace, 284
  values, 305
  variability, 293, 312, 314
Spectral
  power distribution (SPD), 249
  sensitivity to deposited material, calculation of, 372f
Spectroscopic
  ellipsometer, 373
  sensing applications, 364
Spectrum, 284, 286, 293–294
Spherical
  aberrations, 60
  reconstruction boundary, 117
  voxel elements, 113
  voxel USART, 117
Stages, 139

Standard
    deviation, 195, 198, 195t–197t, 201t,
        277
    Euclidean metric, 277
Stationary
    objects, 142
    target density, 189
Statistical
    based fusion approaches, 245
    methods, 245
    significance test, 127–128
    testing, 114
Stereo matching algorithm, 159,
        163f, 171
Stereoscopical techniques, 158–159
Subpixel target detection, 293, 314
Surface
    mines, 283
    profilometry, 61
    temperatures, 262
Surveillance applications, 283
Synthetic discriminant function
        filter, 15
Synthetic reference wave (SRW), 356

T
Talbot effect, 73
Target detection in multiple views
    range to target, 206
    resolution, 206
    sensor type, 206
    various target detection and recognition
        methods, 206
Target detection, 319
    algorithms, 283–285
Target tracking
    KNIGHT tracking system, 209
    multiple FOVs, 209
    surveillance system, 209
Target
    abundance measure, 290
    classification operation, 190
    detectors, 284
    phenomenology, 204
    spectra, 293–294
Taylor
    expansion, 354
    series, 339
Temporal registration, 256

Temporal templates
    motion energy images, 210
    Motion History, 210
Terrian and Waxman, methods by, 245
Thermal Physical Model, 249–250
    blackbody, 249
    conduction, 249
    convection, 249
    convective heat transfer function, 249
    heat flux, 249
    laminar flow, 249
    radiation, 249
    Stephen–Boltzman law, 249
    thermal resistance, 250
Thermal
    camera, 256
    expansion, 69
Thick holograms, 383
Thin holograms, 383–384
Three-dimensional object recognition, 223
    correlation of the reconstructed 3D
        scenes, 165–171
    direct correlation of integral images,
        164–165
Three-dimensional scene, digital
        reconstruction of
    correction of the depth-dependent
        magnification ratio, 161
    digital visualization of the 3D scene,
        163–164
    example of 3D reconstruction, 161–163
    retrieval of the 3D scene, 158–161
Time of flight, 325
Tolerance angle, 16, 18
Tolerance to
    displacements, 7
    distortions, 7
    in-plane rotation, 14
    longitudinal shifts, 14
    out-of-plane rotations, 14
    scaling, 14
Tomography, 156
Total internal reflection, 337
Tracking operation, 183, 189
    moving target velocities (Eigen
        velocities), 189
    tracking dwell time, 189
Tracking, 176
Training set, 267–269, 274, 278, 280

Transmission bandwidth, 183, 184f, 201
Transmission distance, 342
Transversal locations, 13
Tree area, 305
TRUE IMAGE® LABEL, 390
Tunable laser, high speed wavelength, 175
Tunable Ti–sapphire laser, 374
Turn Face operator, 272f
Twin or conjugate image, 49–50
Two-dimensional (2D) projections, 223
Two-layer waveguide-grating device, 374

U
Uniform lighting, 273
Uniform Simultaneous Algebraic
    Reconstruction Technique
    (USART), 113
  iterations, 128
  implemention by the spherical voxel
    model, 116–117
Unmanned Aerial Vehicles (UAV),
    183–185, 184f, 187, 189, 191–198,
    201,
  altitude, 189, 191, 193, 198, 199f
  flight patterns, 183
  long endurance surveillance, 183
  mission scenarios, 183
  sensor parameters, 183, 184f
UV Ar$^+$ laser interferometer, 373

V
Vacuum chuck, 69
VanderLugt correlator, 343–344, 348
VCSEL arrays, 344
Vector quantization network, *see*
    Kohonen competitive network
Velocity vector, 189, 195
  average, 189
  projected, 189
Vertical cavity surface emitting laser
    diodes (VCSELs), 344
Video
  animation, 66

image, 50
optical phase mask, 85
optical system, 86
phase mask (VPM), 86
Virtugram® technique, 387
Visualization and recognition of
    filamentous algae, experimental
    results, 146
Volume object recognition, performance
    analysis, 130–132
Volume objects reconstructed by the
    spherical-voxel USART, 117
  classes, 117
Voxel density, 113, 115

W
Wavefront curvature, 60, 65
Waveguide optics, 337
Wavelength tunable filter, 175–176
Wavelength-hopped
  laser coding and decoding
    system, 175
  spread spectrum sequence, 176
Weather station
  anemometer, 255
  barometer sensor, 255
  humidity sensor, 255
  temperature sensors, 255
  wind direction, 255
Weight vectors, 32
Wide-sense stationary random processes,
    320, 332
Window Size, 274, 275f

X
X-ray imaging technique, 113

Y
Yale Face Database B, 273

Z
Zernike's phase contrast method, 354
Zero-mean stationary random process, 321